Opportunities in 5G Networks

A Research and Development Perspective

OTHER TELECOMMUNICATIONS BOOKS FROM AUERBACH

AUERBACH PUBLICATIONS
www.auerbach-publications.com
To Order Call: 1-800-272-7737 • Fax: 1-800-374-3401 •E-mail: orders@crcpress.com

Opportunities in 5G Networks

A Research and Development Perspective

Edited by Fei Hu

CRC Press
Taylor & Francis Group
Boca Raton London New York

CRC Press is an imprint of the
Taylor & Francis Group, an **Informa** business

MATLAB® is a trademark of The MathWorks, Inc. and is used with permission. The MathWorks does not warrant the accuracy of the text or exercises in this book. This book's use or discussion of MATLAB® software or related products does not constitute endorsement or sponsorship by The MathWorks of a particular pedagogical approach or particular use of the MATLAB® software.

CRC Press
Taylor & Francis Group
6000 Broken Sound Parkway NW, Suite 300
Boca Raton, FL 33487-2742

First issued in paperback 2020

© 2016 by Taylor & Francis Group, LLC
CRC Press is an imprint of Taylor & Francis Group, an Informa business

No claim to original U.S. Government works

ISBN 13: 978-0-367-57489-5 (pbk)
ISBN 13: 978-1-4987-3954-2 (hbk)

Library of Congress Cataloging-in-Publication Data

Names: Hu, Fei, 1972-
Title: Opportunities in 5G networks : a research and development perspective
/ editor, Fei Hu.
Description: Boca Raton : Taylor & Francis, CRC Press, 2016. | Includes
bibliographical references and index.
Identifiers: LCCN 2015038786 | ISBN 9781498739542 (alk. paper)
Subjects: LCSH: Global system for mobile communications--Research. | Global
system for mobile communications--Technological innovations.
Classification: LCC TK5103.483 .O67 2016 | DDC 621.3845/6--dc23
LC record available at http://lccn.loc.gov/2015038786

Visit the Taylor & Francis Web site at
http://www.taylorandfrancis.com

and the CRC Press Web site at
http://www.crcpress.com

To my lovely children…

Contents

SECTION III 5G PHYSICAL LAYER

SECTION IV CM AND MM WAVE FOR 5G

Preface

Fifth generation (5G) is the future of information networks. It is not an incremental advance of fourth-generation (4G) cellular networks. Instead, 5G is a dramatic paradigm shift that supports high frequencies (such as 60 GHz), extreme node densities, large-scale antenna array, massive bandwidth, and so on. 5G networks require high flexibility and intelligence in the following aspects: spectrum sharing, millimeter-wave (mmWave) communications, integrated "Internet of Things" access, massive MIMO, smart antennas, Big Data, cloud computing, and many other disruptive technologies. 5G means a new transmission scheme, a virtualized SDN-like control, extreme energy efficiency, and new regulatory and standardization issues. It should support a 1000× higher data rate than 4G.

In recent years, because of the unprecedented growth in the number of connected devices and mobile data, and the ongoing developments in technologies to address this enormous data demand, the wireless industries have initiated a roadmap for transition from 4G to 5G. It is reported that the number of connected devices (the Internet of Things) is estimated to reach 50 billion by 2020, while the mobile data traffic is expected to grow to 24.3 exabyte per month by 2019. The higher cell capacity and end-user data rate are required due to ultra-high-definition multimedia streaming and the extremely low latency requirement for cloud computing and storage/retrieval. 5G can be expected to support immersive applications that demand exceptionally high-speed wireless connections and a fully realized IoT, to experience lower latency, and to promote both spectrum and energy efficiency.

It is expected that the 5G system will support data rates of 10–50 Gbps for low mobility users. The 5G system will provide gigabit-rate data services regardless of a user's location. 5G will also provide an end-to-end latency of less than 5 ms and air latency of less than 1 ms, which is one-tenth compared with the latency of the 4G network. In the 5G system, the simultaneous connections are expected to be over 10^6 per unit square kilometer, which is much higher than that of the legacy system. 5G systems are targeted to be 50 times more efficient than 4G by delivering reduced cost and energy usage per bit. 5G technologies will provide mobility on demand based on each device's and service's needs. On the one hand, the mobility of user equipment should be guaranteed to be at least the same level as the 4G system. On the other hand, the 5G system will support mobility at speeds ranging from 300 to 500 km/h. The cell spectral efficiency is set to the 10 bps/Hz level (in contrast to the 1–3 bps/Hz on 4G networks).

Differences from other 5G books: Although some books on 5G have been in the market for some time, most of them target general introductions and focus on fundamental knowledge. This book targets detailed research and development technical design. We have invited worldwide experts to focus on the detailed engineering design of 5G networks, especially its core components such as mmWave communications, massive multiple-in multiple-out, cloud-based networking, software-defined networking support, Big Data running, energy-efficient protocols, cognitive spectrum management, and other standardization issues.

Targeted audiences: This book targets both academics and industrialists. Researchers will have a profound understanding of the challenging issues in 5G and can thus easily find an unsolved research issue to pursue. Industry engineers can use the principles and schemes provided in the chapters for their practical 5G product design. Administrators will be able to understand the future trend of this exciting area.

Book Architecture: The book consists of four parts:

- *Section I. 5G Fundamentals*: This part includes a few chapters that cover the big picture of 5G. Especially, we will cover the following aspects:
  - *Fundamentals of 5G*: We will answer the following questions: What is 5G? What are the real uses of 5G in cases? What are the effects of 5G for mobile operators?
  - *Evolution from 4G to 5G:* Overview of the 4G networks (Long-Term Evolution/Long-Term Evolution-Advanced) with the dawn of 5G, services visions, requirements, and an overview of key enabling technologies for 5G network.
  - *5G trend and open issues*: Overview of the impact of the wireless backhaul traffic in forthcoming 5G mobile networks by examining the so-called stepping stone architecture toward 5G based on cloud radio access network.
- *Section II. 5G Design*: We will cover the nuts and bolts of 5G design. Specifically, we will explain:
  - *Deployment guidelines and principles for 5G radio access network*: The cellular network deployment policies; directional antennas for cellular networks; vertical sectoring; and so on.
  - *Quality of service:* The development of quality-of-service management principles at the network level in the new Third Generation Partnership Project releases and their implementation in 5G networks.
  - *Massive multiple-in multiple-out*: Cover massive multiple-in multiple-out systems—a key enabling technology for 5G; the issues of channel estimation and channel feedback in massive multiple-in multiple-out.

- *Optical and wireless*: Converged management of radio and optical resources, for example, management of multiple frequency bands with respect to radio capacity demand (current and estimated future demand) and available optical transport capacity and vice versa.
■ *Section III. Physical Layer:* An overview of candidate physical layer technologies for 5G systems; nonorthogonal multiple access; faster than Nyquist signaling, in which the symbol period is increased beyond the Nyquist rate, and so on.
■ *Section IV. cm and mm Wave for 5G*: (1) Centimeter-wave (cmWave) concept (below 30 GHz); 5G cmWave concept for small cells; fundamental technology components such as optimized frame structure, dynamic scheduling of uplink/downlink transmission, interference suppression receivers, and rank adaptation. (2) mmWave models and medium access control design: 5G mmWave communications, including channel modeling, beam tracking, network architecture, and so on; high directional, new medium access control mechanisms for directional mmWave wireless systems, and others.

The chapters contain detailed technical descriptions on the models, algorithms, and implementations of 5G networks. There are also accurate descriptions on the state-of-the-art and future development trends of 5G applications. Each chapter also includes references for readers' further studies.

Thank you for reading this book. We believe that this book can help you with the scientific research and engineering design of 5G systems. Due to time limitations, there might be some errors. Please let us know if you find any.

MATLAB® is a registered trademark of The MathWorks, Inc. For product information, please contact:

The MathWorks, Inc.
3 Apple Hill Drive
Natick, MA 01760-2098 USA
Tel: 508 647 7000
Fax: 508-647-7001
E-mail: info@mathworks.com
Web: www.mathworks.com

Editor

Dr. Fei Hu is currently a professor in the Department of Electrical and Computer Engineering at the University of Alabama, Tuscaloosa, Alabama. He obtained his PhD degrees at Tongji University (Shanghai, China) in the field of signal processing (in 1999), and at Clarkson University (Potsdam, New York, USA) in electrical and computer engineering (in 2002). He has published over 200 journal/conference papers and books. Dr. Hu's research has been supported by the U.S. National Science Foundation, Cisco, Sprint, and other sources. His research expertise can be summarized as *3S: Security, Signals, Sensors:* (1) *Security:* This is about how to overcome different cyberattacks in a complex wireless or wired network. Recently, his focus has been on cyberphysical system security and medical security issues. (2) *Signals:* This mainly refers to *intelligent signal processing*, that is, using machine learning algorithms to process sensing signals in a smart way to extract patterns (i.e., pattern recognition). (3) *Sensors:* This includes microsensor design and wireless sensor networking issues.

Contributors

Heba Abd-El-Atty
Department of Electrical Engineering
Portsaid University
Portsaid, Egypt

Ozgur Baris Akan
Next-Generation and Wireless
   Communications Laboratory
Department of Electrical and
   Electronics Engineering
Koc University
Istanbul, Turkey

Mohammed Alnuem
Department of Electrical and
   Computer Engineering
King Saud University
Riyadh, Saudi Arabia

Pablo Ameigeiras
Department of Signal Theory
University of Granada
Granada, Spain

Meisam Khalil Arjmandi
Department of Electrical and
   Computer Engineering
University of Alabama
Tuscaloosa, Alabama

Busra Gozde Bali
Next-Generation and Wireless
   Communications Laboratory
Department of Electrical and
   Electronics Engineering
Koc University
Istanbul, Turkey

Gilberto Berardinelli
Department of Electronic Systems
Aalborg University
Aalborg, Denmark

János Bitó
Department of Broadband
   Infocommunications and
   Electromagnetic Theory
Budapest University of Technology and
   Economics
Budapest, Hungary

Grigory Bochechka
LLC Icominvest
Moscow, Russia

Davide Catania
Department of Electronic Systems
Aalborg University
Aalborg, Denmark

Symeon Chatzinotas
Interdisciplinary Centre for Security,
  Reliability, and Trust
University of Luxembourg
Luxembourg

Khaled Elsayed
Department of Electronics and
  Communications
Cairo University
Cairo, Egypt

Gerhard Fettweis
Vodafone Chair Mobile
  Communications Systems
Dresden University of Technology
Dresden, Germany

Atílio Gameiro
Department of Electronics,
  Telecommunications, and
  Informatics
Institute of Telecommunications
University of Aveiro
Aveiro, Portugal

Ivan Gaspar
Vodafone Chair Mobile
  Communications Systems
Dresden University of Technology
Dresden, Germany

Moneeb Gohar
Department of Information and
  Communication Engineering
Yeungnam University
Gyeongsang, South Korea

Kazi M.H. Huq
Institute of Telecommunications
University of Aveiro
Aveiro, Portugal

Franciso Javier Lorca Hernando
Distrito Telefónica
Madrid, Spain

Muhammad Ali Imran
Department of Electrical and
  Electronics Engineering
Institute for Communication Systems
University of Surrey
Guildford, United Kingdom

Ahmed Kamal
Department of Electrical and
  Computer Engineering
Iowa State University
Ames, Iowa

Péter Kántor
Department of Broadband
  Infocommunications and
  Electromagnetic Theory
Budapest University of Technology and
  Economics
Budapest, Hungary

Joongheon Kim
Intel Corporation
Santa Clara, California

Mads Lauridsen
Department of Electronic Systems
Aalborg University
Aalborg, Denmark

Jukka Lempiäinen
Department of Electronics and
  Communications Engineering
Tampere University of Technology
Tampere, Finland

Shang Liu
School of Information and
  Communication Engineering
Beijing University of Posts and
  Telecommunications
Beijing, China

Nurul H. Mahmood
Department of Electronic Systems
Aalborg University
Aalborg, Denmark

Jose Gabriel Martinez Martin
Distrito Telefónica
Madrid, Spain

Maximilian Matthé
Vodafone Chair Mobile
  Communications Systems
Dresden University of Technology
Dresden, Germany

Luciano Leonel Mendes
Department of Telecommunications
  Engineering
National Institute of
  Telecommunications
Minas Gerais, Brazil

Nicola Michailow
Vodafone Chair Mobile
  Communications Systems
Dresden University of Technology
Dresden, Germany

Preben Mogensen
Department of Electronic Systems
Aalborg University
and
Nokia Networks
Aalborg, Denmark

Paulo P. Monteiro
Department of Electronics,
  Telecommunications, and Informatics
Institute of Telecommunications
University of Aveiro
Aveiro, Portugal

Shahid Mumtaz
Institute of Telecommunications
University of Aveiro
Aveiro, Portugal

Oluwakayode Onireti
Department of Electrical and
  Electronics Engineering
Institute for Communication Systems
University of Surrey
Guildford, United Kingdom

Björn Ottersten
Interdisciplinary Centre for Security,
  Reliability, and Trust
University of Luxembourg
Luxembourg

Kari Pajukoski
Nokia Networks
Oulu, Finland

Iftikhar Rasheed
Department of Telecommunication
  Engineering
University College of Engineering and
  Technology
The Islamia University of Bahawalpur
  Pakistan
Bahawalpur, Pakistan

Jonathan Rodriguez
Department of Electronics,
  Telecommunications, and
  Informatics
Institute of Telecommunications
University of Aveiro
Aveiro, Portugal

Joonas Säe
Department of Electronics and
Communications Engineering
Tampere University of Technology
Tampere, Finland

Mohamed Selim
Department of Electrical and
Computer Engineering
Iowa State University
Ames, Iowa

Muhammad Usman Sheikh
Department of Electronics and
Communications Engineering
Tampere University of Technology
Tampere, Finland

Zhenyu Shi
Huawei Technologies Co., Ltd.
Shanghai, China

Fernando M. L. Tavares
Department of Electronic Systems
Aalborg University
Aalborg, Denmark

Valery Tikhvinskiy
LLC Icominvest
Moscow, Russia

Anestis Tsakmalis
Interdisciplinary Centre for Security,
Reliability, and Trust
University of Luxembourg
Luxembourg

Manuel Violas
Department of Electronics,
Telecommunications, and
Informatics
Institute of Telecommunications
University of Aveiro
Aveiro, Portugal

Yi Wang
Huawei Technologies Co., Ltd.
Shanghai, China

Turker Yilmaz
Next-Generation and Wireless
Communications Laboratory
Department of Electrical and
Electronics Engineering
Koc University
Istanbul, Turkey

Syed Fahad Yunas
Department of Electronics and
Communications Engineering
Tampere University of Technology
Tampere, Finland

Dan Zhang
Vodafone Chair Mobile
Communications Systems
Dresden University of Technology
Dresden, Germany

5G
FUNDAMENTALS

1

Chapter 1

Basics of 5G

Iftikhar Rasheed

Contents

1.1 Brief History

In the twentieth century, the introduction of first-generation mobile technology (1G) in 1981 changed human thinking. Using 1G, the Nordic Mobile Telephone was the latest technology in its era. At the time, scientists, researchers, and engineers were working on improvements to the telecommunication system to increase user reliability and reduce wastage of precious time so that users could communicate and transport data effectively under high encryption. Ten years later, the concept of the second generation (2G) was introduced. It was launched by Global System for Mobile Communications (GSM) in Finland in 1991. The major benefit of 2G is its introduction of the short message service (SMS), picture messages, and the multimedia messaging service (MMS), which are basic concepts for transporting data around the world under privacy or encryption. Unlike 1G, which was analog, 2G was digital. Later, 2G was expanded into second and a half generation (2.5G) and second and three-quarters generation (2.75G). 2G technology is still used in some countries, and is still important due to its high market share. However, 2G has relatively low spectral efficiency as compared with the new access technologies in GSM. The orthogonal subchannel (OSC) and dynamic frequency and channel allocation (DFCA) are two new functionalities introduced in GSM technologies [1].

2.5G uses general packet radio service (GPRS) technology, which has been implemented in the packet switching domain as well as the circuit switch domain. The introduction of GPRS is a major step in the evolution of GSM to the third-generation (3G) networks. Enhanced data rates for GSM evolution (EDGE) technology, or enhanced GPRS, developed by the introduction of 8PSK encoding, are used in 2.75G. This digital mobile phone technology improved the data rate, and was introduced in the United States on GSM in 2003. In 3G, the wireless communication data rate reaches up to 2 Mbps. To achieve this data rate, the core network and the access network are significantly changed to cater for the high data requirement. 3G mobile telecommunications is a generation of standards for mobile phones and mobile telecommunication services that fulfills the International Mobile Telecommunications-2000 (IMT–2000) disclaimers by the International Telecommunication Union. Application services include wide-area wireless voice telephone, mobile Internet access, video calls, and mobile TV, all in a mobile environment.

In 2009, fourth generation (4G) became the wireless cellular standard. This uses the International Mobile Telecommunication Advanced (IMT-Advanced) technology at a required peak speed for 4G of 100 Mbps for high-mobility communication and 1 Gbps for low-mobility communication. The world's first publicly accessible Long-Term Evolution (LTE) service was opened in the two Scandinavian capitals, Stockholm (Ericsson, Nokia, and Siemens network systems) and Oslo (a Huawei system), on December 14, 2009.

Fifth-generation (5G) mobile networks or fifth-generation wireless systems is the name used in some research papers and projects to denote the next major phase of mobile telecommunications standards beyond the 4G/IMT-Advanced standards effective since 2011. It is expected that 5G will be introduced in the 2020s. According to researchers, scientists, and engineers, 5G will provide users with 1000 times greater bandwidth as well as a 100 times larger data rate to cover the huge applications of future mobile stations [2]. It is also expected that various techniques will be used in 5G to fulfill user requirements, one of which will be terahertz band mobile communication [3]. The 5G wireless mobile Internet networks are real wireless world, which will be supported by large area synchronized code division multiple access (LAS-CDMA), orthogonal frequency-division multiplexing (OFDM), multicarrier code-division multiple access (MC-CDMA), ultrawide band (UWB), network-local multipoint distribution service (LMDS), and Internet protocol version 6 (IPv6). 4G and 5G share the basic IPv6. A classification of the five generations is given in Table 1.1.

1.2 Introduction

The purpose of this chapter is to familiarize the reader with the coming 5G technology. 5G will be essential worldwide in the future due to the increasing traffic rates of data, voice, and video streaming in this modern era. The present 3G and 4G technologies cannot fulfill the future increasing capacity requirements of Internet data traffic. There is no unique definition of 5G [4,5]. Basically, we need to know "What is 5G?" in real terms. First of all, we have to clarify the real meaning of 5G in a technological sense. Three questions arise that are most important for clarifying the term *5G*:

1. What is 5G?
2. What are real example uses of 5G?
3. What are the effects of 5G for mobile operators?

A number of generation changes have been experienced by mobile technologies, which have transformed the cellular background into a global set of inter-related networks. By 2020, 5G will support voice and video streaming and a very complex range of communication services over more than nine billion subscribers, as well as billions of devices that will be connected to each other. But, what is 5G? 5G provides a new path for thinking. It includes a radical network design for installing machine-type communication (MTC). Also, 5G networks will be able to provide efficient support applications with widely varying operational parameters, providing greater elasticity for installing services. As for the previous generations, 5G is a combination of developed network technologies. The coming 5G technology will have the ability to share data everywhere, every time, by everyone and everything, for the benefit of individuals, businesses, and society, as well as the technological environment, by using a bandwidth of unlimited access for carrying information. It is expected that specific and standard activities will begin in 2016, leading to commercial availability of the equipment and machinery around 2020. The future 5G technology is much more than a new set of technologies and will require enormous upgrades of equipment/devices or machinery as compared with the previous generations. The purpose of this technology is to build on the developments already achieved by telecommunication systems. The complementary technologies (a combination of core and cloud technologies) employed in much of the existing radio access will be used in 5G to cater for higher data traffic

Table 1.1 Comparison between 1G, 2G, 2.5G, 3G, 3.5G, 4G, and 5G Technologies

Generation	Definition	Throughput/ Speed	Technology	Time Period	Features
1G	Analog	14.4 Kbps (peak)	AMPS, NMT, TACS	1981–1990	Wireless phones are used for voice only
2G	Digital narrowband circuit data	9.6/14.4 Kbps	TDMA, CDMA	1991–2000	Multiple users on a single channel via multiplexing. Cellular phones are used for data also along with voice
2.5G	Packet data	171.2 Kbps (peak) 20–40 Kbps	GPRS	2001–2004	Internet becomes popular. Multimedia services and streaming start to show growth. Phones start supporting web browsing
3G	Digital broadband packet data	3.1 Mbps (peak) 500–700 Kbps	CDMA 2000 (1 × RTT, EVDO) UMTS, EDGE	2004–2005	Multimedia services support along with streaming. Universal access and portability

			HSPA	2006–2010	*Higher throughput and speeds to support higher data*
3.5G	Packet data	14.4 Mbps (peak) 1–3 Mbps			
4G	Digital broadband packet, all IP, very high throughput	100–300 Mbps (peak) 3–5 Mbps 100 Mbps (Wi-Fi)	WiMAX LTE Wi-Fi	Now (transitioning to 4G)	*High speed and definition streaming. New phones with HD capabilities surface. Portability is increased further. Worldwide roaming*
5G	Not yet	Gigabits	LAS-CDMA, OFDM, MC-CDMA, UWB, Network-LMDS	Soon (probably 2020)	*Currently there is no 5G technology deployed. It will provide very high speeds and efficient use of bandwidth when deployed*

AMPS, Advanced Mobile Phone System; EVDO, evolution data-optimized; HSPA, high-speed packet access; NMT, Nordic Mobile Telephone; RTT, round-trip time; TACS, Total Access Communication System; TDMA: time-division multiple access; UMTS: Universal Mobile Telecommunications System.

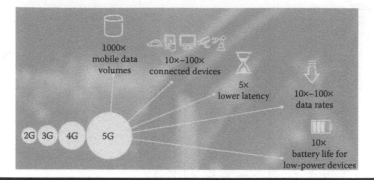

Figure 1.1   Estimated requirement levels of 5G.

and more types of devices under different operating requirements in different situations. Figure 1.1 shows the estimated performance levels of 5G technology needed to meet these requirements.

Universal agreement is building around the idea that 5G is simply the integration of a number of techniques, scenarios, and use environments rather than the origination of a new single radio access technology. The estimated performance levels that 5G [6] technologies will need to cater are

- 10–100 times higher typical user data rate
- 10 times longer battery life for low-power devices
- 10–100 times higher number of connected devices
- Five times reduced end-to-end latency
- 1000 times higher mobile data volume per area

Now, the problem is "How will we get there?" The next generation (5G) will mostly allow connectivity. But this technology is not developed in isolation. The developing next generation will play a significant role in shaping various factors such as long-term sustainability, cost, and security, and will need to provide connectivity to billions of subscribers. While the comprehensive conditions for 5G have nevertheless to be set, it is clear that flexibility to accommodate thousands of applications is the key to 5G and what it will enable. The parameters on which 5G technology will be developed include

- Data integrity
- Latency
- Smart communication
- Traffic capacity
- Data throughput
- Energy consumption
- Technology convergence

1.3 Next-Generation (5G) Concept in Wireless Technology

The 5G concepts correspond to the open system interconnected (OSI) layers. Four basic layers are used in 5G. Figure 1.2 shows a comparison between the OSI and 5G layers.

1.3.1 Open Wireless Architecture

Open wireless architecture in 5G corresponds to the physical layer and the data link layer or medium access control (MAC) layer of the OSI model, which are commonly known as Layer 1 and Layer 2, respectively [7].

1.3.2 Network Layer

Layer 2 of 5G is subdivided into upper and lower layers, as shown in Figure 1.2. The network layer of 5G technology corresponds to the OSI Layer 3, which is the network layer. This layer is based on IP. Currently, there is no competition on this level. IP version 4 (IPv4) is widespread globally. It has various problems, such as limited address space and no real possibility for quality-of-service (QoS) support per flow. These issues are solved in IPv6, but with the trade-off of a significantly bigger packet header. Also, mobility still remains a problem. There is a Mobile IP standard as well as many micromobility solutions. All mobile networks will use Mobile IP in 5G, and each mobile terminal will be a foreign agent (FA), maintaining care of address (CoA) mapping between

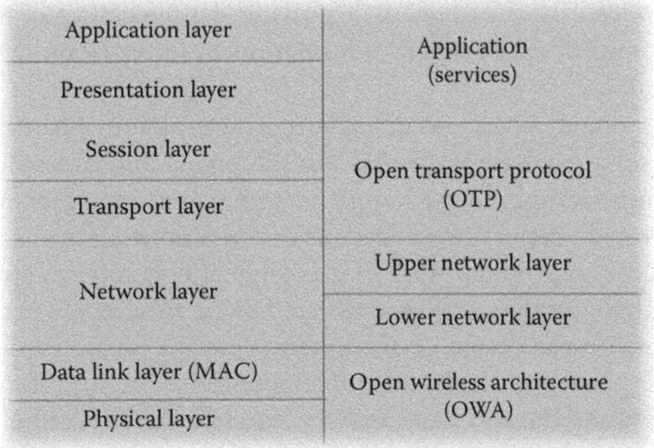

Application layer	Application (services)
Presentation layer	
Session layer	Open transport protocol (OTP)
Transport layer	
Network layer	Upper network layer
	Lower network layer
Data link layer (MAC)	Open wireless architecture (OWA)
Physical layer	

Figure 1.2   Compression between OSI and 5G layers. (A. Gohil et al., 5G technology of mobile communication: A survey, *International Conference on Intelligent Systems and Signal Processing (ISSP)*, IEEE, Gujarat, pp. 289–290, 2013.)

its fixed IPv6 address and the CoA address for the current wireless network. However, a mobile can be attached to several mobile or wireless networks at the same time [8].

1.3.3 Open Transport Protocol

The open transport protocol layer is the third layer of 5G technology, which corresponds to the transport and session layers of the OSI model. Wireless networks and mobiles differ from underwired networks regarding the transport layer. In all transmission control protocol (TCP) versions, it is assumed that the packet loss is due to network congestion. But due to a higher bit error ratio in the radio interface, losses may occur in wireless technology. Therefore, TCP amendments and alterations are anticipated for the mobile and wireless networks, which retransmit the damaged TCP segments over the wireless link only. For 5G mobile terminals, it will be suitable to have a transport layer that can be downloaded and installed. Such mobiles will have the ability to download a version that is targeted to a specific wireless technology installed at the base stations (BS). This is called an *open transport protocol* (OTP) [9].

1.3.4 Application

The application layer is the last layer of 5G as well as the OSI model. Regarding applications, the ultimate request from the 5G mobile terminal is to provide intelligent QoS management over a variety of networks. Today, the users of mobile phones manually select the wireless interface for a particular Internet service without having the ability to use the QoS history to select the best wireless connection for a given service. The 5G phone will provide the possibility of service quality testing and storage of measurement information in information databases in the mobile terminal.

The QoS parameters, such as delay, jitter, losses, bandwidth, and reliability, will be stored in a database in the 5G mobile phone and can be used by intelligent algorithms running in the mobile terminal as system processes, which in the end will provide the best wireless connection according to the required QoS and personal cost constraints. With 5G, a range of new services and models will be available. These services and models need to be further examined regarding their interface with the design of 5G systems [10]. In future wireless networks, there must be a low complexity of implementation and an efficient means of negotiation between the end users and the wireless infrastructure. The Internet is the driving force for higher data rates and high-speed access for mobile wireless users. This will be the motivation for an all-mobile IP-based core network evolution.

1.4 Disruptive Technologies for 5G

In the last few years, mobile and wireless networks have undergone remarkable development. The fifth era (5G) is advancing. What advances will characterize it? Will 5G be simply an advance on 4G, or will developing technologies cause an interruption, obliging a wholesale reconsideration of settled cellular standards? We believe that the following five potential advances could prompt both architectural and component design changes for 5G [11]:

1. Device-centric architecture
2. Millimeter wave (mmWave)
3. Massive multiple-in multiple-out (MIMO)
4. Smarter devices
5. Essential support for machine-to-machine (M2M) communication

1.4.1 Device-Centric Architecture

In 5G, the BS architecture for cellular systems may change. For better information flow routes within the network with different purposes and preference toward different sets of nodes, we have to reconsider the concepts of control and data channels, as well as uplink and downlink. Cellular designs have generally depended on the hardware part of *cells* as crucial units inside the radio access system. A device obtains service by building a downlink and further, an uplink connection carrying control and data traffic with the BS commanding the cell. In the last few years, a disruption of this cell-centric structure has been indicated by different trends:

■ The BS density is expanding swiftly, determined by the ascent of heterogeneous systems. Heterogeneous systems were institutionalized in 4G, but the architecture was not locally intended to support them. System densification could necessitate some real changes in 5G. The organization of BSs with vastly different transmitted power and coverage areas, for example, requires a decoupling of downlink and uplink in a manner that permits the signalling data to move through diverse sets of nodes [12].
■ The requirement for an extra spectrum will inexorably prompt the coexistence of frequency bands with profoundly distinctive propagation attributes inside the same framework. In this situation, [13] proposes the idea of a phantom cell in which the information and control planes are differentiated: high-power nodes send the control data at microwave frequencies, while low-power nodes pass on the payload information at mmWave frequencies.

- Another idea termed *incorporated baseband*, has been developed in the cloud-based wireless access systems [14]. Such an idea allows the virtualization of the network topology, which means that the virtualized network node and the actual equipment assigned to that node could be in different physical locations of the network. Equipment assets in a pool, for example, could be variably assigned to distinctive nodes relying on measurements characterized by the system administrator.
- The use of smarter devices could affect the radio access system. Specifically, both device-to-device (D2D) and smart caching calls require a design redefinition whereby the center of gravity moves from the system center to the outskirts (devices, relays).

In view of these trends, our vision is that the cell-centric architecture needs to develop into a device-centric one. A given device (human or machine) should have the capacity to communicate by trading numerous information flows through a few conceivable sets of heterogeneous nodes. The set of system nodes providing integration into a given device and the objectives of these nodes in a specific communication session should be custom-made for that particular device and session. Under this vision, the ideas of uplink/downlink and a control/information channel need to be reevaluated (Figure 1.3). While the requirement for a disruptive change in architectural design seems clear, real research endeavors are still expected to change the subsequent vision into a cognizant and practical suggestion. Since the historical backdrop of developments [15] shows that design changes are frequently the drivers of major innovative discontinuities, we accept that the above-listed four trends will have a significant impact on the advancement of 5G.

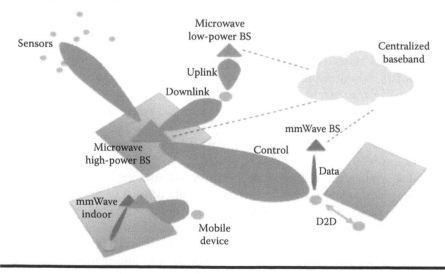

Figure 1.3 Device-centric architecture example. (F. Boccardi et al., *IEEE Comm Mag* 52: 76, 2014.)

1.4.2 Millimeter Wave

Conquering the worldwide bandwidth shortage created due to the rapid increase in the use of smartphones as well as mobile data growth presents exceptional challenges [16,17]. A precious, narrow microwave cellular range of around 600 MHz is divided among operators [18]. For modern communication, a wider microwave spectrum is required. There are two ways to increase the microwave spectrum for cellular communication:

- ▪ To refarm the spectrum. This has been done worldwide by repurposing the TV spectrum for rural broadband access applications; however, repurposing has not provided much more spectrum: only around 80 MHz, at a high cost.
- ▪ To impart the spectrum using, for example, cognitive radio techniques. The initial high hopes for cognitive radio have been dashed by the fact that an occupant who is not completely ready to cooperate is a real hindrance to spectrum efficiency for secondary users.

On the whole, it appears that at microwave frequencies, doubling the current cellular bandwidth is the best approach. More importantly, the mmWave frequency spectrum available for cellular communication ranges from 3 to 300 GHz. It could be expected that several gigahertz could be made accessible for 5G. The sensitivity to blockage is the major difference between microwave and mmWave frequencies: the results in [18] indicate that for line-of-sight propagation, the path loss exponent is 2, but for nonline of sight it is 4 (plus additional power loss). mmWave cellular research will need to incorporate affectability by blockages and more intricate channel models into the investigation, and furthermore, to study the impact of enablers, for example, relays and high-density infrastructure.

In an mmWave system, antenna arrays are the key features. The impact of interference is also reduced by adaptive arrays with a narrow beam, with the proviso that interference-limited conditions are not suitable for mmWave systems, which are better able to operate in noise-limited conditions. Essential communication may be performed by acceptable array gain, requiring new random access protocols that work when transmitters emit in certain directions and receivers only receive from certain directions. Adaptive array-processing algorithms are needed that can respond quickly when antenna beams are blocked by people or objects, or when the device antenna is hidden by the user's own body. mmWave systems also need some modifications to their hardware.

From the preceding discussion, and referring to the Henderson–Clark model [15], we come to the conclusion that radical changes in the system are required for the mmWave, and note that the mmWave is a potential disruptive technology for 5G network implementation, which, provided the abovementioned difficulties can be overcome, could provide excellent data rates and a totally distinctive user experience.

1.4.3 Massive MIMO

Massive MIMO refers to large-scale MIMO, or large-scale antenna systems, in which antennas at the BS are much larger in number than devices connected per signaling resource [19], with numerous BS antennas delivering channels to different devices. We contend in the Henderson–Clark system that massive MIMO is a disruptive technology for 5G because

- It is a scalable technology at a node level. This is completely different from 4G, which is not scalable: further sectorization is not possible due to (i) the limited space for heavy azimuth directional antennas and (ii) the imminent angle spread of propagation; thus, the single MIMO user is limited by the predetermined number of antennas that can fit into certain cell phones. Conversely, there are very nearly no restrictions on the number of antennas mounted on BSs in massive MIMO, given that time-division duplexing is used to empower channel estimation through uplink pilots.
- It empowers new architectures and deployments. While one can imagine macro BSs being directly replaced by arrays of low-gain antennas, other deployments in rural areas are also possible. Besides, the same massive MIMO standards that oversee the use of assembled arrays of antennas apply additionally to disperse deployments in which a campus or a whole city could be covered by a huge number of dispersed antennas that in aggregate, serve numerous users.

Massive MIMO still has various research challenges. Critical channel estimation and user movement force a limited continuity interval during which channel knowledge must be gained and used, and subsequently, there are a limited number of orthogonal group sequences that can be accredited to devices. From an implementation perspective, massive MIMO can be realized with low-power, low-cost hardware, with each antenna working semi-independently, yet significant effort is still needed to demonstrate the cost-effectiveness of this solution. From the foregoing discussion, we come to the conclusion that massive MIMO is a disruptive technology for 5G, but before employing it, we would have to overcome the challenges of massive MIMO implementation.

1.4.4 Smarter Devices

The cellular systems of prior eras were built with the configuration rationale of having complete control at the infrastructure side. Here, we discuss some possibilities that permit the devices to play a more vital role, and consequently, how 5G design could bring about an increment in device smartness. We concentrate on three distinct examples of technologies that could be combined into much smarter devices: D2D, advanced interference rejection, and local caching.

■ D2D has the capability of taking care of local communication more proficiently. Local high data rates could also be attained by other technologies, such as Bluetooth or Wi-Fi direct. Applications requiring a mixture of local and nonlocal content, or a mixture of high data rate and low latency limitations, could represent more convincing explanations for the use of D2D. Specifically, we imagine D2D as an imperative empowering influence for applications requiring low latency, particularly in the deployment of future systems using baseband centralization and radio virtualization. D2D is currently being examined by the 3rd Generation Partnership Project (3GPP) as a 4G, with the main focus being vicinity detection for public safety [20].
■ The idea of caching a massive amount of information at the edge of the wireline system just before the wireless node only applies to delay-tolerant traffic, and for this reason, it has no place in voice-centric systems. Caching may at long last have a future in data-centric systems [21]. Thinking ahead, it is not difficult to imagine cell phones with a huge amount of memory. Local caching is an imperative option both at the radio access system edge and cell phones, also thanks to empowering agents such as mmWave and D2D.
■ Notwithstanding D2D's capabilities and its huge amount of memory, future cell phones might likewise have varying form factors. In a few cases, the devices may necessitate a few antennas, with the resulting possibility of active interference rejection alongside beam forming and spatial multiplexing.

From this discussion, we regard smarter devices as having all the properties of a disruptive technology for 5G implementation.

1.4.5 Essential Support for M2M Communication

The essential consideration for M2M communication in 5G includes fulfilling three generally distinctive necessities that are related to diverse classes of low data rate services: backing of countless low rate devices, sustaining a negligible data rate in more or less all circumstances, and low-latency data transfer. Attending to these necessities in 5G requires new techniques and ideas at both the architectural and the component level.

As with electricity or water, wireless communication is becoming a commodity [22]. This commoditization offers a huge range of new services with new types of prerequisites:

■ A massive number of connected devices
■ Very high link reliability
■ Low latency and real-time operation

References

1. S. Lasek, D. Tomeczko, and J. T. J. Penttinen, GSM refarming analysis based on orthogonal sub channel and interference optimization, 8th IEEE IET International Symposium of Communication System, Networks and Digital Signal Processing, IEEE, Poznań, Poland, 2012.
2. Nokia Siemens Networks 2011, 2020: Beyond 4G radio evolution for the gigabit experience, White Paper, February 2011.
3. B. S. Rawat, A. Bhat, and J. Pistora, THz B and nano antennas for future mobile communication. In *Signal Processing and Communication* (ICSC), 2013 International Conference on, pp. 48–52, December 12–14, 2013.
4. J. G. Andrews, S. Buzzi, W. Choi, S. V. Hanly, A. Lozano, A. C. K. Soong, and J. C. Zhang, What will 5G be? *IEEE J Sel Areas Comm* 32: 1065–1082, 2014.
5. P. Pirinen, A brief overview of 5G research activities, In *Proceedings of the 1st International Conference on 5G for Ubiquitous Connectivity (5GU)*, IEEE, Akaslompolo, pp. 17–22, 2014.
6. P. Popovski, V. Braun, H.-P. Mayer, P. Fertl, Z. Ren, D. GozalvesSerrano, E. Strom, et al. ICT-317669-METIS/D 1.1 V I scenarios, requirements and KPls for 5G mobile and wireless system, Technical Report, May 2013. Available from: https://www. metis2020.com/wp-content/uploads/deliverables/METIS_D1.1_v1.pdf.
7. J. Govil and J. Govil, 5G: Functionalities development and an analysis of mobile wireless grid, *First International Conference on Emerging Trends in Engineering and Technology*, IEEE, Nagpur, pp. 270–275, 2008.
8. 5G mobile technology abstract. Available from: http://www.seminarsonly.com/ Labels/5g-Mobile-Technology-Abstract.php.
9. M. Hata, Fourth generation mobile communication systems beyond IMT–2000 communications, *Proceedings of 5th Asia Pacific Conference on Communication and 4th Optoelectronics Communications Conference*, Vol. 1, IEEE, Beijing, pp. 765–767, 1999.
10. A. Gohil, H. Modi, and S. K. Patel, 5G technology of mobile communication: A survey, *International Conference on Intelligent Systems and Signal Processing (ISSP)*, IEEE, Gujarat, pp. 289–290, 2013.
11. F. Boccardi, R. W. Heath Jr, A. Lozano, T. L. Marzetta, and P. Popovski, Five disruptive technology directions for 5G, *IEEE Comm Mag* 52(2): 76, 2014.
12. J. Andrews, The seven ways HetNets are a paradigm shift, *IEEE Comm Mag* 51(3): 136–144, 2013.
13. Y. Kishiyama, A. Benjebbour, T. Nakamura, and H. Ishii, Future steps of LTE-A: Evolution towards integration of local area and wide area systems, *IEEE Wireless Comm* 20(1): 12–18, 2013.
14. C-RAN: The road towards green RAN, China Mobile Research Institute, Beijing, White Paper, Vol. 2.5, 2011.
15. A. Afuah, *Innovation Management: Strategies, Implementation and Profits*. London: Oxford University Press, 2003.
16. T. S. Rappaport, J. N. Murdock, and F. Gutierrez, State of the art in 60 GHz integrated circuits & systems for wireless communications, *Proc IEEE* 99(8): 1390–1436, 2011.
17. Z. Pi, and F. Khan, An introduction to millimeter-wave mobile broadband systems, *IEEE Comm Mag* 49(6): 101–107, 2011.

18. T. S. Rappaport, S. Sun, R. Mayzus, Z. Hang, Y. Azar, K. Wang, G. N. Wong, J. K. Schulz, M. Samimi, and F. Gutierrez. Millimeter wave mobile communications for 5G cellular: It will work! *IEEE Access* 1: 335–349, 2013.
19. T. L. Marzetta, Noncooperative cellular wireless with unlimited numbers of base station antennas, *IEEE Trans Wireless Comm* 9(11): 3590–3600, 2010.
20. 3GPP TR 23.703 v.0.3.0, Study on architecture enhancements to support proximity services (ProSe), 2013.
21. N. Golrezaei, A. F. Molisch, A. G. Dimakis, and G. Caire, Femtocaching and device-to-device collaboration: A new architecture for wireless video distribution, *IEEE Comm Mag* 51(1): 142–149, 2013.
22. M. Weiser, *The Computer for the 21st Century*. New York: Scientific American, 1991.

Chapter 2

5G Overview: Key Technologies

Meisam Khalil Arjmandi

Contents

2.1 Why 5G?

Before discussing the structure and characteristics of fifth generation (5G), it seems that the necessity for designing such a network should be clearly explained. Therefore, it is informative to review previous network generations. 1G is the first generation of wireless telephone technology, which provides a speed of up to 2.4 Kbps. The voice calls provided by this network are limited to one country and the network is based on using an analog signal. There are many pitfalls with 1G, such as poor voice quality, poor battery life, large phone size, limited capacity, and poor hand-off reliability. The second generation is 2G, which is based on the global system for mobile communication (GSM). This network uses digital signals and its data speed is up to 64 Kbps. This network provides services such as text messages, picture messages, and multimedia messages (MMS). The quality and capacity of the network is also better compared with 1G. The high dependency of this network on strong digital signals and its inability to handle complex data such as video are its most important drawbacks. The technology between 2G and third generation (3G) was called second and a half generation (2.5G), which was a combination of 2G cellular technology with general packet radio service (GPRS). The characteristics of this network are providing phone calls, sending and receiving e-mail messages, enabling web browsing, and providing a network speed of 64–144 Kbps. With the introduction of 3G in 2000, the data transmission speed increased from 144 to 2M Kbps. The prominent features of 3G are that it provides faster communication, enables sending and receiving large e-mails, and provides high-speed web, videoconferencing, TV streaming, and mobile TV. However, the license services for 3G are expensive and building the infrastructure is challenging. Requiring a high bandwidth, large cell phones, and expensive 3G phones are other drawbacks of 3G. The new generation, which is called *fourth generation* (4G), provides communication with higher data rates and high-quality video streaming in which Wi-Fi and WiMAX are combined together. This network is able to provide speeds of 100 Mbps to 1 Gbps. 5G is expected to be a significant advance on previous networks especially 4G. The quality of services (QoS) and security are significantly promoted in 4G while the cost per bit is low. In comparison with previous network generations, there are some issues with 4G such as greater power consumption (battery use), it is difficult to implement, the hardware required is too complicated, and the high cost of the equipment needed to implement the next-generation network. 5G, which is the subject of this chapter, is going to be the next generation. In brief, it aims to provide a complete wireless communication with almost no limitations. Considering all the advances in different areas, 5G is going to be responsible for providing a unique network that is able to broadcast large amounts of data in gigabits per second (Gbps), enabling multimedia newspapers and TV programs with high-definition (HD) quality. Improving the dialing speed and the clarity of audio and video, and supporting interactive multimedia are other advantages of the 5G network. Table 2.1 compares different network generations. Figure 2.1 shows how 5G will collect all possible networks to establish a single thorough network.

There are some main expectations from the 5G network to present a better telecommunication network. First, the 5G network aims to provide a very high data rate for huge number of users. It should also be able to support several simultaneous connections for deploying massive numbers of sensors. Compared with 4G, the spectral efficiency of the 5G network should be greatly enhanced. This network should also be compatible with 4G Long-Term Evolution (LTE) and Wi-Fi to provide high-rate coverage and smooth communication with low latency. Figure 2.2 shows the variation in the volume of data traffic per month for the Internet protocol (IP) network in petabits. From Figure 2.2, it is clear that with this huge growth in demand for data transmission, we need a new generation network with high ability.

Table 2.1 Characteristics of Different Network Generations

Network	1G	2G/2.5G	3G	4G	5G
Deployment	1970/1984	1980/1999	1990/2002	2000/2010	2014/2015
Bandwidth	2 Kbps	14–64 Kbps	2 Mbps	200 Mbps	>1 Gbps
Technology	Analog cellular	Digital cellular	Broadband width/CDMA/IP technology	Unified IP and seamless combo of LAN/WAN/WLAN/PAN	4G+wwww
Service	Mobile telephony	Digital voice, short messaging	Integrated high-quality audio, video, and data	Dynamic information access, variable devices	Dynamic information access, variable devices with AI capabilities
Multiplexing	FDMA	TDMA/CDMA	CDMA	CDMA	CDMA
Switching	Circuit	Circuit/circuit for access network and air interface	Packet except for air interface	All packet	All packet
Core network	PSTN	PSTN	Packet network	Internet	Internet
Hand off	—	Horizontal	Horizontal	Horizontal	Horizontal and vertical

Note: FDMA, frequency-division multiple access; LAN, local area network; PAN, personal area network; PSTN, public switched telephone network; TDMA, time-division multiple access; WAN, wide area network; WLAN, wireless local area network.

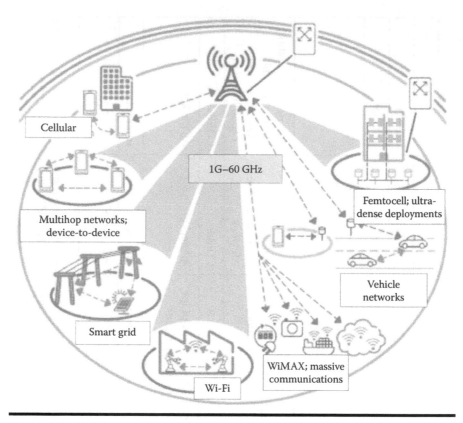

Figure 2.1 Multiple integrated wireless/access solutions that enable a long-term networked society.

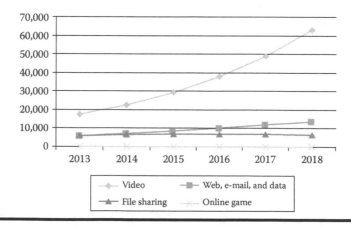

Figure 2.2 Demand for transmission over IP in different years. Values are in petabit per month.

To achieve these goals, the 5G network must have the following characteristics: (1) it should be highly flexible and intelligent; (2) it should have a significant spectrum management scheme; (3) it is expected to improve efficiency while decreasing the cost; (4) it should be able to provide an Internet of Things (IoT), including billions of devices from different sources; (5) it should introduce flexible bandwidth allocation based on the demands of users (what and how much people want to buy); and (6) it should be able to integrate with previous and current cellular and Wi-Fi standards, which give a high rate of communication and decrease delays. In general, for 5G technologies, issues that are under discussion are network densification and millimeter wave (mmWAV) cellular systems, and developing multiple-input, multiple-output (MIMO).

5G networks are required to handle multiple technologies including Wi-Fi and LTE, provide multiple frequency bands, and support greater numbers of users compared with previous networks. Considering the implementation of the LTE system in 4G and its maturity, researchers usually attempt to answer the question of whether this is needed for the next generation. The annual Visual Network Index (VNI) has made it clear that relying only on incremental advances on 4G will not satisfy the ever-increasing demands for more network capacity by a huge number of users [1]. During the last decade, there has been a remarkable increase in the use of smartphones, tablets, video streaming, and online games, so that establishing a new network with better performance is of great value. Besides the growing volume of data, the number of devices and the data rate that is related to channel capacity will increase dramatically. Since using applications for personal communication attracts individuals of different ages and many companies are working on designing new ones, 5G will be responsible for covering all related issues [2–4]. These facts show how much engineers are responsible for innovating new technology to meet the requirements. There are several projects such as METIS [5] and 5GNOW [6] in which academic researchers are involved in designing and establishing the new network. Also, industry is working on 5G standardization activities.

The requirements for a 5G system can be described by different terms. Table 2.2 shows the current state and what is expected from the 5G network. Although all the terms are important, satisfying all of them simultaneously may not be possible and depends on the application. For example, in applications such as HD video streaming, latency and reliability can be ignored to some extent; however, in driverless cars or public safety applications, these parameters cannot be compromised. The first parameter in Table 2.2 is the data rate, which determines the degree to which a network is able to support the mobile data traffic explosion. Table 2.2 measures this term in several ways: (1) area capacity (aggregate data rate), which is the total amount of data served by a network in bits/s; (2) edge rate (%), which is the worst data rate that is expected by a user within the range of a network; and (3) peak rate, which is the highest expected data rate. In general, 5G aims to increase the aggregate data rate and the edge rate, respectively, by factors of 1000 and 100 compared with 4G. Latency is another issue with which the network can

Table 2.2   Expected Improvement from 5G Network

	Data Rate		Latency	Cost
	Area Capacity	Edge Rate		
The needed improvement from 4G to 5G	1000× 4G	100× 4G	15 ms in 4G to 1 ms in 5G	5G ≪ 4G

be evaluated. Although the current round-trip latencies of 4G are sufficient for providing services, 5G is anticipated to support a network containing new cloud-based technologies, and practical applications such as Google Glass and many other wearable devices. With this aim, the researchers involved in designing 5G should provide a round-trip latency of about 1 ms, which is remarkably lower than 4G (15 ms). Reducing the cost and energy of the network is another fact that is going to be feasible through 5G. The data rate will increase by 100× in 5G, therefore the cost per bit should decrease by 100×. This implies that a cheaper mmWAV spectrum should be provided for 5G.

2.2   What is 5G?

5G, also known as the fifth-generation mobile network or fifth-generation wireless systems, applies the next generation of mobile telecommunication standards. The following are some of the main expectations from the 5G network to improve the telecommunication network. First, the 5G network aims to provide a very high data rate for a huge number of users. Second, it also aims to support several simultaneous connections for deploying massive numbers of sensors. Compared with 4G, there should be a noticeable enhancement in the spectral efficiency of the 5G network. The telecommunication area has been experiencing a new generation of mobile networks almost every 10 years since the advent of 1G. Introducing any new mobile generation comes about by assigning new frequency bands and a wide spectral bandwidth per frequency channel. Table 2.3 shows the progress of different telecommunication systems and their corresponding spectral bandwidth [7].

Other parameters that are expected to be enhanced in 5G include higher peak bit rate, handling more simultaneously connected devices, higher spectral efficiency, lower battery consumption, lower outage probability (better coverage), high bit rate in larger portions of the coverage data, lower latencies, higher numbers of supported devices, lower infrastructure deployment costs, and more reliable communication. The expected deployment for this network is 2020.

A challenging issue is that the available networks will not support such an increasing number of network usages, which will increase the need to establish a

Table 2.3   Different Mobile Generations and Their Corresponding Bandwidth

Network	Year of Appearance	Bandwidth
1G	1981	<30 kHz
2G	1991	<200 kHz
3G	2001	<20 MHz
4G	2012	<100 MHz

flatter and more distributed network. The increasing demand for sharing and transferring several file formats such as video, audio, image, and data through networks indicates that we need new source coding such as H.264. Another thing that should be considered is using advanced radio access networks (RANs) such as heterogeneous networks (HetNets), and higher technologies for radio access (RATs) such as the new wireless wide area network (WWAN). Considering 5G for the future network necessitates the need to improve technologies related to transportation at cell sites corresponding to needed change in the network speed and its interoperability. Generally, this optimization will be on the network, devices, and applications. 5G wireless technologies will provide a very high bandwidth by changing the way we use wireless gadgets. Another fact about 5G is that it will interconnect the entire world without limits by employing intelligent technology. It will be based on a new concept of a multipath data path scheme for providing a real worldwide wireless web (wwww). To design such a wireless world, the integration of networks is required. The final design is expected to be a multi-bandwidth data path, which is designed through collecting the current and future networks and introducing the new network architecture of 5G in reality. Figure 2.3 shows this structure, which integrates the present and future networks.

Therefore, in such a real wireless world (5G), code-division multiple access (CDMA), orthogonal frequency-division multiplexing (OFDM), multicarrier code-division multiple access (MCCDMA), ultrawide band (UWB), and Internet protocol version 6 (IPv6) will support the network. As a result of such an extensive architecture, by using 5G it will be possible to have remarkable data capabilities and connect unlimited call volumes and infinite data broadcast. This ability necessitates that the applied technology for router and switch in 5G should be able to provide a high connectivity for the network. Another anticipation of 5G is its ability to distribute Internet access to nodes across the world at a smooth speed. Using 5G, the provided resolution for a wireless network will be high and there will be bidirectional large bandwidth shaping. A great characteristic of 5G technology will be its ability in remote diagnostics. Users will experience a network that gets better and fast solutions via remote management.

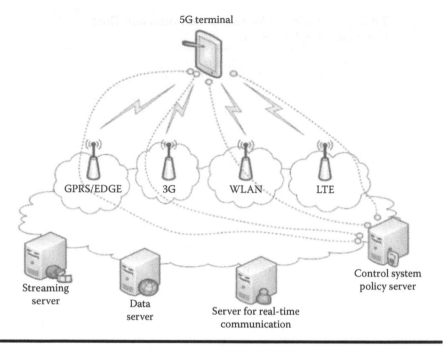

Figure 2.3 **5G structure based on combining the current and future networks.**

2.3 Applications for 5G

With the advent of 5G, every type of communication will be affected to a great degree. Let us look at the motivations for seeking the new network called 5G. It is clear that an increasing demand for high throughput connections, the need to increase the volume of data over wireless networks, the demand for better quality service, and a lower price are among the factors that have led to the 5G network. Mobile networks, health care, video and audio steaming over the Internet, games, security monitoring, and various aspects of our lives will take advantage of the 5G network. It will also play a significant role in business, industry, schools, and colleges, in the lives of doctors, pilots, and the police, in vehicles and many other areas of our lives. One of the greatest advantages of 5G is its ability to establish a global network. This global network is based on using all the available communications. Think about the availability of wearable devices with artificial intelligence capabilities, which can help us to monitor our body's activities such as heart rate variability, blood pressure, and the brain's activities, and setting up online communication with a central health-care center. 5G aims to make such a great contribution.

2.4 5G Specifications

In 3G and 4G, improving the peak rate and spectral efficiency are the primary goals. 5G aims to increase the efficiency of a network based on one of the most helpful low-cost architectures called the *dense HetNet*. This is to satisfy all the demands of industries and provide consistent connectivity. In 5G, the architecture of HetNet will be such that a diverse set of frequency bands will be incorporated. This range of frequency bands includes macrocells in a licensed band such as LTE and small cells in a licensed or an unlicensed band such as Wi-Fi. Another possibility is using a higher-frequency spectrum such as mmWAV in small cells, which will provide ultrahigh data rate services.

2.5 Challenges

The mechanisms for integrating various standards and providing a common platform and a suitable infrastructure are among the most important challenges in designing and establishing 5G networks. In establishing 5G wireless networks, the requirements can be addressed under three main categories. First, from what is expected of the 5G network, it should be capable of providing huge capacity and large connectivity. Second, the 5G network is going to support a vast variety of services, applications, and users related to different areas of life. The third point in establishing the 5G network is its flexibility and efficiency in utilizing all of the available capacity in the spectrum for deploying different networking scenarios. Mobile networks have been increasingly covering all aspects of our daily communications. Therefore, these networks should be able to deliver a connection with a suitable QoS and be highly reliable and fairly secure. To achieve these goals, the designed technology for establishing the 5G should consider the ability to support visual communications with ultra-high-quality and alluring multimedia interactions.

The ultimate goal of the 5G is a network that will support many devices from cars to wearable devices to household appliances and many more. The performance of such an extensive network can be termed as *unlimited* so that multiple gigabits per second are needed. One of the primary goals of the 5G network is building smart cities that provide the required infrastructure. These smart cities would provide mobile industrial automation, vehicular connectivity, and other IoT applications, with the network providing a connection with low latency and high reliability.

As mobile services become increasingly diverse with a wide range of services, different performance requirements are needed. Figure 2.4 shows an overview of the requirements of the 5G network, such as network throughout, latency, and number of connections.

According to Figure 2.4, there are several important challenges in designing the 5G network to satisfy all the aforementioned service requirements. To meet

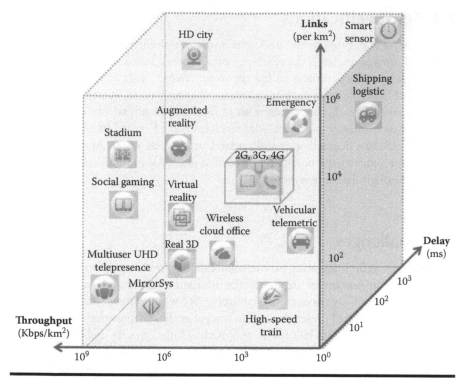

Figure 2.4 **The requirements for 5G service and scenario.**

the requirements for providing ultra-HD video and virtual reality applications, 5G should be able to support at least 1 Gb/s or more data rate. Figure 2.4 clearly demonstrates how much 5G is expected to improve to meet all the requirements in terms of data rate, latency, switching time between different radio access technologies, and energy consumption. In general, the potential requirements for 5G networks include increasing their capacity by a factor of nearly 1000 in traffic load, a peak data rate of 5–10 Gbps, a spectral efficiency of 10 bps/Hz, and latency of 1 ms for the user plane and 50 ms for the control plane. It should also consider mmWAV and unlicensed bands for spectrum usage. Another requirement is mobility with a maximum speed higher than 350 km/h and a hand-off switch time lower than 10 ms. The reliability of the designed 5G networks is expected to be very high.

2.6 Key Technology for 5G Networks

Figure 2.5 shows the requirements for 5G networks. In 5G networks, it is desired to provide a multi-gigabit-per-second-based data rate for communication by using massive MIMO, mmWAVs, and new waveforms. There is a great demand for a radical increase in the capacity and bandwidth of different cellular and wireless networks.

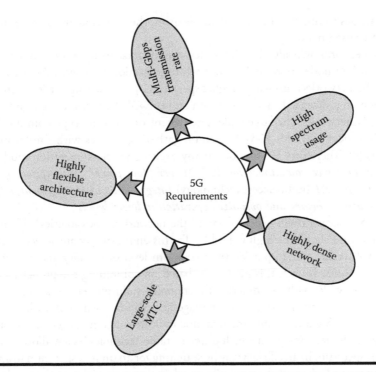

Figure 2.5 Key technology trends for 5G networks and their requirements.

The data rates in a future wireless generation 5G network must increase up to several gigabit per second. This high data rate can be processed by using mmWAV spectrum steerable antennas. This smaller millimeter wavelength can be integrated with directional antennas for higher throughput because massive MIMO as a spatial processing technique can provide orthogonal polarization and beam-forming adaptation.

Figure 2.6 shows the available mmWAV bands for a mobile access network. Carrier aggregation will be applied to offer considerably higher data rates, which create a larger virtual bandwidth by combining a separate spectrum band. One of

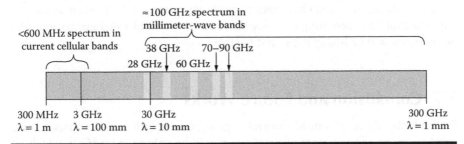

Figure 2.6 Millimeter waveform bands for mobile access networks.

the strategies to improve the bandwidth is using the carrier aggregation of licensed and unlicensed bands.

5G networks will also be highly dense networks, using advanced small cells, advanced internode coordination, and self-organization networks. Another advantage in 5G networks is utilizing a higher spectrum by considering carrier aggregation, operation on unlicensed bands, operation on mmWAV bands, and cognitive radio.

In 5G networks, the large-scale deployment of machine-type communication (MTC) devices will be achieved based on gathering devices with similar mobility patterns [8]. Therefore, 5G supports many exciting wireless operation modes such as *device-to-device communication* (D2D), *very low power consumption operation modes, multi-RAT* (radio access technology) *integration and management, advanced multiple-access schemes,* and *optimized operation in lower bands.*

The 5G network will benefit from all the networking possibilities and therefore its architecture should be highly flexible. To this end, using context-aware networking (CAN) is one way to provide the maximum level of stability and reliability for digital networks. In fact, this network is based on combining the properties of two different networks with two different functionalities: dumb networks and intelligent networks. A dynamic radio resource management will be used in 5G, which is based on software-defined radio [9]. The dynamic radio resource management is based on cognitive radio technology in which different radio technologies are allowed to share the same spectrum in an efficient way by searching for an unused spectrum in an adaptive manner and adapting the transmission scheme corresponding to the requirements of the technologies that share the spectrum. Network function virtualization (NFV) is another way to make the 5G more flexible. Through using this function, we can decouple network functions from dedicated devices, thereby allowing network services to be hosted on a virtual machine. By using NFV, 5G will result in a decrease in the amount of proprietary hardware needed to launch and operate network services.

MIMO technology, which is based on multi-input and multi-output, combines multiple transmitters and receivers or antennas and can be considered as a smart antennas array group. This will be used in 5G networks since it provides higher performance than partial multiuser MIMO. Through massive MIMO, which is based on arrays with up to hundreds of elements and its typical operation is in higher frequencies larger than 10 GHz, with enough elements the capacity can be increased. There are some challenges in applying massive MIMO such as mutual antenna coupling, designing complex RF hardware, and channel estimation. The 5G network will be highly dense and flexible.

2.7 Conclusion and Future Works

5G with the abovementioned features is going to revolutionize the market for a wireless system. The concept of a super core will be enhanced by 5G in which all the network operators will be connected through one single core and have one

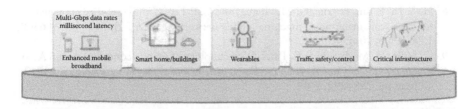

Figure 2.7   5G as a platform for implementing any future wireless application.

single infrastructure no matter what their access technologies are. The 5G network will be a combination of several improved technologies to meet the requirements for establishing a more efficient network with higher capacity and better QoS, and with green technology. To design a network with such great quality and ability, a network that is denser with small cells is key. Spectrum sharing is still a challenge for the wireless industry in 5G networks.

5G is going to address all the issues related to progressing from today's wired communication to a wireless one. Safety and security are among other important issues of the available network generation. It is expected that 5G will establish an extensive and reliable network with the ability to provide security. It is anticipated that the 5G network will be established by 2020 since the growth in data traffic necessitates having such a strong network. The 5G network is where, as Figure 2.7 shows, any future wireless application can be implemented.

References

1. Cisco, Visual Networking Index, 2014, White Paper. Available from www.cisco.com.
2. M. S. Corson, R. Laroia, L. Junyi, V. Park, T. Richardson, and G. Tsirtsis, Towards proximity-aware internetworking, *IEEE Wireless Comm Mag* 17(6): 26–33, 2010.
3. A. Maeder, P. Rost, and D. Staehle, The challenge of M2M communications for the cellular radio access network. In *Proceedings of Würzburg Workshop IP, Joint ITG Euro-NF Workshop "Visual Future Generation Networks" EuroView*, Würzburg, Germany, pp. 1–2, 2011.
4. Analysis Mason Inc. (Forecast Report), Machine-to-machine device connections: Worldwide forecast 2010–2020, 2010. Available from: http://www.analysysmason.com/Research/Custom/Reports/RRE02_M2M_devices_forecast/.
5. FP7 European Project 317669 METIS (Mobile and Wireless Communications Enablers for the Twenty-Twenty Information Society), 2012. Available from: https://www.metis2020.com/.
6. FP7 European Project 318555 5G NOW (5th Generation Non-Orthogonal Waveforms for Asynchronous Signalling), 2012. Available from: http://www.5gnow.eu/.
7. Interview with Ericsson CTO: There will be no 5G: We have reached the channel limits, *DNA India*. May 23, 2011. Available from: http://www.dnaindia.com/money/interview-there-will-be-no-5g-we-have-reached-the-channel-limits-ericsson-cto-1546408. Retrieved September 27, 2013.

8. H.-L. Fu, P. Lin, H. Yue, G.-M. Huang, and C. P. Lee, Group mobility management for large-scale machine-to-machine mobile networking, *IEEE Trans Veh Tech* 63(3), 1296–1305, 2014.
9. G. Asvin, H. Modi, and S. K. Patel, 5G technology of mobile communication: A survey, *International Conference on Intelligent Systems and Signal Processing (ISSP)*, IEEE, Gujarat, pp. 288–292, 2013.

Chapter 3

From 4G to 5G

Moneeb Gohar

Contents

3.1 Introduction

Over the last few years, the global demand for mobile data services has experienced phenomenal growth because of the rapid proliferation of smart devices. Mobile data traffic is predicted to increase anywhere from 20 to 50 times over the next 5 years. Almost 80% of mobile data traffic is being generated indoors, which requires increased link budget and coverage extension to provide a satisfactory end-user experience. Indoor performance is poorer than outdoor performance because of radio signals, which are seriously attenuated, distorted, and redirected by walls, ceilings, floors, and so on. Thus, current cellular architectures that were originally designed to serve large coverage areas are no longer able to efficiently cope with such dominant indoor traffic.

The increased usage of smartphones creates the problem of frequent short on/off connections and mobility, generating heavy signaling traffic load in the network. This consumes a disproportionate amount of network resources, compromising the network throughput and efficiency, and in extreme cases causing the third-generation (3G) or fourth-generation (4G; Long-Term Evolution [LTE] and LTE-Advanced [LTE-A]) cellular networks to crash [1].

As the conventional approaches to improving the spectral efficiency or the allocation of additional spectrum or both are fast approaching their theoretical limits, there is a growing consensus that current 3G and 4G (LTE/LTE-A) cellular radio access technologies (RATs) will not be able to meet the anticipated growth in mobile traffic demand. To address these challenges, the wireless industries have initiated a roadmap for transition from 4G to fifth generation (5G). Based on the 4G network, it is generally understood that 5G must address the challenges that are not adequately addressed by the state-of-the-art deployed 4G network (LTE/LTE-A).

The need for a new 5G mobile system is based on a few remarkable changes in mobile network environments, such as the avalanche of overwhelming Internet traffic, the explosive growth in the number of various connected devices, and the large diversity of use cases and requirements [2,3].

The 5G mobile system will be designed to effectively cope with such environmental changes. Among these changes, the most crucial factor to be considered is how to handle the explosive growth of mobile data (Internet) traffic. A report says that overall mobile data traffic is expected to grow up to 24.3 exabytes per month by 2019 [4].

To address the mobile data traffic explosion issue, many ideas are being proposed, including small cell approach, device-to-device (D2D) communication, and so on. However, we note that such efforts are mainly focusing on how to increase the capacity of wireless radio links. The 5G system consists of a radio link part and a mobile core network part. It is believed that the effective design of a mobile core network, as well as the radio link part, is also very crucial to achieve the goals of the 5G system.

This chapter provides a brief overview of 4G networks (LTE/LTE-A) with the dawn of 5G, 5G services visions, 5G requirements, and an overview of key enabling technologies for the 5G network.

3.2 Overview of LTE

3.2.1 LTE Basics

The ongoing growth of the transmission bandwidth is challenging the limits of 3G networks, hence it was decided by the 3rd Generation Partnership Project (3GPP) in 2005 to start work on the next-generation network. LTE is the latest standard that is being implemented within the 3GPP to ensure the competitiveness of 3G for the next 10 years and beyond. LTE supports both time-division duplex (TDD) and frequency-division duplex (FDD) [5,6].

An LTE base station (BS) is referred to as an enhanced NodeB (eNB) to differentiate it from a Universal Mobile Telecommunication System (UMTS) BS, which is known as NodeB. eNB are made more intelligent than NodeB by removing the radio network controller (RNC) and transferring the functionality to eNB and partly to the core network gateway. eNB can also perform handovers through the X2 interface and forward the downlink data from the source eNB to the target eNB. The X2 interface uses the tunneling protocol for the control plane (GTP-C). eNB connects to the gateway nodes through the S1 interface. All eNBs are connected to at least one mobility management entity (MME) over the S1-MME interface. The MME is the control plane, which mainly handles mobility management, authentication, bearer management, selection of gateway, session management signaling, and location tracking of mobile devices. The MME relies on the existence of subscription-related user data for all users trying to establish Internet protocol (IP) connectivity. For the purpose of user subscription information, the MME is connected to the home subscriber server (HSS) over the S6a interface. Subscription data includes credentials for authentication and access authorization. The HSS also supports mobility within LTE as well as between LTE and other access networks. The data packets flowing to and from the mobile devices are handled by two nodes, called the *serving gateway* (S-GW) and the *packet data network* (PDN-GW). The S-GW terminates the S1-U interface toward the eNBs, and works as a local mobility anchor (LMA) for intra-3GPP handover. The S-GW also buffers downlink packets when terminals are in idle mode. For roaming users, the S-GW always resides in the visited network and supports accounting functions (billing settlements). The S-GW and MME can be implemented on the same hardware or separately. If implemented separately, the S11 interface is used to communicate between them. PDN-GW is the point at which the LTE network interconnects with the external IP network through the SGi interface. PDN-GW provides the IP multimedia services, charging, packet filtering, policy-based control, and the allocation of IP addresses. The S-GW and PDN-GW are connected through the S5 interface (if a user resides in a home network) or S8 (if a user resides in a visited network). It uses a general packet radio service (GPRS) tunneling protocol-user plane (GTP-U) to tunnel user data from/to the S-GWs and the GPRS tunneling protocol-signaling protocol (GTP-S) for the initial establishment of a user data tunnel and subsequent tunnel modifications when the user moves between cells that are managed by different S-GWs. The policy and charging rules function (PCRF) determines policies, such as quality of service and charging rules, and connects to the S-GW/PDN-GW through the Gx interface. When Proxy Mobile IPv6 (PMIPv6) is used on the S5, the PCRF connects to the S-GW through the Gxc interface. The home PCRF (nonroaming) and visited PCRF (roaming) interconnects through the S9 interface. The S10 interface interconnects two MMEs, when the MME that is serving a user has to be changed either due to maintenance or node failure or when a terminal moves between two pools [7]. The basic LTE network architecture [6,7] is shown in Figure 3.1.

Orthogonal frequency-division multiple-access (OFDMA) technology is used for LTE downlink transmission, that is, from BSs to the end-users' devices. The basic concept of OFDMA is that the total available channel spectrum (e.g., 10 MHz) is subdivided into a number of 15 kHz channel, each carrying one subcarrier. The subcarriers' spacing provides orthogonality among carriers. The transmission speed of each subcarrier can be much lower than the overall data rate, because many bits of data are transmitted in parallel. This not only minimizes the multipath fading but also the effect of multipath fading and delay spread become independent of the channel bandwidth used. This is because the bandwidth of each subcarrier remains the same and only the number of subcarriers is changed for a different achievable overall bandwidth. The most common modulation techniques used are binary phase-shift keying (BPSK), quadrature phase-shift keying (QPSK), and quadrature amplitude modulation (QAM).

For OFDMA downlink transmission, the inverse fast Fourier transform (IFFT) is used to transform the signal from the frequency domain to the time domain. The resulting signal is transmitted in the air after modulation and amplification. Firstly, when the signal is received by the receiver, it demodulates and amplifies the signal. After this, the fast Fourier transform (FFT) is used to convert the signal back from the time domain to the frequency domain. The multiple access (MA) in OFDMA refers to the downlink data that are received by several users simultaneously. By using control messages, the mobile devices send information regarding waiting for

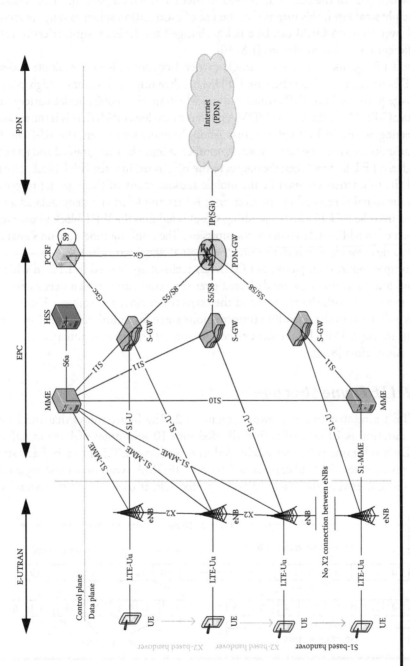

Figure 3.1 LTE network architecture.

data, which part of the data is addressed to them and which part they can ignore. On the physical layer, this means that the use of modulation schemes ranging from QPSK over 16 to 64 QAM can be quickly changed for different subcarriers to fulfill different reception conditions [1,8–10].

For LTE uplink transmission, single-carrier frequency-division multiple access (SC-FDMA) is used. This is because OFDMA inherently suffers from a high peak-to-average power ratio (PARP), which can quickly drain the mobile device battery. In general, SC-FDMA is similar to OFDMA but has much lower PARP. This is the reason for selecting SC-FDMA for uplink transmission. In many subcarriers, the SC-FDMA also transmits data over the air interface. A number of input bits are grouped and passed through the FFT first and then the output of the FFT is fed into the IFFT block. Since not all the subcarriers are used by the mobile station, many of them are set to zero. When the signal is received by the receiver, it is first amplified and demodulated and then fed into the FFT block. The resulting signal is fed into the IFFT block to counter the effect of an additional step in the transmission. The resulting time-domain signal is fed into a detector block, which recreates the original signal bits [1,8–10].

Multiple-input, multiple-output (MIMO) technology is used in LTE in which two and four transmissions are delivered over the same band, which needs two or four antennas at both the receiver and the transmitter side, respectively. However, LTE is only used in the downlink transmissions since for uplink transmissions it is difficult to use MIMO for mobile devices because of the limited antenna size and power constraints [8,11].

3.2.2 LTE Frame Structure

The LTE frame structure is shown in Figure 3.2. The figure shows that the LTE frame duration is 10 ms and is then divided into 10 subframes of 1 ms duration each. Each subframe is further subdivided into two slots of 0.5 ms each. Each slot of 0.5 ms consists of 12 subcarriers and 6 or 7 OFDMA symbols depending on if either the standard or the extended cyclic prefix (CP) is used. When the extended

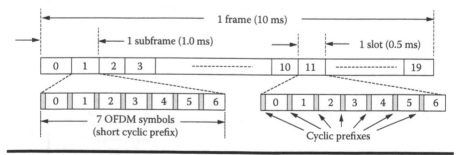

Figure 3.2 LTE frame structure. (S. Hussai, An innovative RAN architecture for emerging heterogamous networks: The road to the 5G era, *Dissertation and Thesis,* **2014.)**

CP is used, then the number of OFDMA symbols is reduced to 6. The grouping of 12 subcarriers results in a physical resource block (PRB) bandwidth of 180 kHz. Two slots group together to form a subframe, which is also known as the *transmit time interval* (TTI). In the case of the TDD operation, the subframe can be used to downlink or uplink. This is decided by the network frames used to downlink or uplink. However, in LTE, most networks are likely to use FDD in which separate bands are used to uplink and downlink [1,8–10].

3.2.3 eNB, S-GW, and MME Pools

eNB, S-GW, and MME pools are shown in Figure 3.3 [1].

> *Tracking Area (TA):* A group of BSs providing radio services for a wider area. Each area is identified by a TA identity (TAI). User equipment (UE) does not need to send a TA update as long as it is roaming in a TA.
> *Pool Area:* Can be one or more TAs, served by one or more MME/S-GW pools.
> *MME Pool:* One or more MMEs can serve other (RAN) pool areas.
> *S-GW Pool:* One or more S-GWs.

3.2.4 Protocol Stack

Figure 3.4 shows the protocol stack for the GTP-based data delivery in a 4G evolved packet core (EPC). The radio access uses the protocols media access control (MAC), radio link control (RLC), and packet data conversion protocol (PDCP). The GTP is used between eNB and S-GW/PDN-GW. The GTP encapsulates the original IP packet into an outer IP packet [7].

3.2.5 Initial Registration

Figure 3.5 describes the initial procedures in 4G-EPC: network attachment and binding update by UE, and the data delivery from one UE to another UE. When the UE establishes a radio link with the eNB, it sends an *Attach Request* to the MME. Then, security-related procedures are performed between the UE and the MME. The MME will update the associated HSS. To establish a transmission path, the MME sends a *Create Session Request* to the S-GW. When the S-GW receives the request from the MME, it will send a *Modify Bearer Request* message to the PDN-GW. The PDN-GW responds with a *Modify Bearer Response* message to the S-GW. Then, the S-GW will respond with a *Create Session Response* to the MME. Now, the MME sends the information received from the S-GW to the eNB within the *Initial Context Setup Request* message. This signaling message also contains the *Attach Accept* notification, which is the response to the *Attach Request*. The eNB duly responds with an *Initial Context Setup Response* to the MME. Then,

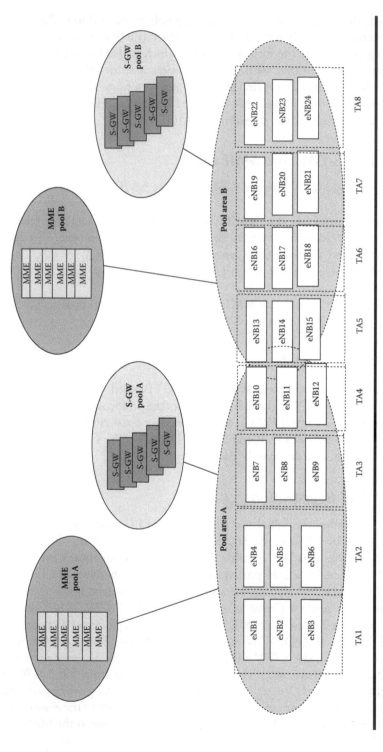

Figure 3.3 eNB, S-GW, and MME pools.

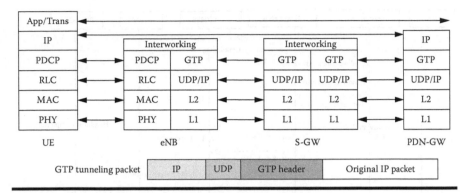

Figure 3.4   Protocol stack for data delivery.

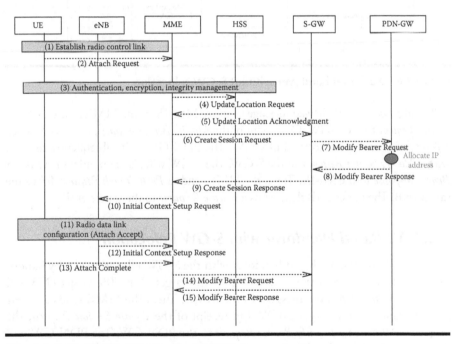

Figure 3.5   Initial registration in 4G-EPC.

the UE sends an *Attach Complete* message to the MME. Consequently, the MME sends a *Modify Bearer Request* message to the S-GW, and the S-GW will respond with a *Modify Bearer Response* to the MME [7].

3.2.6 X2-Based Handover without S-GW Relocation

Figure 3.6 shows the X2-based handover without the S-GW relocation of 4G-EPC [7]. By handover, the UE moves from the source eNB to the target eNB. The target eNB

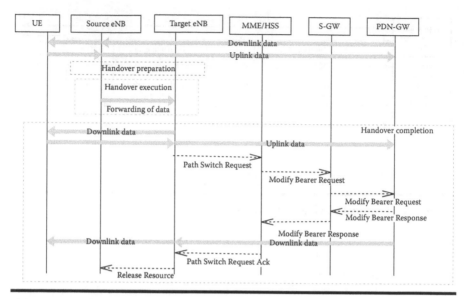

Figure 3.6   X2-based handover without S-GW relocation.

will send a *Path Switch Request* message to the MME. Then, the MME sends a *Modify Bearer Request* to the S-GW. On receipt of the *Modify Bearer Request*, the S-GW sends a *Modify Bearer Request* to the PDN-GW. The PDN-GW will subsequently respond with a *Modify Bearer Response* to the S-GW. The S-GW will also respond with a *Modify Bearer Response* to the MME. Then, the MME sends a *Path Switch Request Ack* to the target eNB. The target eNB then sends a *Release Resource* to the source eNB.

3.2.7 X2-Based Handover with S-GW Relocation

Figure 3.7 shows the X2-based handover with the S-GW relocation [7]. By handover, the UE moves from the source eNB to the target eNB. The target eNB will send a *Path Switch Request* message to the MME. Then, the MME sends a *Create Session Request* to the target S-GW. On receipt of the *Create Session Request*, the target S-GW sends a *Modify Bearer Request* to the PDN-GW. The PDN-GW will duly respond with a *Modify Bearer Response* to the S-GW. The S-GW will also respond with a *Create Session Response* to the MME. Then, the MME sends a *Path Switch Request Ack* to the target eNB. Consequently, the target eNB sends a *Release Resource* to the source eNB.

3.2.8 S1-Based Handover

Figure 3.8 shows the S1-based handover with S-GW and MME relocation [7]. The source eNB decides to initiate an S1-based handover to the target eNB. This can be triggered, for example, by no X2 connectivity to the target eNB, or by an error

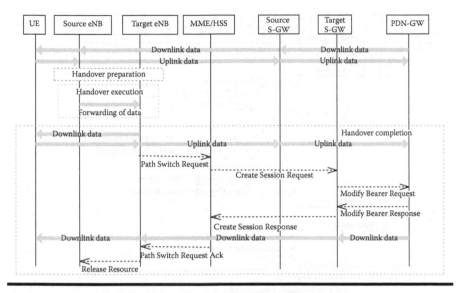

Figure 3.7 X2-based handover with S-GW relocation.

indication from the target eNB after an unsuccessful X2-based handover, or by
dynamic information learned by the source eNB (Step 1). Then, the source eNB
sends *Handover Required* to the source MME (Step 2). The source MME selects
the target MME and if it has determined to relocate the MME, it sends a *Forward
Relocation Request* message to the target MME. If the MME has been relocated,
the target MME verifies whether the source S-GW can continue to serve the UE. If
not, it selects a new S-GW (Step 3). If the MME has not been relocated, the source
MME decides on this S-GW reselection. If the source S-GW continues to serve
the UE, no message is sent in this step. In this case, the target S-GW is identical
to the source S-GW. If a new S-GW is selected, the target MME sends a *Create
Session Request* message per PDN connection to the target S-GW. The target S-GW
sends a *Create Session Response* message back to the target MME (Step 4). The target
MME sends a *Handover Request* message to the target eNB. This message creates
the UE context in the target eNB, including information about the bearers, and the
security context. The target eNB sends a *Handover Request Acknowledge* message to
the target MME (Step 5).

If indirect forwarding applies and the S-GW is relocated, the target MME
sets up forwarding parameters by sending a *Create Indirect Data Forwarding
Tunnel Request* to the S-GW. The S-GW sends a *Create Indirect Data Forwarding
Tunnel Response* to the target MME (Step 6). If the MME has been relocated,
the target MME sends a *Forward Relocation Response* message to the source MME
(Step 7). If indirect forwarding applies, the source MME sends a *Create Indirect
Data Forwarding Tunnel Request* to the S-GW. The S-GW responds with a *Create
Indirect Data Forwarding Tunnel Response* message to the source MME (Step 8).

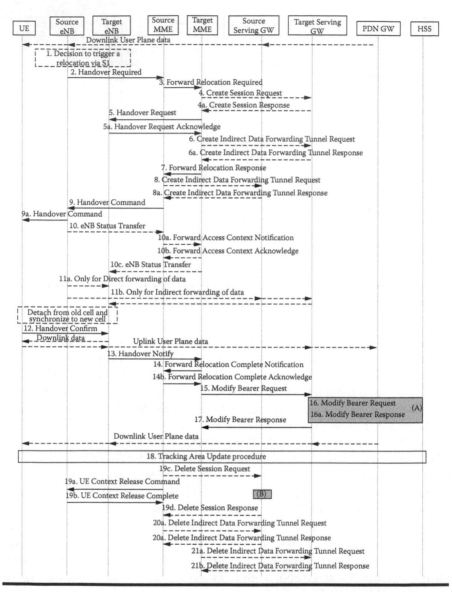

Figure 3.8 S1-based handover with S-GW and MME relocation. (3GPP TS 23.401. General Packet Radio Service (GPRS) enhancements for Evolved Universal Terrestrial Radio Access Network (E-UTRAN) access, Rel-12 Ver. 12.4.0, 2014.)

The source MME sends a *Handover Command* message to the source eNB. The *Handover Command* is constructed using the target to source transparent container and is sent to the UE (Step 9). The source eNB sends the eNB *Status Transfer* message to the target eNB via the MME(s). If there is MME relocation, the source MME sends this information to the target MME via the *Forward Access Context*

Notification message, which the target MME acknowledges with a *Forward Access Context Notification Acknowledge* message. The target MME sends the information to the target eNB via the *eNB Status Transfer* message (Step 10). The source eNB should start forwarding the downlink data from the source eNB to the target eNB (Step 11). After the UE has successfully synchronized with the target cell, it sends a *Handover Confirm* message to the target eNB. Downlink packets forwarded from the source eNB can be sent to the UE. Also, uplink packets can be sent from the UE, which are forwarded to the target S-GW and on to the PDN-GW (Step 12). The target eNB sends a *Handover Notify* message to the target MME (Step 13). Then, the target MME sends a *Forward Relocation Complete Notification* message to the source MME. The source MME in response sends a *Forward Relocation Complete Acknowledge* message to the target MME (Step 14). After that, the target MME sends a *Modify Bearer Request* message to the target S-GW for each PDN connection (Step 15). If the S-GW is relocated, the target S-GW sends a *Modify Bearer Request* message per PDN connection to the PDN-GW. The PDN-GW updates its context field and returns a *Modify Bearer Response* message to the target S-GW. The PDN-GW starts sending downlink packets to the target S-GW. These downlink packets will use the new downlink path via the target S-GW to the target eNB (Step 16). The target S-GW sends a *Modify Bearer Response* message to the target MME. The message is a response to a message sent at Step 15 (Step 17). The UE initiates a *Tracking Area Update* procedure when one of the conditions listed in the clause "Triggers for tracking area update" applies (Step 18). When the timer started in Step 14 expires, the source MME sends a *UE Context Release Command* message to the source eNB. The source eNB releases its resources related to the UE and responds with a UE *Context Release Complete* message (Step 19). If indirect forwarding was used, then the expiry of the timer at the source MME started at Step 14 triggers the source MME to send a *Delete Indirect Data Forwarding Tunnel Request* message to the S-GW to release the temporary resources used for indirect forwarding that were allocated at Step 8 (Step 20). If indirect forwarding was used and the S-GW is relocated, then the expiry of the timer at the target MME started at Step 14 triggers the target MME to send a *Delete Indirect Data Forwarding Tunnel Request* message to the target S-GW to release temporary resources used for indirect forwarding that were allocated at Step 6 (Step 21).

3.2.9 *Proxy Mobile Internet Protocol–Long-Term Evolution*

PMIPv6 [12] has been considered to support IP mobility in the LTE/system architecture evolution (SAE) [13–15]. In the study, to support the PMIPv6 in the LTE/SAE architecture, the PDN-GW is used for the LMA of PMIPv6 and the S-GW is used as the mobile access gateway (MAG) of PMIPv6. The X2- and S1-based handovers are the same as conventional LTE/SAE. PMIP-LTE uses generic routing encapsulation (GRE) tunneling between the S-GW and the P-GW instead of GTP tunneling, and *Proxy Binding Update* (PBU) and *Proxy Binding Ack* (PBA) messages are exchanged between the S-GW and the P-GW instead of *Modify Bearer Request* and *Response* messages. The PMIP-LTE network architecture is shown in Figure 3.9.

3.3 Overview of LTE-Advanced

LTE-A is the next major milestone in the evolution of LTE and is a crucial solution for addressing the anticipated 1000× increase in mobile data. It incorporates multiple dimensions of enhancements including the aggregation of carriers and advanced antenna techniques. But most of the gain comes from optimizing heterogeneous networks (HetNets), resulting in better performance from small cells.

The benefit of small cells in providing capacity where needed, is well understood. So are the challenges and solutions for managing the interference. Enhancements such as "range expansion," introduced in LTE-A, increase the overall network capacity much more than by merely adding small cells. The interference management techniques of LTE-A make adding more small cells possible without affecting the overall network performance.

LTE-A will meet or exceed International Mobile Telephony (IMT)-Advanced (IMT-A) requirement within the International Telecommunication Union Radiocommunication Sector (ITU-R) time plan. Extended LTE-A targets are adopted in LTE Release 11 and Release 12, for example, additional carrier aggregation band combinations. LTE-A also supports new frequency bands. LTE-A is backward compatible with LTE Release 8. An LTE Release 8 UE can operate in an LTE-A network. Also, an LTE-A UE (R10 or higher) can operate in an LTE Release 8 network. LTE-A deployment uses increased deployment of indoor eNB and home eNB (HeNB), which is a type of femtocell with a very small coverage area, typically less than a 50 m radius [1,16].

3.4 Dawn of 5G Era

In recent years, with the unprecedented growth in the number of connected devices and mobile data, and the ever-fast approaching 4G technologies to address this enormous data demand, the wireless industries have initiated a roadmap for transition from 4G to 5G. It is reported that the number of connected devices (Internet of Things [IoT]) is estimated to reach 50 billion by 2020 [17], while the mobile data traffic is expected to grow to 24.3 exabyte per month by 2019 [4], as shown in Figure 3.10.

Furthermore, the impact of higher cell capacity and end-user data rate requirements due to ultra-high-definition (UHD) multimedia streaming and extremely low latency requirements for cloud computing and storage/retrieval, 5G can be expected to support immersive applications that demand exceptionally high-speed wireless connections, a fully realized IoT, and experience lower latency and promote

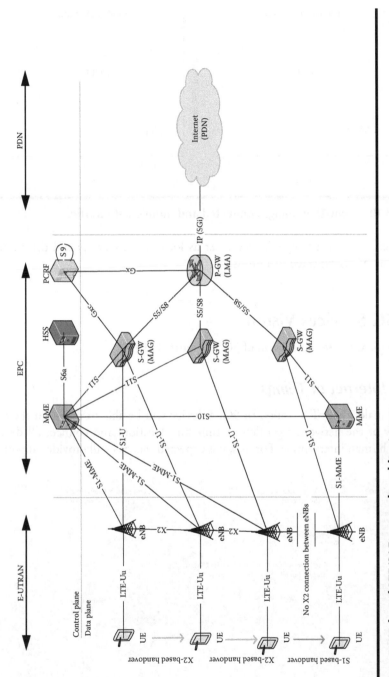

Figure 3.9 PMIP-based LTE/SAE network architecture.

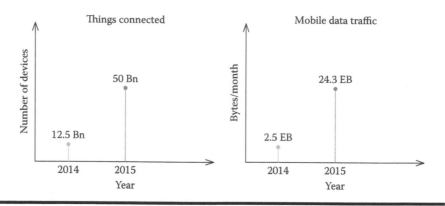

Figure 3.10   Growth in things connected and mobile data traffic.

both spectrum and energy efficiency. Let us look at the services and the requirements that 5G is expected to address.

3.5 5G Services Vision

Let us look at the services vision of 5G in Figure 3.11 [18].

3.5.1 Internet of Things

5G will make the IoT a reality. In 5G, a device will be able to maintain the connectivity of the network regardless of time and location, and connect all devices without human intervention. For this, it is expected that 5G will provide support for

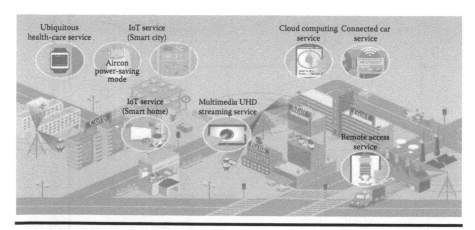

Figure 3.11   5G services vision. (DMC R&D Center, Samsung electronics "5G Vision", White Paper [Online] 2015.)

up to a million simultaneous connections per square kilometer, enabling a variety of machine-to-machine services, including wireless metering, mobile payment, smart grid, connected home, smart home, smart transportation, fitness/health care, smart store, smart office, and connected car. Intelligent devices will communicate autonomously and freely share information with each other in the background.

3.5.2 Immersive Multimedia Experience

In 5G, users will experience lifelike multimedia streaming anytime and anywhere. Users will feel as if they are part of the scene when they watch videos on their smart devices. To provide such an immersive experience, it is expected that UHD video streaming will provide a lifelike experience in the 5G system. Currently, UHD services are already standardized in some countries. Some smartphones in the market are now equipped with a camera that can record with 4K UHD video. It is expected that UHD services will likely be mainstreamed by 2020. Other examples are virtual reality (VR) and augmented reality (AR). VR provides a world where physical presence is simulated by computer graphics and users can interact with the simulated elements as in immersive sports broadcasting. Other scenarios are interactive 360 movies, online games, remote education, and virtual orchestra. In an AR, the computer-aided, real-time information based on user context is graphically augmented to the display. AR will help to inform the price, popularity, and details of a given product. Another AR service is navigation on a windshield, where navigation information and other helpful notifications are displayed on the windshield of a car.

3.5.3 Everything on the Cloud

5G will provide a desktop-like experience based on cloud computing for users. Everything is stored and processed on the cloud and immediately accessed with low latency. As an example, when you go shopping, the smart device can notify you about the arrival of a new coat that you might like or it will let you know how well the coat in the new inventory matches with your liking based on your purchase history. This notification can be triggered as you step into a shop.

3.5.4 Intuitive Remote Access

In the 5G environment, users will be able to control remote machines (heavy industrial machines) and appliances and access hazardous sites remotely, as if they were right in front of them, even from thousands of miles away.

3.6 5G Requirements

5G requirements consist of seven key performance indices as shown in Figure 3.12 [18].

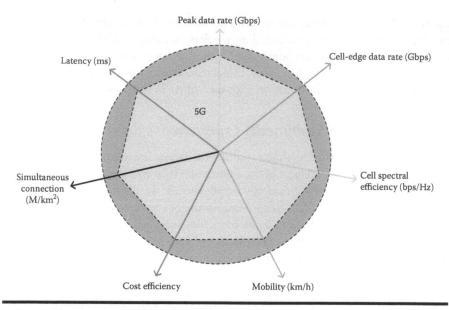

Figure 3.12   5G requirements.

3.6.1 Cell-Edge Data Rate and Peak Data Rate (Gbps)

It is expected that the 5G system will support data rates of 10–50 Gbps for low-mobility users. The 5G system will provide gigabit-rate data services regardless of a user's location, as shown in Figures 3.13 and 3.14.

3.6.2 Latency

5G will provide an end-to-end latency of less than 5 ms and air latency of less than 1 ms, as shown in Figure 3.15, which is one-tenth the latency of the 4G network.

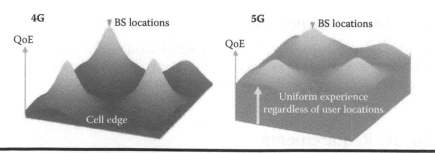

Figure 3.13   Edgeless RAN-1 Gbps anywhere.

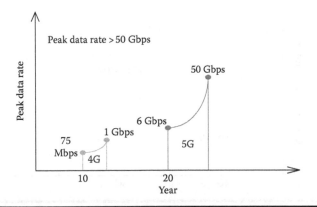

Figure 3.14   Data rate comparison of 5G with 4G.

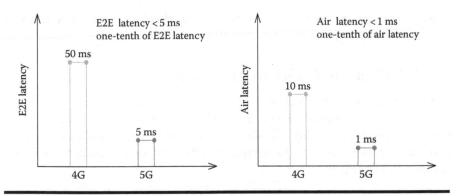

Figure 3.15   Latency comparison of 5G with 4G.

3.6.3 *Simultaneous Connection (M/km²)*

The simultaneous connections in the 5G system are expected to be over 10^6 per unit square kilometer, which is much higher than that of the legacy system.

3.6.4 *Cost Efficiency*

5G systems are targeted to be 50 times more efficient than 4G by delivering reduced cost and energy usage per bit. This sequentially requires low-cost network equipment, lower deployment costs, and enhanced power-saving functionality on the network and UE sides.

3.6.5 *Mobility*

5G technologies will provide mobility on demand based on the needs of each device and service. On the one hand, the mobility of the UE should be guaranteed to be

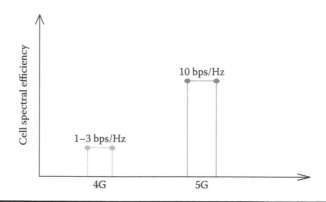

Figure 3.16   Cell spectral efficiency comparison of 5G with 4G.

at least the same level as the 4G system. On the other hand, the 5G system will support mobility at speeds ranging from 300 to 500 km/h.

3.6.6 Cell Spectral Efficiency (bps/Hz)

The cell spectral efficiency is set to 10 bps/Hz level (in contrast to the 1–3 bps/Hz on 4G networks) as shown in Figure 3.14. 5G is also expected to deliver an efficient use of the spectrum by using MIMO, advanced coding and modulation schemes, and a new waveform design. The cell spectral efficiency comparison of 5G with 4G is shown in Figure 3.16.

3.7 Overview of 5G Key Enabling Technologies

The 5G enabling technologies will meet unprecedented speeds, near-wireline latencies, and ubiquitous connectivity with uniform quality of experience (QoE), and they will have the ability to connect massive numbers of devices with each other [18]. 5G technologies will provide an immersive experience, even while the user is on the move. The future 5G system will boast wireless capacity utilizing new frequency bands, advanced spectrum efficiency enhancement methods in the legacy bands, and seamless integration of licensed and unlicensed bands. Table 3.1 summarizes the enabling technologies in terms of the 5G requirements.

Figure 3.17 shows an overview of the 5G key enabling technologies. In the 5G system, the massively higher capacity needs will be addressed by new millimeter-wave (mmWave) systems, which provide 10 times more bandwidth than the 4G cellular bands; advanced small cell in which it is necessary to deploy a large number of cells in a given area and to manage them intelligently; advanced multiple input, which experiences small interuser and intercell

Table 3.1 Enabling Technologies in Terms of 5G Requirements

Enabling Technologies	Latency	Simultaneous Connection	Cost Efficiency	Mobility	Cell Spectral Efficiency	Cell-Edge Data Rate	Peak Data Rate
mmWave system	√	√	√	—	√	√	√
Multi-RAT	√	—	—	√	√	√	—
Advanced network	—	—	—	√	√	√	√
Advanced MIMO	—	√	√	—	√	√	—
ACM and multiple access	—	√	√	—	—	√	—
Advanced D2D	—	—	√	—	—	√	√
Advanced small cell	—	√	—	√	√	—	—

Source: DMC R&D Center, Samsung electronics "5G Vision", White Paper (Online) 2015.

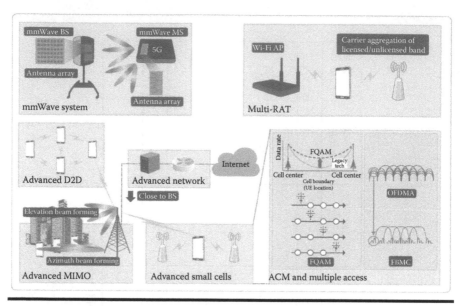

Figure 3.17 Overview of the 5G key enabling technologies. (DMC R&D Center, Samsung electronics "5G Vision", White Paper [Online] 2015.)

interferences and consequently achieve significantly higher throughput than the state-of-the-art MIMO system, and multiple output (MIMO); and new MA schemes such as filter bank multicarrier (FBMC). Adaptive coding and modulation such as frequency and quadrature amplitude modulation (FQAM) can significantly improve the cell-edge performance and, combined with higher density deployments with multi-BS cooperation, will help to deliver on the promise of "Gbps anywhere" and uniform QoE. Multi-RAT integration including carrier aggregation of licensed and unlicensed bands will inevitably help in increasing the available system bandwidth. On the network side, novel topologies including application servers placed closer to the network edge will contribute to significantly reducing the network latency. Advanced D2D technology can help reduce the communications latency and support larger numbers of simultaneous connections in a network.

3.8 Conclusion

This chapter provided a brief overview of 4G networks (LTE/LTE-A) with the dawn of 5G, 5G services visions, requirements that are not adequately addressed by the state-of-the-art deployed 4G network, and an overview of the key enabling technologies for the 5G network.

References

1. S. Hussain, An innovative RAN architecture for emerging heterogamous networks: The road to the 5G era, *Dissertation and Thesis*, 2014.
2. FP7 METIS project. Mobile and wireless communications enablers for the 2020 information society. 2013. Available from: http://www.metis2020.com.
3. H. Benn, Vision and key features for 5th generation (5G) cellular: Samsung, Technical Report, 2014.
4. Cisco, Cisco visual networking index: Global mobile data traffic forecast update: 2013–2018, Cisco, 2014.
5. D. Calin, H. Claussen, and H. Uzunalioglu, On femto deployment architectures and macrocell offloading benefits in joint macro-femto deployments, *IEEE Comm Mag* 48: 26–32, 2010.
6. Qualcomm, A comparison of LTE advanced HetNets and WiFi, White Paper (Online) 2013. Available from: https://www.qualcomm.com/documents/comparison-lte-advanced-hetnets-and-wi-fi.
7. 3GPP TS 23.401. General Packet Radio Service (GPRS) enhancements for Evolved Universal Terrestrial Radio Access Network (E-UTRAN) access, Rel-12 Ver. 12.4.0, 2014.
8. Telesystem Innovations Inc., LTE in nutshell: The physical layer, White Paper (Online) 2010. Available from: http://www.tsiwireless.com/.
9. Anritsu Corporation, LTE resource guide, 2009. Available from: http://www.us.anritsu.com/.
10. J. Zyren, Overview of the 3GPP long term evolution physical layer, Freescale Semiconductor, White Paper (Online) 2007. Available from: http://www.freescale.com/.
11. Juniper, WiFi and femtocell integration strategies 2011–2015, White Paper (Online) 2011. Available from: http://www.juniperresearch.com/.
12. S. Gundavelli, K. Leung, V. Devarapalli, K. Chowdhury, and B. Patil, Proxy Mobile IPv6, *IETF RFC 5213*, 2008.
13. J. Laganier, T. Higuchi, and K. Nishida, Mobility management for all-IP network, *NTT DOCOMO Tech J* 11(3): 34–39, 2009.
14. Y-S. Chen, T-Y. Juang, and Y-T. Lin, A secure relay-assisted handover protocol for Proxy Mobile IPv6 in 3GPP LTE systems, *Wireless Pers Comm* 61(4): 629–656, 2011.
15. A. R. Prasad, J. Laganier, A. Zugenmaier, M. S. Bargh, B. Hulsebosch, H. Eertink, G. Heijenk, and J. Idserda, Mobility and key management in SAE/LTE, *CNIT Thyrrenian Symposium on Signals and Communication Technology*, Springer, New York, pp. 165–178, 2007.
16. S. Abeta, T. Abe, and T. Nakamura, Overview and standardization trends of LTE-advanced, *NTT Tech Rev* 10(1): 1–5, 2012.
17. UMTS, Mobile traffic forecasts 2010–2020 report, *UMTS Forum*, 2011.
18. DMC R&D Center, Samsung electronics "5G Vision", White Paper (Online) 2015.

Chapter 4

Communication Haul Design for 5G Radio: Challenges and Open Issues

Kazi M.H. Huq, Shahid Mumtaz, Jonathan
Rodriguez, and Manuel Violas

Contents

4.1 Introduction

In a data communication system, the segment that connects the core and the access networks is termed the *backhaul*. The edges of any telecommunication network are connected through backhauling. Backhaul links have been one of the building blocks for the next-generation mobile networks. Research communities around the world make tremendous efforts in their research on efficient and enhanced backhaul technology and topology. The importance of backhaul research is spurred by the need for increasing data capacity and coverage to cater for the ever-growing population of electronic devices—smartphones, tablets, and laptops—which is foreseen to hit unprecedented levels by 2020. The backhaul is anticipated to play a critical role in handling large volumes of traffic, with stringent demands placed on it from both mobile broadband and the introduction of heterogeneous networks (HetNets). In fact, broadband has been evolving rapidly for the last decade and this has prompted the accompanying backhaul technologies to adjust invariably to satisfy both users and operators. The evolution in radio access network (RAN) backhaul is being triggered by the adoption of the Ethernet as the physical interface and the proven benefits of Internet protocol (IP) in unraveling the network layer.

Mobile traffic and the use of more sophisticated broadband services have been steadily increasing, pushing the limit on current mobile standards to provide tighter integration between wireless technologies and higher speeds, thus moving further toward a new generation of mobile communications: fifth generation (5G) [1]. 5G is foreseen as the convergence of Internet services with mobile networking, leading to the term *mobile Internet* over HetNets [2], in the context of personal adaptive, global networks, expanding the availability of a true broadband connection beyond the home and the office. However, more sophisticated services along with the higher capabilities of mobile devices increase their power requirements. "Green communications," therefore, also play a pivotal role in the 5G evolution with key mobile stakeholders driving the momentum toward a greener society through cost-effective design approaches. Small cells are becoming a clear solution for energy-efficient, high-speed wireless Internet connection.

In addition to novel radio conception, it is pretty evident that the stringent requirements of 5G have a substantial impact on the network behind the radio transmitter. Researchers generally anticipate that both the mobile and the fixed access networks will need to be optimized collectively to fulfill the demands of 5G. Future access networks integrate a multiplicity of fixed and wireless technologies (e.g., fiber, copper, digital subscriber line [DSL], microwave and millimeter wave [mmWave], and free-space optics) by using the Ethernet and IPs, forming an integrated communication platform. Moreover, by using these protocols, it is feasible by means of virtualization and cloud-based radio technology that multiple operators share a common physical infrastructure. The support of multivendor technology and the use of the network by multiple operators becomes possible in so-called open networks [3].

4.2 Influences of Backhaul/Fronthaul for 5G

The well-known femtocell [4] represents the indoor version of the small cell solution, while picocell deployment caters mainly for outdoor coverage. The former represents a cost-effective solution, but is limited to indoor scenarios; while the latter provides a more generic solution for outdoor coverage, but is subject to high capital expenditure (CAPEX) and operation expenditure (OPEX) on the operator side due to radio networking infrastructure and network planning. It is clear that if we can break the mold of current femto applications and extend its accessibility to the outdoor world, we would perhaps stumble on the next generation of femtocell technology for future 5G networks.

The foreseen increase in the number of connected mobile devices coupled with the ever more stringent quality-of-service (QoS) requirements from emerging broadband services means that employing today's technologies and strategies for network expansion will fail to deliver competitive tariffs as the transmission cost per bit will rocket. Unless new disruptive techniques are exploited, just opting to "buy more spectrum or infrastructure" to accommodate extra users will no longer solve the issue of operators meeting customer demand effectively in an era where spectral resources are at a premium. It is clear that a new proactive stance is needed if we are going to meet today's requirements in a cost-effective way.

In designing and planning small cell deployments in HetNets, mobile operators encounter two significant problems [5]:

■ How to transport traffic from the small cell at the edge to the core of a mobile network.
■ How to manage the RAN. Specifically, interference and resource management.

However, these are two separate research problems that are tightly coupled; since the extent to which traffic is managed and routed back to the mobile core network will influence the design requirements for the interference coordination strategy, and may push the operators toward particular solutions.

The edge link that connects small cells to the rest of the network may use different technologies: wired (e.g., fiber) or wireless (e.g., line of sight [LOS] or nonline of sight [NLOS]); on licensed or license-exempt spectrum; point-to-point (PTP) or point-to-multipoint (PMP); microwave or mmWave. mmWave has already been standardized for short-range services and deployed for application such as small cell backhaul; therefore, it could lead to unrivaled data rates and a completely different user experience if deployed also for broadband applications.

Traditionally, the backhaul segment connects the RAN to the rest of the network where the baseband processing takes place at the cell site. However, the notion of "fronthaul" access is gaining interest since it has the potential to support remote baseband processing based on adopting a cloud-RAN (C-RAN) architecture [6] that aims to mitigate interference in operator-deployed infrastructures; this significantly eases the requirements in interference-aware transceivers. The emergence of wireless fronthaul solutions widens the appeal of fronthaul for small cell deployments because fiber—the technology typically used for fronthaul—is too expensive or just not available at many small cell sites.

In wireless cellular radio networks, the backhaul contribution to the total power consumption is generally overlooked since it falls outside the scope with regard to minimizing the power consumption on the radio access domain. However, satisfying the almost exponential increase in mobile data traffic demands a prominent number of (mainly small) base stations (BSs) or macrocells along with remote antenna elements such as remote radio units (RRU). Therefore, we can easily deduce that the deployment of the backhaul links will increase the network expenditure (both CAPEX and OPEX) including greater power consumption in the highly anticipated future 5G wireless system. Obviously, the latency and synchronization of the backhaul links will vary for different kinds of backhaul technology and topology.

4.3 Scenarios and Their Respective Challenges

In this chapter, we summarize two reference scenarios for C-RAN along with one benchmark scenario. Three system scenarios will be discussed as follows:

1. Backhaul-based baseline small cell scenario according to the 3rd Generation Partnership Project (3GPP)
2. Fronthaul-based C-RAN scenario for interference management in a multitier infrastructure network
3. Mobile small cells with imperfect backhaul for ubiquitous high-speed data services on demand

To evaluate the performance of these scenarios, it is worthwhile defining how small cells are deployed and managed in today's mobile networks.

4.3.1 Baseline Scenario of Small Cell Deployment According to 3GPP

When using a backhaul architecture, an integrated small cell (antenna, wireless transceiver, plus baseband) is connected to an aggregation point—that is, a macrocell or other location that is connected (typically by fiber) to the mobile core. Since the picocell processes the RAN traffic, the operator can use many solutions for backhaul, including fiber or other wireline technologies, or wireless links. Wireless links may include LOS (mmWave) and NLOS (microwave) bands; PTP, PMP, or mesh topologies; and licensed or license-exempt bands. To emphasize, this is the deployment network operators' use in today's networks to deliver data services at low cost, as shown by Figure 4.1.

In this "infrastructure-based" multitier deployment, the picocell overlay network exploits a backhaul service to deliver high-speed services at relatively low cost, while the typical macrocell network continues to deliver a standard lower data rate service over wide area coverage. The limitation with such systems is the effect of co-channel interference, not only between tiers in the infrastructure deployment, but also due to the random deployment of small cells, which we typically refer to as femtocell technology. This current deployment scenario has spurred interest in technology enhancement techniques such as coordinated multipoint (CoMP) transmission [7] and interference management approaches such as the almost blank space (ABS) approach, both of which are already standardized and form a pivotal part of the Long-Term Evolution-Advanced (LTE-A) architecture. In the former, CoMP is used to control the interference between clusters of macrocells as well as providing coverage at the cell edge, while the latter approach is used to manage interference between tiers by switching the radio resource space in the macrocell while the picocell (small cell) is transmitting.

However, all current approaches are still limited in terms of spectral efficiency and complexity, and therefore how to effectively manage interference in multitier cellular environments that include the random deployment of small cells still remains an open research challenge.

4.3.2 Reference Scenario 1: C-RAN for Interference Management in Multitier Infrastructure Networks

Future emerging scenarios in small cell deployment are heading toward the notion of cloud radio. C-RAN is a novel mobile technology that separates baseband processing units (BBUs) from radio front-ends such as RRU. In this technology, BBUs of several BSs are positioned in a central entity where the radio front-ends of those BSs are deployed at the cell sites [8–10]. Therefore, this new framework unfolds a new paradigm for algorithms/techniques that need centralized and cooperative processing. However, the deployment of this new technology faces several potential research challenges, for instance, latency, efficient fronthaul design, and radio resource management for converged networks.

Fronthaul enables a C-RAN architecture, in which all the BBUs are placed at a distance from the cell site. The fronthaul transports the unprocessed radio-frequency (RF) signal from the antennas to the remote BBUs. While the fronthaul requires a higher bandwidth, lower latency, and more accurate synchronization than the backhaul, it enables a more efficient use of RAN resources, which, coupled with legacy interference and mobility management tools, can significantly minimize interference in the structured part of the network, including picocell–macrocell interference.

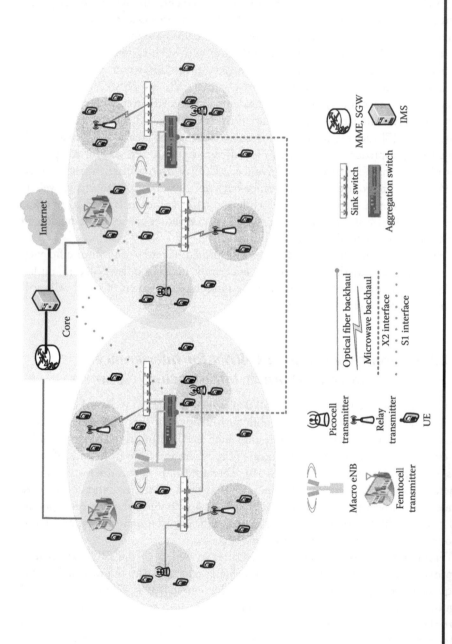

Figure 4.1 Legacy small cell deployment.

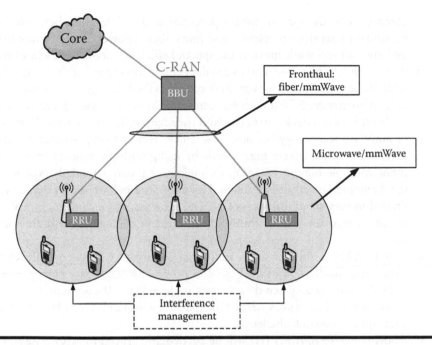

Figure 4.2 Operator's perspective of fronthaul-based reference scenario.

The general architecture of the aimed fronthaul-based reference Scenario 1 is illustrated in Figure 4.2, which consists of three main components [10], namely (1) a centralized BBU pool, (2) RRUs with antennas, and (3) a transport link, that is, a fronthaul network that connects the RRUs to the BBU pool.

In downlink, the RRUs transmit the RF signals to user equipment (UE), or in uplink, the RRUs carry the baseband signals from the UE to the BBU pool for further processing. The BBU pool is composed of BBUs that operate as virtual BSs to process baseband signals and optimize the network resource allocation for one RRU or a set of RRUs. The fronthaul links can be made of different technologies, namely, wired (fiber→ideal) and wireless (mmWave→nonideal).

4.3.2.1 Research Challenges

Introducing a C-RAN has the potential for several new advantages, in terms of high-speed connectivity to network-deployed small cells, as well as a means for mitigating interference. However, there are still several challenges to overcome in terms of interference management, fronthaul design, and mobility.

■ Although interference can be significantly controlled through coordinated resource management, this will require *coordinated scheduling* between small cells, and the macrocell network. A coordinated approach also affects the complexity of the system, with a potential trade-off foreseen between the coordination set size, complexity, and transmission power. How to attain this delicate trade-off while maximizing spectral efficiency needs to be addressed.

■ The *fronthaul design* is pivotal to the data connection speed of the small cell, with the potential to deliver very high speeds to the cell edge. Until now, optical and microwave PTP links have been viable options providing a reasonable trade-off between deployment cost and throughput. However, with the onset of mmWave technology, we now have a new avenue to explore that has the potential to provide very high speeds by using multiple-antenna (multiple-input, multiple-output [MIMO]) technology that can potentially exploit up to 64 antennas on a handset design. However, studies have shown that this is limited to very small distances of NLOS. How we can include an mmWave fronthaul segment, and its feasibility in terms of deployment cost are open issues.

■ True C-RAN could provide an additional opportunity for energy efficiency since the centralization of the baseband processing might save energy, especially if advances on green data centers are leveraged. The amount of energy consumed by the circuitry needed for C-RAN is still unknown. How to deal with this is a research challenge.

■ *mmWave energy efficiency* [11] will be particularly crucial to investigate since mmWave offers unprecedented bandwidths.

■ The *trade-off* between having fewer macrocells and a large number of small cells taking into account their distinctive types of energy consumptions is also of considerable interest.

However, there is still *interference from femtocell* technology/random deployment of small cells that can cause issues. How to deal with this is a research challenge.

4.3.3 Reference Scenario 2: Mobile Small Cells for Ubiquitous High-Speed Data Services on Demand

We introduce the notion of mobile small cells, where the mobile handset adopts the role of the access point or RRU. This shifts the philosophy of legacy mobile networks from purely being network centric toward being device centric, where mobile devices are now seen as a pool of additional network resources to be used by the operator to extend network coverage on demand. This new paradigm gives way to research challenges in terms of incentives for cooperation. On a broader perspective, the separation allows new ways of sharing infrastructures owned and deployed by different operators that can be managed/operated by the control plane according to the specific set of commercial rules under a common agreement [12].

This reference Scenario 2 is characterized by Figure 4.3, where we firstly split the control/data plane where the macro BS proves the signaling service for the

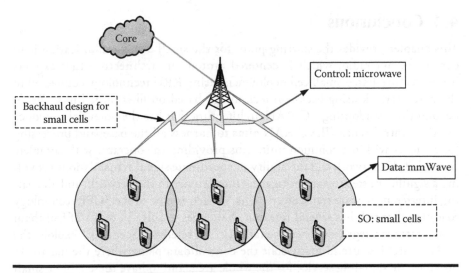

Figure 4.3   Device-centric scenario for on-demand small cells.

whole area, and to have these mobile small cells specialized to delivering data services for high-rate transmission with a light control overhead and appropriate air interface (mmWave could be the best option), which is illustrated in Figure 4.2.

4.3.3.1 Research Challenges

This scenario is an evolution of the second scenario. Introducing on-demand small cells has the potential for several new advantages, in terms of reducing the signaling overhead to network-deployed small cells, as well as a means of enhancing mobility management. However, several challenges remain in terms of interference management and mobility management.

- *Self-organizing (SO) interference management* between randomly created on-demand small cells.
- How to deal with *SO mobility management* in on-demand small cells.
- *SO node discovery* mechanism is also a research challenge.
- The problem of *jointly optimizing* resources in the communication haul and access network for on-demand small cells has not yet been investigated. How to optimize this problem is a research challenge.

In wireless networks, traffic may be off-loaded for a different reason: spatial and temporal demand fluctuations. Such fluctuations will be greater in small cell networks. *A bargaining approach* for data off loading from a cellular network onto a collection of Wi-Fi or femtocell networks could be considered an open research issue.

4.4 Conclusions

This chapter provides the starting point for the small cell communication haul design for 5G radio, which is centered around an architecture that exploits infrastructure-based small cell deployment using RRU technology connected to the core network using backhaul technology based on fiber-optic links. We go beyond this by adopting a C-RAN architecture that has the potential to reduce multitier interference. The C-RAN aims to harness all the baseband processing for all users within a common unit, thus providing the operator with complete control of the network and the ability to coordinate signal transmissions providing a significant step toward mitigating interference in the network, and alleviating interference-aware transceivers. This, in fact, is part of the 3GPP technology roadmap that still has several research challenges to solve in terms of fronthaul deployment strategy (reference Scenario 1). The reference Scenario 2 exploits the C-RAN architecture to firstly split the control/data plane where the macro BS provides the signaling service for the whole area, and to have these mobile small cells specialized toward delivering data services for high-rate transmission with a light control overhead and appropriate air interface (mmWave). However, we shift the networking philosophy of legacy mobile networks from purely being network centric toward being device centric, where mobile devices are now seen as a pool of additional network resources to be used by the operator to extend network coverage on demand. We utilize the notion of mobile small cells (reference Scenario 2), where user devices are able to emulate RRU services and provide high-speed data services on the fly. In this scenario, we consider ultradense small cell deployment with imperfect backhaul. This raises significant research challenges in terms of mobile small cell coexistence, and mobility management. A self-organizing network (SON) is considered an indispensable technique for the success of small cell networks. To enhance the capacity of small cell users, a SON small cell network should adopt an automatic SO scheme that adapts transmit power control and dynamic frequency selection, guaranteeing the performance of macrocell users.

Acknowledgment

This work was carried out under the E-COOP project (PEst-OE/EEI/LA0008/2013—UID/EEA/50008/2013), funded by national funds through FCT/MEC.

References

1. J. G. Andrews, S. Buzzi, W. Choi, S. V. Hanly, A. Lozano, A. C. K. Soong, and J. C. Zhang, What will 5G be? *IEEE J Sel Area Comm* 32(6): 1065–1082, 2014.

2. A. Damnjanovic, J. Montojo, Y. Wei, T. Ji, T. Luo, M. Vajapeyam, T. Yoo, O. Song, and D. Malladi, A survey on 3GPP heterogeneous networks, *IEEE Wireless Comm* 18(3): 10–21, 2011.
3. V. Jungnickel, K. Habel, M. Parker, S. Walker, C. Bock, J. Ferrer Riera, V. Marques, and D. Levi, Software-defined open architecture for front- and backhaul in 5G mobile networks, in *2014 16th International Conference on Transparent Optical Networks (ICTON)*, Graz, IEEE, pp. 1–4, 2014.
4. V. Chandrasekhar, J. Andrews, and A. Gatherer, Femtocell networks: A survey, *IEEE Comm Mag* 46(9): 59–67, 2008.
5. M. Paolini, Fronthaul or backhaul for micro cells? A TCO comparison between two approaches to managing traffic from macro and micro cells in HetNets, Technical Report of Senza Filli Consulting, 2014.
6. China Mobile Research Institute. C-RAN: The road towards green RAN, Technical Report, 2011.
7. K. M. S. Huq, S. Mumtaz, F. B. Saghezchi, J. Rodriguez, and R. L. Aguiar, Energy efficiency of downlink packet scheduling in CoMP, *Trans Emerging Tel Tech* 26(2): 131–146, 2015.
8. A. Checko, H. L. Christiansen, Y. Yan, L. Scolari, G. Kardaras, M. S. Berger, and L. Dittmann, Cloud RAN for mobile networks: A technology overview, *IEEE Comm Surv Tutor* 17(1): 1, 2014.
9. Y. Beyene, R. Jantti, and K. Ruttik, Cloud-RAN architecture for indoor DAS, *IEEE Access* 2: 1205–1212, 2014.
10. R. Wang, H. Hu, and X. Yang, Potentials and challenges of C-RAN supporting multi-RATs towards 5G mobile networks, *IEEE Access* 2: 1187–1195, 2014.
11. X. Ge, H. Cheng, M. Guizani, and T. Han, 5G wireless backhaul networks: Challenges and research advances, *IEEE Network* 28(6): 6–11, 2014.
12. R. J. Weiler, M. Peter, W. Keusgen, E. Calvanese-Strinati, A. De Domenico, I. Filippini, A. Capone, et al., Enabling 5G backhaul and access with millimeter-waves, in *2014 European Conference on Networks and Communications (EuCNC)*, Bologna, IEEE, pp. 1–5, 2014.

5G DESIGN

Chapter 5

Planning Guidelines and Principles for 5G RAN

Syed Fahad Yunas, Joonas Säe,
Muhammad Usman Sheikh, and Jukka Lempiäinen

Contents

5.1 Brief History of Cellular Concept/ Cellular Networks

In the brief history of mobile networks, the concept of *cellular networks* or a *cellular system* is not novel, having already been informally introduced by D. H. Ring in 1947 [1,2]. The concept proposed a new idea for increasing the network capacity by more efficient reuse of the spectrum resources through densifying the existing macrocellular network elements. Several years and publications later (e.g., [3–7]), patents for mobile communication systems with cellular network capabilities started to appear. Among the first of these was a patent by a Bell Labs engineer, Amos E. Joel, Jr., for a mobile communication system [8]. Joel proposed the idea of cell switching, that is, transferring the call connection from the coverage area of one base station (BS) cell to the coverage area of the neighboring cell. This enabled automatic cell reselection and released the used spectrum resources back to the other users in the first cell. It also suggested a way to locate a user within the cellular network using location areas, thus making it possible, for example, to route a call to a mobile network user to the correct cell in the network by knowing roughly the location of the mobile phone in the network.

5.1.1 From Omnidirectional Antennas to Six Sectors

Another way that the cellular concept proposed to increase coverage and capacity was the idea of cell sectorization, with the use of directional antennas instead of the traditional omnidirectional antennas along with different cell deployment layouts, as presented in the late 1970s [9,10]. The first BS sites were three-sectored BS sites with a 120° separation between the antennas in the horizontal direction. Nowadays, three-sectored BS sites are initially used in radio network design for planning new BS sites in the network. Therefore, to add more capacity or enhance the area spectral efficiency in a particular area in the network, the densification is taken even further, with more sectors per BS site. This has resulted in adding three additional sectors to sites, ending up with six-sectored sites. This will be discussed in more detail in Section 5.3.1.

5.1.2 Macro, Micro, and Mini

The concept of street-level cellular deployment was proposed in the mid-1980s [11–14]. It took the idea of macrocellular network element densification a step further, and is nowadays known as *microcell deployment*. It is based on the simple idea of adding more capacity to hot-spot areas in urban environments by having smaller cells, that is, cells with smaller coverage areas, below the rooftop level.

It is worth noticing that prior to 1990, all cellular deployments used unshielded twisted pair or coaxial cables for backhaul transmissions. This limited the spread of microcell deployment, as the backhaul connections were rather expensive at the time. A solution for this came from the advancements in optical fiber technology, leading to the fiber-optic microcellular concept. This happened at the beginning of the 1990s and provided a cost-effective, high-capacity, and low-latency transmission medium solution [15–17]. Along with microcellular deployment came new research problems related to microcell placement, location, and size, and resource management between the macro- and microcell layers, as addressed in [18–24].

When cellular networks were built to contain both macro- and microcells, a mixture of these different layers was formed. These are called *heterogeneous networks*, or *HetNets*, and they usually contain a macrocell layer for coverage purposes and a microcell layer for added capacity. The transition from pure macrocellular or microcellular network parts to HetNets was slow at first, but gained momentum when data capabilities were introduced in the second-generation (2G) mobile networks. Even more attention was paid to HetNets with the deployment of third-generation (3G) networks. These HetNets were soon also known as *hierarchical cellular structures* (HCSs) and attracted serious attention from operators.

During the last decade, the mobile data traffic volume increased at an exponential rate. This happened mostly because of the availability of high-speed mobile broadband services, thanks to evolved 3G networks having flat-rate pricing, and the proliferation of smartphones. This led to a situation in which the telecom industry realized that it could not fulfill the demand using legacy wireless infrastructure. Hence, efforts were put into finding a cost-effective solution for the problem. The "solution," or a proposal for the solution, began to take shape when a new, compact, self-optimizing home cell site was first reported by Alcatel [25] in 1999. This new concept of an operator-managed, self-configured, and stand-alone home BS came to be known as a *femtocell*, and was adopted by the industry around 2005. The actual standardization related to femtocells started in 2007 with the start-up Femto Forum (now known as the Small Cell Forum, which originally included mainly microcells and slightly smaller picocells) [26]. Initially, the standardization activities focused on residential femtocells; however, heterogeneous small cell deployments with a wider focus have recently been gaining ground. These include, for example, enterprise femtocells, outdoor urban femtocells, and rural femtocells [27–29].

Over the years, a trend toward decreasing the physical size of cells has been observed. Macrocells have the largest coverage area, followed by microcells with smaller coverage areas, and finally, picocells and femtocells, which have the smallest coverage. Nowadays, some network vendors have even proposed *attocells*, which are considered to be mini-femtocells, providing coverage over an area as small as 1 m². Hence, for simplification, this chapter categorizes the different cells into three broad classes: macro, micro, and mini (including all cells smaller than micro).

5.1.3 Performance Bottleneck of Universal Mobile Telecommunications System High-Speed Packet Access (UMTS/HSPA) and Long-Term Evolution (LTE) Networks

One common or well-known "fact" about the performance of cellular networks is that the actual data rate achieved will be much lower than the "promised" one. The 3rd Generation Partnership Project (3GPP) HSPA, which is available in most countries around the globe (over 500 operators in over 200 countries [30]), can theoretically offer up to 14.4 Mbps in the downlink direction. The evolved HSPA (HSPA+) can offer up to 28 Mbps with dual-carrier HSPA+ (DC-HSPA+) for quadrature amplitude modulation having 16 constellation points (16-QAM) and 42 Mbps with 64 constellation points (64-QAM) or 21 Mbps with only one carrier. However, these bit rates are not achieved in practical situations. Thus, the "normal" throughputs for the previous data rates are notably lower than these, and even the best data rates are only available in very good radio channel environments, that is, very close to the BS antennas. One study in [31] shows that the practical user experience in a commercial network reached data rates of more than 5 Mbps over 50% of the time in an HSPA+ network (with a theoretical maximum data rate of 21 Mbps) and recorded a peak data rate of 17 Mbps.

The same bottleneck can also be seen for 3GPP LTE. The theoretical peak data rate of the initial Release 8 presented 100 Mbps for downlink with a 20 MHz bandwidth. However, the actual average data rate is far less, roughly 15–20 Mbps [32]. Thus, it should be highlighted that the practical data rates are always below the "promised" ones. This affects the performance of user equipment (UE), especially in the cell-edge areas.

5.2 5G Sandbox

5.2.1 Definition of 1000× Capacity Increase

Recent data analytics [33–35] on the mobile data traffic trends have predicted a 1000× increase in capacity demand in the near future. With such a massive increase in data traffic, it is envisioned [35] that the capacity of fourth-generation (4G) networks, which are currently being deployed, will soon reach its limit. To tackle the 1000× challenge, the industry is already working toward the fifth generation (5G) of mobile cellular networks, which is conceived to address the growing capacity demand in a sustainable, energy-efficient, and cost-effective manner. It is currently in the prestandardization phase, and a great deal of debate is ongoing in the industry on what the 5G standard will really encompass. Nevertheless, a general consensus among the experts reveals [36,37] that the upcoming technology will not be just about enhancements in the radio access technology, but will, rather, represent an ecosystem of interoperable technologies and network layers, working together to provide ubiquitous high-speed connectivity, scalable network capacity, and seamless user experience. One clear enabler toward substantially improved network area capacities is the increasing levels of network densification at different layers of the overall heterogeneous radio access system. Ultradense networks or *DenseNets* aim to take network densification to a whole new level, where extreme spatial reuse is deployed. This chapter looks at two different deployment methodologies based on DenseNets for the successful realization of beyond 4G (B4G) networks.

5.2.2 Outdoor versus Indoor Traffic?

Contemplating the relative share of today's indoor/outdoor data traffic, there is a general agreement within the telecom industry that the majority of the data traffic, approximately 80%–90%, is generated by indoor users [38]. Assuming that this share of the outdoor/indoor data traffic will persist in the future as well, the projected capacity demand from indoor locations can then be estimated to increase by approximately 800–900× (assuming 1000× overall increase in data traffic demand), while the outdoor data traffic will increase by a factor of 100–200×. Such a massive increase in the capacity demand cannot be delivered in a cost- and energy-efficient manner by dense outdoor deployments, due to associated indoor capacity inefficiencies. Furthermore, poor indoor coverage from outdoor deployments, in buildings with high building penetration loss (BPL), has been and still is the foremost complaint with which mobile operators struggle. Hence, for B4G networks, the operators will need to deploy an extremely dense network of indoor small cells. Such massive-scale deployment of indoor cells will result in shifting the current "outside-in" deployment approach toward a new one based on an "inside-out" approach, where not only indoors, but potentially also some low-mobility outdoor neighborhood users, can be served by indoor BSs.

5.2.3 Major Service Provisioning Layer

One of the fundamental questions of deploying a cellular network is how to actually implement it. At first, networks were deployed having only the macro layer, but nowadays the standard is a heterogeneous network. However, the question of the major layer remains, that is: "What should be the number one technique in expanding the mobile network (coverage)?" "Would it be reasonable to assume that the best practice for expanding the coverage would be to use the macro layer?" or "Would it be more suitable, taking into account the capacity limits at the very beginning, to enhance the network only with a micro layer?" Either way, the cost aspects also need to be taken into account.

Whichever the major layer for cellular networks, there is also a need for the other layer and possibly even other layers to support the major layer. The major layer must be able to provide continuous coverage so that a ubiquitous cellular network service is possible.

5.3 Proposed 5G Cellular Techniques/Technologies

5.3.1 Higher-Order Sectorization

As mentioned at the start of the chapter, network densification can be achieved either by increasing the site density (also called *site splitting*) or by site sectorization. Site sectorization involves increasing the number of physical sectors or cells within a BS serving area. Each of the sectors then serves a portion of the cell site coverage area. The main idea behind sectorization is the higher use of space division multiple access (SDMA), that is, the available spectrum is reused more heavily than before. Traditionally, mobile operators have used three-sector cell sites in their networks to fulfill the coverage and capacity requirements. However, with the advent of mobile broadband networks such as UMTS, LTE, and so on, one might nowadays even find six-sector cell sites in some capacity-strapped locations such as downtowns. The concept of higher-order sectorization takes network densification a step further by increasing the sector count, for example, the use of 12-sector cell sites investigated in [39]. This, in general, would mean using antennas with narrower horizontal beamwidths to reduce intracell overlapping, which may cause intrasite interference. In a three-sector cell site, the antenna horizontal half power beamwidth (HPBW) is usually 65°–70°, which reduces to approximately 30°–32° HPBW with six-sector cell sites. Now, when upgraded to a 12-sector cell site, the HPBW of each antenna needs to be already as low as 15°–16°. The gain of using higher-order sectorization would at first seem obvious: doubling the amount of sectors in one BS site would yield a proportional capacity increase, that is, the available capacity would also double. However, this is not the case, since the amount of cell overlapping also increases. This results from the fact that decreasing the HPBW results in more directivity, which means higher gain in the main beam direction. This leads to a situation in which cell ranges increase and more overlapping occurs. Nevertheless, the capacity gain achieved through higher-order sectorization is still remarkable. This helps the operators to provide their subscribers with more capacity than with lower-order sectorization. Concerning higher-order sectorization, it is also required to keep in mind the tessellation used. Thus, when the number of sectors is increased in the cellular network BS sites, the antenna directions should also be optimized. In [39,40], the best suitable tessellation for three-sector sites, using 65°, was found to be the *cloverleaf* layout, as shown in Figure 5.1a, while for six-sector BS sites, the best tessellation was found to be the *snowflake* layout, as shown in Figure 5.1b. Likewise,

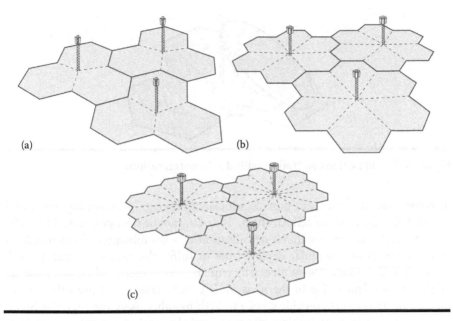

(a)

(b)

(c)

Figure 5.1   Different layouts for different levels of sectorization.

in a 12-sector case, the most suitable tessellation was the so-called *flower* layout, as shown in Figure 5.1c [39,40].

5.3.2 Vertical Sectoring

A relatively new concept called *vertical sectorization* is another possible way of enhancing the capacity of cellular networks. As the name suggests, instead of adding more sectors in the horizontal direction, the vertical direction is used to separate more than one cell from another. This has been studied in [41,42], where the impact of vertical sectorization has been taken into account in system-level simulations with the help of advanced antenna systems (AAS). To differentiate the normal horizontal sectorization from vertical sectorization, the "traditional" three-sector BS site using horizontal sectorization is marked as 3×1, while a BS site employing three sectors in the horizontal direction and two sectors in the vertical direction is denoted as 3×2, as shown in Figure 5.2. The findings of [41,42] reveal that in vertical sectorization, not only does the capacity of the network increase, but also the coverage is improved. For further reading on vertical sectorization, the reader may refer to [41–43] and the references therein.

5.3.3 Densification of Legacy Deployment Solutions

Macrocellular networks have been, and still continue to be, the basis for cellular network deployments globally. High-power transmitters with highly elevated and

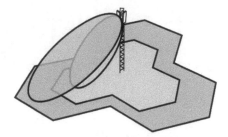

Figure 5.2   Vertical sectorization with 3×2 configuration.

directive antenna structures provide wide-area coverage provisioning and fulfill the mobility demands of cellular users. Thus far, mobile operators have been relying on either pure macrocellular densification or a combination of macrocellular and street-level microcellular deployments to fulfill the network capacity, both outdoors and indoors. This is especially true in urban centers, where the majority of the mobile data traffic in the network is concentrated. In dense urban areas, macrocellular networks provide service to high-mobility users and high-rise buildings, whereas microcellular sites offer the needed capacity in hot-spot locations. However, as mentioned in Section 5.2.1, due to the rapid proliferation of mobile broadband services, a 1000× increase in capacity demand is expected in the near future. Thus, keeping in view the exponentially rising capacity demand, coupled with the fact that the majority of this traffic will originate from indoor locations in the future, the question that needs to be answered is whether the current deployment solution will be able to sustain such high traffic growth; in other words, whether densifying the legacy deployment solutions would be enough to cater for future capacity requirements. In [44,45], it was shown that under a full network load scenario, the densification of macrocellular and microcellular solutions suffers from poor indoor capacity saturation in dense urban areas. As an example (taken from [44]), Figure 5.3 shows the performance of a macrocellular network densification from both the network spectral-efficiency and the network energy-efficiency point of view for both outdoor and indoor user locations in an urban environment. It is clear that the indoor environment suffers from capacity inefficiency with increasing cell density. As the network is densified, the sites are brought closer together, which increases the intercell interference (ICI). To limit the ICI in the macro layer, the antennas have to be downtilted to a greater degree [11]. This results in poor coverage on some floor levels inside high-rise buildings, which degrades the radio channel conditions, thereby affecting the overall achievable cell spectral efficiency. Furthermore, as a result of the reduction in cell-level spectral efficiency, the network spectral efficiency starts to saturate in the indoor environment, implying that macrocellular network densification becomes less efficient from the indoor service provisioning point of view. From the network energy-efficiency perspective, the indoor capacity inefficiency has a direct impact on the energy-efficiency performance of the network, as lower spectrum efficiency in the indoor environment results in higher energy consumption per bit, in other words, lower network throughput per unit power consumed.

The authors of [44] argue that the fundamental behavior seen in Figure 5.3 necessitates a shift from the current outside-in approach to a new deployment paradigm that puts more focus on the service provisioning from an indoor

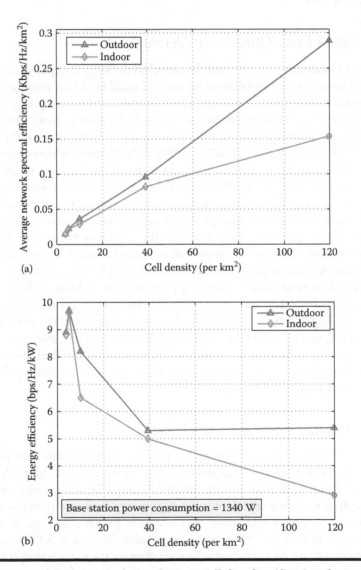

Figure 5.3 Performance analysis of macrocellular densification from outdoor and indoor receiver's perspective. (a) Average network spectral efficiency and (b) network energy efficiency, versus cell density.

perspective. Hence, for future networks, mobile operators will be required to opt for solutions that provide better indoor coverage as well as fulfilling the future capacity requirements. As such, the indoor deployment of small cells has been identified as a cost-efficient solution that offers wireless carriers a sustainable evolutionary pathway to meet the indoor capacity demands of the future [36]. An increasing trend for operators to opt for such small cell deployments has been recently observed.

5.3.4 Small Cells/Ultradense Networks

For B4G networks, experts envision that to fulfill the surging capacity demands of 1000× or more, an extremely dense network of small cells that provides seamless coverage and mobility will be essential, thus giving rise to the concept of DenseNets. Network densification, based on the ultradense deployment of small cells, is being considered as one of the key aspects of the emerging 5G cellular networks that will truly address the 1000× data challenge [36]. A large share of these deployments will be indoors, as this is believed to be the arena where the majority of the data traffic will originate in future. While the outdoor basic macro layer is still always needed for high-mobility outdoor users, such massive-scale indoor deployments may shift the current outside-in deployment strategy toward a new paradigm based on an inside-out approach, where not only the indoor users but potentially also some low-mobility outdoor neighborhood users can be served by indoor BSs [46]. It is believed that the majority of these small cells will be purchased and deployed either by end users in their homes in a plug-and-play fashion, or by enterprises in commercial buildings with no or minimal assistance from mobile operators, thereby enabling significant savings for the operators in terms of capital expenditure (CAPEX) and operational expenditure (OPEX). Moreover, due to a very small coverage footprint, the extreme density of these small cells will enable very tight frequency reuse, resulting in large network capacity gains, thus fulfilling the indoor capacity demands in a cost-efficient manner.

5.3.4.1 Indoor Femtocell-Based Solutions

From the outdoor service provisioning perspective, mobile operators can also leverage the dense indoor small cell deployment to their advantage to bring their operating costs down. Indoor small cells, despite having a limited coverage footprint, tend to radiate/spill their signals into the neighboring outdoor environment. These signals usually originate from small cells located in nearby buildings. By enabling the indoor small cells to operate in an open subscriber group (OSG) mode and thus provide service in their immediate outdoor vicinity, mobile operators can significantly lower their infrastructure costs by benefiting from zero site rental, radio-frequency (RF) engineering, and backhaul connectivity costs, thereby offering connectivity to outdoor users/customers with lower incurred costs. A similar concept of indoor-to-outdoor service provisioning (IOSP) has been presented by Qualcomm as "neighborhood small cells (NSC)" in [36,46]. Furthermore, key challenges related to the deployment, mobility management, and radio resource management (RRM) of NSC have been discussed quite nicely in [46]. However, one key item that is still missing from the studies is how the IOSP concept will perform in modern constructions with high wall penetration losses (WPL), reported, for example, in [47–49].

Recently, due to the increased level of awareness of global warming and the resulting requirements to save energy and cut down CO_2 emissions, the construction industries have started to develop, manufacture, and use modern construction materials that provide a greater degree of thermal insulation. Unfortunately, these types of materials significantly impact radio propagation in the form of high BPL. Traditionally, the values have been in the range of 5–15 dB; however, a more recent study has reported BPL of up to 35 dB in modern constructions [48]. High WPL attenuate signals penetrating through them, resulting in signal quality deterioration, which in turn affects the network capacity and data throughput.

In [50], the performance of three different mobile network deployment strategies has been evaluated; the pure macrocellular network, the pure densely deployed indoor femtocellular network, and the heterogeneous co-channel macro–femto deployment in a suburban environment, with different BPL recently encountered in modern buildings. Figure 5.4 shows the performance of different deployment strategies, considered in [50], for outdoor and indoor users in a suburban environment with different WPL. It can be seen that for all three BPL scenarios of 10/20/30 dB, the indoor femtocell-based solutions prove to be performing quite consistently well in terms of *coverage, cell-* and *network-level capacities*, as well as *energy efficiency*. Hence, the authors conclude that to counter the growing concerns of the mobile operators related to "zero-energy" and other modern buildings, the best solution is to deploy dedicated indoor solutions such as femtocells and outdoor small cell (e.g., metro femtocells) solutions to overcome the inherent capacity and energy inefficiencies of traditional macrocells. In this way, the exponentially increasing amounts of mobile data can still be supported with sustainable energy efficiency and scarce spectral resources. Furthermore, mobile operators can provide certain services to outdoor users from indoor access points; however, to guarantee higher bit rates, the operators will be required to deploy dedicated outdoor installations as well. It is pertinent to mention that the proposed IOSP will work only in small streets or neighborhoods, as the indoor small cells, due to lower transmitted power levels, will only be able to cover areas in the vicinity of the buildings. Any outdoor location wider than a few tens of meters might experience coverage limitation if there is no dedicated outdoor access layer available. The IOSP solution using indoor DenseNets can thus be considered as a good complement to the outdoor network, as a means of, for example, offloading capacity at times when the outdoor layer is overloaded with users during busy hours.

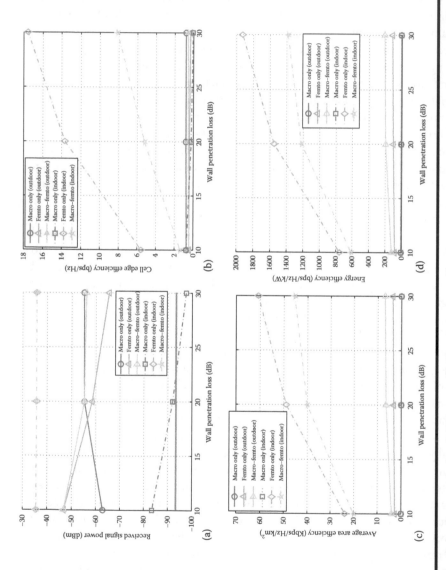

Figure 5.4 Performance analysis of different deployment strategies for outdoor and indoor users in a suburban environment with different wall penetration losses. (a) Cell-edge coverage conditions, (b) cell-edge cell spectral efficiency, (c) average network area spectral efficiency, and (d) network energy efficiency.

5.3.4.2 Wi-Fi Consideration?

The Institute of Electrical and Electronics Engineers (IEEE) 802.11 standards for a wireless local area network (WLAN) are more often referred to as "Wi-Fi." This abbreviation is a trademark of Wi-Fi Alliance, which used to be falsely interpreted as "Wireless Fidelity." However, the term *Wi-Fi* or *hot spot* is nowadays used everywhere, and has also attracted a lot of attention from mobile network operators. It was originally considered as a private wireless network, which was easy to implement and maintain, since only a fixed connection point was needed to set up a wireless network, for example, in offices or home locations. Thus, some might think that Wi-Fi could be a "threat" to mobile operators, since one can usually connect to it for free. In fact, by the end of 2014, there were 47.7 million Wi-Fi access points available worldwide [51], and not just indoors. Outdoor Wi-Fi coverage areas have also been deployed, which include areas in the vicinity of universities, coffee shops, hotels, airports, and so on. It is also expected that the number of these Wi-Fi hot spots will reach over 340 million by 2018 [51].

LTE for unlicensed spectrum (LTE-U) or license-assisted access (LAA) usually refers to some level of cooperation between Wi-Fi and LTE. The idea is to use LTE in the dual-band Wi-Fi, or more precisely, in the unlicensed 5 GHz band. It is expected to be possible to use this unlicensed band for extra bandwidth or to help in offloading the LTE traffic to Wi-Fi in places where the LTE is congested. However, many fear that there will be some problems in terms of interference when these two technologies are used at the same frequency band.

Time will show whether LTE-U or LAA will gain any momentum, but from the 5G development point of view, some sort of interconnectivity between these two technologies is bound to occur. In many other aspects, the interconnectivity between the existing wireless technologies is more likely to move toward "one transparent network" to ease up the "connection jungle of different cellular network generations" for the basic mobile network user.

5.3.5 *Distributed Antenna System (DAS)/Dynamic DAS*

For outdoor service provisioning, due to relatively low traffic volume and high-mobility users, mobile operators may still continue for some time to rely on the macrocellular layer to provide wide area coverage. This trend, however, will not last long, as the recent advancements in wireless connectivity, for example, for vehicles, supporting different applications ranging from infotainment and security to navigation and so on, will make stringent demands on the infrastructure of the mobile operators outdoors as well. Such innovations will demand high bit rates with "anywhere anytime availability," which the legacy outdoor deployments, such as macrocells, inherently lack. Furthermore, in a typical cellular network, the overall traffic fluctuates depending on the time of day. An idle or low load is usually experienced, for example, during late night or early morning, while high/peak load is observed during the daytime or early evening, as shown in Figure 5.5. Also, the traffic pattern

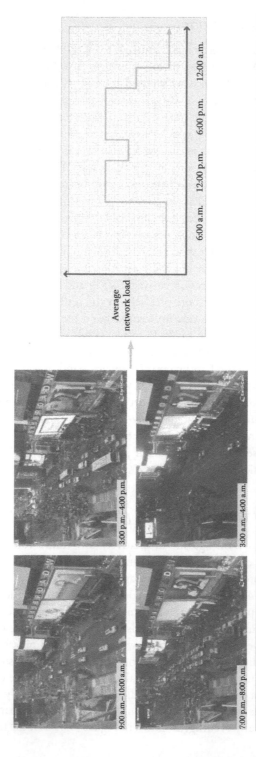

Figure 5.5 Outdoor traffic conditions during different time periods on Broadway in New York, New York, USA. (Copyright EagleCam.)

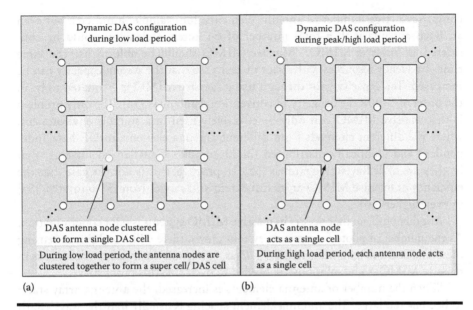

(a)   (b)

Figure 5.6 Dynamic distributed antenna system configurations during (a) low network load time period and (b) peak/high network load time period.

varies separately for the outdoor (after office hours—on highways, boulevards, etc.) and indoor environments (during office hours and evenings). Hence, there is an inherent need to provide capacity dynamically outdoors whenever needed.

In [52,53], an advanced form of DAS, dynamic DAS, is proposed, which can provide such solutions for the mobile operators. The dynamic DAS solution dynamically configures all the remote nodes into a "super microcell serving" or individual small cells based on the outdoor traffic conditions, as shown in Figure 5.6. During a high network load period, when many users are simultaneously accessing the network, the dynamic DAS configures the remote radio nodes to act as individual small cells. As a result, the frequency reuse increases within the area, thereby allowing more users to be served in that specific location. When the load in the network decreases, the remote nodes are configured into one "super" cell, wherein each participating node transmits the same radio signal, thereby providing consistent service to outdoor users as they move across the cell. Hence, the authors of [52] conclude that the dynamic DAS concept may be a good complementary solution for mobile operators to efficiently serve the outdoor traffic demand in future.

5.3.6 Massive MIMO

One of the major proposals for 5G is massive multiple-input multiple-output (MIMO) techniques. Massive MIMO consists of using an antenna system that comprises a large number of antenna elements, for example, hundreds, but even an 8×8 antenna with a total number of 64 antenna elements could be considered as "massive MIMO." Massive MIMO should be able to use the same time–frequency resources with space diversity, so that the system capacity can be increased. The capacity gain through using massive MIMO is estimated to be in the order of 5–20× the capacity of conventional antennas [54]. The only problem is that massive MIMO can only be effectively used in a suitable environment, where the different channels from different antenna elements to UE have independent and identically distributed (i.i.d.) complex Gaussian coefficients, that is, they are i.i.d. Rayleigh channels [55]. In practice, this is not the case, but the efficiency of massive MIMO at its full extent still ranges from 55% to 90% [55] or even higher [56].

Another problem that occurs in massive MIMO systems is pilot contamination. As the number of pilots per antenna system grows, there will be a limit to reusing the pilots in the neighboring cells [57]. Thus, to expand the number of channels in massive MIMO, some solutions are needed to overcome this problem.

When the number of antenna elements is increased, the antenna array starts to become too large. The antenna element spacing is usually half the wavelength of the frequency band used; for example, for a 900 MHz signal, the wavelength is roughly 30 cm. Thus, half of this corresponds to 15 cm spacing for the elements, resulting in a 9.6 m long uniform linear array (ULA) antenna with 64 antenna elements. Even for a 2.6 GHz frequency range, the length of a ULA antenna would be 7.3 m [56] with 128 antenna elements. One solution to overcome this problem would be to use a different kind of antenna array shape, such as a uniform cylindric array (UCA). This would reduce the size of the previous example of a 2.6 GHz antenna to roughly 30 cm in height and diameter when 64 dual-polarized patch antennas result in the same 128 antenna elements [56]. Another solution would be to use massive MIMO only with much higher frequencies, resulting in lower wavelengths, meaning lower antenna spacing and finally smaller antenna arrays. One such technology is the so-called millimeter-wave (mmW) communications.

5.3.7 Millimeter-Wave Communication

Another major proposal for 5G is the use of the so-called mmW communication. As the name suggests, it is based on using the millimeter wavelength part of the electromagnetic spectrum. This corresponds to frequencies of around 30–300 GHz. At those frequencies, much wider bandwidths are possible, which enables more capacity. In fact, the capacity gain for mmW communications is considered to be in the order of 20× the capacity of the existing cellular systems [58].

The idea of using these frequencies for communication purposes is not new, as they are already in use for point-to-point line-of-sight (LOS) microwave links for high-bandwidth communication links. These are used, for example, for connecting relay BSs in hard-to-reach environments such as hilly terrain or mountains. However, the "novelty" of using this mmW communication in cellular networks comes from using it as the access method for UE. Traditionally, it has been thought that the problem of mmW communication is the high propagation loss, since these frequencies have a lot of atmospheric attenuation in certain frequencies, such as 60 GHz. However, the effects of this attenuation are not so dramatic in all the frequencies in the 30–300 GHz frequency range. In fact, when used only in short ranges, it is possible to have even non-line-of-sight (NLOS) mmW communication [59]. In fact, the idea of using mmW communication as an access technique for cellular networks has been widely studied recently. The authors of [60] have carefully worked out the possibilities of mmW communication. Several studies [58,59] have also already performed some field tests to find out how well mmW communication is able to operate in different environments.

Even if mmW communication proves not to be feasible for cellular technology, it is still being developed for Wi-Fi. IEEE 802.11ad or microwave Wi-Fi, originally known as Wireless Gigabit, developed by Wireless Gigabit Alliance (WiGig), is the next step for WLAN techniques. It consists of using the mmW frequency band around 60 GHz for fast throughputs of a few gigabits per second, as the original name suggests.

5.4 Noncellular Approach for "Real 5G/6G?"

5.4.1 Cellular versus Noncellular?

Nowadays, the cellular concept is the basic idea behind mobile networks, and it seems that this will remain the case, at least for a while. However, the problem with the cellular concept is that there will always be interference areas in the network that lower its quality and capacity, especially in the cell-edge/border areas. Traditionally, it has been thought that to decrease the effect of the interference, the cell sizes need to be reduced so that overlapping areas with the neighboring cells decrease as well. However, some studies [45,61] have shown that this is not actually happening. It is true that the physically overlapping areas are smaller with smaller cells, but the problem comes with the increased number of those cells. Thus, the percentage of interfering areas, that is, the overlapping areas between cells, is actually increasing when the cell sizes are reduced and more cells are deployed.

Traditional mobile networks are based on the cellular approach, where radio resources are reused in every cell or after every few cells. In legacy cellular networks, sites are generally deployed with wide, 65° and 32° HPBW antennas. Therefore, there is interference from the neighboring cells of the same and other sites. A new possibility to overcome the severe problem of interference is to use antennas with extremely narrow beams, as suggested in [62]. The novel concept of single path multiple access (SPMA), in which interference was limited by adopting needle beams, was proposed in [62]. These needle beams enable users spatially separated by a few meters to reuse the radio resources (spectrum). In this way, the same spectrum can be reused after every few meters, which will drastically increase the capacity of the system. SPMA does not follow the conventional cellular approach; rather, in SPMA, each user is assumed to have its own *virtual cell.* This approach will not only radically increase the frequency reuse for centralized macro sites, but it will also bring a revolutionary change in traditional/conventional cellular concept thinking. More detail and assumptions of SPMA can be found in the next section.

5.4.2 *Innovative Concept of SPMA*

SPMA, proposed in [62], is an "innovative deployment paradigm" based on advanced antenna solutions. SPMA can be considered as an evolved and enhanced version of SDMA. SPMA exploits the spatial characteristics of the radio channel and uses the characteristics of independent propagation paths at particular geographical locations. In the traditional cellular approach, the received power at the mobile station (MS) is the sum of independent multipaths between the BS and the MS. Each multipath component follows a unique path (trajectory) and experiences a different number of reflections, diffractions, and losses. In the case of SPMA, instead of using numerous signal paths, the target is to use only a single independent multipath component to establish a link between a BS and an MS.

The essence of the SPMA concept relies on the strong assumption that future novel antennas will be able to form simultaneously several narrow adaptive antenna beams. These extremely narrow beams, also known as needle beams, will have a horizontal beamwidth of around 0.5° and a vertical beamwidth of around 0.2°. In [63], it is shown that a beam with these dimensions should be able to distinguish users that are spatially separated by a few meters in the macrocellular environment. It is also assumed that there will be an independent adaptive beam for each active user, and that the beam will precisely track the user. To have a narrow radiation pattern in an azimuth and elevation plane, generally an antenna array is needed. Therefore, to form such an extremely narrow beam, either a very large antenna array is required, such as massive MIMO, which will make the physical size of the antenna gigantic, or some other antenna manufacturing solution is needed. A key assumption for the SPMA concept is based on the expectation that new electrical materials, for example, artificially structured metamaterials such as graphene, carbon nanotube (CNT), and carbon nanoribbon (CNR), will be used for antenna manufacturing. The potential antenna and RF applications of graphene at microwave and terahertz frequencies have already been proposed in [64], but much advancement and hard work is still required from the research and development sector to make it feasible at the cellular band.

A centralized macro-site approach is suggested for SPMA, in which a traditional BS with a finite number of sectors is replaced by an SPMA node. This will allow existing macro-site locations to be used for SPMA nodes [65]. A single SPMA node is expected to have numerous needle beams with a predefined beamwidth in an azimuth and elevation plane. As in the case of SPMA, each user is assumed to have its own virtual cell; therefore, in the downlink direction, every user is acting as an interferer to another user, no matter whether the other user is served by its own serving SPMA node or any other SPMA node. Therefore, its own serving SPMA node will also be a source of interference, along with other SPMA nodes.

In SPMA, the signal-to-interference ratio (SIR) at the mth receiver point (Γ_m) can be calculated as follows:

$$\Gamma_m = \frac{S_m}{\sum I_{n,\text{own}} + \sum I_{p,\text{other}}}, \; n \neq m \text{ and } p \neq m \qquad (5.1)$$

where:

S_m is the received signal power at the mth receiver point coming from the serving SPMA node intended for the mth receiver point

$I_{n,\text{own}}$ is the received interference power at the mth receiver point coming from the serving SPMA node intended for the nth receiver point

$I_{p,\text{other}}$ is the received interference power at the mth receiver point coming from the other SPMA node intended for the pth receiver point

The difference between the coverage of the conventional cellular approach and state-of-the-art needle beams is depicted in Figure 5.7 (geometry is arbitrary).

In [63], the SPMA performance was evaluated in real-world urban and dense urban environments, and compared with respect to traditional 3-sector and higher-order sectored sites, that is, 6- and 12-sector sites. The metrics considered for the analysis were SIR, cell spectral efficiency, and area spectral efficiency. Figure 5.8a shows the cumulative distribution function (CDF) plot of the SIR achieved with 3-, 6-, and 12-sector sites. It can be seen that despite using narrow antenna patterns for higher-order sectored sites, the mean cell SIR degrades with higher-order sectorization. Due to the presence of a greater number of sectors at a site, the spatial separation between the sectors is reduced, which results in increased interference from the neighboring cells. Figure 5.8b presents the CDF plot of the SIR achieved with SPMA. It shows that SPMA outperforms other sectored antenna configurations. Simulation results validate the assumption made for SPMA that directive narrow needle beams can help avoid interference with nearby users. The SIR is significantly improved by SPMA compared with traditional sectored sites. It was

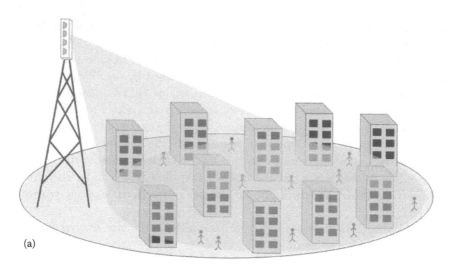

(a)

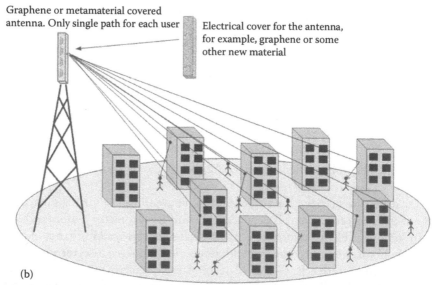

Graphene or metamaterial covered antenna. Only single path for each user

Electrical cover for the antenna, for example, graphene or some other new material

(b)

Figure 5.7 **Comparison of the traditional wide beam antenna and SPMA. (a) Cell coverage with traditional wide beam antenna and (b) demonstration of service provision with state-of-the-art needle beams. (M. U. Sheikh and J. Lempiäinen, *Wireless Personal Communications*, 78(2), 979–994, 2014.)**

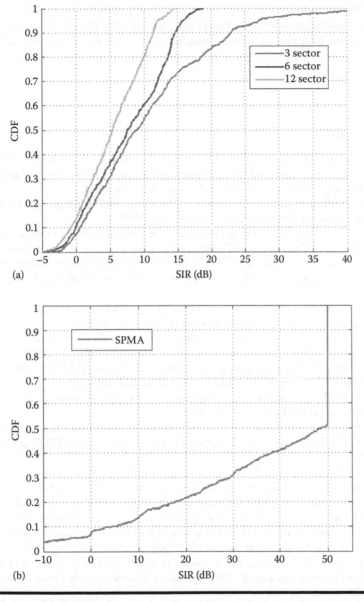

(a)

(b)

Figure 5.8 Signal-to-interference ratio (SIR) performance of (a) average network spectral efficiency and (b) network energy efficiency, versus cell density. (M. U. Sheikh, et al., *Wireless Networks,* **1–13, 2015.)**

found that the samples with high SIR were homogeneously found over almost the entire area under consideration, except for a few receiver points. These bad samples were scattered over the whole area, possibly due to the nonavailability of distinct paths between the two nearby users, or due to the lack of a dominant (LOS or reflected) path.

References

1. W. C. Jakes, *Microwave Mobile Communications*. New York: IEEE Press, 1972.
2. W. R. Young, Advanced mobile telephone service: Introduction, background and objectives, *The Bell System Technical Journal* 58: 1–14, 1979.
3. H. J. Schulter and W. A. Cornell, Multi-area mobile telephone service, *IRE Transactions on Communications Systems* 9: 49, 1960.
4. K. Araki, Advanced mobile telephone service: Introduction, background and objectives, *Review of the Electrical Communication Laboratory* 16: 357–373, 1968.
5. R. Frenkiel, A high capacity mobile radiotelephone system model using a coordinated small zone approach, *IEEE Transactions on Vehicular Technology* 19(2): 173–177, 1970.
6. P. T. Porter, Supervision and control features of a small zone radiotelephone systems, *IEEE Transactions on Vehicular Technology* 20(3): 75–79, 1971.
7. N. Yoshikawa and T. Nomura, On the design of a small zone land mobile radio system in UHF band, *IEEE Transactions on Vehicular Technology* 25(3): 57–67, 1976.
8. E. Amos, Mobile communication system, U.S. Patent 3,663,762 (Online) 1972. Available from: https://www.google.com/patents/US3663762.
9. V. Donald, Advanced mobile phone service: The cellular concept, *The Bell System Technical Journal* 58(1): 15–41, 1979.
10. V. Palestini, Evaluation of overall outage probability in cellular systems, in *Proceedings of the IEEE 39th Vehicular Technology Conference (VTC)*, Italy, IEEE, pp. 625–630, 1989.
11. R. Steele, Towards a high-capacity digital cellular mobile radio system, *IEEE Communications, Radar and Signal Processing* 132(5): 405–415, 1985.
12. R. Steele and V. Prabhu, High-user-density digital cellular mobile radio systems, *IEEE Communications, Radar and Signal Processing* 132(5): 396–404, 1985.
13. K. Wong and R. Steele, Transmission of digital speech in highway microcells, *Journal of the Institution of Electronic and Radio Engineers* 57(6): 246–254, 1987.
14. S. El-Dolil, W.-C. Wong, and R. Steele, Teletraffic performance of highway microcells with overlay macrocell, *IEEE Journal on Selected Areas in Communications* 7(1): 71–78, 1989.
15. T.-S. Chu and M. Gans, Fiber optic microcellular radio, *IEEE Transactions on Vehicular Technology* 40(3): 599–606, 1991.
16. W. C. Y. Lee, Efficiency of a new microcell system, in *Proceedings of the IEEE 42nd Vehicular Technology Conference (VTC)*, Denver, CO, IEEE, pp. 37–42, 1992.
17. L. Greenstein, N. Amitay, T.-S. Chu, L. Cimini, G. Foschini, M. Gans, I. Chih-Lin, A. Rustako, R. Valenzuela, and G. Vannucci, Microcells in personal communications systems, *IEEE Communications Magazine* 30(12): 76–88, 1992.

18. X. Lagrange, Multitier cell design, *IEEE Communications Magazine* 35(8): 60–64, 1997.
19. J. Sarnecki, C. Vinodrai, A. Javed, P. O'Kelly, and K. Dick, Microcell design principles, *IEEE Communications Magazine* 31(4): 76–82, 1993.
20. I. Chih-Lin, L. Greenstein, and R. Gitlin, A microcell/macrocell cellular architecture for low- and high-mobility wireless users, *IEEE Journal on Selected Areas in Communications* 11(6): 885–891, 1993.
21. M. Murata and E. Nakano, Enhancing the performance of mobile communications systems, in *Proceedings of the IEEE 2nd International Conference on Universal Personal Communications: Gateway to the 21st Century*, vol. 2, Ottawa, ON, IEEE, pp. 732–736, 1993.
22. A. Yamaguchi, H. Kobayashi, and T. Mizuno, Integration of micro and macro cellular networks for future land mobile communications, in *Proceedings of the IEEE 2nd International Conference on Universal Personal Communications: Gateway to the 21st Century*, vol. 2, Ottawa, ON, IEEE, pp. 737–742, 1993.
23. J. Shapira, Microcell engineering in CDMA cellular networks, *IEEE Transactions on Vehicular Technology* 43(4): 817–825, 1994.
24. D. M. Grieco, The capacity achievable with a broadband CDMA microcell underlay to an existing cellular macrosystem, *IEEE Journal on Selected Areas in Communications* 12(4): 744–750, 1994.
25. S. A. Ahson and M. Ilyas, *Fixed Mobile Convergence Handbook*. Boca Raton, FL: CRC Press, 2010.
26. Small Cell Forum, http://www.smallcellforum.org (Online: accessed 5 May 2015).
27. Small Cell Forum, Enterprise femtocell deployment guidelines, Technical Report, SCF032.05.01, 2014.
28. Small Cell Forum, Deployment issues for urban small cells, Technical Report, SCF096.05.02, 2014.
29. Small Cell Forum, Deployment issues for rural and remote small cells, Technical Report, SCF156.05.01, 2014.
30. GSA, Global mobile broadband market update, http://www.gsacom.com, 2014.
31. Signals Research Group, The real-world user experience in a commercial hspa+ network, 2009.
32. UMTS Forum & IDATE, Mobile traffic forecasts 2010–2020 report, 2011.
33. Qualcomm, The 1000x mobile data challenge, White Paper, 2013.
34. Nokia Networks, Enhance mobile networks to deliver 1000 times more capacity by 2020, White Paper, 2014.
35. Review Ericsson, 5G radio access: research and vision, White Paper, 2013.
36. N. Bhushan, J. Li, D. Malladi, R. Gilmore, D. Brenner, A. Damnjanovic, R. Sukhavasi, C. Patel, and S. Geirhofer, Network densification: The dominant theme for wireless evolution into 5G, *IEEE Communications Magazine* 52(2): 82–89, 2014.
37. B. Bangerter, S. Talwar, R. Arefi, and K. Stewart, Networks and devices for the 5G era, *IEEE Communications Magazine* 52(2): 90–96, 2014.
38. Analysys Mason, Global mobile network traffic: A summary of recent trends, Report, 2011.
39. M. U. Sheikh and J. Lempiäinen, A flower tessellation for simulation purpose of cellular network with 12-sector sites, *Wireless Communications Letters, IEEE* 2(3): 279–282, 2013.

40. M. U. Sheikh, J. Lempiäinen, and H. Ahnlund, Advanced antenna techniques and high order sectorization with novel network tessellation for enhancing macro cell capacity in DC-HSDPA network, *International Journal of Wireless & Mobile Networks* 5(5): 65–84, 2013.
41. O. Yilmaz, S. Hamalainen, and J. Hamalainen, System level analysis of vertical sectorization for 3GPP LTE, in *Wireless Communication Systems, 2009. ISWCS 2009. 6th International Symposium on*, Tuscany, IEEE, pp. 453–457, 2009.
42. F. Youqi, W. Jian, Z. Zhuyan, D. Liyun, and Y. Hongwen, Analysis of vertical sectorization for HSPA on a system level: Capacity and coverage, in *Vehicular Technology Conference (VTC Fall), 2012 IEEE*, Quebec City, IEEE, pp. 1–5, 2012.
43. Y. Fengyi, Z. Jianmin, X. Weiliang, and Z. Xuetian, Field trial results for vertical sectorization in LTE network using active antenna system, in *Communications (ICC), 2014 IEEE International Conference on*, Sydney, NSW, IEEE, pp. 2508–2512, 2014.
44. S. F. Yunas, T. Isotalo, J. Niemelä, and M. Valkama, Impact of macrocellular network densification on the capacity, energy and cost efficiency in dense urban environment, *International Journal of Wireless and Mobile Networks (IJWMN)* 5(5): 99–118, 2013.
45. S. F. Yunas, T. Isotalo, and J. Niemelä, Impact of network densification, site placement and antenna downtilt on the capacity performance in microcellular networks, in *Proceedings of the 6th Joint IFIP/IEEE Wireless and Mobile Networking Conference (WMNC)*, Dubai, IEEE, pp. 1–7, 2013.
46. Qualcomm, Neighborhood Small Cells for Hyper Dense Deployments: Taking HetNets to the Next Level, White Paper, 2013.
47. A. Asp, Y. Sydorov, M. Valkama, and J. Niemelä, Radio signal propagation and attenuation measurements for modern residential buildings, in *Proceedings of the IEEE Globecom Workshops (GC Wkshps)*, Anaheim, CA, IEEE, pp. 580–584, 2012.
48. A. Asp, Y. Sydorov, M. Keskikastari, M. Valkama, and J. Niemelä, Impact of modern construction materials on radio signal propagation: Practical measurements and network planning aspects, in *Proceedings of the IEEE 79th Vehicular Technology Conference (VTC)*, Seoul, IEEE, pp. 1–7, 2014.
49. I. Rodriguez, H. Nguyen, N. Jorgensen, T. Sorensen, and P. E. Mogensen, Radio propagation into modern buildings: Attenuation measurements in the range from 800 MHz to 18 GHz, in *Proceedings of the IEEE 80th Vehicular Technology Conference (VTC)*, Vancouver, BC, IEEE, pp. 1–5, 2014.
50. S. F. Yunas, A. Asp, J. Niemelä, and M. Valkama, Deployment strategies and performance analysis of macrocell and femtocell networks in suburban environment with modern buildings, in *Proceedings of the IEEE 39th Conference on Local Computer Networks (LCN'14)*, Edmonton, AB, IEEE, pp. 643–651, 2014.
51. K. Varia, iPass Wi-Fi Growth Map Shows 1 Public Hotspot for Every 20 People on Earth by 2018, http://www.ipass.com/press-releases/ipass-wi-fi-growth-map-shows-one-public-hotspot-for-every-20-people-on-earth-by-2018/.
52. S. F. Yunas, M. Valkama, and J. Niemelä, Spectral efficiency of dynamic DAS with extreme downtilt antenna configuration, in *Proceedings of the IEEE 25th International Symposium on Personal, Indoor and Mobile Radio Communications (PIMRC'14)*, Washington, DC, IEEE, pp. 1–6, 2014.
53. S. F. Yunas, M. Valkama, and J. Niemelä, Spectral and energy efficiency of ultra-dense networks under different deployment strategies, *IEEE Communications Magazine* 53(1): 90–100, 2015.

54. I. Hwang, B. Song, and S. Soliman, A holistic view on hyper-dense heterogeneous and small cell networks, *Communications Magazine, IEEE* 51(6): 20–27, 2013.
55. X. Gao, O. Edfors, F. Rusek, and F. Tufvesson, Massive MIMO in real propagation environments, *CoRR*, vol. abs/1403.3376 (Online) 2014. Available from: http://arxiv.org/abs/1403.3376.
56. G. Xiang, F. Tufvesson, O. Edfors, and F. Rusek, Measured propagation characteristics for very-large MIMO at 2.6 GHz, in *Signals, Systems and Computers (ASILOMAR), 2012 Conference Record of the Forty Sixth Asilomar Conference on*, Pacific Grove, CA, IEEE, pp. 295–299, 2012.
57. F. Rusek, D. Persson, B. K. Lau, E. G. Larsson, T. L. Marzetta, O. Edfors, and F. Tufvesson, Scaling up MIMO: Opportunities and challenges with very large arrays, *CoRR*, vol. abs/1201.3210 (Online) 2012. Available from: http://arxiv.org/abs/1201.3210.
58. S. Rangan, T. Rappaport, and E. Erkip, Millimeter-wave cellular wireless networks: Potentials and challenges, *Proceedings of the IEEE* 102(3): 366–385, 2014.
59. W. Roh, J.-Y. Seol, J. Park, B. Lee, J. Lee, Y. Kim, J. Cho, K. Cheun, and F. Aryanfar, Millimeter-wave beamforming as an enabling technology for 5G cellular communications: Theoretical feasibility and prototype results, *Communications Magazine, IEEE* 52(2): 106–113, 2014.
60. T. Rappaport, S. Sun, R. Mayzus, H. Zhao, Y. Azar, K. Wang, G. Wong, J. Schulz, M. Samimi, and F. Gutierrez, Millimeter wave mobile communications for 5G cellular: It will work! *Access, IEEE* 1: 335–349, 2013.
61. M. U. Sheikh, J. Säe, and J. Lempiäinen, In preparation for 5G networks: The impact of macro site densification and sector densification on system capacity, *Springer Journal of Telecommunication Systems* (submitted and under review).
62. M. U. Sheikh and J. Lempiäinen, Will new antenna materials enable single path multiple access (SPMA)? *Wireless Personal Communications* 78(2): 979–994, 2014 (Online). Available from: http://dx.doi.org/10.1007/s11277-014-1796-x.
63. M. U. Sheikh, J. Säe, and J. Lempiäinen, Evaluation of SPMA and higher order sectorization for homogeneous SIR through macro sites, *Wireless Networks* 1–13, 2015 (Online). Available from: http://dx.doi.org/10.1007/s11276-015-1019-8.
64. J. M. Jornet and I. F. Akyildiz, Graphene-based nano-antennas for electromagnetic nanocommunications in the terahertz band, in *Antennas and Propagation (EuCAP), 2010 Proceedings of the Fourth European Conference on*, Barcelona, pp. 1–5, 2010.
65. M. U. Sheikh, J. Säe, and J. Lempiäinen, Arguments of innovative antenna design and centralized macro sites for 5G, *International Journal of Electronics and Communications* (submitted and under review).

Chapter 6

Quality of Service in 5G Network

Valery Tikhvinskiy and Grigory Bochechka

Contents

6.1 QoS Management Model Evolution in Mobile Networks

By developing the previous generations of mobile networks, the 3rd Generation Partnership Project (3GPP) has successfully standardized the principles and models of service quality management at the network level. Moreover, the new feature of service quality management has been introduced in 3GPP networks.

Ensuring the quality of service (QoS) in 3GPP networks by their evolution from high-speed packet access (HSPA) technology to Long-Term Evolution (LTE)–Advanced technology is based on the following principles [1]:

■ Provision of service management by operator
■ Differentiation of service quality and users

- Minimal involvement of the user terminal in the service quality management process
- Support of QoS for client applications that are invariant to the access network
- Rapid establishment of sessions
- Continuity of quality management function with mobile networks of previous generations
- Convergence of services in the interaction of mobile networks with fixed-access networks
- Rapid introduction of new services to the market

The implementation of QoS management principles at the network level suggests a steady increase in the number of mobile applications that control QoS based on the service quality requirements and the creation of the necessary high-level data exchange by bearer services.

Fourth-generation (4G) networks based on QoS model management at the network level have implemented new types of QoS management that can use QoS network model. In these cases, these old applications have to be refreshed. However, we can find some terminals where QoS terminal model management was used. This means that in some years, two QoS management models coexisted in mobile terminals. The situation when two QoS management models were used and the evolution of these models are shown in Figure 6.1. The period of 2008–2010 is considered to have been the point of transition from a model of QoS management based on user terminals to the network QoS management model.

To date, there are 3GPP requirements, in particular for general packet radio service (GPRS) networks and packet switched networks, to maintain QoS management at both the user terminal layer and the network layer to provide a smooth transition to QoS management only at the network level.

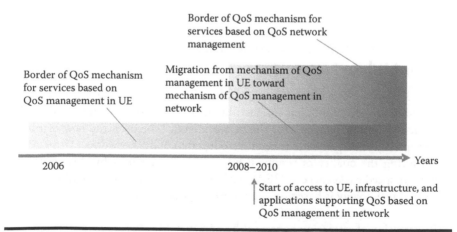

Figure 6.1   Two QoS management models in mobile networks.

The implementation of 3G users' requirements to ensure QoS in the "end user–end user" (E2E) chain begins with the activation of the QoS parameter negotiation procedure in the network. This procedure depends on the parameters of the user subscription to services stored in the home subscriber server (HSS) database and the current availability of network resources to 3G subscribers, which allows the final compound for a subscriber to be guaranteed. The procedure of QoS parameter approval and QoS management in the 3G network begins with sending a signaling message of session control by the user terminal at the nonaccess stratum (NAS) layer.

In 4G networks, unlike packet connections in second-generation (2G)/3G networks, a typical service of data exchange with a predetermined class of QoS is ready to form a connection to the packet network when a subscriber terminal is connecting to the network. QoS options for data exchange services are determined by the QoS parameters in the user profile, which are stored in the Subscription Profile Repository (SPR) database. This situation is very similar to the QoS management in GPRS/3G networks. However, in 4G networks, after transmission of the first data packet from the user terminal, this packet is routed to the packet data network (PDN), where the policy and charging rules function (PCRF) node that manages network policies and billing analyzes the quality class of the requested service in the E2E chain. Depending on the requested service class, the PCRF node can use different modifications of the QoS parameters to all nodes involved in the management of QoS data services. An LTE user terminal, unlike a 2G/3G user terminal, has no opportunity to request a particular QoS class, and only the LTE network is responsible for managing QoS. Similarly, a 4G network subscriber cannot request information about the QoS parameters, as is done, for example, through the use of a secondary context in the 3G network.

One feature of QoS management in the 4G network is that one user terminal can simultaneously support a variety of active services in the E2E chain, and each of these services will have its own individual QoS profile. A 4G user terminal may have up to 256 Evolved UMTS Terrestrial Radio Access Network (E-UTRAN) Radio Access Bearers (E-RAB) (communication services between the access terminal (AT) and the serving gateway [S-GW] service connections) by using E-UTRAN protocols, while in 3G networks only 15 different RAB-IDs are identified.

Thus, the 5G QoS management mechanism must be supported by network function virtualization (NFV) software solutions.

For the realization of QoS management in the network, we have to define the main QoS parameters for future 5G networks that will enable quality management for the new technology.

6.2 QoS as a Factor of the Trust in 5G Networks

Currently, leading organizations in the international standardization and development of telecommunication technologies, such as the International

Telecommunication Union (ITU), 3GPP, the Institute of Electrical and Electronics Engineers (IEEE), and the European Telecommunications Standards Institute (ETSI), have not formulated a strict definition of *trusted network*. However, the trust in the communication network significantly affects consumers' choice of communication operator, the regulation of operators' activities by state bodies, and the market demand for communication services and equipment.

The trust in network or communication technology has market and regulatory aspects that can contribute to the development of the network and the technology and increase the attractiveness of the services. Therefore, networks and communication technologies should correspond to both market and regulatory requirements of trust.

Given the many factors affecting the trust in 5G networks, in this chapter we will briefly review the major factors and examine in detail the impact of service quality on the trust in 5G networks.

The existing understanding of a trusted network is based on the concepts adopted by the developers of computer networks, which traditionally include [2]:

- Secure guest access: Guests obtain restricted network access without threatening the host network.
- User authentication: A trusted network integrates user authentication with network access to better manage who can use the network and what they are allowed to do.
- End point integrity: A trusted network performs a health check for devices connecting to the network. Devices out of compliance can be restricted or repaired.
- Clientless end point management: A trusted network offers a framework to assess, manage, and secure clientless end points connected to the network, such as Internet protocol (IP) phones, cameras, and printers.
- Coordinated security: Security systems coordinate and share information via the Interface for Metadata Access Points (IF-MAP) standard, improving accuracy and enabling intelligent response.

According to the Kaspersky Internet Security Company definition [3], a trusted network is a network that can be considered absolutely safe, within which your computer or device will not be subjected to attacks or unauthorized attempts to gain access to your data.

The proposed comprehensive review of the issue of trusted communication networks complements the concepts of computer network developers with the views of consumers, which also include the QoS provided by trusted networks. Subscribers' and regulators' views on the quality aspects of a trusted network are not always taken into account when creating a new mobile technology that reduces trust in the network.

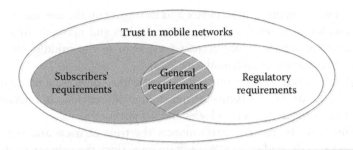

Figure 6.2   Domains of trust in mobile networks.

To implement a systematic approach to the trusted communication network, the trust of two major players in the telecommunications market should be considered: consumers, who provide market demand for communication services, and regulators, who monitor the effectiveness of operators' network infrastructure. As can be seen from Figure 6.2, consumers' and regulators' requirements for a trusted mobile communication network may either coincide or differ. The main factors affecting the trust of the subscriber and the regulator are shown in Table 6.1, taking into account their importance in descending order.

Most factors are the same for consumers and regulators, but factors determining consumer trust, according to the authors' evaluation, have the dominant influence on the mobile network.

Traditional factors influencing consumers' and regulators' trust in 5G networks are information security of confidential user data, security of subscribers' devices, and network infrastructure. The basis for such security is resistance to physical attacks on subscriber devices, such as illegal substitution of identification modules (Universal Subscriber Identity Module [USIM] cards), installation

Table 6.1   Main Factors Affecting the Trust of the Subscriber and the Regulator in the Network

Consumer	Regulator
Quality of service	Network security
Quality of experience	Information security
Information security	Network performance
Network performance	Network reliability
Network reliability	Quality of service
Convenience and security of subscriber's equipment	

of malicious software on the user device and its impact on the user device configuration, resistance to network attacks on user devices and network infrastructure, such as denial of service (DoS) attacks and "man in the middle" attacks, and resistance to attacks on confidential user data.

Ensuring the safe functioning of 5G networks, devices, and applications, including the security of transmission and storage of user data, is a major priority for developers of future 5G technologies and networks.

In addition to the security performance, the trust of users and regulators in 5G networks will depend on quality performance, since the security of the mobile network itself does not guarantee that the communication service will be provided without interruption and with the stated quality. A reduced quality in 5G networks will lead to a decrease of trust in them, and as a result, in an outflow of subscribers. Also, given that the 5G network will be used in a variety of financial systems, public safety systems, and traffic and energy management systems, a deterioration in their quality could lead to the loss of human life, environmental disasters, and financial fraud.

The quality parameters of 5G networks can be divided into three levels: network performance (NP), QoS, and quality of experience (QoE), as shown in Figure 6.3. NP and QoS are objective indicators that can be measured using specialized analyzers, while QoE indicators are subjective, estimated by users on the basis of their personal experience. The deterioration of QoS and NP will primarily lead to lowering the trust of regulators and business-to-business (B2B) and business-to-government (B2G) customers in 5G networks, while QoE deterioration will lead to lowering the trust of the mass market.

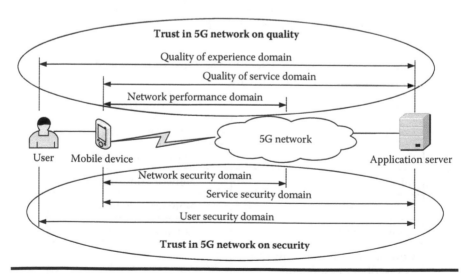

Figure 6.3   Quality and security levels of trust in a mobile network.

6.3 Services and Traffic in 5G Networks

To meet the QoS requirements in future 5G networks, we can take advantage of the analysis of future requirement levels for services such as high-density video and machine-to-machine communications, and then transfer these levels into 5G QoS requirements.

Mobile and Wireless Communications Enablers for the Twenty-Twenty Information Society (METIS) projects consider three basic business models of 5G services: extreme Mobile BroadBand (xMBB), massive machine-type communications (M-MTC), and ultrareliable machine-type communications (U-MTC) [4].

Forecasts from the leading specialists working on international 5G projects [5–7] show that video services, such as high-definition (HD) and ultra-high-definition (UHD) video, with high-quality resolution will have a dominant position among services rendered in 5G networks. According to reports by leading 4G network operators, video services dominate in the subscribers' traffic and will continue to dominate in the content of 5G networks.

For instance, the current traffic volume of video services is estimated by different operators [5] to be from 66% to 75% of the total traffic in 4G networks, including 33% for YouTube services and 34% for clear video, as well as closed-circuit television (CCTV) monitoring (video surveillance) in machine-to-machine (M2M) networks. In addition, by 2020, the volume of mobile M2M connections will grow with a compound annual growth rate (CAGR) index of 45% [8], up to 2.1 billion connections. Given the growing mass scale of M2M services in all industries, they will dominate over basic services (voice and data) in 4G and 5G networks.

The 5G European development strategy also aims to enable subscribers by 2025 to choose how to connect to TV broadcasts, via a 5G modem or antenna with digital video broadcasting–terrestrial (DVB-T), so this will require appropriate quality management mechanisms.

Therefore, the efforts of developers to improve quality management mechanisms will focus on video and M2M services traffic, the improvement of quality checking algorithms, and the creation of new quality assessment methods.

When developing requirements for QoS in 5G networks, two key traffic models should be firstly considered: high-speed "server–subscriber" video flow and massive M2M. Video transmission services will be an important stimulus to development and a rapidly growing segment of 5G network traffic. In 2013, the volume of video services in the total traffic of 4G network subscribers already exceeded 50%, and by 2019, it is forecast to increase by at least 13 times [9]. Thus, we can already observe the first wave of the oncoming "tsunami" of subscribers' traffic in 4G networks. The monthly consumption of data transmission traffic in 4G networks has already reached 2.6 GB, and the monthly consumption of traffic in 5G networks will exceed 500 GB.

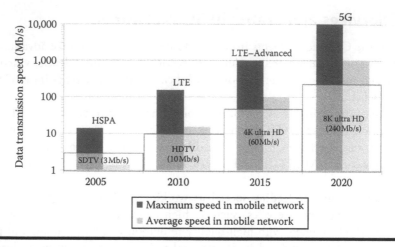

Figure 6.4   Technological capabilities of video transfer for mobile networks of various generations.

The growth of video services traffic volume will be associated with the implementation of various technologies of video services image quality, from standard-definition (SD) TV to UHD TV (8K), which in its turn requires a data transmission speed of up to 10 Gb/s in the network. The technological capabilities of mobile networks of various generations to broadcast video with various qualities of video image are shown in Figure 6.4 [10,11]. The capability of video broadcasting depends on the data transmission speed in the radio access network (RAN).

According to the forecasts shown in Figure 6.5, in 2018, the number of M2M connections in the networks of mobile operators will exceed 1.5 billion [12], which is five times higher than the current rate, and in 2022, mobile operators will have more than 2.6 billion M2M connections. At the same time, the share of M2M

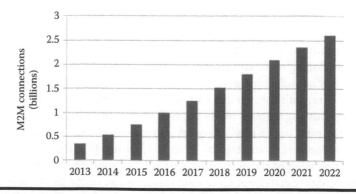

Figure 6.5   Number of M2M connections in mobile networks. (From Machina Research, *The Global M2M Market in 2013*, London, 2013.)

connections out of the total number of connections in the mobile operators' networks will increase from the current 5% to 15% in 2018 and to 22% in 2022.

The strategies of M2M operators are aimed at creating universal M2M platforms capable of operating in multiple vertical economic sectors. This will lead to the possibility of implementing approaches, tools, and processing methods for structured and unstructured Big Data derived from M2M networks.

According to ABI Research forecasts, the M2M Big Data and analytics industry will grow by a robust 53.1% over the next 5 years, from US$1.9 billion in 2013 to US$14.3 billion in 2018. This forecast includes revenue segmentation for the five components that together enable analytics to be used in M2M services: data integration, data storage, core analytics, data presentation, and associated professional services.

M2M services require much smaller data rates than video services, and generally do not require a guaranteed data rate. However, many M2M services, especially those used in the management of industrial systems, are critical of delays in mobile networks. Therefore, M2M services will also affect the quality of 5G networks.

6.4 QoS Parameters

Quality control and management in mobile networks are based on the use of the key QoS parameters, such as bit rate, latency, and packet loss.

In the current generation of mobile networks, there are two major types of network bearers: guaranteed bit rate (GBR) and nonguaranteed bit rate (non-GBR). GBR bearers are used for real-time services, such as rich voice and video. A GBR bearer has a minimum amount of bandwidth that is reserved by the network, and always consumes resources in a radio base station regardless of whether it is used or not. If implemented properly, GBR bearers should not experience packet loss on the radio link or the IP network due to congestion. GBR bearers will also be defined with the lower latency and jitter tolerances that are typically required by real-time services.

Non-GBR bearers, however, do not have a specific network bandwidth allocation. Non-GBR bearers are used for best-effort services, such as file downloads, e-mail, and Internet browsing. These bearers will experience packet loss when a network is congested. A maximum bit rate for non-GBR bearers is not specified on a per-bearer basis. However, an aggregate maximum bit rate (AMBR) will be specified on a per-subscriber basis for all non-GBR bearers.

Packet Delay Budget (PDB): This parameter identifies a maximum acceptable end-to-end delay between the user equipment (UE) and the packet data network gateway (PDN-GW). The purpose of using the PDB parameter is to support the queues of planning process and network functions at the connection level. The maximum delay budget (MDB) parameter is interpreted as the maximum packet delay with a confidence level of 98%. The PDB parameter defines the time limit for packet delay, for which the "final" package of the session will be transmitted with a delay of not greater than a predetermined value of the PDB. In this case, the packet should not be dropped.

Packet Error Loss Rate: This is the proportion of packets lost due to errors when receiving data packets. The maximum value of this parameter specifies the largest number of data packets lost during transmission over the network.

According to the assumptions of this analysis, these QoS parameters will be used in the process of developing 5G QoS requirements supporting the three main business models of 5G.

6.5 Quality Requirements in 5G Networks

5G mobile technologies that are expected to appear on the market in 2020 should significantly improve customers' QoS in the context of the snowballing growth of data volume in mobile networks and the growth in the number of wireless devices and the variety of services provided [5]. It is expected that mobile communication networks built on the basis of 5G technologies will provide a data transfer speed of more than 10 Gb/s.

Figure 6.6 shows the evolution of the QoS class from 2G to 4G, which has doubled the QoS class. These trends raise the question of how many QoS classes will be enough for 5G.

Previous 4G technologies (LTE/LTE–Advanced) provide flexible QoS management based on the division of data transfer characteristics into nine classes. These classes cover both of the 4G quality principles: services provision without quality assurance (best effort or non-GBR) and guaranteed QoS provision (GBR) [1].

Unfortunately, these LTE technological advances in the field of QoS management cover only part of the E2E chain, in particular 5G–5G and 4G–4G intranetwork connections. The quality management system does not extend to connections

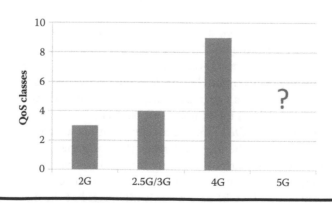

Figure 6.6   Evolution of QoS classes in mobile networks.

between 5G subscribers and other mobile 2G/3G/4G and fixed networks. The absence of the possibility of coordinated and flexible quality management in fixed IP and mobile networks of previous generations will for a long time remain a brake on the new level of subscribers' service quality in 5G networks.

The METIS project has identified 12 use cases for 5G networks: virtual reality office, dense urban information society, shopping mall, stadium, teleprotection in smart grid network, traffic jam, blind spots, real-time remote computing for mobile terminals, open-air festival, emergency communications, massive deployment of sensors and actuators, and traffic safety and efficiency, and has developed QoE requirements for them [13]. The QoE performance requirements that provide trust in network 5G are presented in Table 6.2. The highest requirements for experienced user throughput are developed for the "virtual reality office" use case. End users should be able to experience data rates of at least 1 Gbps in 95% of office locations and for 99% of the busy period. Additionally, end users should be able to experience data rates of at least 5 Gbps in 20% of office locations, for example, at their actual desks, for 99% of the busy period. The highest requirements for network latency are developed for the "dense urban information society" use case, in which device-to-device (D2D) latency is less than 1 ms. The highest requirements for availability and reliability of the 5G network are identified for the "traffic safety and efficiency" use case: 100% availability with a transmission reliability of 99.999% is required to provide services at every point on the road.

During the evolution of QoS management mechanisms in 3GPP (Global System for Mobile Communications [GSM]/Universal Mobile Telecommunications System [UMTS]/LTE) networks, there was a migration from QoS management at the UE level to QoS management at the network level. This approach to QoS management will be maintained in 5G networks as well.

QoS management mechanisms in 5G networks should provide video and voice over IP (VoIP) traffic prioritization toward web-search traffic and other applications tolerant to quality.

The service of streaming video transfer without buffering is very sensitive to network delay, so one of the most important parameters that determine QoS requirements is the total PDB, which is formed on the RAN air interface and is treated as the maximum packet delay with a confidence level of 98%.

Table 6.2   QoE Performance Requirements for 5G Networks

QoE Indicators	Requirements
Experienced user throughput	5 Gbps in data level (DL) and user level (UL)
Latency	D2D latency less than 1 ms
Availability	≈100%
Reliability	99.999%

Table 6.3   Requirements for Delay in 3G/4G/5G Networks

QoS Terms	Packet Delay Budget (ms)		
	3G	4G	5G
Without quality assurance	Not determined	100–300	Not determined
With guaranteed quality	100–280	50–300	1

Table 6.3 lists the requirements for delay in 3G/4G/5G networks formed in 3GPP [14] and the METIS project [15]. These data demonstrate that with the increase in mobile network generation, the requirements for the lower boundary of the total data delay across the network decline. Also, an analysis of the requirements for the overall 5G network delay revealed that given the accumulation effect, the delay in the 5G RAN network should be less than 1 ms.

A comparison of the requirements for delay in the control and user planes for signaling traffic and user traffic, respectively, presented in Figure 6.7, shows that the requirements for 5G networks will be twice as rigid for traffic in the user plane and 10 times more rigid in the subscriber traffic plane [7].

Another parameter is the proportion of packets lost due to errors when receiving data packets (IP packet error rate). The values for this parameter, which determines the requirements for the largest number of IP packets lost for video broadcasting through 3G/4G/5G mobile networks, are shown in Table 6.4 [16].

For M2M services, the quality will also be determined by the proportion of packets lost when receiving in 3G/4G/5G networks. Given the service conditions of M2M subscriber devices determined for both cases, with a guaranteed QoS and without guarantees, the requirements for the share of lost packets differ by three

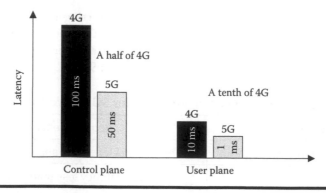

Figure 6.7   Requirements for delay in control and user planes for 4G/5G networks. (From Yongwan, P., 5G Vision and Requirements of 5G Forum, Korea, 2014.)

Table 6.4   Requirements for the Packet Error Loss Rate for Video Broadcasting

	Packet Error Loss Rate			
QoS Terms	*SDTV*	*HDTV*	*4K UHD*	*8K UHD*
Possibilities of mobile communication generation	3G/4G	4G	4G	5G
Video broadcasting with guaranteed quality	10^{-6}	10^{-7}	10^{-8}	10^{-9}

Table 6.5   Requirements for the Packet Error Loss Rate for M2M Services

	Packet Error Loss Rate		
QoS Terms	*3G*	*4G*	*5G*
Without guaranteed quality (non-GBR)	10^{-2}	10^{-3}	10^{-4}
With guaranteed quality (GBR)	10^{-2}	10^{-6}	10^{-7}

orders. Requirements for the packet error loss rate for M2M services are shown in Table 6.5.

The development of the near-field communication (NFC) concept will lead to a virtualization of the quality management function that could be introduced in the form of two main functions: the cloud QoS management function (CQMF) and the cloud QoS control function (CQCF), shown in Figure 6.8.

The CQCF function of QoS control provides real-time control of traffic flows in a 5G network on the basis of QoS levels established during the connection. Basic QoS control mechanisms include traffic profiling, planning, and management of data flows.

The CQMF function of QoS management provides QoS support in a 5G network in accordance with service level agreement (SLA) contracts, as well as providing monitoring, maintenance, review, and scaling of QoS.

The implementation of algorithms for traffic prioritization in 5G networks will be based on traffic classification procedures with a focus on video traffic priorities and M2M traffic. The traffic classification procedure should take into consideration the possibility of adaptation, as the traffic characteristics will dynamically change with the emergence of new applications, both in the M2M area and in the field of video services.

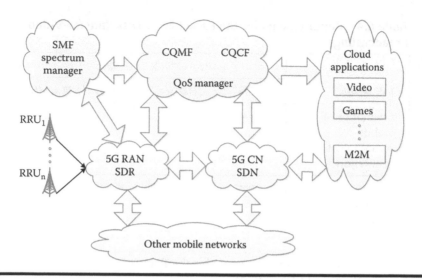

Figure 6.8   Virtualization of control and management functions in 5G network.

In addition to QoS management functions in the 5G network related to traffic management and prioritization, the scope of service quality management also includes the management of radio-frequency resources used by mobile networks (the Spectrum toolbox). The capabilities of access to the radio spectrum on the principles of licensed share access (LSA) in 5G networks require QoS guarantees to operators who grant other operators access to their spectrum [17,18].

The spectrum management function (SMF) in the 5G network will be designed as a Spectrum Manager entity. Spectrum Manager, as a network entity, is responsible for deciding how resources should be shared between mobile networks or other radio networks of the same regulatory priority.

In the case of a shortage of frequency resources to provide service with the required QoS, the 5G network must decide to use additional frequency channels for aggregation and select the channel from the frequency ranges that use a spectrum based on LSA or licensed exempt (LE) principles [18]. Therefore, the QoS manager must exchange information with the Spectrum Manager to manage the spectrum resources effectively in the interest of 5G network QoS. The process of deciding about spectrum resource allocation is based on a policy that takes into consideration the 5G QoS requirements of the primary spectrum user in LSA.

6.6 Conclusion

The emergence of 5G networks in the market in 2020 will focus on a significant improvement in the characteristics of mobile networks, including QoS, which will provide a high level of trust in these networks.

A one-sided view of trusted 5G networks from the security position alone will limit the growth in trust of customers and regulators. Developing high-level requirements in the QoS field will allow 5G developers to obtain trust in 5G at an early stage.

Given that the principles of QoS control will be preserved during the transition from 4G to 5G, the main effort of 5G developers should be focused on the virtualization of network functions that are responsible for the management and control of QoS in the network. Also, the QoS architecture of 5G should provide information exchange between the QoS manager and the Spectrum Manager for the effective management of spectrum resources, with the benefit of ensuring QoS and trust in 5G networks.

References

1. V. O. Tikhvinskiy, S. V. Terentiev, and V. P. Visochin, *LTE/LTE Advanced Mobile Communication Networks: 4G Technologies, Applications and Architecture*, Moscow: Media, 2014.
2. Network Access & Identity. Trusted Computing Group. Available from: http://www. trustedcomputinggroup.org/solutions/endtoend_trust/, 19 June 2015.
3. Trusted Network. Kaspersky Internet Security, 2005. Available from: http://support. kaspersky.com/6423.
4. ICT-317669-METIS/D6.6. Final report on the METIS 5G system concept and technology roadmap, *Project METIS Deliverable D6.6*, 2015.
5. V. O. Tikhvinskiy and G. Bochechka. Perspectives and quality of service requirements in 5G networks, *Journal of Telecommunications and Information Technology* 1: 23–26, 2015.
6. W. Yin. No-Edge LTE, Now and the Future 5G World Summit, 2014. Available from: http://ws.lteconference.com/.
7. P. Yongwan. 5G Vision and Requirements of 5G Forum, Korea, 2014.
8. V. O. Tikhvinskiy, G. S. Bochechka, and A. V. Minov. LTE network monetization based on M2M services, *Electrosvyaz* 6: 12–17, 2014.
9. B. Sam. Delivering New Revenue Opportunities with Smart Media Network. 5G World Summit 2014, Amsterdam, 2014.
10. Series H: Audiovisual and Multimedia Systems. Infrastructure of audiovisual services—Coding of moving video. High efficiency video coding. Recommendation ITU-T H.265, October 2014.
11. P. Elena. HDTV and beyond. ITU Regional Seminar Transition to digital terrestrial television broadcasting and digital dividend, Budapest, 2012.
12. Machina Research. *The Global M2M Market in 2013*, London, 2013.
13. ICT-317669-METIS/D1.1, Scenarios, requirements and KPIs for 5G mobile and wireless system, *Project METIS Deliverable D1.1*, 2013.
14. Adrian Scrase. 5G ETSI Telecoms Standards Workshop. The future of telecoms standards, London, 2015.
15. Project METIS Deliverable D2.1 Requirements and general design principles for new air interface. *Project METIS Deliverable D2.1*, 2013.

16. ETSI Technical Specification. Digital Video Broadcasting (DVB). Transport of MPEG-2 TS Based DVB Services over IP Based Networks. ETSI TS 102 034 V1.4.1, 2009.
17. V. O. Tikhvinskiy and G. Bochechka. Spectrum occupation and perspectives millimeter band utilization for 5G networks, *Proceedings of ITU-T Conference Kaleydoscope-2014*, St Petersburg, 69–72, 2014.
18. ICT-317669-METIS/D5.4. Future spectrum system concept, *Project METIS Deliverable D5.4*, 2015.

Chapter 7

Massive MIMO for 5G

Shang Liu

Contents

At the end of 2010, Thomas L. Marzetta proposed the massive multiple-in multiple-out (MIMO) concept. Massive MIMO (also termed *large-scale MIMO* or *large-scale antenna systems*) is a form of multiuser MIMO in which the number of antennas at the base station (BS) is much larger than the number of devices per signaling resource. Massive MIMO brings the following benefits. First, massive MIMO has significantly enhanced spatial resolution compared with conventional MIMO. It can deeply mine the resources of the spatial dimension without the need for increased BS densification. Second, beam forming can be focused with extreme sharpness into small regions, thus greatly reducing interference. Third, massive MIMO can improve the spectral and energy efficiency by orders of magnitude compared with a single-antenna system. With these benefits, we can envision the great potential of massive MIMO systems to be a key enabling technology for fifth generation (5G).

Massive MIMO is a new research area in communication theory, propagation, and electronics, as a result of a paradigm shift in the way of thinking with regard to theory, systems, and implementation. To realize massive MIMO, several challenges must be overcome. One example is pilot contamination: performance of analysis of transmission schemes tend to assume that a large-scale MIMO channel is ideal for an independent and identically distributed (IID) channel. In this condition, pilot pollution is considered to be a "bottleneck" problem of large-scale MIMO systems. To fully exploit the advantages of massive MIMO, we need to design accurate channel models that fit the practical scenarios with large-scale MIMO, as well as the corresponding channel measurement models. Meanwhile, the issues of channel estimation and channel feedback in massive MIMO still require extensive research and investigation.

7.1 MIMO Basics

We will discuss the basic theory of MIMO here, and we will then extend the discussion to massive MIMO. Massive MIMO is a popular potential enabling technology for simultaneously increasing capacity and peak data rates while reducing energy consumption and latency. Conventional MIMO solutions are already being deployed in networks today, typically using four to eight antennas. The 5G massive MIMO approach proposes a tremendous increase in the number of antennas at a BS, potentially scaling to arrays of 100×100 or more. There are a number of key technical problems to be addressed in terms of system design and engineering deployment.

7.1.1 MIMO Technology and Its Theoretical Basis

This chapter introduces the basic theoretical knowledge of MIMO systems, which includes the influencing factors of the transmission channel, the signal processing techniques at the transceiver, and the channel capacity and multiuser MIMO under several emission mechanism (multiantenna diversity, multiplexing, and beam forming), which form the basis of the following discussion of massive MIMO.

7.1.1.1 Wireless Transmission Environment

The characteristics of the wireless transmission channel are influenced by several factors, including free-space propagation loss, the shadow effect, the multipath effect, and the Doppler effect [1,2].

7.1.1.2 Free-Space Propagation Loss

In the strict sense, free space refers to a full vacuum condition, but in reality, an ideal space in general will be referred to as free space. In a free-space environment, there are no wave reflection, refraction, or diffraction phenomena; therefore, the propagation speed of a radio wave is equal to the speed of light c. In these channel conditions, the received signal power, P_r, can be expressed as

$$P_r = P_t \left(\frac{\lambda}{4\pi d} \right)^2 g_t g_r \tag{7.1}$$

where:
P_t is the transmission power
g_r and g_t are the antenna gains of the transceiver
λ is the transmission wavelength
d is the corresponding transmission distance

The transmission loss, L_s, is the ratio of the transmission power and the received power, which is expressed as

$$L_s = \frac{P_t}{P_r} = \left(\frac{4\pi d}{\lambda}\right)^2 \frac{1}{g_t g_r} \tag{7.2}$$

From Equation 7.2, we know that the received signal power and the square of the transmission distance are an inverse relationship:

$$P_r \propto \frac{1}{d^2} \tag{7.3}$$

7.1.1.3 Shadow Fading

Shadow fading is also known as slow fading. Due to blockage caused by buildings and the topography at the receiving antenna, an electromagnetic shadow is generated, and the electromagnetic field strength of the received signal will change. In comparison with fast fading, it changes macroscopically and is usually measured on a larger spatial scale. In addition, the decline in the slow fading speed has nothing to do with frequency; it mainly depends on the environment. The most common shadow fading statistical model is the lognormal shadow model, which is represented as

$$p(\psi) = \frac{\xi}{\sqrt{2\pi}\sigma_{\psi_{dB}}\psi} \exp\left[-\frac{\left(10\log\psi - \mu_{\psi_{dB}}\right)^2}{2\sigma^2_{\psi_{dB}}}\right], \quad \psi > 0 \tag{7.4}$$

where:

$\quad\quad \xi = 10/\ln 10$

$\quad\quad \psi = p_t/p_r$, is the ratio of the transmission power and the received power

$\quad\quad \psi_{dB}$ represents the unit ψ' in decibels

$\quad \mu_{\psi_{dB}}$ and $\sigma_{\psi_{dB}}$ are the mean and variance, respectively, of ψ'_{dB}

7.1.1.4 Multipath Effect

The multipath effect is a major feature of wireless communication. In the wireless mobile environment, the signals at the receiving end are not spread over a single path, but are the superposition of many reflected signals. Since the arrival time and phase of the different reflected waves are not the same, the superimposed signals may result in either signal enhancement or signal fading.

Assuming that the transmitter transmits a pulse signal unit, $s_0(t) = \delta(t)$, to the receiver, and because of the multipath effect, the received signal can be expressed as

$$s(t) = \sum_{n=1}^{N} a_n \delta(t - \tau_n) e^{j\theta_n} \tag{7.5}$$

where:
- N is the multipath number
- τ_n is the path delay of the nth path
- a_n is the reflection coefficient
- $e^{j\theta_n}$ represents the phase change after the signal goes through the nth multipath channel

7.1.2 Multiantenna Transmission Model

A point-to-point multiantenna transmission system diagram is shown in Figure 7.1.

Assuming that the channel between the sending and receiving ends is the flat fading channel, the $N_r \times N_t$ dimensional channel matrix of the MIMO system H is expressed as

$$H = \begin{Bmatrix} h_{11} & \cdots & h_{N_t 1} \\ \vdots & \ddots & \vdots \\ h_{1N_r} & \cdots & h_{N_t N_r} \end{Bmatrix} \tag{7.6}$$

Where $h_{ij} \in C$ expresses the channel fading coefficients between the sender antenna i and the receiver antenna j, which is subject to the complex Gaussian random variable, the relationship between input and output signals of the corresponding multiantenna system can be written as

$$y = Hx + n \tag{7.7}$$

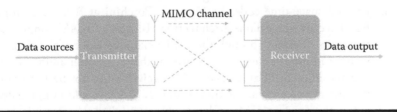

Figure 7.1   MIMO system model diagram.

where:

$y = \left[y_1, \cdots, y_{N_r}\right]^T$ is the received signal vector

$x = \left[x_1, \cdots, x_{N_t}\right]^T$ is the transmission signal vector

$n = \left[n_1, \cdots, n_{N_r}\right]^T$ is the additive noise

The noise variables $n_i \sim CN(0, \sigma^2)$ are IID zero mean complex circular symmetric Gaussian variables.

7.1.2.1 Transmitter Technology

To improve the overall performance of the MIMO system and reduce the error rate, typically space-time code (STC) and linear precoding can be combined at the transmitter.

1. Space-time code: The STC obtains diversity gain and coding gain through space and time joint coding under the condition of bandwidth that is not increasing. STC is a coding technique that can achieve higher transmission rates, resist fading, improve energy efficiency, and achieve parallel multiplex transmission, which results in promoting spectrum efficiency. Several relatively common ATCs are described in this section.

    a. Space-time block code: Alamouti proposed the space-time block code (STBC) in 1998 [3]. In essence, it is based on a simple transmission diversity technology about two transmission antennas. The basic idea is similar to the maximum ratio receive merger of the receive diversity. Alamouti's double antenna transmission policy [3] is

$$S = \begin{Bmatrix} s_1 & -s_2^* \\ s_2 & -s_1^* \end{Bmatrix} \tag{7.8}$$

    where at the nth time, the first symbol vector $(s_1\, s_2)^T$ is sent, then at the $(n+1)$th time, the second symbol vector $(-s_2^*\, s_1^*)^T$ is sent; that is, the system transmits two symbols during two moments, and the transmission speed is the same as in the single-antenna transceiver system.

    b. Layered space-time code: In 1996, G. Foschini at Bell Labs proposed Bell Layered Space-Time Architecture (BLAST). BLAST improves the spectrum efficiency through the spatial characteristics of the channel. It requires multiple antennas to be used at the transmitter and the receiver, and the number of antennas at the receiver must not be less than the number of antennas at the transmitter ($N_r \geq N_t$). In addition, the decoding at the receiver needs to know the exact channel state information (CSI). The block diagram is shown in Figure 7.2.

Figure 7.2 BLAST system block diagram.

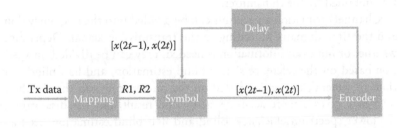

Figure 7.3 Block diagram of differential space-time code.

   c. Differential space-time coding: Tarokh proposed space-time block coding based on an orthogonal design according to differential encoding technology under a single antenna. The basic principle is shown in Figure 7.3.

        From Figure 7.3, the current transmission signal of two symbol periods $[x(2t+1), x(2t+2)]$ is not only related to the current emission data, but it is also associated with the signal of the first two symbol periods, as shown in Equation 7.9.

$$\left[x\left(2t+1\right),\ x\left(2t+2\right)\right]=R_1\left[x\left(2t-1\right),x\left(2t\right)\right]+R_2\left[-x\left(2t\right)^*,x\left(2t-1\right)^*\right] \quad (7.9)$$

2. Precoding technique: Precoding at the transmitter increases the system channel capacity by using the CSI. Common precodings can be divided into linear and nonlinear precoding. Linear precoding at the receiver can decode through various linear detection means, including zero-forcing (ZF), minimum mean square error (MMSE), and scalable video coding (SVD). Typical nonlinear precoding comprises Tomlinson–Harashima precoding (THP) [4] and vector precoding (VP) [5].

7.1.2.2 Receiver Technology

The wireless communication channel has great randomness, which results in the amplitude, phase, and frequency of the receiver signal being distorted, with a resulting major impact on the system performance. Channel estimation and reception detection, as the key components of the wireless communication receiver, are important fields in wireless communications research.

1. Channel estimation technique: The channel estimation of communication systems is generally completed at the receiver, and the estimation accuracy directly affects the overall system performance. For example, the precoding performance previously described is heavily dependent on the accuracy of the CSI obtained by the transmitter.

   Channel estimation algorithms can be divided into the frequency domain and the time domain, according to the transmission signal. Depending on whether or not prior information is needed, they can be divided into estimation based on the reference signal, blind estimation, and half blind estimation. In general, estimation based on the reference signal, that is, the pilot sequence, gives a more accurate estimation result, but the disadvantage is the lower spectrum efficiency. Blind and half blind estimation need a very short sequence, if any, to obtain high spectral efficiency. But the corresponding disadvantages are low estimation accuracy, serious phase ambiguity, and estimated nonconvergence problems. The receiver uses the estimated channel to detect, and also requires the CSI to precode. Both the transmitter and the receiver need the CSI to improve the channel capacity. The CSI at the transmitter is fed back and transmitted through the receiver, as shown in Figure 7.4.

   Common feedback modes include mean feedback, variance feedback, and limited feedback based on the codebook. Due to the simple structure and small amount of limited feedback, it is attracting a great deal of attention and research.

2. Reception detection technology: Reception detection of the MIMO communication system is a key step in the practical process of MIMO technology. Signal detection algorithms can be divided into linear and nonlinear detection algorithms according to the reception signal processing. In linear

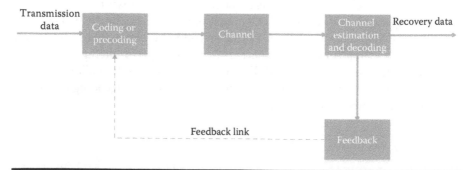

Figure 7.4 The receiver feedback model.

detection, the received signal x can completely recover the original transmitted signal s by linear operations:

$$\hat{s} = Fx \tag{7.10}$$

where F is the processing matrix of the receiver. Common linear detectors include ZF and MMSE.

Compared with linear detection algorithms with low complexity, the characteristic of a nonlinear detector is higher processing complexity, but with significantly improved system performance. In the simplest nonlinear detector, a feedback decision mechanism is added on the basis of the linear detector, namely, the interference cancellation (IC) algorithm. Another type of nonlinear detector approximates optimum detection; it needs to search the constellation point set to obtain optimum test results. The representative algorithm is called a *sphere decoding algorithm*, which has low computational complexity due to the use of the maximum likelihood estimation model. The solution to the sphere decoding algorithm is shown in Equation 7.11:

$$
\begin{aligned}
\hat{s} &= \arg\min_{s\in\Omega^n} \left\| x' - Rs \right\| \frac{n!}{r!(n-r)!} \\
&= \arg\min_{s\in\Omega^n} \left(\sum_{i=1}^{N_t} \left| x' - \sum_{j=i}^{N_t} R_{ij}s_j \right|^2 + \sum_{i=N_t+1}^{N_r} \left| x_i' \right|^2 \right) \\
&= \arg\min_{s\in\Omega^n} \left(\sum_{i=1}^{N_t} \left| x' - \sum_{j=i}^{N_t} R_{ij}s_j \right|^2 \right) \\
&= \arg\min_{s\in\Omega^n} \left[f_{N_t}\left(s_{N_t}\right) + \cdots + f_1\left(s_{N_t},\ldots,s_1\right) \right]
\end{aligned} \tag{7.11}
$$

The channel matrix carries out QR decomposition to obtain $H=QR$, the received signal is processed to obtain $x'=Q^H x$, $f_k\left(s_{N_t},\ldots,s_k\right)=\left| x' - \Sigma_{j=k}^{N_t} R_{kj}s_j \right|^2$ and Ω^n represents the set of constellation points.

7.1.3 Channel Capacity of Multiantenna Diversity, Multiplexing, and Beam Forming

According to Equation 7.12, assuming that the transmission signals are dependent and not associated, and their total power is limited, which is denoted as P_t, the covariance matrix of the transmission signal is expressed as

$$R_{xx} = E\left(xx^H\right) = \frac{P_t}{N_t} I_{N_t} \tag{7.12}$$

Assuming that the signal and noise are independent and unrelated, the covariance matrix of the received signal can be written as

$$R_{yy} = E\left(yy^H\right) = HR_{xx}H^H + R_{nn} \tag{7.13}$$

where $R_{nn} = N_0 I_{N_r}$.

According to the information theory, the channel capacity of the model can be expressed as

$$C = \log_2 \left| I_{N_r} + HR_{xx}H^H R_{nn}^{-1} \right|$$

$$= \log_2 \left| I_{N_r} + \frac{P_t}{N_t N_0} HH^H \right| \tag{7.14}$$

7.1.3.1 Spatial Multiplexing

Spatial multiplexing considers the transmission rate as the primary factor. The transmission information at the transmitter, which requires serial to parallel conversion, is divided into multiple parallel data streams. Each data flow corresponds to different antennas, which causes the final transmission rate to increase linearly with the number of data streams.

An eigenvalue decomposition is performed for HH^H:

$$HH^H = U \begin{Bmatrix} \lambda_1 & \cdots & 0 \\ \vdots & \ddots & \vdots \\ 0 & \cdots & \lambda_N \end{Bmatrix} U^H \tag{7.15}$$

where:

U is the square matrix of $N \times N$

ith column is the eigenvector q_i of the matrix HH^H

λ_i is the corresponding eigenvalue of q_i

The channel capacity can be obtained under the spatial multiplexing C_{SM}:

$$C_{SM} = \log_2 \left| I_{N_r} + \frac{P_t}{N_0 N_t} \left\{ \begin{bmatrix} \lambda_1 & \cdots & 0 \\ \vdots & \ddots & \vdots \\ 0 & \cdots & \lambda_N \end{bmatrix} \right\} \right|$$

$$= \sum_{i=1}^{N_r} \log_2 \left(1 + \frac{P_t}{N_0 N_t} \lambda_i \right) \tag{7.16}$$

where:

N_r denotes the number of antennas at the receiver
N_t denotes the number of antennas at the transmitter
p_t denotes the transmitter power
N_0 denotes the noise power

7.1.3.2 Spatial Diversity

MIMO technology essentially provides the gains of both spatial multiplexing and spatial diversity. Compared with spatial multiplexing, spatial diversity considers the quality of the transmission signal as the primary factor and takes full advantage of the multipath components of the wireless channel. The receiver can obtain many original copies of the transmitted signal with different fading to improve the accuracy of decisions.

The typical technology of the spatial diversity is STC. If the transmitter uses the STBC, then the channel can be expressed as

$$y = \|H\|^2 x_{SD} + n \tag{7.17}$$

Each element of the vector x_{SD} is p_t/N_r. The channel capacity of the STBC with the rate of 1 C_{SD} is expressed as

$$C_{SD} = \log_2 \left(1 + \frac{p_t}{N_0 N_t} \|H\|^2 \right)$$

$$= \log_2 \left(1 + \frac{p_t}{N_0 N_t} \sum_{i=1}^{N} \lambda_i \right) \tag{7.18}$$

where:

N_r denotes the number of antennas at the receiver
N_t denotes the number of antennas at the transmitter
p_t denotes the transmitter power
N_0 denotes the noise power

7.1.3.3 Beam Forming

The main principle of beam forming generates a signal-directed beam by the digital signal processing method to make the main beam point toward the target user and the side lobe or zero point toward the interference direction, thereby improving the signal-to-noise ratio (SNR) of the system. If the transmitter can obtain the desired CSI, the system can be modeled as

$$y = Hwx + n \tag{7.19}$$

where:
  y denotes the received signal
  H denotes the channel matrix
  n denotes the additive noise

The beam-forming matrix w is the N_t-dimensional vector and $\|w\| = 1$. At this time, the channel capacity of the beam-forming system is denoted as

$$C_{BF} = \log_2 \left\| I_{N_r} + \frac{P_t}{N_t} Hww^H H^H \right\| \tag{7.20}$$

Designing the optimal beam-forming matrix, the optimal weights are the main eigenvalues of HH^H. Then, the capacity equation can be expressed as

$$C_{BF} = \log_2 \left(1 + \frac{P_t}{N_t} \lambda_{max} \right) \tag{7.21}$$

where λ_{max} is the maximum eigenvalue of the eigenvalue matrix HH^H.

7.1.4 Multiuser MIMO

Compared with the traditional single-input single-output (SISO) system, MIMO technology can greatly improve the spectrum efficiency of the system to meet the growing demand rate. With the in-depth study of MIMO technology, the focus of study has gradually shifted from the original single-user MIMO system to a multiuser MIMO system, in particular downlink broadcast transmission.

Compared with single-user MIMO, the most important feature of multiuser MIMO is that there are no restrictions on the number of antennas. Multiuser MIMO transmission systems are divided into uplink multiple access (multi-access channel [MAC]) and downlink broadcast transmission (broadcasting [BC]).

7.1.4.1 Multiuser MIMO System Model

Assume that M denotes the number of antennas at the BS, N denotes the number of antennas at the user equipment (UE), and K denotes the total number of users. Then, the multiuser MIMO system block diagram about the single cell can be represented as in Figure 7.5.

Assuming that ψ_k denotes the transmission signal of the Kth user, the received signal of the BS is denoted as

$$y = \sum_{k=1}^{K} H_k \psi_k + v \tag{7.22}$$

where:

H_k denotes the uplink transmission channel matrix between the Kth user and the BS

$v \sim CN(0, N_0)$ denotes the uplink Gaussian white noise

Since the BS needs to transmit to multiple users simultaneously in the downstream broadcast transmission, the data received by the user will not only be related to its own transmission channel but it will also be affected by other users in the cell. In the multiuser MIMO system, eliminating the interference between multiple users is one of the most important means of improving the system performance.

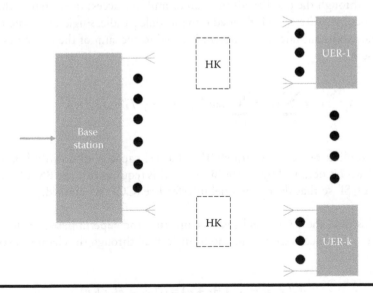

Figure 7.5 Multiuser MIMO system diagram.

If the BS obtains the CSI completely, the interference between multiple users is eliminated completely through dirty paper coding (DPC), compensating for the effects caused by interference.

7.1.4.2 Channel Capacity of Multiuser MIMO System

1. Uplink multiple access: For the uplink system model described in Section 7.1.4.1, if parallel multi-access interference suppression is adopted, the received signal at the uplink BS is expressed as

$$y = H_k \psi_k + \sum_{i \neq k} H_i \psi_i + v \tag{7.23}$$

where $\sum_{i \neq k} H_i \psi_i$ denotes the multiple-access interference. If Equation 7.23 is multiplied by the processing matrix T, which is expressed as

$$Ty = TH_k \psi_k + T \sum_{i \neq k} H_i \psi_i + Tv \tag{7.24}$$

to eliminate the multiple-access interference, a processing matrix T is designed that meets $T\sum_{i \neq k} H_i = 0$. According to [6], the processing matrix T can be constituted by the null-spaces vector of the matrix $\hat{H}_k \hat{H}_k^H$, $\hat{H}_k = [H_1, \ldots, H_{k-1}, H_{k+1}, \ldots, H_k]$.

Through the parallel elimination of multiple-access interference, the multiuser MIMO system is divided into multiple parallel single-user systems, and the maximum rate of the system is equal to the sum of the capacities for all users:

$$C = \sum_{k=1}^{k} C_k = \sum_{k=1}^{k} \max \log_2 \left| I + \frac{p_r}{N_0} T_k H_k (T_k H_k)^H \right| \tag{7.25}$$

2. Downlink broadcast channel: The channel capacity of downlink multiuser MIMO is obtained by means of DPC. This requires that the BS fully obtains the CSI, so that the co-channel interference (CCI) is obtained.

In this case, the CCI signal that is converted and superimposed in this way is the same as the unconverted signal nontransformed through the channel, expressed as

$$C(x_k = H_k s_k + \theta_k + n) = C(x_k = H_k s + n) \tag{7.26}$$

where:

> x_k denotes the received signal of the Kth user in the downlink broadcast link
>
> s_k denotes the received signal of the Kth user from the BS
>
> θ_k denotes the sum of the transmission signal interference caused by other users

$n \sim CN(0, N_0)$ denotes the downlink Gaussian white noise

Equation 7.26 shows that in the case when the BS fully obtains the CSI, the effect caused by the downlink user channel capacity can be eliminated completely, resulting in noninterfering channel capacity. Thereby, the MIMO broadcast channel capacity is obtained, expressed as

$$C_k = \log_2 \frac{\left| I + H_k \left(\sum_{j \geq k} Q_j \right) H_k^H \right|}{\left| I + H_k \left(\sum_{j > k} Q_j \right) H_k^H \right|} \tag{7.27}$$

where Q_j denotes the covariance matrix of the transmission signal s_j.

7.2 Antennas and Beam Forming

The use of very large antenna arrays at the BSs brings new issues, such as the significantly increased hardware and signal processing costs. Installing large-sized antenna arrays may be difficult in practice, which stimulates the design and implementation of antenna arrays to adapt flexibly to the complex environment. Beam forming can be focused with extreme sharpness into small regions, thus greatly reducing the interference.

7.2.1 Antennas

Generally, a single antenna with a relatively wide radiation pattern has relatively poor directivity due to its fixed radiation pattern. With increasing developments in wireless communications, antennas with high directivity are strongly required. An antenna array is constructed as an assembly of antenna elements in a proper electrical and geometrical configuration. Usually, the array elements are identical; this is not necessary, but it is practical and simpler to design. This gives us the freedom to choose (or design) a certain desired array pattern from an array without changing its physical dimensions. One of the basic methods for controlling the overall antenna pattern is to adopt proper geometrical configuration of the antenna array (linear, circular, rectangular, etc.). Antenna arrays that use active antennas can provide beam-forming capability. They can provide a diversity gain in multipath signal reception and high gain (array gain) by using simple antenna elements.

7.2.1.1 Antennas of Massive MIMO

Massive MIMO is based on a multiuser MIMO scheme in which the BSs are equipped with a large number of antennas to cover the number of active users. In each time–frequency resource block, BSs use the spatial multiplexing provided by large-scale array antennas to enhance the multiplexing capability of spectrum resources and the spectral efficiency of each user, while taking advantage of MIMO with its multiantenna diversity and beam-forming technology, greatly enhancing the use of spectrum resources and improving the transmission rate. If the number of antennas at a BS is significantly larger than the number of users served, the channel of each user to/from the BS is nearly orthogonal to that of any other user. This allows very simple transmit or receive processing techniques, such as matched filtering, to be nearly optimal with enough antennas even in the presence of interference.

However, the use of very large antenna arrays at the BSs brings new issues, such as significantly increased hardware and signal processing costs. Installing the large-sized antenna arrays may be difficult in practice, which stimulates the design and implementation of antenna arrays to adapt flexibly to the complex environment.

1. Combining with millimeter wave: A large-scale antenna array system can combine with millimeter wave, allowing the system to work at the higher carrier frequency. For example, at 25 GHz, the length of the uniform linear array (ULA) is reduced from 5.94 to 0.594 m, which can meet the requirements of practical installation
2. Proper antenna array configuration: Also, due to the constraints on array aperture, very large MIMO arrays are expected to be implemented in a two-dimensional (2-D) or three-dimensional (3-D) array structure, such as uniform planar array (UPA) and uniform circular array (UCA).

   One issue to consider is that different antenna array topologies will lead to a change in channel characteristics and will have an impact on the performance of large-scale antenna array systems. Intuitively, with the same number of antennas, an antenna array in which more antennas are placed in the horizontal direction can achieve higher spectrum efficiency. The more antennas placed in an array, the lower the spectrum efficiency obtained, mainly due to vertical antennas having small angular spread and the terminal angle distribution being small in the vertical direction, so that more antennas would be needed to obtain enough gain in the vertical direction to distinguish among these angles. In addition, with a fixed number of antennas, increasing the number of antennas in the vertical direction will lead to reducing the number of the antennas in the horizontal direction, resulting in performance degradation in the horizontal direction.

3. Reducing adjacent element spacing: Reducing antenna spacing also meets the requirements for installation, but if the adjacent element spacing is less than a half wavelength of the antenna, it will increase the correlation between the antennas, so that massive MIMO cannot form a precise beam to distinguish users, which will lower system performance.

7.2.1.2 Issues of Very Large Antennas

These include the configuration and deployment of the arrays.

7.2.1.2.1 Conventional Antenna Arrays

1. 2D-MIMO: In a large-scale antenna array system, increasing the number of antennas will lead to the actual array area increasing rapidly, which brings challenges for the antenna installation at the BS. For example, in a BS containing a ULA of 100 antennas, the array antenna spacing is half a wavelength. For the carrier frequency of 2.5 GHz, the length of the linear array needs to be 5.94 m, which is not feasible for many BSs that have only limited room on the tower. In traditional communication systems, with the constraints of antenna structure at the BS, the transmitted beam is adjusted only on the horizontal cross section of the wireless channels, and in the vertical dimension, once the BS has made the optimization, all cell users are under a fixed angle, so all kinds of beam forming and precoding are based mainly on the horizontal dimension channel. In fact, due to the propagation channel being a 3-D space, a fixed-angle method often cannot allow the system to achieve optimal throughput. With the number of cell users increasing, and distributed in different regions of the cell, including the cell center and the cell edge, using traditional 2-D beam forming can only distinguish in the horizontal direction with the horizontal dimension of channel information, but not in the vertical dimension. When the azimuth angles of the users are the same, it will inevitably cause interference, as shown in Figures 7.6 and 7.7.

2. 3-D MIMO: This deficiency of conventional 2-D beam forming is closely related to the antenna structure. As we know, the current BS antenna ports are equipped with a linear array arranged in the horizontal direction, so that adjusting the amplitude and phase of each antenna port can only control signal distribution in the horizontal dimension. With the emergence of the active antenna system (AAS), which changes the original antenna structure, an AAS array consisting of a plurality of low power, relatively independent

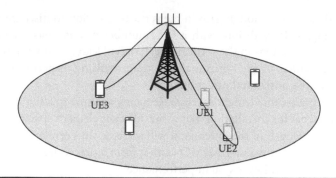

Figure 7.6   2-D beam forming. (Interference exists between UE1 and UE2, which are at the same azimuth angle.)

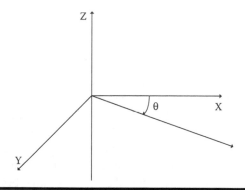

Figure 7.7   2-D MIMO with only azimuth angle (ø).

of dipoles, radio frequency (RF)-integrated modules can independently control each active antenna, resulting in the capacity for efficient and flexible beam control. Therefore, the antenna structure of massive MIMO should be designed with 3-D spatial channel models taking into account both the azimuth and elevation directions of signal propagation between the BS and users.

As shown in Figure 7.8, a 2-D antenna array based on AAS can enable a MIMO system to make full use of the vertical dimension of the propagation space. Compared with 2-D MIMO, which can only adjust the beam angle with the azimuth angle, 3-D MIMO can adjust the beam angle with both azimuth and elevation angles.

3. Antenna modeling: The 3rd Generation Partnership Project (3GPP) 3-D antenna models [7], as shown in Figure 7.9, deploy a 2-D planar antenna array, including cross-polarized array and ULA, where N is the number of columns, M is the number of antenna elements in each column.

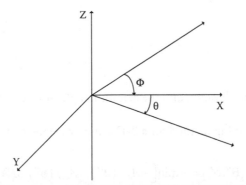

Figure 7.8   3-D MIMO with azimuth angle (Φ) and elevation angle (θ).

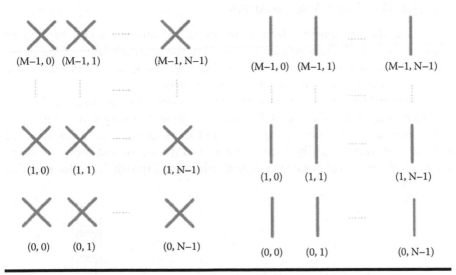

Figure 7.9   2-D planar antenna arrays with cross-polarization (a) and co-polarization (b). (From L. You, et al., *IEEE Trans.*, 14, 3352–3366, 2015.)

From [1], each antenna element has a vertical radiation pattern:

$$A_{E,V}\left(\theta''\right) = -\min\left[\left(\frac{\theta'' - 90°}{\theta_{3dB}}\right)\right]$$

and also a horizontal radiation pattern:

$$A_{E,V}\left(\phi''\right) = -\min\left[-12\left(\frac{\phi''}{\phi_{3dB}}\right), A_m\right]$$

where:

$SLA_V = 30$ dB
$A_m \quad = 30$ dB
$\theta_{3dB} \quad = 65°$
$\phi_{3dB} \quad = 65°$
θ'' and ϕ'' are the elevation and the azimuth angle, respectively

Then, the combining method for a 3-D antenna element pattern is

$$A''(\theta'',\phi'') = -\min\left[-A_{E,V}(\theta'') + A_{E,V}(\phi''), SLA_V\right]$$

7.2.1.2.2 Very Large Antenna Arrays

As the number of antenna elements increases, the antenna arrays need to be extended to a 2-D/3-D antenna array model, and the hardware design of the antennas should consider the different configurations and deployment scenarios that use the actual antenna arrays in a massive MIMO system.

As Figure 7.10 shows, antenna array topologies include ULA (antenna elements arranged along a straight line), UPA (antenna elements arranged over some planar surface, such as a rectangular array), and UCA (antenna elements arranged around a circular ring). The most common and most analyzed geometry is the ULA, which consists of N antenna elements placed on a straight line. However, this

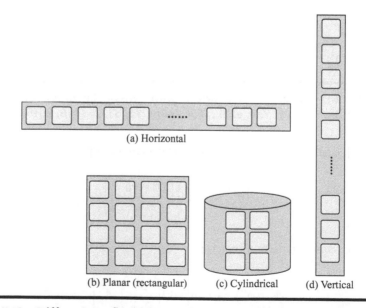

Figure 7.10   Different configurations and deployment for actual antenna arrays.

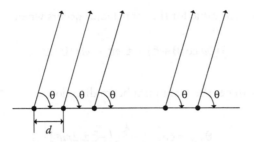

Figure 7.11 ULA with adjacent element spacing *d*.

configuration, when adopted for large-scale antennas, is unsuitable for BSs having limited space that is unable to accommodate the structure of a ULA.

1. ULAs: ULAs, with their elements uniformly separated along a straight line, generally tend to achieve high directivity in a limited direction. The ULA can be easily analyzed and designed due to its simple geometry. For ULAs, as depicted in Figure 7.11, the channel vector of a single user is represented by

$$a\left(0_{i}\right)=\left[1,e^{((j2\pi d\cos(\theta))/\lambda)}+\xi,\ldots,e^{j(N-1)(((2\pi d\cos(0))/\lambda)+\xi)}\right]^{T}$$

where:
   λ is the carrier wavelength
   d denotes the adjacent element spacing

Here, we assume that the excitation currents have the same amplitude, but the phase difference between adjacent elements is ξ.

From $h=2\pi/\lambda$, we can get the array factor for a ULA:

$$\mathrm{AF}=1+e^{jhd\cos(\theta)+\xi},+\cdots+,+e^{j(N-1)hd\cos(\theta)+\xi}$$

$$\mathrm{AF}=\sum_{n=1}^{N}e^{j(n-1)hd\cos(\theta)+\xi}$$

$$\mathrm{AF}=\frac{\sin\left((N/2)hd\cos\theta+\xi\right)}{\sin\left((1/2)hd\cos(\theta)+\xi\right)}e^{j1/2(N-1)\left(hd\cos(\theta)+\xi\right)}$$

The maximum value for the array factor occurs when

$$\tfrac{1}{2}\left(hd\cos\theta+\xi\right)=\pm m\pi,\quad m=0,1,2\dots$$

Then we can get the maximum beam direction:

$$\theta_{max}=\cos^{-1}\left[\frac{\lambda}{2\pi d}\left(-\xi\pm2m\pi\right)\right]$$

2. UCAs: UCAs are uniformly spaced along the circumference of the circle. Equipped with such a configuration, they generally tend to achieve wide angle radiation direction (Figure 7.12).

The array factor for UCA is

$$AF=e^{j\left[hd\sin\phi\cos(\theta-\theta_1)+\xi_1\right]}+\cdots+e^{j\left[hd\sin\phi\cos(\theta-\theta_m)+\xi_m\right]}+\cdots+e^{j\left[hd\sin\phi\cos(\theta-\theta_M)+\xi_M\right]}$$

$$\sum_{m=1}^{M}e^{j\left[hd\sin\phi\cos(\theta-\theta_m)+\xi_m\right]}$$

The maximum value for the array factor occurs when

$$hd\sin\phi\cos\left(\theta-\frac{2\pi}{M}\right)+\xi=\pm2n\pi$$

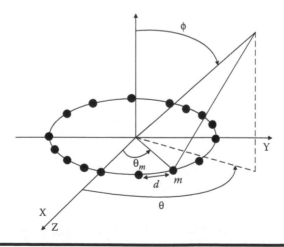

Figure 7.12   Uniform circular array.

7.2.1.2.3 Adjacent Element Spacing and Mutual Coupling Effect

In practical implementation of large numbers of antennas required for massive MIMO, it is found that mutual coupling has a substantial impact on capacity, as the increased number of antennas, especially for a fixed array aperture, results in reduced adjacent element spacing, adversely impacting system performance. With reduced adjacent element spacing, it is intuitive that the array effectively reduces to only one antenna, which can be characterized as a single antenna. When two or more antennas are placed close to each other, there is current or electric field coupling among the antenna elements, which will produce mutual interference. This interference is called the *mutual coupling effect*. Mutual coupling effects between the antennas are usually characterized by the coupling coefficient matrix.

Figure 7.13 shows the array mutual coupling network model: V_{st} and Z_{Lt} are the source voltages and load impedances, respectively; Z_{st} is the impedance of the arrays; i_i are the excitation and received currents (at the ith port) of the antenna system; and V_t is the terminal voltage across the ith transmit antenna port.

From the derivation in [8], the coefficient matrix is an identity matrix. If we ignore the mutual coupling, the coupling coefficient matrix is

$$C_r = \left(I_n + Z_L^{-1}Z\right)^{-1}\left(I_n + Z_L^{-1}Z_S\right)$$

where:

Z_L $= \mathrm{diag}(z_{Li})$, $(I = 1, 2,\ldots, n)$ is the load diagonal matrix
I_n is the n-dimensional identity matrix
Z $= (z_{ij})$, $(i, j=1, 2,\ldots, n)$ is the mutual impedance matrix
Z_s $= \mathrm{diag}(z_{si})$, $(I = 1, 2,\ldots, n)$ is the impedance matrix of the arrays

Generally, we can assume that the impedance of each antenna has identical values, and hence the conjugation value is equal to the load impedance. Under such an assumption, the Z matrix elements can be calculated based on [9].

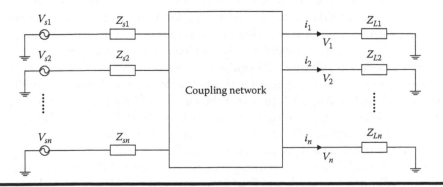

Figure 7.13   Coupling network of antenna arrays.

The mutual coupling effect ultimately has an impact on the entire array, which can be expressed as [8]

$$A = C_r a_i$$

where:

$a = [a_1, a_2, \ldots, a_n]^T$ are the signal vectors of each antenna with mutual coupling effect

$a_t = [A_{t1}, A_{t2}, \ldots, At_n]^T$ are the signal vectors of antennas without mutual coupling effect

7.2.1.3 Massive MIMO Test Bed

In this section, we will introduce some development work on prototypes of large-scale antenna systems recently launched by some research institutes.

Linköping University, Lund University, and Bell Labs have developed an ASS with 128 antennas, which works on 2.6 GHz and contains circular array and line array configurations [10]. The circular array consists of 128 antenna ports, while the antenna array consists of 16 dual-polarized patch antenna elements arranged in a circle, with four such arrays stacked to form a cylindrical array antenna. This array has a simple structure, and may solve the scattering problem at different elevation angles. However, due to its limited aperture, its azimuth resolution is weak. The linear array consists of 128 antenna ports.

Channel measurement results show that when the number of antennas is 10 times the number of users, they can get close to the ideal theoretical performance, even in difficult propagation environments [10]. The results in [10] also confirm that when the number of antennas increases a certain amount, the multiuser channels show certain orthogonality among them, and by using linear precoding we can possibly approach the best link capacity. This helps to explore the potential of massive MIMO.

Argos [11], developed by Rice University, Bell Labs, and Yale University, and working at the 2.4 GHz band, is a prototype for a large-scale multiuser beamforming antenna test-bed system that can simultaneously serve multiple users. It adopted a hierarchical and modular design, which makes the system more scalable. Argos V1 made it possible for a BS equipped with 64 antennas, in an indoor environment, to serve 15 users equipped with single antennas simultaneously. In 2013, Argos V2, based on Argos V1, increased the number of BS antennas to 96. Argos consists of Wireless Open-Access Research Platform (WARP) boards, a laptop, an Ethernet switch, and an AD9523-based clock distribution board.

In the original prototype, the system includes a central controller, an Argos hub, and 16 modules, each containing four radios. Each WARP module consists of four radio daughter cards and four antennas, and a field programmable gate array (FPGA) mainly converts digital signal to RF signal, or vice versa; the clock distribution board is for the realization of clock synchronization between components; the central controller is responsible for baseband data processing and analysis; and the Ethernet switch is responsible for gathering the collected signal from all RF front-end and forwarding to the central controller baseband.

7.3 Beam forming

7.3.1 Overview of Beam Forming

Beam forming is a kind of signal-pretreatment technology based on the antenna array. Beam-forming technology produces a directional beam by adjusting the weighting coefficient of each beam element in the antenna array; thereby, obvious array gain can be obtained. Therefore, beam-forming technology has great advantages in expanding coverage, improving edge throughput, suppressing interference, and so on. Because of the space selection resulting from beam forming, it has a close relationship with space division multiple access (SDMA).

The purpose of beam forming is to accomplish the best combination or distribution of baseband (intermediate-frequency) signal according to the system performance index. Specifically, its main task is to compensate signal fading and distortion introduced by factors such as free-space loss and multipath effect in the process of wireless transmission and reduce the interference among co-channel users at the same time. So, firstly, a system model needs to be established to describe all kinds of signals in the system, and then it will become possible to express the combination or distribution of a signal as a mathematical problem based on the system performance requirements and seek the optimal solution.

Beam-forming technology in multiple-antenna systems has been widely investigated for its effective suppression of multiuser interference, which can greatly improve the system capacity. By forming the emitting beam, it is possible direct the main lobe direction of a signal toward the intended users as far as possible, and beam null-forming to other users at the same time to reduce interference to other users. And by receiving the formed beam, users can avoid interference from others, and match the main lobe direction of the desired signal as far as possible at the same time.

As shown in Figure 7.14, an antenna array adopting beam-forming technology can form a strong beam in the preferred directions (generally the communication directions) and a weak beam in undesired directions (generally the interference directions). Beam forming can improve the SNR in the direction of the strong beam and suppress CCI in a multiuser environment in the direction of the weak beam, thus improving the overall signal-to-interference-plus-noise ratio (SINR). So, the system performance can be further improved by combining beam-forming technology with space multiplexing technology or space diversity technology.

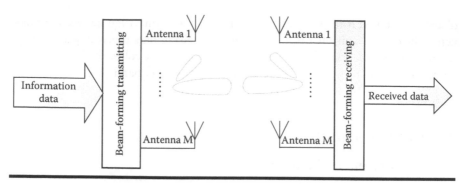

Figure 7.14 **Wireless MIMO beam-forming technology principle diagram.**

7.3.2 Beam-Forming System

Figure 7.14 shows the general structure of the beam-forming system, which consists of three sections: the antenna array section, the analog/digital (A/D) conversion section, and the weight calculation section.

1. Antenna array section: According to the rule of array combination and array element arrangement, an array of antennas can be divided into a linear array, an area array, a circle array, and so on. The number of antenna array elements also has a direct impact on smart antenna performance. In practice, the usual number of antenna array elements is 8 or 16. For example, the structure of the smart antenna in time-division synchronous code-division multiple access (TD-SCDMA) is an eight-array element circle array.
2. A/D conversion section: Each channel of the smart antenna on the BS side has an A/D converter. In the receiving (transmitting) mode, the antenna converts the received analog signals into digital signals, or vice versa.
3. Weight calculation section: This part is the core of the beam forming. A digital signal processor can adjust the weight value coefficient W_1, W_2,... adaptively to obtain the appropriate beam-forming network, or choose the optimal values in a set of preset weights to get the best main beam direction. In this way, the beam-forming device can track users or choose beam figures intelligently.

A receiving beam-forming structure is shown in Figure 7.15. The transmitting beam-forming structure is slightly different from the receiving beam-forming structure. Its weighted device or weighted network is set to be in front of the antenna, and without an add combiner.

7.3.3 Basic Principle of Beam Forming

Beam forming is a method of spatial filtering in terms of signal processing, which can pick up the signal we are concerned with from the input signal, which is a mixture of signal, interference, and noise.

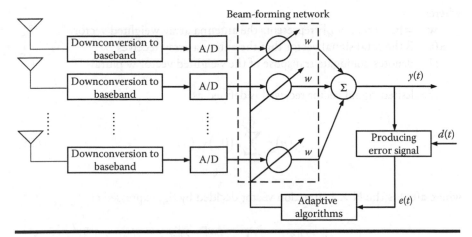

Figure 7.15   General beam-forming device structure.

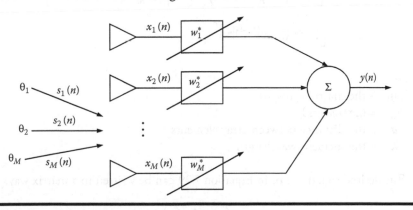

Figure 7.16   Adaptive beam-forming schematic diagram.

A diagram of adaptive beam forming is shown in Figure 7.16. Assume that *n* denotes the number of transmitting array elements, *M* the number of receiving array elements, and d the distance between array elements. Every antenna array receives the desired signals, interference signals, and noise (Gaussian white noise). The desired signal $x_s(n)$ is received from the direction of the arrival angles $\theta_1, \theta_2,..., \theta_M$, and the interference signal $x_i(n)$ from the direction of the arrival angles $\theta_1, \theta_2,..., \theta_J$. Assuming that the received signal has been converted through downconversion and A/D conversion, the output of the array system *y(n)* is the weighted sum of components that the received signal vector *x(n)* has on each array element, expressed as

$$y(n) = \mathbf{w}^{H}\mathbf{x}(n) = \sum_{i=1}^{M} w_i^* x_i(n) \qquad (7.28)$$

where:

$\mathbf{w}$ $= [w_1, w_2,..., w_M]^T$ represents the antenna array weighted vector

$\mathbf{x}(n)$ is the total signal vector that the antenna array receives

H denotes conjugate transpose of the weighted vector $\mathbf{w}$ plural

The desired signal vector received is defined as

$$x_s(n) = \sum_{m=1}^{M} a(\theta_m) s_m(n) \tag{7.29}$$

where $\mathbf{a}(\theta_m)$ is the N × 1 direction vector decided by θ_m, expressed as

$$\mathbf{a}(\theta_m) = \left\{ 1, \exp\left(j\frac{2\pi d}{\lambda}\sin\theta_m \right),...,\exp\left[j\frac{2\pi d}{\lambda}(M-1)\sin\theta_m \right] \right\}^T$$

$$= \left[1, e^{j\beta_m},..., e^{j(M-1)\beta_m} \right]^T$$

where:

$[(\bullet)]^T$ is the matrix transpose

β_m $= (2\pi/\lambda)d\sin\theta_m$

d is the distance between array elements

λ is the incident wavelength

The desired signal vector in Equation 7.29 can be written in a matrix way:

$$\mathbf{x}_s(n) = \mathbf{A}_s \mathbf{s}(n) \tag{7.30}$$

where:

$\mathbf{A}_s$ is the N×M vector and direction vector of the desired signal, defined as

$$\mathbf{A}_s = \left[\mathbf{a}(\theta_1), \mathbf{a}(\theta_2),..., \mathbf{a}(\theta_M) \right] \tag{7.31}$$

$\mathbf{s}(n)$ is the M×1 vector and the desired signal vector, defined as

$$\mathbf{s}(n) = \left[s_1(n), s_2(n),..., s_M(n) \right]^T \tag{7.32}$$

Similarly, we can get the interference signal vector:

$$\mathbf{x}_i(n) = \left[\mathbf{A}_i \mathbf{i}(n) \right] \tag{7.33}$$

where:

A$_i$ is the ×I vector and the direction vector of the interference signal, defined as

$$\mathbf{A}_i = \left[\mathbf{a}(\theta_1), \mathbf{a}(\theta_2), \ldots, \mathbf{a}(\theta_I) \right] \tag{7.34}$$

i(n) is the I×1 vector and the interference signal vector, defined as

$$\mathbf{i}(n) = \left[i_1(n), i_2(n), \ldots, i_I(n) \right]^T \tag{7.35}$$

Hence, the total signal vectors the array receives can be expressed as the sum of desired signal, interference signal, and noise signal:

$$\mathbf{x}(n) = \mathbf{x}_s(n) + \mathbf{x}_i(n) + \mathbf{n}(n) \tag{7.36}$$

Equation 7.28 can be rewritten based on the derived result in Equation 7.36:

$$y(n) = \mathbf{w}^H \mathbf{x}(n) = \mathbf{w}^H \left[\mathbf{x}_s(n) + \mathbf{x}_i(n) + \mathbf{n}(n) \right] \tag{7.37}$$

The array amplitude beam diagram is defined as

$$F(\theta) = \left| w^H a(\theta) \right| \tag{7.38}$$

To make the beam point in the normal direction ($\theta = 0°$), setting $\mathbf{w}$ as

$$\mathbf{w} = \left[1, 1, \ldots, 1 \right]^T \tag{7.39}$$

At this time, the beam diagram is

$$F(\theta) = \left| \mathbf{w}^H \mathbf{a}(\theta) \right| = \left| \sum_{i=1}^{M} e^{j(i-1)\beta} \right| = \left| \frac{\sin(M\beta/2)}{\sin(\beta/2)} \right| = \left| \frac{\sin\left[(M\pi d/\lambda)\sin\theta \right]}{\sin\left[(\pi d/\lambda)\sin\theta \right]} \right| \tag{7.40}$$

where the distance between array elements $d = \lambda/2$.

A ULA is shown in Figure 7.17, while Figure 7.18 is an eight-element ULA beam diagram. From Figure 7.18, the following conclusions can be obtained:

1. The direction of maximum antenna gain, that is, the main lobe, is located in $\theta = 0°$.

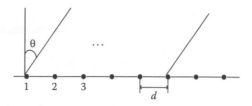

Figure 7.17 Uniform linear array.

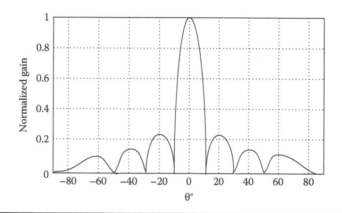

Figure 7.18 Eight-element uniform linear array beam diagram.

2. The angle between the first zero points on both sides of the antenna main lobe is referred to as zero point lobe width BW_0:

$$BW_0 = 2\arcsin\frac{\lambda}{Md} \tag{7.41}$$

If $Md \gg \lambda$, then

$$BW_0 \approx \frac{2\lambda}{Md} \tag{7.42}$$

The width of the main lobe half power points (where the intensity drops to half of maximum power) can be approximated as

$$BW_{0.5} = 0.886\frac{\lambda}{Md}(\text{rad}) = 50.8\frac{\lambda}{Md}(°) \tag{7.43}$$

3. The electrical level of the first side lobe in the directional diagram is highest, and decreases with the increase of the array element number M.

4. When $d < \lambda/2$, there is only one main lobe in the antenna directional diagram. When $d > \lambda/2$, a larger side lobe, called the grating lobe, may be produced. There will be two or more large outputs in different directions when the grating lobe appears, so that it will be impossible to ensure from which direction the signal enters the array. To avoid the grating lobe, usually $d \leq \lambda/2$.

7.3.4 Classification of Wireless MIMO System Beam-Forming Technology

The initial research on beam-forming technology mainly focused on receiving beam forming, such as main beam forming and depth, width of beam nulling, and constraints on the beam side lobe. It is in the application of multiple transmit antennas and the multiuser environment that MIMO beam forming is different from traditional beam forming, so with the increased focus on receiving beam forming, transmitting beam-forming technology is now attracting attention. MIMO technology has become the core of the wireless communication system. One of the key problems affecting MIMO system performance is interference suppression problems between users in the multiuser MIMO environment. Beam generator design is an important way to limit multiuser interference.

In the field of wireless communications, beam forming is mainly divided into three categories: receiving beam forming, transmitting beam forming, and relay beam forming.

7.3.4.1 Receiving Beam-Forming (Uplink) Technology

Compared with transmitting beam forming, receiving beam forming in wireless communication is in theory more direct and easier to implement. So, receiving beam forming made great progress in the last century, with analytical methods similar to those of radar research. Receiving beam forming is generally in the uplink, so it is also referred to as *uplink beam forming*.

At present, the main direction of receiving beam-forming technology development is robust design. Traditional receiving beam-forming design usually needs to estimate the guidance signal vector (data and channel) or the received signal vector (steering vector and noise). In the practical system, for reasons such as channel change, quantitation, and calculation precision, the signal estimation error is almost inevitable. In this case, beam formers are generally designed using limited samples and employing minimum variance (MV) technology based on sample matrix inversion (SMI). However, SMI-based MV estimation will affect the beam-forming design significantly, because a slight estimation error will lead to large changes in the design results, thus affecting receiving performance. To solve this problem and improve the robustness of SMI-based MV, the diagonal loading (DL) method can be used. However, it is difficult to obtain the DL coefficient using traditional DL, which is its main drawback. In recent years, a robust design method has been developed based on the optimization of the worst-case system performance, which obtains the beam former by optimizing the worst performance. This method is derived from strict theory and can guarantee robust performance, which makes it currently an important research direction for the robust design of receiving beam forming. Many applications for robust design are being promoted, such as introducing the ellipsoid model to robust design, robust design of broadband systems, and robust design combined with other technologies (such as STC).

7.3.4.2 Transmitting Beam-Forming (Downlink) Technology

Traditional beam forming often forms a strong beam in only one direction and a weak beam in other directions. In multiuser wireless communication, however, multiple beam-forming vectors are required to form a strong beam in the direction of all users. These beam-forming vectors need joint optimization, strengthening the beam in the direction of the target users and reducing the beam in the direction of the other users at the same time. Transmitting beam forming is also called *downlink beam forming*. Currently, research on beam forming is mainly focused on the downlink. The development of downlink beam forming can be roughly divided into three stages: multiuser MIMO downlink beam forming, multiuser MIMO downlink robust beam forming, and multicell multiuser (cellular) MIMO downlink link cooperative beam forming.

7.3.4.2.1 Multiuser MIMO Downlink Beam Forming

The optimization variables of receiving beam forming can be independently optimized in some cases, but those of transmitting beam forming usually need joint processing. Therefore, the design of transmitting beam forming is relatively complex, and it is even more complex in a multiuser environment. One design criterion is for the purpose of eliminating interference between users. This can be eliminated (or close to eliminated) by the ZF method or the block diagonal (BD) method. Another design criterion is measured by quality of service (QoS), generally converting the beam-forming problem into a QoS maximization problem under the restriction of resource consumption (such as transmitting power) and a resource consumption minimization problem under the restriction of QoS. Uplink and downlink duality is an important and effective method for solving QoS problems. Research shows that a power minimization beam-forming problem under the restriction of SINR in the downlink can be solved by an iterative method in the uplink, because it is easier to solve uplink problems. The optimality of the uplink and downlink duality method has been proved. In a MIMO system, either the combination of receiving and transmitting beam forming or the situation of multiple data streams for each user should be considered. In addition to the uplink and downlink duality method, convex optimization is also an effective method for power minimization problems in the design of multiuser MIMO QoS beam forming.

7.3.4.2.2 Multiuser MIMO Downlink Robust Beam Forming

Much of the research into transmitting beam forming assumes that the ideal CSI is known to the transmitter. In practice, due to estimation error, channel change, quantization, and feedback delay, among other reasons, the transmitter usually receives a nonideal channel estimation matrix, and the calculation of the beam former is very sensitive to the channel matrix. Therefore, the robust design of beam forming in the case of nonideal CSI has received much attention in recent years. There are generally two methods for the analysis of robust design of beam forming. One of them is the statistical analysis method, based on the channel matrix statistics error (SE) model, which assumes that the channel error meets certain statistical properties. This model is applicable in the case when channel errors are caused by channel estimation processing. Similarly to the ideal CSI case, uplink and downlink duality still holds true to the mean square error (MSE) of the SE model, and can be used in the design of the downlink beam forming. The other method is a worst-case design method based on the norm bounded error (NBE) model, which assumes that the channel error is limited to a certain bounded range. The model does not need a statistical hypothesis, and is suitable in the case when channel error is caused by the quantization of channel coefficient. In the actual system, there is little possibility for worst case to appear. Similarly to receiving beam forming, it is necessary to introduce probability factors, and the robustness of beam-forming design can then be quantified in the form of probability.

7.3.4.2.3 Multicell Multiuser (Cellular) MIMO Downlink Link Cooperative Beam Forming

Because of the existence of interference among cells, the capacity gain obtained by multiuser MIMO processing will be seriously reduced in a multicell environment. To solve this problem, MIMO processing strategies or algorithms are proposed by using the collaboration between the BSs in the multicell environment. For cellular MIMO downlink beam forming, there are two main collaboration strategies. One strategy assumes complete collaboration for all BSs: that is, all stations fully share the channel information and data and common service mobile terminal, and thus, all BSs can be viewed as one big BS. The cellular network beam-forming problems can be considered as multiuser MIMO beam forming under the restriction of total power. Although this strategy can achieve optimal performance, complete collaboration has a high requirement for system coordination and is difficult to implement. The other kind of collaborative strategy is called *interference collaboration*: that is, all BSs are assumed to obtain only channel information or local channel information, and each mobile terminal is served by only one BS.

These two kinds of collaborative strategy are aimed at all BSs, which is almost impossible to realize in practice. Because the interference among cells is mainly from neighboring cells, cells can be divided into groups and allowed to collaborate with cells in the same group. In addition, in the process of designing cellular network MIMO downlink beam forming, some practical factors should be taken into consideration, such as user distribution, distributed design, and robust design.

7.3.4.3 Relay Network Beam-Forming Technology

The study of the theoretical information capacity of relay channels and relay networks shows broad prospects for relay collaboration. The key idea of relay network beam forming is to combine the antennas of a multiple relay node, build a virtual antenna array, and then transform information to the target communication relay. To some extent, relay beam forming can be understood as a combination of receiving beam forming and transmitting beam forming. Relay beam forming, however, is very different from receiving and transmitting beam forming. The main difference is that due to the space distribution, it is quite difficult for a relay node to obtain receiving signals from other relay nodes; thus, the beam-forming design must be distributed. In relay beam-forming technology, a simple relay strategy is the amplify-and-forward (AF) protocol. At present, most studies are based on this strategy.

In the actual system, a beam-forming strategy should also consider the intersymbol interference (ISI) suppression problem. In addition, there are other development directions for relay network beam-forming research, such as robust beam-forming design, joint beam-forming design of transceiver node and relay node design, and multihop beam-forming design.

7.3.5 MIMO Beam-Forming Algorithms

There are various beam-forming algorithms and classification methods. For example, these algorithms can be divided into DOA and non-DOA estimation algorithms on the basis of whether a DOA estimation is needed; adaptive nonblind and adaptive blind algorithms according to whether a reference signal is needed; uplink and downlink beam-forming algorithms according to different applications; or time domain or space algorithms and space-time joint processing algorithms according to the signal domain classes.

As shown in Figure 7.19, the main factors affecting MIMO system and rate performance are noise, the same user interference between different data streams, and multiuser interference. With low SNR, system performance is mainly influenced by noise. In this case, algorithm design mainly focuses on improving the power gain of the desired signal; when SNR is high, multiuser interference and the same user interference between different data streams play a dominant role in influencing system performance. So, the algorithm design in a case of high SNR mainly considers how to suppress multiuser interference and interference between data streams.

There are several main beam-forming algorithms, such as matched filtering (MF), minimizing interference, maximizing the signal-to-interference and noise

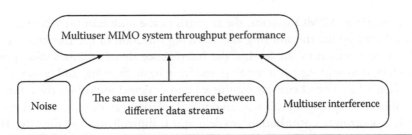

Figure 7.19 Main factors influencing the throughput performance of multiuser MIMO systems.

ratio (Max SINR), and maximizing the signal to leakage and noise ratio (Max SLNR).

The MF algorithm, designed for noise and interference between data streams, implements singular value decomposition (SVD) on the channel matrix. Derived from SVD, the left singular vector denotes the receiving beam-forming matrix, and the conjugate transpose of the right singular vectors denotes the transmitting beam-forming matrix, to match the channel. However, when used in a multiuser MIMO system, system performance will be bad under the condition of high SNR, because the algorithm only considers how to match the desired channel, regardless of the existence of multiuser interference.

The minimizing interference algorithm is a receiving beam-forming design only intended for multiuser interference suppression. Under the condition of low SNR, the performance of the algorithm is poor, because in this case, the influence of noise on the performance plays a leading role; but when SNR is high, because the interference can be suppressed effectively, the algorithm can achieve better performance.

The Max SINR algorithm is a trade-off between the MF and minimizing interference algorithms, and considers the influence of noise, interference between data streams, and multiuser interference on system performance at the same time. So, in the case of both low SNR and high SNR, the performance is better than with the former two algorithms. It has been shown by analysis that these three algorithms only optimize the performance of a single user, rather than improving the capacity of the whole system from the angle of the system. Therefore, in a multiuser MIMO communication system, these algorithms are not optimal.

The Max SLNR algorithm also considers the influence of noise, interference among data streams, and multiuser interference, and can completely eliminate interference among the data streams. But because the SLNR defined in the transmitter cannot match the SINR completely, the effect of interference suppression is less obvious than with the minimizing interference and Max SINR algorithms. However, the Max SLNR algorithm can simplify the problem of multiuser joint design and then obtain the closed-form solution of the transmitting beam-forming matrix by the Rayleigh theorem. Obviously, it is a suboptimal algorithm.

In multiuser MIMO systems, the strength of the multiuser interference is not only affected by the transmitting and receiving direction of the transmitting signal, but also related to the level of the interference signal's transmission power. Therefore, transmitting beam forming, receiving beam forming, and power allocation should be comprehensively considered and designed jointly for the optimization of system capacity. In this way, an optimal solution can be obtained. However, this is a nonconvex optimization problem and is difficult to solve directly. There are generally two ideas for solving a nonconvex optimization problem. One is the iteration method; the other is to find the tight lower bounds of the nonconvex problem (i.e., to estimate the tight upper bound of the performance), converting the nonconvex problem into a convex optimization problem, and then to design an optimization scheme that can come close to or actually achieve the tight lower bounds of performance.

References

1. L. You, X. Gao, X. Xia, N. Ma, and Y. Peng, Pilot reuse for massive MIMO transmission over spatially correlated Rayleigh fading channels, *Wireless Communications, IEEE Transactions on* 14(6): 3352–3366, 2015.
2. C. Xiao, Y. R. Zheng, and N. C. Beaulieu, Novel sum-of-sinusoids simulation models for Rayleigh and Rician fading channels, *Wireless Communications, IEEE Transactions on* 5(12): 3667–3679, 2006.
3. S. M. Alamouti, A simple transmit diversity technique for wireless communications, *IEEE Journal of Selected Areas in Communications* 16(8): 1451–1458, 1998.
4. K. Kusume, M. Joham, W. Utschick, and G. Bauch, Efficient Tomlinson-Hiroshima preceding for spatial multiplexing on flat MIMO channel, *International Conference on Communication* 3: 2021–2025, 2005.
5. R. R. Muller, D. Guo, and A. L. Moustakas, Vector precoding for wireless MIMO systems and its replica analysis, *IEEE Journal of Selected Areas in Communications* 26(3): 530–540, 2008.
6. G. Caire and S. Shamai, On achievable rates in a multi-antenna broadcast downlink, *38th Annual Allerton Conference on Communication, Control and Computing*, Monticello, IEEE, pp. 1188–1193, 2000.
7. 3GPP TR 36.873, Study on 3D channel model for LTE (Release 12).
8. X. Li and Z-P Nie, Mutual coupling effects on the performance of MIMO wireless channels, *IEEE Antennas and Wireless Propagation Letters* 3: 344–347, 2004.
9. H. E. King, Mutual impedance of unequal length antennas in echelon, *Antennas and Propagation, IRE Transactions on*, Los Angeles, CA, IEEE, vol. 45, pp. 306–313, 1957.
10. F. Rusek, D. Persson, B. K. Lau, E. G. Larsson, T. L. Marzetta, O. Edfors, and F. Tufvesson, Scaling up MIMO: Opportunities and challenges with very large arrays, *IEEE Signal Processing Magazine* 30: 40–60, 2013.
11. W. Calyton, *Shepard, Argos: Practical Base Stations for Large-Scale Beamforming*, Houston: Rice University, 2012.

Chapter 8

Self-Healing in 5G HetNets

Mohamed Selim, Ahmed Kamal, Khaled Elsayed,
Heba Abd-El-Atty, and Mohammed Alnuem

Contents

The main requirements of the fifth generation (5G) are emerging through the efforts of diverse groups such as 4G America in the United States, the IMT-2020 (5G) promotion group in China, and the 5G Private Public Partnership (5G PPP) in Europe. The 5G requirements will greatly increase the network complexity, which will demand auto-integration and self-management capabilities that are well beyond today's self-organizing network (SON) features. Additionally, ultrareliable communications impose very stringent latency and reliability requirements on the architecture.

The main challenges for 5G networks are the continued evolution of mobile broadband and the addition of new services and requirements, for example, anything-to-anything communication, very low latency (<1 ms), as well as reduced signaling overhead and energy consumption (greener network). Future mobile networks will not only have significantly increased traffic volumes and data transmission rates, but also many more use cases. These include not only traffic between humans and between humans and the cloud, but also between humans, sensors, and actuators in their environment, as well as between sensors and actuators themselves.

Small access nodes, with low transmitting power and no planning requirements, are conceived to be densely deployed, resulting in an ultradense network (UDN). UDNs are also called *heterogeneous networks* (HetNets), that is, multilayered networks with high-power macrocells and very dense small cells (SCs) with low power [1]. This approach improves spectral efficiency per area by reducing the distance between transmitters and receivers, and improves macrocell service by offloading wireless traffic. UDNs are a step further toward low-cost, plug-and-play, self-configuring, and self-optimizing networks. 5G will need to deal with many more base stations (BSs), deployed dynamically and in a heterogeneous manner, combining different radio technologies that need to be flexibly integrated. Moreover, a massive deployment of small access nodes introduces several challenges, such as additional backhaul and mobility management requirements, which 5G needs to address (CROWD project: http://www.ict-crowd.eu/).

The Mobile and Wireless Communications Enablers for the Twenty-Twenty Information Society (METIS) provides a consolidated 5G vision. According to this vision, the most prominent requirements of 5G are: (1) total capacity increased 1000-fold; (2) 10–100 times more connected devices; (3) end-user data rates expected to increase by 10–100 times; (4) latency reduced by five times; and (5) requirements (1)–(4) at today's cost or less. The road toward 5G is gradual. But perhaps, a key 5G qualification that dominates all these requirements is network flexibility and reliability. These qualifications can be achieved by integrating SONs in upcoming 5G networks.

The purpose of this chapter is to cover the topic of self-healing in 5G networks. Self-healing is an important functionality of SONs, and it means that networks migrate from manual operation to automated operation (minimizing human interaction). SON defines three areas: self-configuration (plug-and-play network elements [NEs]), self-optimization (automatically optimizing NEs and parameters), and self-healing (automatically detecting and mitigating failures in NEs).

This chapter is structured as follows. Section 8.1 captures SON and addresses it both before 5G and within 5G networks. Section 8.2 presents a detailed introduction to self-healing and its two main categories, cell outage detection (COD) and compensation, and then illustrates the state of the art in self-healing in Section 8.3. Section 8.4 presents a backhaul self-healing case study in detail. Section 8.5 concludes the chapter.

8.1 Introduction to SONs

SON is a paradigm defined under the auspices of the 3rd Generation Partnership Project (3GPP) and the Next Generation Mobile Networks (NGMN) Alliance, aiming to automate mobile network operation, administration, and management (OAM). Via SON, operators seek to achieve optimum performance at minimum cost. Figure 8.1 shows a comparison between network operation with SON functions and the conventional OAM, which relies on human intervention or service tools. SON aims to leapfrog the performance of future networks to a higher level of automated operation by self-managing the planning, configuration, optimization, and healing of networks, which act in three main areas [2]:

Self-configuration: The plug-and-play capabilities of NEs, including self-planning, in which the selection of the new site location is determined automatically depending on dead coverage zones or dense mobile user areas. Also, the authentication and verification of each new site are done automatically.

Self-optimization: The adjustment and auto-tuning of parameters during the operational life of the system, making use of the measurements and performance indicators collected by the user equipment (UE) and the NEs.

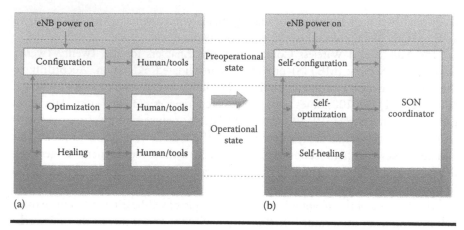

Figure 8.1 **Network operation (a) without and (b) with SON function.**

Self-healing: Failure detection, diagnosis, compensation, and recovery of the network. This is done by the execution of certain actions that force the network to work in normal or near-normal conditions even in the presence of failures.

8.1.1 SON Architecture

There are three different architectures for SON functionality:

1. Centralized SON: In centralized SON, optimization algorithms are executed in the OAM system. In such solutions, SON functionality resides in a small number of locations, at a high level in the architecture. Thus, all SON functions are located in OAM(s), so it is easy to deploy them. However, as this architecture adds latency to simple optimization cases due to centralized processing, it is not suitable for the UDN, and thus not suitable for 5G networks.
2. Distributed SON: In distributed SON, optimization algorithms are executed in macrocells (evolved node B [eNB] in Long-Term Evolution [LTE]). In such solutions, SON functionalities are found in many locations at a relatively low level in the architecture. Thus, all SON functions are located in macrocells. This adds overhead to deployment. It is also difficult to support complex optimization schemes that require coordination among many BSs. However, it is easy to support quick optimization responses among a small number of BSs.
3. Hybrid SON: In hybrid SON, some of the optimization algorithms are executed in the centralized OAM system, while others are executed in macrocells. Thus, simple and quick optimization schemes are implemented in macrocells, and complex optimization schemes are implemented in OAM. Thus, it has the flexibility to support different kinds of optimization cases.

8.1.2 SON before 5G

8.1.2.1 SON in Global System for Mobile Communications and Universal Mobile Telecommunications System

Deploying and operating conventional cellular networks (Global System for Mobile Communications [GSM] and Universal Mobile Telecommunications System [UMTS]) was a complex task that comprised many activities, such as planning, dimensioning, deployment, testing, launching and operating optimization, performance monitoring, failure mitigation, failure correction, and general maintenance. The GSM network was much simpler than today's networks and had a low degree of automation and operational efficiency. This is why the SON trend was developed parallel to the evolution and increased complexity of cellular networks. When GSM was deployed, it required a great deal of manual operational effort, which was then gradually reduced in the evolved systems, such as UMTS and LTE, which became more sophisticated and needed more automation to operate with high performance.

The NGMN Alliance (the reader is referred to www.ngmn.org for more details) identified excessive reliance on manual operational effort as a main problem in conventional mobile networks and defined operational efficiency as a key target. A project was started in 2006 by the NGMN with its main focus on SON. The project's main objective was to solve the manual operation problem of the conventional mobile networks (GSM and UMTS). The deliverables of this project were adopted during the development of the 3GPP. At that time, the 3GPP implemented some functions for the SON, such as minimization of drive test, energy saving, handover optimization, and load balancing.

8.1.2.2 SON in 3GPP

The 3GPP mandate is the development of 3G and 4G networks based on the GSM standards. The 3GPP was originally concerned with technical requirements and specifications for 3G systems; then the maintenance and development of GSM systems were added to its responsibilities, and it is currently responsible for the evolution of LTE and LTE-Advanced. The structure of 3GPP consists of four technical specification groups (TSGs), each divided into several working groups (WGs). The WGs responsible for the evolution of SON are mainly radio access network (RAN) WG3, and SA WG5 [3].

The SON has been defined by 3GPP in different releases, starting with Release 8. In Release 9, the main self-configuration functionalities were standardized, and a framework was described that covers all the necessary steps to put a new eNB in an operational state (self-configuration). Also, manual configuration was still an option. Self-optimization in Release 9 included the following use cases: neighbor list optimization with 2G and 3G neighbors, mobility load balancing, mobility robustness optimization, and interference control. Finally, there are a few use cases of self-healing standardized in Release 9: self-recovery of NE software failures, self-healing of board failures, and self-healing of cell outage.

In Release 10, more and more SON functions were standardized, such as management aspects of interference control, capacity and coverage optimization, interworking and coordination between different SON functions, and the definition of inputs and outputs of self-healing. In 3GPP Release 10, the study of applying SON to HetNets was initiated in order to be standardized in later revisions or releases. In each of the releases after Release 10, more functions were added to each SON category. Also, white papers from vendors and deliverables from different projects added new functions and solutions to the SON. The SOCRATES project was one of the major projects dealing with self-configuration and self-optimization aspects. It ran for three years from January 2008. The project's main objectives were to validate the developed concepts through extensive simulations, and to evaluate the implementation and the impact of the proposed and standardized schemes [4].

8.1.2.3 SON in 4G

The explosive growth in mobile broadband services has resulted in great demands on wireless radio networks. HetNet technology was developed for LTE-Advanced systems to overlay low-power nodes within a macrocell to improve capacity and extend coverage. In HetNets, SON is a critical technology. On the one hand, the use of SON in HetNets allows operators to streamline their operations, not only reducing the complexity of managing co-channel interference in HetNets, but also saving operational costs for all macro and heterogeneous communication entities, to harmonize the whole network management approach and improve the overall operational efficiency. The availability of SON solutions for HetNets leads to the identification of more powerful optimization strategies that are able to suppress interference and improve energy efficiency. Unfortunately, on the other hand, most HetNets still do not offer SON features because the architectures of SON specified in 3GPP were designed primarily for a homogeneous network topology.

This latter problem, SON incompatibility with HetNets, stimulated standardization and research efforts to focus mainly on the extension of the scope of the SON paradigm to include GSM, general packet radio service (GPRS), enhanced data rates for GSM evolution (EDGE), UMTS, and high-speed packet access (HSPA) radio access technologies as well as LTE. The availability of a multi-RAN SON solution enables more complex optimization strategies that deal with several functionalities simultaneously. Such multi-RAN SON is crucial in case there is a cross-layer restriction in the optimization processes. Also, the multi-RAN SON will leapfrog the performance of the smart load balancing strategies between different technologies, which will therefore increase the overall network grade of service and capacity. However, for implementing the multi-RAN SON, one must take into consideration that SON is not a native technology in 2G and 3G networks, and an external centralized entity will be needed to provide the SON functionalities to all multi-RAN technologies [5].

8.1.3 SON for 5G

A candidate future technology of the 5G network is one that requires a major revolution at different layers, starting with migration toward software-defined wireless networking (SDWN), centralized radio access network (C-RAN), and others, while equipping NEs with SON capabilities where required. The SON in 5G networks will be different from that proposed to be implemented in 4G networks, because those candidate technologies have new features that will directly affect the application of SON functionalities. Most of the new technologies, such as C-RAN and software-defined wireless network controller (SDWNC), will facilitate the implementation of the SON. Both of them will collect a large amount of parameters and information, needed by the SON controller, in a centralized entity.

8.1.3.1 C-RAN in 5G

C-RAN is a cloud computing-based new mobile network architecture that supports current and future wireless communication standards. C-RAN consists of the splitting of the conventional BS into two parts: (1) the antennas and radio-frequency (RF) units will be located on-site and (2) all other BS functions will be migrated to the cloud. Between them are the high-bandwidth and low-latency fronthaul links. The focal concept is to redistribute functions that are traditionally found in BSs toward a cloud-operated central processor. So, the architecture has three main components: (1) remote radio heads (RRHs); (2) a baseband unit (BBU) pool; and (3) fronthaul links (which can be wired or wireless).

Such intelligent centralization would consequently enable cooperative operation among cells for more efficient spectrum use and for greener communication. A fully centralized RAN consists of taking most of the BSs functionalities to the cloud and leaving only the RRHs for the cells. The main function of the RRHs is to transmit the RF signals to the users in the downlink (DL) or to forward the baseband signals to the BBU pool for further processing in the uplink (UL). Consequently, the BBU or the virtual BS, which is traditionally located in the BS equipment, is relocated to the cloud or central processor, hence forming a pool shared by all connected RRHs. The BBU processes baseband signals and optimizes the network resource allocation. Implementing the C-RAN will leapfrog the performance of data rate–boosting radio features such as enhanced intercell interference coordination (eICIC), massive multiple-in multiple-out (MIMO), and coordinated multipoint (CoMP) transmission, which require tight and fast coordination between various cells; hence, SON would benefit from centralized processing in 5G networks [6].

8.1.3.2 SDWN in 5G

Software-defined networking (SDN), characterized by a clear separation of the control and data planes, is being adopted as a novel paradigm for wired networking. By implementing SDN, network operators can run their infrastructure more efficiently with faster deployment of new services while enabling key features such as virtualization.

The SDN paradigm has many advantages, including a global optimal routing function implementation, simplification of networking, enabling programmability and easy deployment of new functions, applications, and protocols that will make the 5G network much more flexible and efficient. Adding the concept of SDN to the 5G architecture will add challenges to the upcoming 5G standard such as operating and controlling a large number of small cells and at the same time reducing the cost of implementation and operation.

SDN is currently being considered as an alternative to classic approaches based on highly specialized hardware executing standardized protocols. Until recently, most of the key use cases used to present the benefits of the SDN paradigm have been limited to wired networks such as Google data centers. The adoption of the SDN concept for wireless access and backhaul environments can be even more beneficial than for wired networks. Indeed, the control plane of wireless networks is more complex than that of wired networks, and therefore, higher gains can be achieved from the increased flexibility provided by SDN. The use of an SDWN architecture would allow the network to offer the service provider an application programming interface (API) to control how the networks behave to serve traffic that matches a certain set of rules. By using SDWN, an API could be offered to external parties (e.g., service providers) to enable them to participate in the decision on which access technology is used to deliver each type of traffic to a specific mobile user or group of users.

The UDN will receive a significant benefit from the decoupling of the user data and control planes. 5G users frequently cross cell borders, generating signaling load from handovers and cell reselections. The concept of decoupling the user data and control planes between the macrocell and SCs is often referred to as *soft cell* or *phantom cell*. Figure 8.2 shows the conventional cellular network versus the 5G network implementing the concept of control/data plane split. The phantom cell sends and receives data to and from the user (data offloading from the macrocell), while the macrocell sends and receives all radio resource control and signaling information to and from the user. The ultradense SCs/phantom cells in the figure are omitted for clarification and figure simplicity. This split will collect all the information needed for SON decisions at the macrocell or the central entity, which will result in reducing the complexity of the SON functionalities in the 5G networks.

8.1.3.3 New Path Loss Models in 5G

By implementing the UDN, the BSs (SCs) will communicate and exchange information with each other. This will introduce new path loss models for different

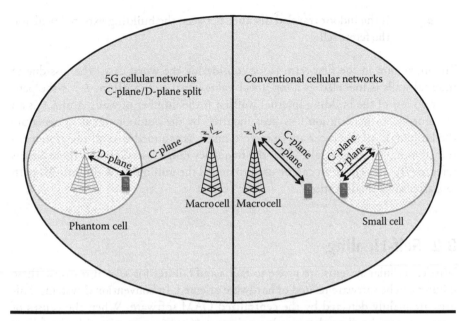

Figure 8.2 Control/data split networks versus conventional networks.

communication scenarios. The conventional communications scenarios, such as outdoor-to-outdoor communications, indoor-to-indoor communications, or even outdoor-to-indoor communications, will be the same. But a new case that will be introduced due to the implementation of the UDN in 5G networks is indoor-to-outdoor-to-indoor communications.

An example of this scenario is a femtocell in a certain building that communicates with another femtocell in a different building. In this case, a new path loss equation is proposed by modifying the conventional equation of the outdoor-to-indoor communication. PL_{I2O2I} is the path loss from indoor-to-outdoor-to-indoor between two femtocells in two different buildings. Here, there are two outdoor-indoor penetration losses at each side, and the indoor distances in the two buildings are considered to be equal. This is equivalent to dividing the total distance between the two femtocells into $(d_{in1} + d_{out1})$ and d_{in2}. For simplicity, considering $d_{in1} = d_{in2} = d_{in}$, $d_{out1} = d_{out}$ and $L_{ow1} = L_{ow2} = L_{ow}$, the path loss equation is given by [7]

$$PL_{I2O2I,dB} = \max(38.46 + 20\log_{10} d_{out},\ \kappa + \nu\log_{10} d_{out} + 0.3(2d_{in}) + qL_{iw} + 2L_{ow}$$

where:

κ and ν correspond to the path loss constant and path loss exponent, respectively

d_{out} is the distance traveled outdoors between the two buildings

d_{in} is the indoor traveled distance between the building external wall and the femtocell

The maximum in the first term is for considering the worst case. The loss due to internal walls is modeled as a log-linear value equal to 0.3 dB/m, L_{iw} is the penetration loss of the building internal walls, q is the number of walls, and L_{ow} is an outdoor-indoor penetration loss (loss incurred by the outdoor signal to penetrate the building). All distances are in meters, and it is assumed that all the aforementioned formulas are generalized for the frequency range 2–6 GHz (for more details refer to [7]). This equation can be used to model the millimeter-Wave (mmW) path loss after some modifications.

8.2 Self-Healing

Wireless cellular systems are prone to faults and failures for several reasons. These failures can be software related or hardware oriented. In conventional systems, failures are mainly detected by the centralized OAM software. When the causes of alarms cannot be cleared remotely, maintenance engineers must visit the failure location. This process could take days before the system returns to normal operation. In some cases, failures are undetected by the OAM and cannot be addressed except when an unsatisfied user files a formal complaint, thus resulting in salient degradation in the network performance. In future SON systems, this process needs to be improved by incorporating self-healing functionality.

Self-healing is the execution of actions that keep the network operational or prevent disruptive problems from arising. In response to failure, self-healing procedures smoothly mitigate it, and the network works near normal conditions, even in the presence of failure. Self-healing is done in two steps: (1) COD and (2) cell outage compensation (COC).

COD and COC provide automatic mitigation of BS failures, especially in the case where the BS equipment is unable to recognize being out of service and has therefore failed to notify OAM of the outage. Detection and compensation are two distinct cases that cooperate to completely mitigate, or at least alleviate, the failure.

Both detection and compensation provide many benefits to the network operator. Conventional cellular networks have experienced failures of which operators had no knowledge until they received notification from customer support of problems in the field. COD ensures that the operator knows about the fault before the end user does, so it detects and classifies failures while minimizing detection time. COC provides temporary alleviation of the main failure problems, such as cell power outage. It executes actions to solve or, at least, alleviate the effect of the problem. If the failure time exceeds a certain threshold, it is considered as a permanent failure, and a site visit by the operator is needed to recover the site. Hence, automatic detection and compensation of failures in 5G networks is mandatory.

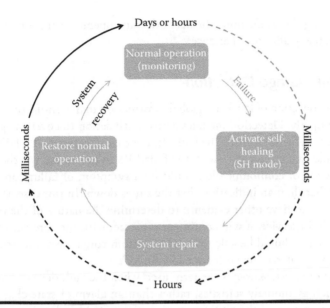

Figure 8.3 Self-healing procedures.

In Figure 8.3, the normal operation of the network means that it is operating without any failures in any BS in the network. The system is monitored for the detection of any failure. In the case of failure, the self-healing functions are activated and implemented in the failure region only, while the rest of the network operates in the normal mode. To mitigate the failure of a cell site, the neighboring cells may increase their power or change their antenna tilt to overcome the coverage problem in the failure region. When the failure is repaired, normal operation mode is recovered, and the neighboring cells restore their original power and antenna tilt configuration.

From a timing perspective, Figure 8.3 shows that normal operation continues for hours or days without any failure or error interruption. In the case of failure, the system will detect and activate the self-healing strategy within a few milliseconds. System repair, which is done by the operator maintenance personnel if the failure is not repaired automatically, can typically be done within 24 h at most, and in this case, the self-healing strategy will provide recovery that satisfies the minimum system requirements in the failure region, even in the worst case (multiple failures). Switching back to normal operation is performed milliseconds after system repair.

8.2.1 Sources of Failure

Wireless cellular systems, like most systems, are prone to failures. These can be classified as software or hardware failures. Software failures can be mitigated automatically by restarting or reloading the failed node software, while most hardware failures have to be manually repaired through a cell site visit, which may take up to 24 h. During the repair time, the network must operate near normal conditions with acceptable quality of experience.

8.2.2 Cell Outage Detection

COD uses a collection of local and global information to determine that a BS is not operating properly. Detection includes active notification to cover the generalized case in which OAM is aware of the fault. In the case of complete BS failure, OAM will be unable to communicate with the failed BS to determine whether its cell is in service. Lack of communication could be a symptom of failure on the OAM backhaul rather than an indication that the site is down. In this case, the network manager needs to have other evidence to determine the nature of the problem. If the cell is still in service, it will continue to interact with the core network, so the network manager should be able to determine from core network metrics whether there is ongoing interaction with a specific BS.

Latent fault determination is the term used when the fault detection is based on evidence, such as anomaly statistics, rather than an alarm or state change. This is the most challenging of the COD scenarios, since OAM indications will suggest that it continues to operate normally. This type of detection may be achieved with a combination of statistics and activity watchdog timers. The operator will typically have a set of generic policies defined, each of which describes the combination of events that are deemed to indicate a cell outage. This may be enhanced using a set of cell type-specific rules, whereby all cells of a specific type use a separate or additional policy (macrocells, picocells, femtocells, or a combination). Perhaps the most valuable of the evidence-based detection mechanisms is the use of time/day profiling on a per cell basis. This is achieved by collecting statistics over a period of time and gradually building a statistical picture of the expected performance for a given time of day, weekday, or weekend. When statistics collected for a cell deviate significantly from values normally seen for that cell, there is a likelihood of a latent fault.

8.2.3 Cell Outage Compensation

COC actions are entirely based on detection by the COD, described in Section 8.2.2. Some software failures can be automatically mitigated, while other software and hardware failures have to be manually cleared by network engineering. Compensation actions are triggered immediately after the detection of failure. If the whole BS fails, the compensation action(s) will be done by neighboring BSs. They initiate reconfiguration actions, such as changing their antenna tilt and increasing transmission power to extend their cell coverage to cover the failed BS footprint.

Transmit power has an immediate impact on the BS coverage; however, conventional cellular systems make maximum use of the available power, without leaving headroom to enable a BS to increase its coverage in the direction of a neighboring outage. This issue must be considered in planning future networks.

Changing the antenna pattern to achieve additional coverage for its neighbor is an effective way to cover the footprint of the failed BS. In most cases, changes are achieved with antenna tilt. However, newer antenna technology, such as massive MIMO, enables many complex adjustments of the coverage pattern on demand. The real challenge with the use of antenna tilt change in support of any SON functions is ensuring that there is a control loop to guarantee that antenna adjustment to improve coverage in the failure area will not affect the coverage of the BS itself. SON needs to collect sufficient measurements to profile the impact of antenna pattern adjustment. Increasing the BS power and changing the antenna tilt are not the only methods used in the COC process, but they may be considered as the conventional techniques for failure compensation.

8.3 State of the Art in Self-Healing

SON is a rapidly growing area of research and development, and in the last decade a plethora of diverse efforts, from different research bodies, have been exerted in this field. In [8], the authors carried out a broad survey covering all three categories of SON, including self-healing. They provided a comprehensive review that aimed to present a clear understanding of this research area. They compared the strengths and weaknesses of existing solutions, and highlighted the key research areas for future development. Imran and Zoha [9] explored the challenges in 5G from the SON and big data point of view. They identified what challenges hinder the current SON paradigm from meeting the requirements of 5G networks. They then proposed a framework for empowering SONs with big data to address the requirements of 5G.

Self-healing has been extensively studied in ad hoc and wireless sensor networks. Recently, little research has addressed self-healing in 4G and 5G networks. Most work done in the self-healing field has addressed COD and COC. In COD, a cell is said to be in outage if it is still operating but suboptimally, still operating with a major fault, or in complete outage, that is, system failure. In [10], the authors presented a novel COD algorithm based on statistical performance metrics that enable a BS to detect the failure of a neighboring cell. Their simulation results indicated that the proposed algorithm can detect the outage problem reliably in real time. The authors in [11,12] focused on COD in the emerging femtocell networks. They proposed a cooperative femtocell outage detection architecture, which consists of a trigger stage and a detection stage. They formulated the detection problem as a sequential hypothesis-testing problem. They achieved improvements in both communication overhead and detection accuracy.

In [13], the authors employed a classification algorithm, K-nearest neighbor (KNN), in a two-tier macro–pico network to achieve automatic anomaly detection.

They proved the efficiency of the proposed algorithm. The authors in [14] considered the SDN paradigm and proposed a data COD scheme for HetNets with separate control and data planes. Then they categorized their data COD scheme into a trigger phase and a detection phase. Their simulation results indicated that the proposed scheme can detect the data cell outage problem in a reliable manner.

COC has received slightly more attention from researchers than COD. In [15], Amirijoo presented a cell outage management description for LTE systems. Unlike previous works, they give a complete overview of both COD and COC schemes, highlighting the role of the operator policies and performance objectives in the design and choice of compensation algorithms. The possibilities of false detection were also highlighted. In another work by the same research group [16], they proposed concrete compensation algorithms and assessed the achieved performance effects in various scenarios. Their simulation results showed that the proposed compensation algorithm is able to serve a significant percentage of the users that would otherwise be dropped, while still providing sufficient service quality in the compensating cells.

In [17], the authors proposed to compensate cells in failure by neighboring cells optimizing their coverage with antenna reconfiguration and power compensation, resulting in filling the coverage gap and improving the quality of service (QoS) for users. The right choice of their reconfigured parameters is determined through a process involving fuzzy logic control and reinforcement learning. The results show an improvement in the network performance for the area under outage as perceived by each user in the system. Moysen and Giupponi [18] also used reinforcement learning for reconfiguring parameters. They proposed implementing a COC module in a distributed manner in eNBs, which intervenes when a fault is detected, compensating for the associated outage. The eNBs surrounding the outage zone automatically and continually adjust their downlink transmission power levels and find the optimal antenna tilt value to fill the coverage and capacity gap. Their results demonstrated that this approach outperforms state-of-the-art resource allocation schemes in terms of the number of users recovered from outage.

In [19], the authors presented a novel cell outage management framework for HetNets with split control and data planes. In their architecture, the control and data functionalities are not necessarily handled by the same node. The control BSs manage the transmission of control information and UE mobility, while the data BSs handle UE data. An implication of this split architecture is that an outage to a BS in one plane has to be compensated by other BSs in the same plane. They addressed both COD and COC, using two COD algorithms to cope with the idiosyncrasies of both the data and control planes and integrating these two COD algorithms with a COC algorithm that can be applied to both planes. The COC algorithm is a reinforcement learning–based algorithm, which optimizes the capacity and coverage of the identified outage zone in a plane. Their results showed that the proposed framework can detect both data and control cell outage, and also compensate for the detected outage in a reliable manner. Fan and Tian

[20] proposed a coalition game–based resource allocation algorithm to enable self-healing and compensate for abrupt cell outage in SC networks. In their proposed algorithm, SCs play coalition games to form a set of coalitions that determines the allocation of subchannels, and each coalition of SCs serves a user cooperatively with optimized power allocation. Their results proved that the proposed algorithm can effectively solve network failure problems.

8.4 Case Study: Backhaul Self-Healing

Backhaul connections in 5G networks are expected to carry over 1000× more traffic than today's networks. So, the question is how to forward hundreds of gigabits per second of backhaul traffic in ultradense cell networks with guaranteed QoS. A wide range of data rates has to be supported in 5G networks, which can be as high as multiple gigabits per second, and tens of megabits per second need to be guaranteed with very high reliability in the presence of failures. In high-speed networks (such as 5G networks), if the backhaul connectivity is lost even for only a few seconds, the data loss will be of the order of hundreds of gigabits. To solve this problem, we proposed a novel preplanned reactive self-healing approach using new added self-healing radios (SHRs) to each BS in the 5G network. These SHRs operate only in the case of backhaul failures in any BS in the network. A new controller is introduced to handle the self-healing procedures.

8.4.1 Backhaul Requirements in 5G Networks

The backhaul solutions for 5G networks need to be more cost-efficient and scalable as well as easily installed with respect to conventional backhaul solutions. To achieve high data rates and low latency in 5G networks, backhaul transport has to provide a native support for this by enabling ultrahigh-capacity data transfer between the end point and core network. Traditionally, copper and fiber have been considered cost-efficient transport alternatives for locations with wired infrastructure. However, for locations without wired infrastructure, the building of wired transport may not be a feasible approach in terms of cost, as well as scalability. Therefore, there is a need to develop very high-capacity, very low-latency (less than 1 ms), and scalable as well as cost-efficient wireless backhaul transport solutions for 5G mobile broadband networks.

To enable cost-efficient deployment of very high-capacity backhaul links, an mmW radio in the 30–80 GHz frequency range can provide an attractive alternative with respect to existing technologies. Naturally, there are several research topics to be addressed for mmW technology-based wireless backhauling; for example, the impact of backhaul network topologies, the impact of mobility for backhauling, spectrum sharing between access and backhaul, and the impact of different duplexing schemes are under investigation. However, the coexistence of mmW with other HetNet layers is not easy. To guarantee coexistence with other HetNet layers, potentially non-mmW, efficient interworking between different HetNet layers should be enabled. Therefore, there is a need to design coexistence methods for mmW communication as an integral part of the overall 5G system.

8.4.2 Proposed Backhaul Self-Healing Architecture for 5G Networks

The proposed 5G heterogeneous network architecture consists of macrocells. Within a macrocell footprint there are a number of SCs, as shown in Figure 8.4. The SCs are categorized into picocells and femtocells. The picocells are backhauled from the macrocell via mmW and microwave connections, depending on the distance between the macrocell and the picocell, assuming that line-of-sight (LOS) is available; otherwise, the picocell can be connected using a non-line-of-sight (NLOS) connection. There are two types of femtocells in this architecture. The first type are called *preplanned femtocells* (PFCs); these are owned and controlled by the 5G network operator and are deployed in large enterprises such as universities, malls, airports, and other public places. They play a vital role in the self-healing process. They are always backhauled using fiber connection to guarantee high-speed

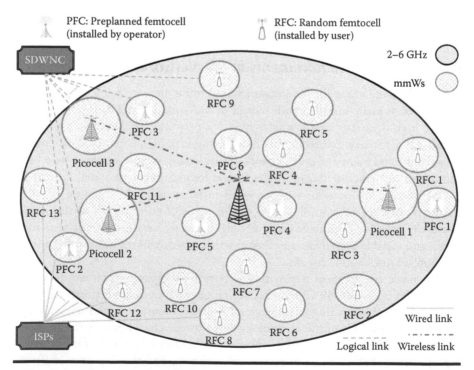

Figure 8.4   The system model.

connectivity to the users and to the BSs that they will heal. The second type are *random femtocells* (RFCs). These are owned by users and are installed in homes or small offices; hence, the deployment of this type of femtocell is totally random. Users are free to purchase regular femtocells (which does not contain SHRs) or self-healing femtocells. In the latter case, users are involved in the self-healing process, and the 5G network operator must use some means of compensation or incentives for such users.

In this proposed network architecture, two 5G candidate technologies are included: mmW and SDWN.

8.4.2.1 mmW Communications

Recently, the mmW band for 5G networks has been proposed in the literature. It is already used in short-range LOS links, for example, between a macrocell and a nearby picocell to provide a gigabits per second backhaul connection within a limited range. This is due to the high attenuation and oxygen absorption of these waves [21–23].

Using mmW with SCs will provide the target data rates for 5G users, and SCs in this case will rely on high-gain beam forming to mitigate path loss [24].

The macrocell situation is more complicated due to the wide area that it covers (up to 2 km). Using mmW with the macrocell is still under intensive research because of the high attenuation associated with the large coverage distance and the high penetration loss. As is well known from 4G networks, 80% of network traffic is used indoors and only 20% is used outdoors [25]. The 20% outdoor traffic will be served by the macrocell and the outdoor SCs (picocells). This trend is expected to continue in 5G, which means that the traffic served by the macrocell in 5G networks will be much lower than that carried by the indoor and outdoor SCs.

In our proposed solution, the mmW is used only in short-range LOS communications (backhauling between different BSs) and in NLOS communications between SCs and UEs. However, the macrocell will use the traditional cellular band (2–6 GHz) for NLOS communications with the UEs. The motivations for using traditional cellular bands for macrocells are better coverage, lower penetration loss, and elimination of the interference issue between the SCs tier and the macrocell tier. This elimination will avoid complex and sophisticated interference cancellation schemes. The only limitation of the traditional cellular band is its limited bandwidth, but using massive MIMO, carrier aggregation/channel bonding, and other technologies can facilitate the achievement of macrocell gigabits per second throughput [26].

8.4.2.2 Software-Defined Wireless Network Controller

The SDN concept separates the network control plane from the forwarding plane. This enables the network controller to become directly programmable. SDN is an emerging architecture that is manageable, dynamic, and cost-effective. These advantages of SDN make it ideal for use in the upcoming 5G network [27].

The concept of SDN has been recently introduced in wireless networks, defining SDWN. For more detailed and exact implementation of the SDWN, the reader is referred to [27]. Using the SDWN concept, heterogeneous radio access networks implement the access, forwarding, and routing functions, and support multiple functionality levels at Layers 2 and 3. The proposed SDWNC is logically connected to all the network BSs through transmission control protocol (TCP) connections. In our 5G architecture, the SDWNC is a required component, as it acts as the supervisor, decision-maker, and administrator for all self-healing procedures applied to all network BSs.

8.4.2.3 Novel Self-Healing Radios

We proposed to add SHRs to the 5G BSs, including macrocells, picocells, and femtocells. The SHRs are activated only in the case of backhaul failure. Each BS can have one or more SHRs, and the number of SHRs is determined by the operator or by optimizing the number of SHRs for each type of BS (macrocell and SCs). The SHRs can be integrated within the BSs antennas or can stand alone, depending on the vendor's/operator's point of view. Also, they can operate in the same band as the BS access antennas, or in another band, as discussed in Section 8.4.2.4.

8.4.2.4 Self-Healing Radio Band

The SHRs can use (1) the mmW band, (2) the traditional cellular band (2–6 GHz), or (3) a new dedicated band. For the mmW band, there are limitations on the distance between NLOS SHRs, and it also suffers from high penetration loss. Dedicating a portion of the 2–6 GHz band, which is also used by the macrocell, to SHR communication will affect the band use, because SHRs are only activated if backhaul failures occur; otherwise, this portion of the band will not be used. Finally, using a dedicated band (purchased for self-healing communication only) will dramatically increase the capital expenditure (CAPEX), and this band also will not be fully used.

Our proposed solution is to use the 2–6 GHz band, as in Solution (2), but to use the cognitive radio (CR) concept for SHR communications to optimize the use of the available spectrum. The band used is the same as in the second solution, whereby a portion of the traditional cellular band (i.e., 20%) will be dedicated to the SHRs.

The main difference is that when SHRs are inactive, that is, in the absence of failure, this portion will be available for the macrocell to use as a secondary user, using the CR concept. Therefore, if the macrocell is starved for bandwidth, it will sense the SHR portion (channels), and if it is free, the macrocell (i.e., the secondary user) will use the vacant channels until the primary users (SHRs) are activated.

Once the primary user (the SHR) is active (this means that a failure has occurred), the macrocell will vacate this channel to be used by the failed BS (the primary user).

Furthermore, in our model, the macrocell can avoid wasting time on spectrum sensing by simply knowing the failure status of the NEs from the SDWNC. If there is no failure, the macrocell will use the reserved portion of the band without sensing. If a failure happens, the SDWNC will immediately request the macrocell to vacate the used channels to be used by the primary users (SHRs).

8.4.2.5 Self-Healing Radio Range

The SHRs in the SCs are omnidirectional, but they are directional in three sectoral macrocells, where there is one or more SHR in each sector of the macrocell. The coverage range of the SHRs in the SCs is larger than the access radios' range. This will allow the SCs to find other SHRs in other cells in the vicinity. The SHR coverage of SCs is shown in Figure 8.5. It is clear from the figure that RFC 4 has activated its SHR(s), and its coverage is much larger than the access radios' coverage. If the SHRs' coverage is the same as the coverage of access radios, it will not find any other SHRs from other BSs to connect to. In the former case, RFC 4 can connect to the macrocell, RFC 5, or RFC 6, depending on the number of SHRs in RFC 4.

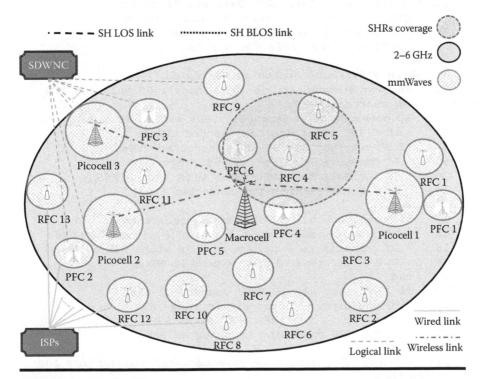

Figure 8.5   Small cells SHRs range.

8.4.3 Novel Self-Healing Approach

The network architecture described in Section 8.4.2 shows the BSs under normal operation, that is, without any backhaul failure. The SDWNC's main functions are to monitor the status of the whole network, detect failures, and apply appropriate procedures to mitigate these failures. When any BS backhaul fails, the BS will automatically activate its SHRs, but it will not be able to inform the SDWNC, because it is totally disconnected from it. In this case, backhaul service providers (other BSs, Internet service provider (ISP) or core network) will report the failure to the SDWNC. The SDWNC will then activate the SHRs of all neighboring BSs.

After activating the SHRs of the failed BS and the nearby BSs, the failed BS will try to connect to these BSs via its SHRs according to a certain priority order, available resources, and the received signal strength (RSS).

The backhaul connection failure may be permanent or transient. Our self-healing approach works with both, and the only difference is that in the case of a permanent failure, the SDWNC, after a certain threshold time, will inform the network operator that there is a permanent failure that requires maintenance personnel to visit the failed site.

```
Algorithm 8.1: Small Cell (SC) Backhaul Algorithm
       Input: SC backhaul status, K (number of SC SHRs)
   1.  if SC backhaul status is failed then
   2.    SDWNC activates SHRs for all BSs
   3.  end
   4.  while 6 Backhaul status is failed do
   5.    Recover communications for picocell SC measures RSS
         from macrocell SHRs
   6.    SC connects to M macrocell SHRs with: RSS > RSS_th
   7.    K=K-M (SC remaining unconnected SHRs)
   8.    if K! =0 (not all SC SHRs connected) then
   9.      SC measures RSS from all in range PFs' SHRs
   10.     SC connects to N PFs' SHRs with: RSS > RSS_th
   11.     K=K-N (SC remaining unconnected SHRs)
   12.   end
   13.   if K!=0 (not all SC SHRs connected) then
   14.     SC measures RSS from all in range picocells SHRs
   15.     SC connects to P picocells SHRs with: RSS > RSS_th
   16.     K=K-P (SC remaining unconnected SHRs)
   17.   end
   18.   if K!=0 (not all SC SHRs connected) then
   19.     SC measures RSS from all in range RFs SHRs
   20.     SC connects to L RFs SHRs with: RSS > RSS_th
   21.     K=K-L (SC remaining unconnected SHRs)
   22.   end
   23.   if other SCs didn't receive backhaul request for 200
         ms then
```

```
24.     SDWNC deactivates these SCs SHRs
25.  end
26.  SDWNC collects the status of all BSs
27. end
28. SDWNC deactivates SHRs for all BSs
```

8.4.3.1 Single Failure Scenario

Single failure means that only one BS backhaul failed in a certain region (the macro-cell region in our model), for example, an SC (femtocell or picocell). In this case, the SC will enter the self-healing mode, in which it will activate its own SHRs and then try to connect to other BS SHRs, which are activated by SDWNC. Following priority order, the SC will first search for macrocell SHRs, then PFC SHRs, then picocell SHRs, and finally RFC SHRs. RFCs are assigned the lowest priority, because they are user owned and the operator will have to compensate the owners of the RFCs.

Algorithm 8.1 shows the procedures followed by the failed SC (picocell or femtocell) to mitigate its backhaul failure. In the first three lines of Algorithm 8.1, the SDWNC monitors the status of the SC backhauling. Line 4 is the beginning of our self-healing procedure, and the while loop terminates when the backhaul failure has been repaired. From Lines 5 to 22, the SC will try to connect up to K SHRs (K is the number of SHRs in the failed SC). As mentioned in the previous paragraph, it will try first to connect to macrocell SHRs. If it succeeds, it will update K as follows: K = K−M, where M is the number of macrocell SHRs connected to the SC SHRs and M is less than or equal to K. If the new updated K is not equal to zero, the same process will be performed for PFCs, and also K will be updated by N, where N is the number of PFCs' SHRs connected to the SC SHRs. With K still not equal to zero and using priorities in selecting the BS type, this process is repeated with picocells and RFCs, respectively. As shown in Lines 23–26, the SDWNC deactivates any SC not participating in the self-healing process. This step is done to minimize the power wasted by the unconnected SHRs. In Line 26, the SDWNC collects the status of all BSs, and if the failed SC backhaul has been repaired, it will terminate the self-healing process and deactivate all SHRs of all BSs, as shown in Line 28.

An example of the failure scenario is shown in Figure 8.5, where the wired backhaul connection of RFC 10 fails. Then RFC 10 will try to connect to the neighboring BSs SHRs using priorities as described in Algorithm 8.1, but because it is out of the SHR coverage of the macrocell, PFCs, and nearby picocells, it will connect to the SHRs of RFC 8 and RFC 12 to mitigate the failure. Once the failure is repaired (RFC 10 backhauling returns to work properly), the SHRs will be deactivated by the SDWNC, and the network will return to normal operation.

The failure of the macrocell backhaul is not considered as a single failure, because this failure will immediately cause the failure of all picocells backhauled through the macrocell, causing a multiple failures scenario in the network.

8.4.3.2 Multiple Failures Scenario

Multiple failures refers to the situation in which two or more backhaul failures occur at the same time, except for the macrocell backhaul failure explained in Section 8.4.3.1, which causes multiple failures in the network. The first case can be seen as the failure of two or more SCs, for example, two femtocells or one femtocell and one picocell in the same region. Algorithm 8.1 can be used to mitigate these failures when implemented for each failed BS, and as the number of failures increases, the probability of healing each failed BS will depend on the nearby BSs, which can provide temporary backhauling using their SHRs. Thus, as expected in 5G networks, the dense deployment of SCs will enhance the performance of our self-healing approach, especially in the multiple failures case.

8.4.3.3 Macrocell Backhaul Failure

The macrocell plays a vital role in HetNets, where in addition to its main function, coverage for outdoor users, it provides wireless backhaul links to other SCs (picocells in our model). Figures 8.6 through 8.8 show step by step the macrocell backhaul failure and the mitigation process. The backhaul of the macrocell is optical fiber based, and it may fail due to either a hardware or a

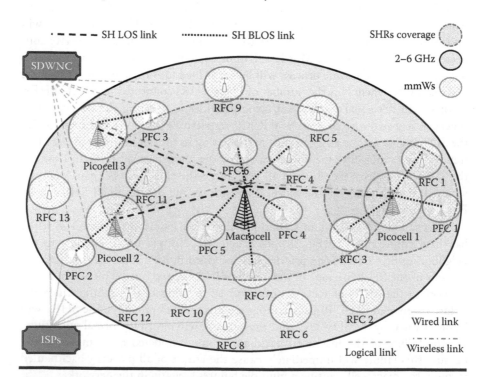

Figure 8.6   Macrocell backhaul failure mitigation Step 1.

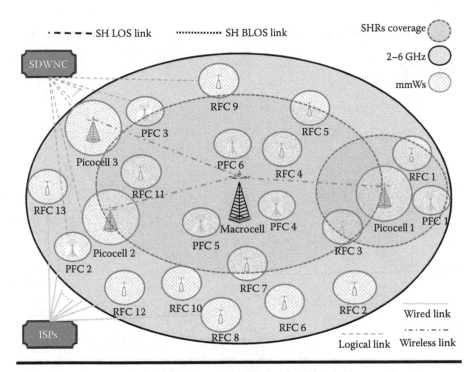

Figure 8.7   Macrocell backhaul failure mitigation Step 2.

software problem with the interfaces at any of the two ends. The fiber may even be damaged or cut due to digging or for some other reason. The SHRs' coverage area is shown only for the macrocell and Picocell 1 for the sake of simplicity.

Figure 8.6 shows the first step, when the backhaul failure has just happened, which is indicated in the figure by the dark gray color on the base of the macrocell. This failure will cause an immediate failure to all BSs backhauled from the macrocell. The macrowave connections from the macrocell to Picocell 1, Picocell 2, and Picocell 3 are indicated with a dashed-dot line to show that those links also failed.

Figure 8.7 shows the second step, when the macrocell and the three failed picocells will activate their SHRs, and the SDWNC will activate the SHRs of all nearby BSs. At this point, the failed BSs' SHRs will search for their temporary backhaul connections from the neighboring BSs. The macrocell will connect to PFC 4, PFC 5, PFC 6, RFC 4, and RFC 7, as is shown in the figure. Picocell 1 will find three neighboring BSs from which to acquire its temporary backhaul connection: PFC 1, RFC 1, and RFC 3. Picocell 2 will connect to PFC 2 and RFC 11, and finally, Picocell 3 will connect only to PFC 3, because it is the only BS that is located in its SHRs coverage.

Figure 8.8 shows the third and final step, when after recovering the macrocell and picocells from the nearby BSs, the picocells will also temporarily backhaul

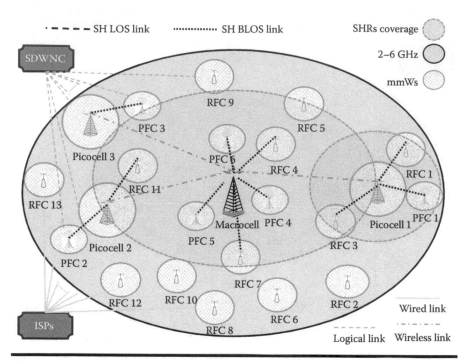

Figure 8.8   Macrocell backhaul failure mitigation Step 3.

the macrocell, if needed, using the LOS microwave links. This step is performed because the macrocell traffic is much larger than any SC traffic and needs many sources of recovery to operate at near-normal efficiency. After the third step, all the failed BSs are now able to serve their users, or will at least be able to deliver the minimum rate requirements to their users until the failure is repaired, regardless of how long the backhaul failure remains. A heuristic algorithm for the detailed macrocell self-healing procedures can be found in [28].

8.4.3.4 Results and Discussion

This section presents the simulation results obtained by implementing the proposed self-healing approach in the macrocell coverage area, considering single and multiple failures of BS backhauling connections.

Extensive simulations for different failure scenarios have been conducted for macrocell, picocells, and femtocells. The RFC input rate is heterogeneous (20, 60, and 100 Mbps) and is distributed between these RFCs with ratios of 50%, 40%, and 10%, respectively, in macrocell and picocell failure scenarios. It is fixed at 100 Mbps in the femtocell failure scenario to evaluate the results in the worst case.

We evaluate our self-healing approach in terms of degree of recovery (DoR) from failure. The DoR of a certain BS is defined, in terms of self-healing, as

$$DoR = \frac{\text{Summation of recovered rates from other BSs}}{\text{Original input rate of the failed BS}}$$

8.4.3.5 Macrocell Failure

As mentioned in Section 8.4.3.2, macrocell failure is a special case. Therefore, we have one failure that implicitly causes multiple failures (all picocells backhauled from the macrocell will fail). Hence, the DoR of the macrocell is assessed with respect to the number of SHRs of the macrocell and the other picocells involved in the self-healing process.

As shown in Figure 8.9, when the number of macrocell SHRs is increased, the DoR increases, but not as much as when the number of picocell SHRs is increased. Using one picocell SHR and increasing the number of macrocell SHRs from one to four can improve the DoR by 26%. However, using one macrocell SHR and increasing the number of picocell SHRs from one to four can improve the DoR by 60%. This means that investing in picocell SHRs will enhance the self-healing performance more than investing in macrocell SHRs. In the next two scenarios, the number of macrocell SHRs will be fixed at three.

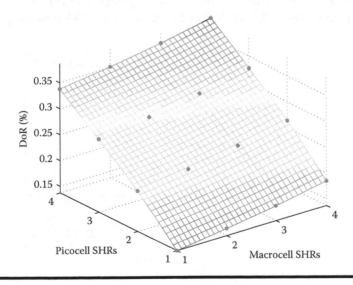

Figure 8.9 DoR of picocells versus number of SHRs.

8.4.3.6 Picocell Failure

Figure 8.10 evaluates the DoR of a picocell when the picocell SHRs are increased from one to seven under single and multiple failure scenarios. In the case of single failure, the DoR increases rapidly from 10% to 20% when the SHRs are increased from one to four. A further increase in the SHRs results in a negligible increase in the DoR, which proves that the recommended number of SHRs to be used by picocells is four.

Also, we can see in Figure 8.10 in the case of multiple failure that all failed picocells can recover their rates by 10% using one SHR. This is because each pico-cell has a dedicated PFC in its SHR's range. Therefore, under multiple failures, each failed picocell can find at least one PFC with which to connect. As the number of SHRs increases, the DoR of the two-failure case and the three-failure case decreases. This is because with the addition of more SHRs in each picocell, the network resources will be consumed, and not all SHRs in each failed picocell will succeed in getting enough resources. Also, it can be seen that a further increase beyond four SHRs per picocell introduces negligible improvement, and the DoR is almost constant. In the next scenario, the number of picocell SHRs will be fixed at four (Figure 8.10).

8.4.3.7 Femtocell Failure

We observe from Figure 8.11 that when femtocells have only one SHR, the DoR of failed femtocells is up to 50%. This is because most of the time the femtocells are recovered from other BSs that have much higher rates. This also explains the reason for exceeding the 100% DoR for femtocells having three SHRs or more. Using two SHRs can recover up to 90% of the failed femtocell rate, which is an acceptable rate

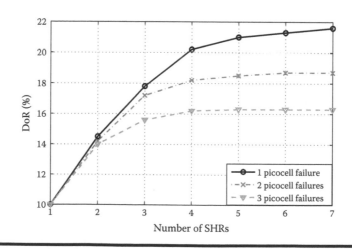

Figure 8.10   DoR of picocells versus number of SHRs.

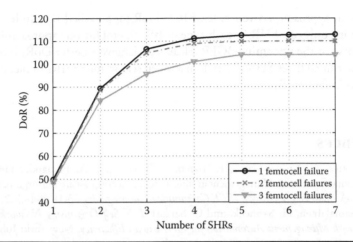

Figure 8.11   The relationship between the DoR of a failed femtocell and its SHR.

in the presence of failure. Recovering more than 100% is not acceptable from the operator point of view. As the number of SHRs increases, the DoR of two failures and three failures decreases, which is similar to the picocell multiple failure case in the previous section. Even in the case of three failures, the DoR is approximately 85% using two SHRs. This indicates that our approach is robust under multiple failures.

8.5 Conclusions

5G networks are expected to be much more complex than 4G networks. This complexity will need more automation and self-X functionalities to be involved in 5G networks. The SON paradigm is one of the candidate technologies that are strongly expected to be integrated in the 5G networks standard. Self-healing is mandatory in 5G networks to guarantee reliability and service continuity even in the presence of failure. The self-healing state of the art shows that there is an ongoing contribution to cell outage management, but this contribution is still not sufficient, and more investigation is needed in this area. Progress from the research community is reducing the gap between 5G backhaul requirements and backhaul capabilities; however, major challenges remain along the way. In the investigated case study, we addressed the problem of backhaul failure in 5G networks, and then we proposed a new backhaul self-healing approach to address unexpected backhaul failures in 4G/5G HetNets. Our approach adds SHRs to the network BSs, the cost of which is negligible compared with the cost of the BS, which used the CR concept to take advantage of the network spectrum and mitigate interference. Simulation results show that our approach can immediately partially recover the failed BS until it returns to normal operation, and this can be done under both single and multiple failures. Our approach recovers at least 10% DoR for all failed BSs under multiple failures using only one SHR in each type of BS. To employ our approach in future 5G networks or the upcoming 3GPP releases to achieve a better DoR, it is recommended to embed three SHRs in each macrocell sector, four SHRs in picocells, and two SHRs in femtocells.

References

1. S. Fortes, A. Aguilar-Garca, R. Barco, F. B. Barba, J. A. Fernández-Luque, and A. Fernández-Durn, Management architecture for location-aware self-organizing LTE/LTE-A small cell networks, *IEEE Communications Magazine* 53(1): 294–302, 2015.
2. S. Hamalainen, H. Sanneck, and C. Sartori, *LTE Self-Organising Networks (SON): Network Management Automation for Operational Efficiency.* New York: John Wiley, 2012.
3. L. Jorguseski, A. Pais, F. Gunnarsson, A. Centonza, and C. Willcock, Self-organizing networks in 3GPP: Standardization and future trends, *IEEE Communications Magazine* 52(12): 28–34, 2014.
4. Socrates Project, *Socrates Final Report (Deliverable D5.9)*, 2012, http://www.fp7-socrates.eu.
5. J. Ramiro and K. Hamied, *Self-Organizing Networks (SON): Self-Planning, Self-Optimization and Self-Healing for GSM, UMTS and LTE.* New York: John Wiley, 2011.
6. R. Wang, H. Hu, and X. Yang, Potentials and challenges of C-RAN supporting multi-RATs toward 5G mobile networks, *IEEE Access* 2: 1187–1195, 2014.
7. J. Meinilä, P. Kyösti, L. Hentilä, T. Jämsä, E. Suikkanen, E. Kunnari, and M. Narandžić, WINNER+Final channel models, *CELTIC CP5-026 WINNER+Project, Deliverable D5.3, v1.0*, 2010.
8. O. G. Aliu, A. Imran, M. A. Imran, and B. Evans, A survey of self organisation in future cellular networks, *IEEE Communications Surveys & Tutorials* 15(1): 336–361, 2013.
9. A. Imran and A Zoha, Challenges in 5G: How to empower SON with big data for enabling 5G, *IEEE Network* 28(6): 27–33, 2014.
10. Q. Liao, M. Wiczanowski, and S. Stanczak, Toward cell outage detection with composite hypothesis testing, *Communications (ICC), 2012 IEEE International Conference on*, Ottawa, ON, IEEE, pp. 4883–4887, 2012.
11. W. Wang, J. Zhang, and Q. Zhang, Cooperative cell outage detection in self-organizing femtocell networks, *INFOCOM, 2013 Proceedings IEEE*, Turin, IEEE, pp. 782–790, 2013.
12. W. Wang, Q. Liao, and Q. Zhang, COD: A cooperative cell outage detection architecture for self-organizing femtocell networks, *Wireless Communications, IEEE Transactions on*, 13(11): 6007–6014, 2014.
13. W. Xue, M. Peng, Y. Ma, and H. Zhang, Classification-based approach for cell outage detection in self-healing heterogeneous networks, *Wireless Communications and Networking Conference (WCNC), 2014 IEEE*, Istanbul, IEEE, pp. 2822–2826, 2014.
14. O. Oluwakayode, I. Ali, I. M. Ali, and T. Rahim, Cell outage detection in heterogeneous networks with separated control and data plane, *European Wireless 2014; 20th European Wireless Conference; Proceedings of*, Barcelona, IEEE, pp. 1–6, 2014.

15. M. Amirijoo, L. Jorguseski, T. Kurner, R. Litjens, M. Neuland, L. C. Schmelz, and U. Turke, Cell outage management in LTE networks, *Wireless Communication Systems ISWCS 2009. 6th International Symposium on*, Tuscany, IEEE, pp. 600–604, 2009.
16. M. Amirijoo, L. Jorguseski, R. Litjens, and L. C. Schmelz, Cell outage compensation in LTE networks: Algorithms and performance assessment, *Vehicular Technology Conference (VTC Spring), 2011 IEEE 73rd*, Yokohama, IEEE, pp. 1–5, 2011.
17. A. Saeed, O. G. Aliu, and M. A. Imran, Controlling self healing cellular networks using fuzzy logic, *Wireless Communications and Networking Conference (WCNC), 2012 IEEE*, Shanghai, IEEE, pp. 3080–3084, 2012.
18. J. Moysen and L. Giupponi, A reinforcement learning based solution for self-healing in LTE networks, *Vehicular Technology Conference (VTC Fall), 2014 IEEE 80th*, Vancouver, IEEE, pp. 1–6, 2014.
19. O. Onireti, A. Zoha, J. Moysen, A. Imran, L. Giupponi, M. Imran, and A. A. Dayya, A cell outage management framework for dense heterogeneous networks, *Vehicular Technology, IEEE Transactions on*, Budapest, 2015.
20. S. Fan and H. Tian, Cooperative resource allocation for self-healing in small cell networks, *Communications Letters, IEEE* 19(7): 1221–1224, 2015.
21. X. Ge, H. Cheng, M. Guizani, and T. Han, 5G wireless backhaul networks: Challenges and research advances, *IEEE Network* 28(6): 6–11, 2014.
22. R. Taori and A. Sridharan, Point-to-multipoint in-band mmwave backhaul for 5G networks, *IEEE Communications Magazine* 53(1): 195–201, 2015.
23. K. Zheng, L. Zhao, J. Mei, M. Dohler, W. Xiang, and Y. Peng, 10 Gb/s HetsNets with millimeter-wave communications: Access and networking—challenges and protocols, *IEEE Communications Magazine* 53(1): 222–231, 2015.
24. J. Singh and S. Ramakrishna, On the feasibility of beamforming in millimeter wave communication systems with multiple antenna arrays, *IEEE Transactions on Wireless Communications*, Canada, IEEE, p. 1, 2015.
25. ETSI TR 101 534 V1.1.1, Broadband Radio Access Networks (BRAN); Very high capacity density BWA networks; System architecture, economic model and derivation of technical requirements, *ETSI TR 101*, 2012.
26. Z. Khan, H. Ahmadi, E. Hossain, M. Coupechoux, L. DaSilva, and J. Lehtomaki, Carrier aggregation/channel bonding in next generation cellular networks: Methods and challenges, *IEEE Network* 28(6): 34–40, 2014.
27. C. Bernardos, A. Oliva, P. Serrano, A. Banchs, L. Contreras, J. Hao, and J. Zuniga, An architecture for software defined wireless networking, *IEEE Wireless Communications & Magazine* 21(3): 52–61, 2014.
28. M. Selim, A. Kamal, K. Elsayed, H. Abd-El-Atty, and M. Alnuem, A novel approach for back-haul self healing in 4G/5G HetNets. (To appear in the proceedings of the IEEE ICC 2015.)

Chapter 9

Convergence of Optical and Wireless Technologies for 5G

Paulo P. Monteiro and Atílio Gameiro

Contents

9.1 Introduction

Data rate demand and quality-of-service (QoS) requirements for wireless access are increasing dramatically. The drivers for this rapid increase include the diversity of multimedia applications and the explosion in the type and number of connections by either human users or connected devices [1,2]. Although estimates vary according to the source, [1] points out that by 2019, aggregate smartphone traffic will be 10.5 times greater than it is today, representing a compound annual growth rate (CAGR) of 60%, while machine-to-machine (M2M) connections will grow from 495 million in 2014 to more than 3 billion by 2019, a 45% CAGR.

Overall by 2020, mobile and wireless traffic volume is expected to increase to reach a 1000 times higher mobile data volume per area, and a 10–100 times higher typical data rate per user device, than in 2010. It is estimated that by 2020, there will be a new class of data-rate-hungry services with low latency requirements. Previous work has shown that applications in the future, such as augmented reality, three-dimensional (3-D) gaming, and "tactile Internet" [3], will require a 100 times increase in the achievable data rate compared with today and a corresponding 5–10 times reduction in latency. The next generation of wireless networks, commonly referred to as fifth generation (5G), must therefore be designed to meet these data rate and latency requirements. Thus, wireless communication systems should support the ever-increasing information capacity requirements, providing a much higher spectral efficiency than today. Moreover, the traffic and system operation requirements for next-generation systems, such as low transport latency, reduced energy consumption, and lower deployment and operation costs, are becoming more stringent.

Till now, the wireless domain has been the main bottleneck to meeting the growing mobile traffic demand, while the fixed optical infrastructure was assumed to be able to transparently accommodate the needs of the wireless domain. In recent years, breakthroughs in the wireless domain (multiple-in multiple-out [MIMO], cognitive radios, cooperation, interference coordination and alignment, etc.) have been considered as the key to providing high-capacity wireless access while at the same time making new, much higher demands on the fixed optical infrastructure. In fact, it should be noted that in existing and emerging systems, the design and management of the two domains (wireless and optical) have been quite independent. However, in future networks, the densification and increasing data rate of wireless access imply increasing demands on the fixed infrastructure. Therefore, it is no longer cost-effective to deploy the optical infrastructure for the most demanding cases. There is an emerging need to handle the information capacity, along with traffic and operation requirements, jointly across the wireless and optical domains in a unified manner. To this end, the convergence of fixed transport networks based on a high-speed optical infrastructure and broadband wireless components has been identified as a key enabler for future networks.

In the wireless domain, new technologies and deployments are currently under investigation to support ultrahigh data rates along with low latency, high reliability, lower energy consumption, and reduced capital/operating expenditure (CAPEX/OPEX) [4]. Ultradense networks are considered as the most promising solution to achieve the ambitious targets set for the next generation of network systems. However, network densification comes with the additional price of infrastructure costs, higher energy consumption, and coordination overheads among multiple access points. These costs are multiplied by the number of different access technologies and the number of operators.

Cloud radio access network (C-RAN) architecture is a promising alternative, which can reduce the cost of ultradense networks by simplifying the small base stations (BSs) to remote antennas (remote radio heads [RRHs]) and moving the baseband processing to a central unit (CU) [5]. C-RAN architecture offers the following features: (1) it improves capacity by jointly processing radio signals received at multiple RRHs (coordinated multipoint [CoMP] gains); (2) it reduces energy consumption by centralized coordination of RRHs and turning off the inactive units; and (3) it improves spectral efficiency by dynamically allocating spectral resources among RRHs depending on the spatial and temporal demand distribution.

However, current C-RAN architectures have certain limitations [5,6]. Namely, most of the existing C-RAN systems target RRHs with single or a limited number of antennas. Similarly, the CUs process signals from only a limited number of RRHs. This chapter will discuss the trends to expand this framework by allowing joint processing of a very high number of RRHs, with multiple antennas per RRH. In this work, we bring together C-RAN with massive MIMO (M-MIMO) and network densification concepts, combining the benefits of all three technologies in a unified and coordinated manner.

The benefits of joint processing on such a large scale can only be achieved through a powerful fiber infrastructure connecting RRHs to the CU (fronthaul) in a reliable manner. Even for a single user, remarkably high data rates are expected in the fronthaul, justified by the radio bandwidth, which can reach an order of up to 2 GHz per end user in the millimeter-wave (mmWave) spectrum. The requirements on the fronthaul are exacerbated by the use of a large number of antenna elements by users (Figure 9.1). This chapter builds on the projected progress in the radio access technologies (RATs) with access to larger bandwidths and spectrally efficient techniques and their simultaneous use. This vision and supporting solutions pave the way for data rates of up to 10 Gbps per user in the near future, and provide a solid basis for systems providing 100 Gbps per user data rates in the long term. Such numbers will lead to an aggregated data rate requirement in the terabits per second range in the RRH–CU links (the fronthaul).

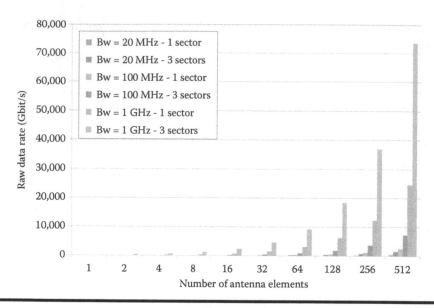

Figure 9.1   Data rates versus number of antenna elements for different technologies.

The provision of such huge capacities for the fronthaul in future radio networks requires a joint approach by the wireless and optical technologies, relying on three main axes:

- *Wireless domain:* Advanced radio-link technologies supporting much higher capacity over the fronthaul links (terabits per second). This can be achieved via advanced compression techniques that take into account the underlying M-MIMO structure, efficient fragmented spectrum use for mobile radio, and high bandwidth for wireless backhaul solution.
- *Optical domain:* Novel optical fronthaul access technologies supporting ultra-high data rate (terabits per second) transport. This is realized through efficient mapping of fragmented radio carriers and antenna signals to optical carriers.
- *Joint wireless and optical domains:* Converged management of radio and optical resources, for example, management of multiple frequency bands with respect to radio capacity demand (current and estimated future demand) and available optical transport capacity, and vice versa.

In this chapter, we address the challenges and the solution for the optical-wireless convergent network. The chapter is organized as follows. In Section 9.2, we outline the trends toward C-RAN architectures, their benefits, and the need to consider convergence between the wireless and optical domains; in Section 9.3, we discuss and illustrate the requirements of the future wireless systems in terms of capacity and latency and their implications for the fronthaul design; in Section 9.4, we address the development of radio-over-fiber (RoF) transceivers for mobile fronthaul (MFH) with the capability to transport analog (A-) and digitized (D-) RoF signals for C-RAN applications. Experimental results on the temperature and bias current behavior of uncooled light sources for potential use in optical fronthauls will be presented. Nonlinear mitigation of A-RoF and D-RoF links by using a memory polynomial based on a simplified Volterra series will be experimentally assessed. The coexistence of the transport of D-RoF with legacy and future Passive Optical Network (PON) systems, such as Next Generation PON 2 (NG-PON2), will be analyzed in Section 9.5.

9.2 Trends and Issues in Wireless and Wired Broadband and Infrastructure Convergence

In recent years, there has been significant research leading to disruptive alternatives that are moving from the hierarchical traditional cellular network toward more centralized processing. The proposals are currently classified under the name *C-RAN*, which is an overused term, publicized and used in several different ways, but essentially designates a network architecture in which the BSs, typically with reduced complexity, are linked to form a cloud of radio heads.

The traditional concept of the C-RAN is shown in Figure 9.2. The radio signals at the RRHs are transparently transmitted/received to/from a CU where all the signal processing is performed. Such an architecture departs from the classical concept of a cell in which each user equipment (UE) is attached to a single BS, as it inherently allows multiple attachments and cooperation among the network elements, thereby fostering distributed MIMO (D-MIMO) and the development of cooperation techniques starting at the physical layer. Considering the high capacities envisioned, optical fiber, due to its low attenuation, electromagnetic interference

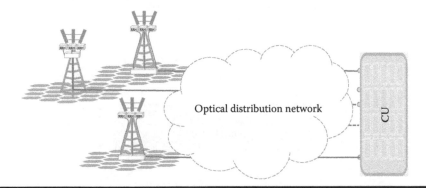

Figure 9.2 C-RAN concept. (From P.P. Monteiro and A. Gameiro, *16th International Conference on Transparent Optical Networks (ICTON)*, Graz, IEEE, 2014.)

immunity, and enormous bandwidth, is the obvious transmission medium of choice to build these transparent interconnections.

The C-RAN concept of Figure 9.2 has since been extended with the deployment of small cells. Small cells have become an integral component in meeting the increased demand for cellular network capacity in what are called *heterogeneous networks* (HetNets). The deployment of small cells enables the densification of the networks, thus paving the way for an increase in the system capacity that complements the benefits related to the joint processing and D-MIMO arising from the C-RAN scenario in Figure 9.2. This is depicted in Figure 9.3, where small cells and macrocells coexist. As shown in the figure, the macrocells overlay the small cells, which are served by RRHs. These RRHs may have different levels of processing, ranging from being limited to radio-frequency (RF) processing to having the processing of a Home Evolved Node B (eNodeB), such as in Long-Term Evolution (LTE). The macro BS may also have the full processing of an eNodeB or may be reduced to the RF functionalities.

In the case where the RRHs or BSs have limited (or null) baseband processing, the RF signals are remoted to a CU where the baseband processing is performed, which enables the solution of several issues of the traditional cellular network regarding the growing complexity of BSs, the need for cooling, and the difficulties in the acquisition of a site. The centralized processing also facilitates the development of

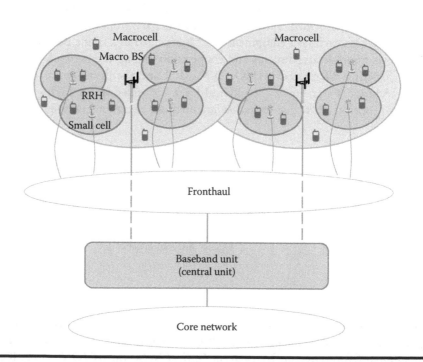

Figure 9.3 Architecture for the coexistence of small cells with macrocells.

coexistence algorithms in the case where all the RF signals are processed at the CU. The idea of distributed BSs can be traced back to the initial work on distributed antennas [7], although under a different name. Relatedly, the concept of small cells has been developed in recent years (e.g., [8]), with similar aims but also with the goal of entering the indoor environment.

As pointed out in Section 9.1, to cope with the very high raw data rates, optical technology is the most obvious choice to link the RRHs or BSs in what is designated as *fronthaul*. Transporting the radio channels to and from the nodes of the antenna networks over the optical transmission and routing lines is one major challenge, due to the massive broadband radio channels of the networks and also due to the latency, jitter, and the tight constraints on synchronization [8,9].

Reception and transmitting techniques with low latency are required, which should also combat the dispersion and nonlinear impairments of transporting such massive lightwave channels over a single-standard, single-mode optical fiber network.

The evolution of mobile networks through the centralization of radio basebands and the introduction of small cells has made significant requirements on the fixed infrastructure, which have led to the introduction of what is called the *fronthaul*. This is mainly in terms of capacity, as shown in Figure 9.1, but there are also significant issues concerning latency, the synchronization of a huge number of RRHs, and joint resource allocation [8,9]. This may in fact represent a significant change in the optical communication industry: up to now, its main challenges were in transport within the core network, but with the densification of the wireless networks, even more stringent requirements in terms of rates are likely to appear within the fronthaul domain.

In the future, the backhaul and fronthaul networks will probably merge into one transport network with baseband pooling, which will be placed closer to the mobile core network.

Furthermore, the move to RRHs instead of full BSs or access points has significant implications in terms of deployment flexibility and upgradeability. The current paradigm in the planning of cellular networks is one of keeping the intercell interference to acceptable levels. This task is becoming more and more complex as the number of heterogeneous services with different requirements increases, and in terms of upgradeability, it implies that significant replanning has to be done each time the market demand requires a network augmentation. Clearly, using simple RRHs does not involve the same constraints when it comes to deciding whether or not the network has to be augmented, and the inclusion of new RRHs and the novel interference patterns it provokes can be controlled dynamically at the algorithmic level.

Also in terms of mobility, the need to cover densely populated areas with small cells (30–100 m in diameter) makes the mobility management of high-mobility users very difficult to handle. Already, at a restricted speed of 40 km/h, it takes only a matter of seconds to move from one cell to the next, and a hierarchical cell structure with high-mobility cells allocated to macro cells is thus called for. Traditional hard-handover mechanisms would cause far too much control signaling between the cells in such a scenario, and more efficient procedures are therefore required [10]. One solution relies on cell grouping in serving areas, with the formation of a virtual large cell; that is, a cluster of cooperating small cells will appear to the user as a single distributed BS [10,11]. In this setting, handovers would occur only at virtual cell boundaries.

Furthermore, in recent years, it has been recognized that the technological complexity, the large number of stakeholders, and the various conflicting goals/ policies will make it impossible to deploy a clean-slate Internet architecture. This has spurred research on network virtualization, with the aim of offering an open and expandable model into which different solutions can fit. An architecture with a massive deployment of simplified RRHs and processing at a CU, conjugated with an efficient optical fronthaul, allows the development of virtualization solutions for radio access for multiple operators and multiple systems. In this way, it may lead to a change of paradigm in the operation of networks (which is already ongoing in several places) whereby the owner of the infrastructure is separated from the traditional telecom operator, and may offer (or enable) network virtualization services allowing multiple operators and different technologies to share the same infrastructure.

9.2.1 Optical Infrastructure and Cognitive Radio

The transparent optical infrastructure can be the enabler of advanced radio concepts such as *cognitive radio*. The massive deployment of simple RRHs with broadband capabilities allows deployment over the same infrastructure by overlaying a Spectrum Sensor Network (SSN) to provide context information enabling the development of cognitive radio algorithms and procedures.

The SSN consists of enabling the RRHs with spectral monitoring capabilities that will be able to detect, in a collaborative manner, the spectrum holes and construct a space-time image of the spectral activities in the serving area. The important aspect is that the transparent transport of the monitoring signals will enable the use of fully collaborative algorithms and therefore provide a reliable image of the spectral activity, even without the use of highly sophisticated sensing algorithms that would be used in localized sensing.

The sensing can be done through a dedicated network of sensors installed at the RRHs or through in-band measurements carried out in the C-RAN wireless systems. The measurements can be made by wireless systems that make use of the fiber infrastructure or provided as a service by the owners of the infrastructure to external systems that would access (through the Internet) a database in which the picture of the spectral activity within the serving area is stored.

The concept is illustrated at a high level in Figure 9.4. The RRHs deployed in the serving area are enhanced with broadband sensors that can monitor the spectral

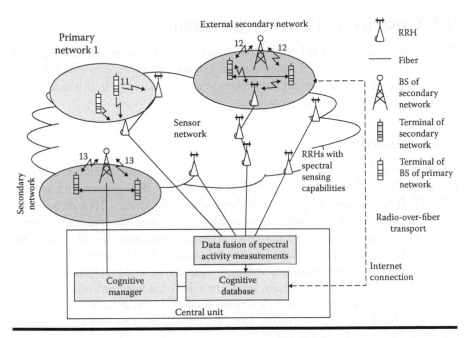

Figure 9.4 Illustration of spectral sensor network concept.

activity over a wide frequency range. The sensors at the RRHs just collect the RF signals that are sent (digitized or in analog format) via the optical infrastructure to the CU. At the CU, the signals from the different RRHs are jointly processed by the data fusion unit, enabling the shadowing effects and hidden terminals problems to be overcome. The output of the processing is stored in the Cognitive Database, and the data stored there is used by the Cognitive Manager to allocate resources or to enable the owner of the infrastructure to provide it to third parties.

Several advantages are provided by the sensor network embedded on the optical infrastructure over conventional approaches relying on terminal localized sensing or wireless cooperative sensing. Localized sensing is obviously limited and cannot cope with hidden terminal problems. Furthermore, to achieve good sensitivity without relying on cooperation requires highly sophisticated algorithms exploiting data features, which may lead to a significant increase in complexity. The sensing performance can be significantly enhanced through wireless cooperation, but then two issues arise: the requirement for overheads to exchange information between the sensing units (or between the sensing units and the data fusion center), which implies that to minimize overheads, only limited side information can be exchanged; and the need for a dedicated channel to exchange this information or eventually resort to ultrawide band. In the proposed architecture, the transmission from the sensing units to the data fusion center is through optical fiber; no additional spectrum is required for such an exchange, and all the information can be transported without any hard-limitation.

9.3 Capacity and Latency Constraints

When designing the optical fronthaul, one faces several issues, which have reported in the previous sections. The most important ones are the need to provide links with very high capacities and the restrictions in terms of latency. In this section, we focus our attention on these two aspects.

9.3.1 Capacity

The current solutions for the optical fronthaul mostly rely on the use of the Common Public Radio Interface (CPRI) [12] and Open Base Station Architecture (OBSAI) [13], and we will give a little detail about the former. Though both the CPRI and OBSAI are standard interfaces, they are currently provided in a "semi-proprietary" way, and they are not designed with interoperability in mind [14]. A European Telecommunications Standards Institute (ETSI) effort named *Open Radio Interface* (ORI) promotes an industry-wide interoperable interface between CUs and RRHs [15]. At the present time, the CPRI has been enhanced by some major mobile operators and industry players, such as the Next Generation Mobile Networks (NGMN) Alliance [14,16]. Later on, we will describe how we can calculate the CRRI rate (Equation 9.1).

In CPRI, the control and user data are organized in a frame hierarchy, as illustrated in Figure 9.5.

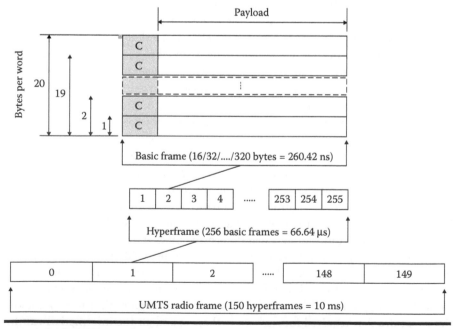

Figure 9.5   CPRI frame hierarchy.

Table 9.1   Specified Line Rates for CPRI and OBSAI Links

CPRI				OBSAI
Option	Bit Rate (Mbit/s)	Line Coding	Bit Rate Calculation	Bit Rate (Mbit/s)
1	614.4	8 B/10 B	$16 \times 1 \times 8 \times 3.84$ MHz $\times 10/8$	768
2	1,228.8	8 B/10 B	$16 \times 2 \times 8 \times 3.84$ MHz $\times 10/8$	1536
3	2,457.6	8 B/10 B	$16 \times 4 \times 8 \times 3.84$ MHz $\times 10/8$	
4	3,072.0	8 B/10 B	$16 \times 5 \times 8 \times 3.84$ MHz $\times 10/8$	3072
5	4,915.2	8 B/10 B	$16 \times 8 \times 8 \times 3.84$ MHz $\times 10/8$	
6	6,144.0	8 B/10 B	$16 \times 10 \times 8 \times 3.84$ MHz $\times 10/8$	6144
7A	8,110.08	64 B/66 B	$16 \times 16 \times 8 \times 3.84$ MHz $\times 66/64$	
7	9,830.4	8 B/10 B	$16 \times 16 \times 8 \times 3.84$ MHz $\times 10/8$	
8	10,137.6	64B/66 B	$16 \times 20 \times 8 \times 3.84$ MHz $\times 66/64$	
9	12,165.12	64B/66 B	$16 \times 24 \times 3.84$ MHz $\times 66/64$	

The basic frame consists of 16 words, 1 for control and 15 for the payload. Each word may contain from 1 to 20 bytes depending on the transmission rate. The transmission rate of the CPRI interface is proportional to the number of bytes per word. The transmission rate of a basic frame is 3.84 MHz, which is equal to the chip rate of the Universal Mobile Telecommunications System (UMTS) technology, with a duration of 260.42 ns.

Table 9.1 presents all the transmission rates supported by CPRI version 6.1 [12] together with those of OBSAI [13]. For each CPRI option, the line coding used is indicated, and the line bit rate calculation is given by the following expression:

$$\text{CPRI rate} = 16 \times N \times 8 \times 3.84 \times 10^6 \times C \qquad (9.1)$$

where:

16	represents the number of words in a basic frame
N	is the number of bytes in a word
8	is the number of bits in 1 byte
3.84×10^6	is the transmission rate of a basic frame
C	is the magnification factor in the transmission rate due to the line coding

CPRI version 6.1 considers C=10/8 for the bit rate options 1–6 using the line coding 8 B/10 B and C=66/64 for options 7A, 8, and 9 using the line coding 64 B/66 B. The variable N may take the values 1, 2, 4, 5, 8, 10, 16, 20, and 24 for the line bit rate options 1–9, respectively.

In the CPRI frame hierarchy, one hyperframe is formed by 256 basic frames, including 256 control words. Each control word has a particular function as there are control words for signaling the start of a hyperframe, synchronization, control, and management, among other things. A radio frame is comprised of 150 hyperframes; the duration of a radio frame is 10 ms, which is equal to the period of one frame in UMTS and LTE. Figure 9.5 shows the hierarchical relationship between the three types of frames.

CPRI options 8 and 9 are specified for a raw rate of 10.378 and 12.165 Gbit/s, respectively. Simple calculations show that they both support LTE connections for the 20 MHz bandwidth sampled at 30.72 MHz for an array of 8 antennas and using 32 quantization bits (16 for the I component and 16 for the Q component). However, impressive as 10 Gbit/s may seem, considering a trisector, a single option 8 (or 9) CPRI connection only allows remoting 20 MHz with 4 antenna elements and 2×12 quantization bits per sample.

In fact, these current capacities will be far from adequate in the very near future. The data rate requirements are likely to increase dramatically with the introduction of M-MIMO. The concept of M-MIMO is simply conventional MIMO [17] pushed to the extreme by using hundreds of antenna elements. This is particularly suitable for the new frequencies intended for use in future systems. To provide high bandwidths that cannot be found in the sub-6 GHz range, the intention is to go up to mmWave [18], where the spectrum is abundant. At such high frequencies, the dimensions of the antenna elements are very small, thus allowing compact array designs. The very large number of antenna elements allows space multiplexing by pointing beams to the desired users, as shown in Figure 9.6.

As the raw data rates required to connect an RRH to the baseband unit (BBU) are given by

$$R = 2 \cdot n_q \cdot n_s \cdot n_A \cdot f_s \tag{9.2}$$

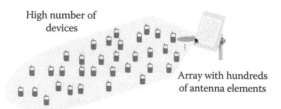

Figure 9.6   Concept of massive MIMO.

where n_q, n_s, and n_A represent the number of quantization bits per component, the number of sectors, and the number of elements in the array, respectively, and f_s is the sampling frequency, it turns out that for bandwidths in the gigahertz range combined with hundreds of antenna elements per array, the requirements per optical link between the BBU and the RRH quickly jump to tens of terabits per second, as shown in Figure 9.1.

As an example, considering a 2 GHz bandwidth in the 60 GHz band, with an antenna array of 256 elements, sampling at 50% above the Nyquist frequency with a 14 bit quantization process leads to 21 Tbit/s only for a single sector. As we may have tens of mmWaves and sub-6 GHz RRHs/km² in ultradense environments, the overall traffic in the fronthaul will be huge, calling for the use of compression algorithms [19–21] and the combination of analog and digitized transmissions [22].

9.3.2 Latency

There is currently significant research concerning low-latency wireless communications to accommodate the requirements of emerging and future applications that demand round-trip delays much lower than those that LTE can currently provide. This is driven by grid applications, the industrial Internet, and also what is called the *tactile Internet* [3], which includes applications related to automotive safety, gaming, and so on. Furthermore, one can envision future M2M applications that form feedback control systems and have stringent latency requirements, such as automatic high-frequency trading (e.g., [23]).

The overall latency of the system has to be minimized, which implies that the delays in the fronthaul have to be kept within stringent limits. As the optical link between RRH and BBU is at the level of the physical radio signals, in a C-RAN architecture, the overall maximum latency has to consider the propagation times of the transmission in the fronthaul.

For LTE radio access technology, the most critical deadline comes from the uplink method Hybrid Automatic Repeat Request (HARQ) on the medium access control (MAC) layer. In case of retransmission, this deadline has on peak data per user. HARQ, which is a retransmission protocol between eNodeB and UE, states that the reception status of every received subframe has to be reported back to the transmitter. This is illustrated in Figure 9.5, which shows that in LTE frequency-division duplexing (FDD)-based networks, the eNodeB has to feed back the ACK/NAK to the fourth subframe following the one where data was received. The HARQ round-trip time (RTT) equals 8 ms, since as shown in Figure 9.7a, the ACK/NAK received at subframe $k + 4$ has to be decoded at the transmitter before subframe $k + 8$ is assembled.

Due to the LTE timing (HARQ) requirement for a round-trip time of 8 ms, there are stringent latency requirements for such a transport system, and according to Figure 9.5, these could be generally expressed as

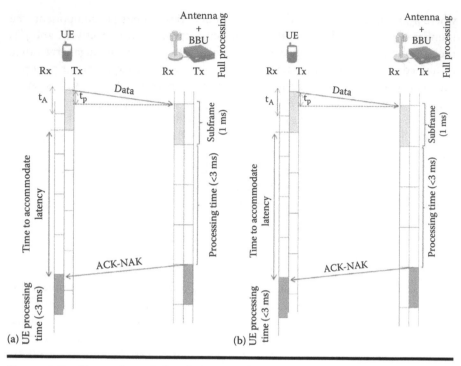

Figure 9.7 Illustration of the timings with LTE: (a) RRH with full processing; (b) RRH with only RF processing and baseband performed at the BBU.

$$\text{Latency margin} = 3 \text{ ms} - T_{p\_NB} + t_A = 3 \text{ ms} - T_{p\_NB} + 2t_{pa} \qquad (9.3)$$

where T_{p_NB} and t_{pa} represent the processing time at the eNodeB and the propagation over the air, respectively (the timing advance t_A is equal to $2t_{pa}$).

In Figure 9.7b, the eNodeB processing time is moved to the BBU, and one has to sum the propagation times between the eNodeB and the BBU. The latency becomes

$$\text{Latency margin} = 3 \text{ ms} - T_{p\_NB} + t_A - 2t_{pf} \qquad (9.4)$$

where t_{pf} is the propagation time between the RRH and BBU, which includes the processing times of all the units between the RRH and BBU.

Considering as an example that t_{pf} includes the RF, CPRI, and fronthaul processing times, and that as a whole they account for a round trip of 120 µs, and that the BBU processing time is 2.5 ms, we are left with the case of minimum time advance with a margin of 380 µs for the round-trip fiber propagation, which represents a length of approximately 38 km. While this may be sufficient for the current frame definition of LTE, things may be significantly affected if the frame duration of the air interface is reduced to reduce the overall latency. Proposals have been forwarded to consider subframes of only 0.25 ms, which would (with everything scaled down by 4) give an overall round trip of 2 ms. Scaling down Equation 9.4 by a factor of 4, one gets

$$\text{Latency margin } (0.25) = 0.75 \text{ ms} - T_{p\_BBU} + t_A - 2t_{pf} \tag{9.5}$$

and then, scaling down the BBU processing time by a factor of 4, it turns out that either the RF and optical transmission times are brought down to very low values or the fiber length will be very limited. Even considering that t_{pf} is only due to the propagation times within the fiber, we are limited to about 12 km. The alternative could be to speed up the processing at the BBU, which because of the centralized nature of the processing, will allow the cost-effective implementation of high-speed processors and buffers. In such a case, one could go to about 40 km fiber length.

9.4 Fronthaul Architectures and Optical Technologies

9.4.1 Fronthaul Architectures

Figure 9.8 depicts different approaches to interconnect the BSs to the RRHs. In the conventional approach, the modulation and demodulation of radio signals (L1 processing) are performed near the antenna site. With this approach, not all details of the radio signals are transported in the backhaul to the Radio Network Controller (RNC) or the CU, thus not providing the benefits of diversity gain or intercell interference cancellation.

In digitized radio, the link between the RRH and the BS transmits digital in-phase (I) and quadrature-phase (Q) samples of the radio signals, and so the RRHs contain digital/analog (D/A) converters, upconverters, and amplifiers in the downlink direction and analog/digital (A/D) converters, downconverters, and amplifiers in the uplink direction. The RRHs are relatively small and simple, since all of the signal modulation functions, including inverse fast Fourier transform (IFFT) processing, take place in the BS, which in C-RAN is located at the CU.

Digital transmission, in the form of digitized radio-over-fiber (D-RoF) links, is popular for third-generation (3G) and fourth-generation (4G) deployments where radio digitized interfaces are available, such as CPRI [12] or OBSAI [13]. As referred to in Section 9.3.1, the main issue with the digitized IQ signal approach is that the increased bandwidth of beyond-4G systems, plus the requirement to transport multiple radio channels for channel monitoring, estimation, feedback, and multi-RRH joint processing requirements, will lead to very high bit rate requirements.

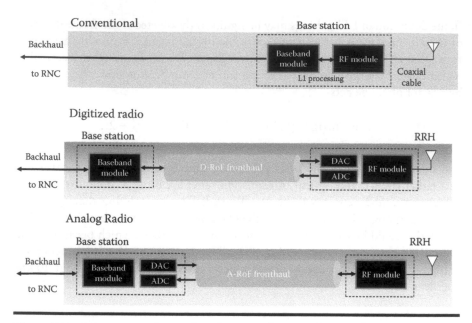

Figure 9.8 Interconnection options between base stations and RRH.

However, the high bandwidth required for D-RoF can be significantly reduced by using high-order modulation formats [24], and also the use of compression can further reduce significantly the bandwidth required and keep the EVM at acceptable levels [14,25]. Moreover, the digital transceivers that are currently massively produced may be used for D-RoF, eventually decreasing the cost and increasing interoperability between manufacturers.

The analog approach depicted in Figure 9.8 offers the possibility for aggregating large numbers of RF channels in a bandwidth similar to the one required by a single digitized channel [26]. The analog radio-over-fiber (A-RoF) links are typically characterized by their transparency, although limited by additional noise and distortion [26]. The RRHs are less complex and have lower power consumption than would be the case for digital transmission, since all digital processing is performed at the CU/BBU. However, analog links suffer from a reduced dynamic range for wide channel bandwidth and high optical loss, but in most cases this can be mitigated using simple techniques such as uplink power control and automatic gain control [26].

A-RoF can increase the capacity of the fronthaul significantly, and it offers a bandwidth-efficient solution that can coexist with D-RoF and comply with elastic optical networking. A suitable fronthaul solution should combine D-RoF for mobile macrocell deployment, due to the robustness and high dynamic range offered by this kind of transmission, with A-RoF transmission for small cells (femtocells, picocells, and microcells) where bandwidth-efficient transport solutions for massive radio channel transmission (such as M-MIMO) or high-speed mmWave services are required [27].

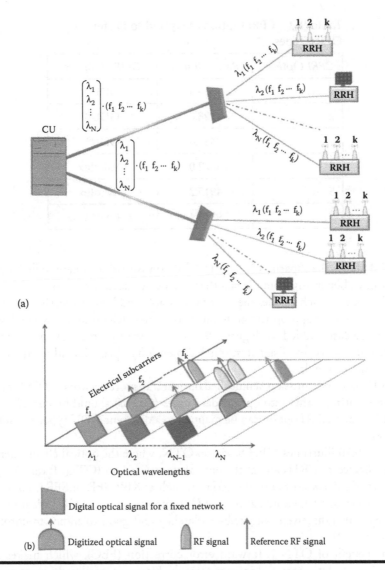

Figure 9.9 **(a) Optical infrastructure and (b) resource allocation.**

One possible solution to achieve these features is to use a combination of RF translation with subcarrier multiplexing (SCM) and low-cost forms of wavelength-division multiplexing (WDM) for the optical distribution network [28], as illustrated in Figure 9.9. A bidimensional space consisting of optical wavelengths and electrical subcarriers can be defined. The RRHs are addressed by wavelength, and individual signals for the different antenna elements of the same RRH (or associated with different wireless systems) are separated in the electrical domain through SCM, as illustrated in Figure 9.9a for a star architecture.

Table 9.2 CPRI Options Mapped to Different ODU Types

CPRI Option	Rate (Mbit/s)	GMP Mapping
1	614.4	ODU0
2	1228.8	ODU0
3	2457.6	ODU1
4	3072.0	ODUflex
5	4915.2	ODUflex
7	9830.4	ODU2

SCM allows the capacity of an optical fiber transmission system to be increased significantly, benefiting from the fact that microwave devices are more mature than optical devices, which enables the use of very stable and highly selective filters [28].

By defining the appropriate granularity, both new RRHs and new wireless systems can be easily added. In Figure 9.9b, one can see that it is possible to accommodate different types of signal in this bidimensional space, including analog and digitized radio signals.

For D-RoF transmission, a modified optical transport network (OTN) solution by encapsulating traffic into optical data units (ODUs) should be considered. In particular, each CPRI option can be mapped to a different ODU type, as Table 9.2 indicates.

Figure 9.10 illustrates MFH based on OTN, where the optical client interfaces (OCIs), located at RRHs and at the optical transport unit (OTU) framer/mapper, can be standard low-cost optical interfaces, such as XFP, SFP, or SFR+. For optical line interfaces (OLIs), located at the OTU framer/mapper and CU, it is important to use optical transceivers with reduced latency and good tolerance to chromatic dispersion (CD).

One benefit of OTN is forward error correction (FEC), which makes links less sensitive to bit errors and improves reach. However, as already mentioned in Section 9.3.2, the fronthaul connection length is generally limited not by technology, but by transport latency. In the case of the fronthaul, FEC will even reduce the achieved distance through introduced latency.

9.4.2 Optical Technologies

Intensity modulation of the optical carrier used for signal transmission, followed by directed detection at the end side of the link, is the widespread choice for RoF applications [26,28]. The simplest intensity modulation technique consists of directly modulating a laser with the electrical signal, to change its output power.

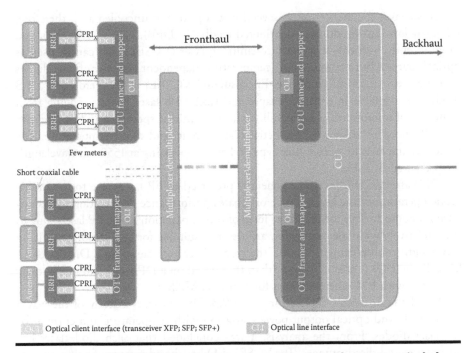

Figure 9.10 Mobile FH based on OTN and coexistence with A-RoF optical channels (OChs). (From P. P. Monteiro and A. Gameiro, *16th International Conference on Transparent Optical Networks (ICTON)*, Graz, IEEE, 2014.)

The laser slope efficiency represents the efficiency with which the RF-modulation current is converted to modulated optical power. The gain of the analog link is therefore dependent, among other parameters, on the laser slope efficiency. The relative intensity noise (RIN), typical of semiconductor lasers, is caused mainly by spontaneous emission. An alternative to direct modulation (DM) is external modulation (EM), in which a laser operates in continuous wave (CW) mode, and an external device is responsible for intensity modulating the laser output power. This option is not limited by the modulation bandwidth of the laser, mainly due to the dependence of the rate of stimulated emission on the carrier number. Among the possible implementations of EM, the preferred option employs a Mach–Zehnder modulator (MZM). MZMs are based on the electro-optic effect, to change the phase of the input optical field, and use an interferometer to convert the phase change into an intensity variation.

Aside from considering which solution is more advantageous to optimize higher transmission data rates and link reach, the quest for the use of low-cost components is most relevant for the commercial implementation of RRHs. For this reason, the use of complex laser structures or temperature controllers in the RRH modules should be avoided, as they increase not only the cost but also the power consumption on the antenna/user side.

Nowadays, the market offers several laser options of uncooled and, therefore, relatively low-cost modules of distributed feedback (DFB), Fabry–Perot (FP), and more recently, vertical-cavity surface-emitting laser (VCSEL) that can be used as optical sources. But the wavelength temperature-dependent behavior of laser diodes is still a great technical challenge. Encapsulated VCSEL, Fabry–Perot, and DFB diodes are available in different shapes and sizes with laser powers ranging from hundreds of microwatts to a few milliwatts. The emitted spectrum and its intensity are influenced by the cavity geometry, bias current, and diode temperature. The typical characteristics of the three type of devices operating at 1550 nm wavelength are presented in Table 9.3.

DFB and VCSEL lasers are generally preferred to FP lasers due to the multi-mode operation of the latter, which originates performance degradation over standard monomode fiber (SMF) in the form of mode-partition noise. FP lasers, due to their relatively wide spectral width, are also not suitable for applications requiring wavelength multiplexing or SCM. The transmitter cost based on a DM VCSEL-laser diode (LD) is normally lower than that based on a DFB-LD and much lower compared with EM schemes using, for example, MZMs.

For the laser diodes depicted in Table 9.3, [29] presents a study of the wave-length shift and optical output power variation with temperature and current for the laser diodes. From the analysis, it was concluded that each diode laser has advantages and disadvantages. The VCSEL-LD has the advantage of being, in general, low cost and has the capability of being directly modulated at high bit rates. However, the studied VCSEL-LD suffered the highest output power variation with temperature and current when compared with the other two devices. The DFB-LD

Table 9.3 Comparison of Uncooled Laser Diodes Emitting at 1550 nm

Characteristics	VCSEL	DFB	FP
Model	RC34051-F	RLD-CD55SF	C1237321423
Manufacturer	RayCan	HGGenuine	Liverage
Data rate (Gbps)	10	10	1.25
Bias current range (25°C) (mA)	2–15	8–120	10–150
Threshold current (mA)	2	8	10
Peak wavelength (nm)	1550	1550	1550
Fiber output power (mW)	0.5	2	1
Side-mode suppression ratio (dB)	35	30	—

is a single-mode laser and it usually has high output power, but the DFB-LD studied showed the highest change of the threshold current over the temperature. The FP-LD is a multimode laser and may not be suitable for DWDM. All three devices require some temperature control for DWDM transmission systems, but this control makes these components more expensive and with higher power consumption.

The RIN, typical of semiconductor lasers, is caused mainly by spontaneous emission. Usually, signal distortion is caused by the modulating device, rather than the photodetector or the fiber. Linearization techniques have been proposed to mitigate the effects of distortion in DM or EM links [26,30–33].

Laser nonlinearities can be classified as either static or dynamic. The main cause of static nonlinearities is laser operation near the threshold, or saturation. Dynamic nonlinearities are related to the nonlinear interaction between photons and electrons in the laser active region. System nonlinearity strongly affects orthogonal frequency-division multiplexing (OFDM) signals due to their high peak-to-average power ratio (PAPR). This limits the maximum average power of the radio signal that can modulate the laser, and consequently the signal-to-noise ratio (SNR) at the output of the RoF link. The average power of the radio signal can be increased if the nonlinearities of the RoF link are compensated. References [34] and [35] present the implementation and experimental results of a predistortion compensation scheme for DM VCSEL with OFDM signals using a memory polynomial [33] that describes the inverse transfer function of an A-RoF link given by the following expression:

$$y(n) = \sum_{K=1}^{K} \sum_{q=0}^{Q} a_{kq} x(n-q) \left| x(n-q) \right|^{k-1} \qquad (9.6)$$

where:

K   is the highest order of the nonlinearity
Q   is the memory length
a_{kq} are the polynomial coefficients
x(n) and y(n) are the input and output signals, respectively, which can assume complex values

To model a system by the memory polynomial, it is necessary (1) to access signals in the system input and output and (2) to estimate the coefficients a_{kq} that best approximate the system. Observing (1), we can conclude that it is y(n) linear with

$$x(n-q)\left| x(n-q) \right|^{k-1} \qquad (9.7)$$

and therefore the coefficients a_{kq} can be calculated using a simple least-squares algorithm.

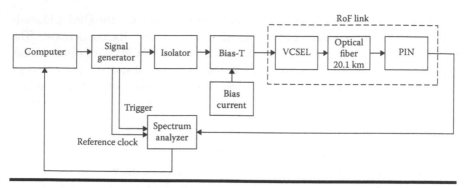

Figure 9.11 Experimental setup.

Figure 9.11 depicts the experimental setup of an RoF link based on OFDM signal transmission by DM of a VCSEL with nonlinearities mitigation using the memory polynomial described in Equation 9.6. An OFDM baseband signal in accordance with the LTE technology was generated using MATLAB®. The signal was generated with a duration of one time slot (0.5 ms), 2048 total subcarriers, where 1201 are used for data transmission using 64 quadrature amplitude modulation (QAM) and the other 847 are used for guard, and a total bandwidth of 20 MHz. Predistortion was applied to the signal, after the signal generator from Rohde & Schwarz (R&S, Munich, Germany) SMW200 A was used to convert the OFDM signal in the baseband (generated by MATLAB) to an analog passband signal. The basic sequence of operations was interpolation, digital-to-analog conversion, IQ modulation, and amplification. A carrier frequency of 900 MHz was used in all the experiments. The RF signal-directly modulated a VCSEL. The VCSEL, with the electrical characteristics shown in Table 9.3, was biased at 6 mA. The signal was propagated though 20.1 km of single-mode fiber. After photodetection by a PIN, the RF signal was amplified by the Nortel Networks (PP-10 G) transimpedance optical front-end (OF). After the OF, the radio signal was demodulated by the R&S FSW spectrum analyzer through the following sequence of operations: conversion of the carrier frequency to an intermediate frequency, filter, sampling, IQ demodulation, and decimation. After that, the data samples were transferred to MATLAB to perform the following signal processing operations: synchronization and compensation of attenuation of the radio signal in the RoF link, and also the phase rotation introduced in the I/Q demodulation.

Figure 9.12a shows the normalized power spectral density of the OFDM signal transmitted and received without and with predistortion. It can be seen that by using the predistorter, the sidebands of the OFDM signal were reduced by more than 5 dB. Figure 9.12b depicts the received constellation without and with predistortion: the measured error vector magnitude (EVM) was 6% and 3.5%, respectively. Since LTE technology requires an EVM lower than 8% for 64 QAM modulation [36], the results demonstrate that the system is operating with an EVM performance well above the limit.

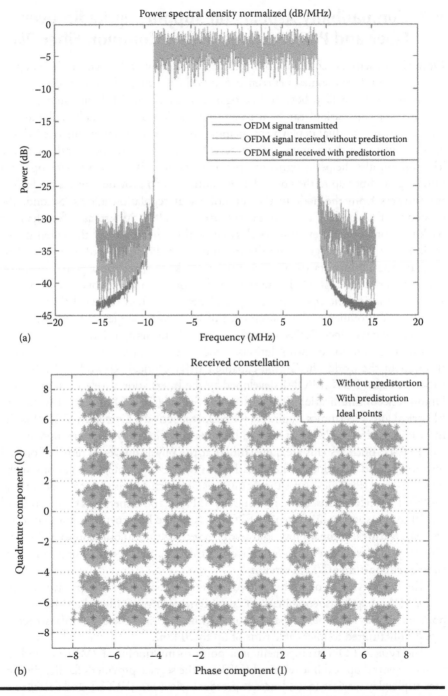

(a)

(b)

Figure 9.12 **(a) Normalized received power spectral densities and (b) received constellation with and without predistortion.**

9.5 Compatibility Aspects on Supporting Radio-over-Fiber and PON Systems with a Common Fiber Plan

Optical and wireless communications have, up to now, taken generally independent paths, and were sometimes seen as competitors, at least for the access segment of the network. In fact, both technologies can have a fruitful coexistence. First, wireless communications can be used as an extension of the fixed broadband network, allowing gradual deployment, helping to phase the investments needed and also to overcome the asymmetries in access between urban and rural scenarios. This option also helps in terms of openness, as even if the incumbent operator only deploys fiber up to the node, the new entrants can provide services by deploying wireless from the node to the subscribers, if regulation allows. Second, the existence of an optical access infrastructure may allow its use as a fronthaul or backhaul, or both, for existing wireless networks. The ability of the wired distribution of broadband signals and the transport of radio signals to/from the RRHs to share the same infrastructure is therefore a key aspect for a global operator, since it allows leverage of the investments needed in the deployment of optical access by sharing the costs through several networks. It is expected that, at least in urban environments, the convergence of both technologies could provide significant benefits, since PONs are widely available in many urban areas. They are mainly based on one of two solutions, which are becoming increasingly popular throughout the world: the Gigabit-capable Passive Optical Network (GPON) and Ethernet-based PON (EPON), developed by the International Telecommunication Union Telecommunication Standardization Sector (ITU-T) and the Institute of Electrical and Electronics Engineers (IEEE), respectively. Currently, GPON is the most popular in the United States and Europe, and EPON systems are more prevalent in Asia. Both systems are quite similar at the physical layer in terms of optical technology and network architecture. However, they use different solutions at the upper layers. GPON leverages the techniques of Synchronous Optical Networking (SONET)/Synchronous Digital Hierarchy (SDH) and generic framing protocol (GFP) to transport, while Ethernet EPON is a native Ethernet solution that leverages the features, compatibility, and performance of the Ethernet protocol [37,38]. GPON provides a better path for delivering the optimum QoS, an important feature for supporting 4G/LTE services.

The PON is based on a fiber network that uses only fiber and passive components, such as splitters, combiners, or coexistent elements, rather than active components such as amplifiers, repeaters, or shaping circuits. The PON infrastructure is typically limited to fiber cable runs of up to 20 km.

The typical PON arrangement is a point-to-multipoint (P2MP) network in which a central optical line terminal (OLT) at the service provider's facility distributes triple play services (data, Internet protocol television [IP-TV], and voice) to as many as 16–128 customers per fiber line. Figure 9.13 exemplifies a PON architecture considering an exemplificative case of three users.

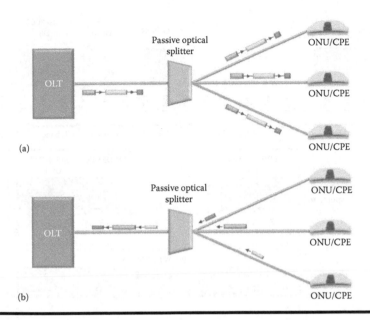

Figure 9.13 Typical PON (GPON and EPON) architecture: (a) downstream; (b) upstream.

In downstream transmission, the passive optical splitter splits a single optical signal into multiple equal but lower power signals to the users. At the user's side, customer-premises equipment (CPE), known as an *optical network unit* (ONU) in IEEE terminology or an *optical network terminal* (ONT) in ITU-T terminology, terminates the fiber-optic line. Its main function is to convert the optical signals into electrical ones (and vice versa) and demultiplex the downstream signal into its component parts. In upstream transmission, the passive power splitter works as a combiner of all data signals transmitted from the CPEs.

EPON and GPON adopt two multiplexing mechanisms. In the downstream direction (i.e., from OLT to CPEs), the optical data packets are transmitted in a broadcast manner, as illustrated in Figure 9.13a, but the data is encrypted using an Advanced Encryption Standard (AES) to prevent eavesdropping. In the upstream direction (i.e., from CPEs to OLT; Figure 9.13b), the optical data packets are transmitted in a time-division multiple-access (TDMA) manner, whereby each user is assigned a time slot using the same upstream optical wavelength, which is different from the downstream one, such that a single fiber can be used for both downstream and upstream data.

GPON delivers 2.488 Gbits/s downstream and 1.244 Gbits/s upstream with an available splitting from 1:64 to 1:128 [39]. EPON uses the same transmission rate for upstream and downstream of 1.25 Gbit/s and a splitting ratio from 1:16 to 1:32 [40]. The wavelength allocation in GPON and EPON systems is currently based on ITU-T G.984.2 [39], defining a downlink band from 1480 to 1500 nm and an uplink from 1260 to 1360 nm, as illustrated in Figure 9.14.

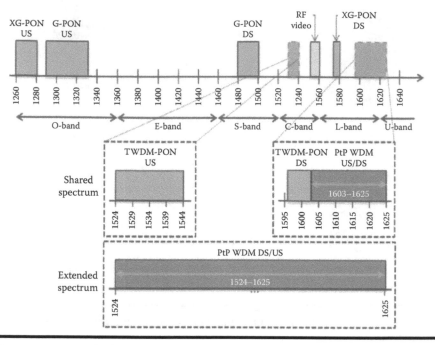

Figure 9.14 Optical spectrum available for different PON technologies. (Proposal under discussion for NG-PON2.)

The latest version of GPON is a 10-gigabit version called XG-PON1, or 10 GPON; XG-PON's maximum rate is 10 Gbits/s (9.95328) downstream and 2.5 Gbits/s (2.48832) upstream [41]. Different wavelength bands are used: 1575–1580 nm for downstream and 1260–1280 nm for upstream, as also depicted in Figure 9.14. This allows 10 Gbit/s service to coexist on the same fiber with standard GPON by using WDM. There is also a 10-Gbit/s Ethernet version, designated 802.3av [42]. The actual line rate is 10.3125 Gbits/s. The primary mode is 10 Gbits/s upstream as well as downstream. A variation uses 10 Gbits/s downstream and 1 Gbit/s upstream. The 10 Gbit/s versions use the same wavelength allocation of XG-PON downstream and 1260–1280 nm upstream, so the 10 Gbit/s can be multiplexed on the same fiber with the standard 1 Gbit/s system.

GPON also considers the transmission of analog video on the same PON infrastructure by adding an overlay channel in the wavelength band 1550–1560 nm [43], also illustrated in Figure 9.14. Analog video services are commonly also deployed by the EPON providers on the same wavelength bands.

9.5.1 D-RoF Transmission in PON Systems

PON is a suitable architecture for fronthaul in C-RAN applications, benefiting from its inherently centralized system, as illustrated in Figure 9.13. However, the native transport of D-RoF signals in EPON and GPON presents several issues. The maximum bit rate transmission of these technologies is not sufficient even for one LTE small cell (1.22 Gbps at 10 MHz 2×2 MIMO); see Section 9.3.1. For XG-PON, it will be possible to capture only one macro cell without compression [14]. However, the most important drawback in the transmission of D-RoF signals (such as CPRI or OBSAI signals) in existing PON systems is the upstream transmission based on TDMA, which significantly increases the latency and jitter, and is usually not compatible with the fronthaul requirements.

To overcome these issues, the ITU-T study group SG15 of the Full Service Access Network (FSAN) Group is now proposing new recommendations for the physical media–dependent (PMD) layer requirements of next-generation PON, known as NG-PON2 [44]. These requirements should provide a flexible optical fiber access network capable of supporting the bandwidth requirements of mobile backhaul, business, and residential services. The NG-PON2 wavelength plan is defined to enable coexistence through wavelength overlay with legacy PON systems. The technology chosen by the industry for the NG-PON2 is the Time and Wavelength Division Multiplexed Passive Optical Network (TWDM-PON), which includes the additional option of a point-to-point wavelength overlay channel (PtP WDM PON). In NG-PON2, the PtP wavelength overlay channel refers to the pair of downstream and upstream wavelength channels providing point-to-point connectivity using the available wavelengths. One of the main PtP WDM applications will be to use NG-PON2 as an MFH to transport D-RoF signals between the CU and the RRHs. However, the wavelength bands available at NG-PON2 for PtP WDM depend on the coexistence requirements with the legacy PON systems (GPON, RF Video, XG-PON) and also with NG-PON2 TWDM [44].

Figure 9.14 depicts the available optical spectrum and a recommendation for proposed shared and extended NG-PON2 spectral options. In the shared spectrum, coexistence with TWDM is required, and the subbands available for PtP WDM are constrained to a relatively narrow bandwidth of 22 nm (1603–1625 nm), which requires the use of modems with the capability to transport D-RoF with very high spectral efficiency to accommodate the high number and high-capacity D-RoF channels foreseen for 5G. However, the bandwidth can be extended from 1524 to 1625 nm if the coexistence with TWDM and also the compatibilities among RF video and XG-PON are lost. This drawback may force operators to use the latter solution only in a green scenario, or to rethink or redraw their networks.

In the framework of the research project Flexicell, flexible optical SCM modems were studied for MFH applications, capable of transmitting D-RoF signals in PON infrastructures with high spectral efficiency. The PtP WDW transmission of D-RoF signals and the coexistence with several access technologies on the same PON network are achieved by using a coexistent element, illustrated in Figure 9.15 [45].

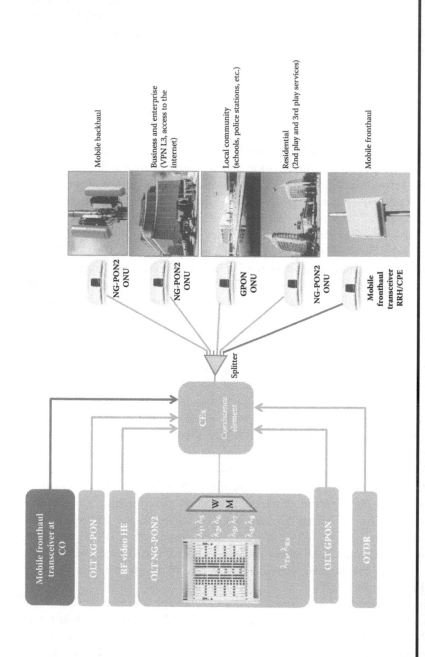

Figure 9.15 Coexistence between different access technologies in a PON system. (Adapted from J. Salgado et al., White Paper by the Deployment & Operations Committee, http://www.ftthcouncil.eu/documents/Publications/DandO_White_Paper_2014. pdf, 2014.)

9.5.2 Mobile Fronthaul Modem for D-RoF Transmission

Figure 9.16 presents the possible transmitter and receiver architectures of a modem for MFH applications. The mapping of each modulated CPRI signal in a subcarrier and the aggregation of all of them in a composed signal are performed entirely on the digital domain, resembling a software-defined radio with a minimal analog radio OF. The digital signal is modulated in a QAM signal with an arbitrary modulation defined by the user; then, synchronization pilots are added to the QAM signal, and the composed signal is predistorted using a predefined transfer function or the memory polynomial described by Equation 9.5. After predistortion, the digital signals are converted into the analog domain by a digital-to-analog converter (DAC) and then low pass filtered. The driver amplifier provides the correct electrical signal amplitude to drive the MZM, which modulates the optical carrier. At the receiver side, illustrated in Figure 9.16b, the optical signal is converted to an electrical signal by a photoreceiver and then converted to the digital domain by the analog-to-digital converter (ADC). After resampling, the selected subcarrier channel

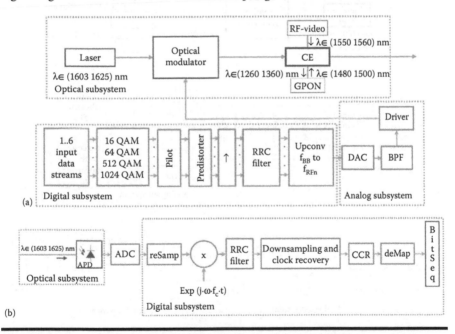

Figure 9.16 Architecture for the mobile fronthaul modem: (a) transmitter; (b) receiver. ADC: analog digital converter; LPF: low pass filter; CE: coexistent element; CCR: clock carrier recovery; CDR: clock data recovery; DAC: digital analog converter; fBB: baseband frequency; fRF: passband frequency; GPON: Gigabit Passive Optical Network; LNA: low noise amplifier; OL: local oscillator; RRC: root raised cosine filter; ↓ down sampling; ↑ up sampling. (From P. P. Monteiro et al., *Transparent Optical Networks (ICTON), 2015 17th International Conference on,* Budapest, IEEE, 2015.)

is downconverted to the baseband and extracted by a root raised cosine matched filter (RRC). After downsampling, clock recovery, and carrier phase recovery, the demodulated subcarrier is ready to be demapped into a bit sequence.

Figure 9.17a presents an alternative, less digital architecture, in which at the transmitter side each subcarrier is digitally generated in the baseband, but upconverted to the corresponding frequency using an analog mixing stage. As shown in Figure 9.17b, the receiver is also different, comprising a two-stage heterodyne receiver. In the analog subsystem, the target subcarrier is first filtered and downconverted to an intermediate frequency. Image rejection is ensured by the two frequency-conversion stages. The resulting signal is then digitalized and transformed to the baseband using a final digital downconverter. This alternative architecture presents the main advantage of using DACs and ADCs with lower sampling rates than in the previous solution, at the cost of requiring more analog devices.

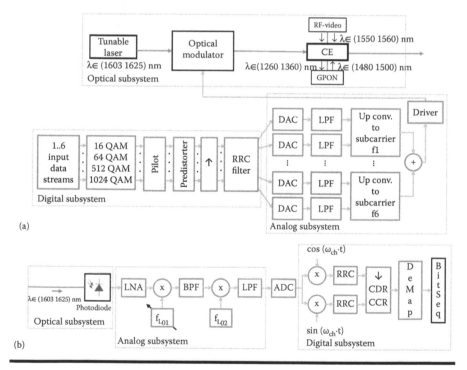

Figure 9.17 Alternative architecture for the mobile fronthaul modem: (a) transmitter; (b) receiver. ADC: analog digital converter; LPF: low pass filter; CE: coexistent element; CCR: clock carrier recovery; CDR: clock data recovery; DAC: digital analog converter; fBB: baseband frequency; fRF: passband frequency; GPON: Gigabit Passive Optical Network; LNA: low noise amplifier; OL: local oscillator; RRC: root raised cosine filter; ↓ down sampling; ↑ up sampling. (From P. P. Monteiro et al., *Transparent Optical Networks (ICTON), 2015 17th International Conference on,* Budapest, IEEE, 2015.)

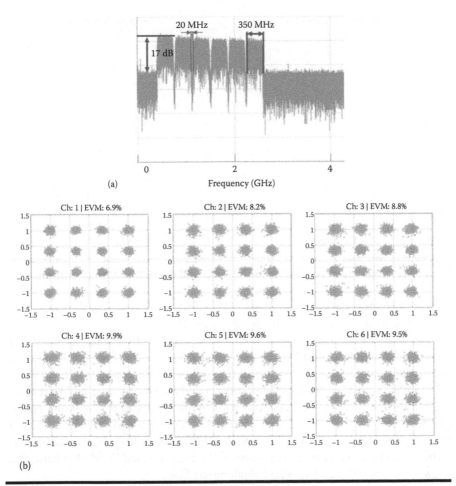

Figure 9.18 **(a) Received spectrum for 6×16 QAM at 1.25 Gbps for 30 km of fiber. (b) Received constellation for 50 km fiber distance with ROP = –11.5 dBm and using a PIN photoreceiver. (From P. P. Monteiro et al.,** *Transparent Optical Networks (ICTON), 2015 17th International Conference on,* **Budapest, IEEE, 2015.)**

However, for the considered sampling frequencies, these analog devices present a much lower cost, and can be integrated into a single chip.

Figure 9.18a shows an SCM spectrum composed of six 16-QAM channels, each at 1.25 Gbps (nearly to CPRI line rate option 2). The spectrum was obtained using a modem architecture, illustrated in Figure 9.16, and for a fiber length of 30 km. A predistortion was performed by using the arccosine function, which is approximately the inverse transfer function of the MZM. Figure 9.18b presents the constellations for all six channels, but considering a fiber length of 50 km and using a PIN photoreceiver for a received optical power (ROP) of –11.5 dBm. All six channels present an EVM lower than 10%. The receiver sensitivity can be improved using a

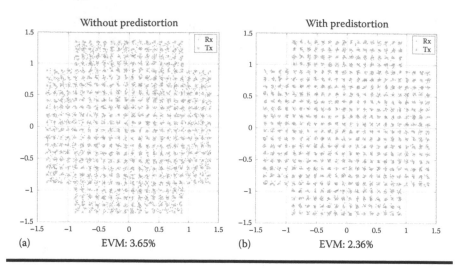

Figure 9.19 Received constellations (a) with and (b) without predistortion.

receiver with an avalanche photodiode (APD) [46]. More details of the experimental setup and additional results are presented in [46].

For higher-order modulation formats, a predistortion equalization using memory polynomials described by Equation 9.5 was used. Figure 9.19 presents the received constellations (a) with and (b) without predistortion for 512 QAM at 1.25 Gbit/s. A memory polynomial of order 3 and a memory order of 7 were used.

A real-time modem was developed with the flexibility to transmit and receive M QAM SCM signals of different orders of modulation and also the possibility of using different forward error correction (FEC) codes [24]. Figure 9.20 shows a 1024 QAM constellation of the worst of the four received SCM channels, after transmission on 20 km of standard single-mode fiber.

An uncoded bit error rate (BER) of 2.9×10^{-4} was achieved for a symbol transmission rate of 100 Mbaud, corresponding to a payload data rate of approximately 900 Mbit/s. Each subcarrier has a bandwidth of only 112 MHz, which represents a very high spectral efficiency. An error free was achieved using a Reed–Solomon (228, 252).

9.6 Conclusions

In this chapter, we have considered the convergence of wireless and optical technologies to provide cost-effective pervasive and broadband networks. The capacity and latency requirements put stringent requirements on the fronthaul design, and it is likely that designing equipment for the fronthaul will be one of the main markets for optical communications in the future. Overall, to provide the features

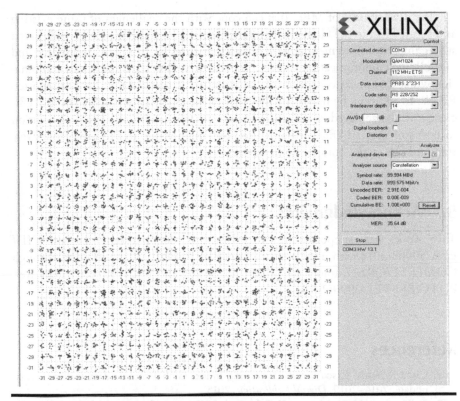

Figure 9.20 Received 1024QAM constellation by a real-time modem and BER performance for the worst SCM channel and after the signal has traveled along a 20 km fiber.

envisioned for a global and pervasive information society, the two technologies, optical and wireless, which in the past have progressed in isolation, need to merge to exploit the synergies and thus provide cost-effective solutions.

We presented hybrid transmission solutions combining digital and analog RoF to transport heterogeneous radio signals efficiently between remote antenna units and a CU. The proposed architectures are well beyond the conventional vision of remoting associated with RoF and will act as an enabler for the development of several wireless technologies: M-MIMO and cognitive radio concepts to achieve broadband wireless transmission; intercell interference cancellation to increase system-level capacity; development of efficient common radio resource management procedures; harmonization of radiation levels in dense urban environments; and smart resource allocation between macrocells and small cells. At the concept level, this represents a shift from the conventional vision of RoF as a remoting technology to one of RoF as an enabler of new wireless architectures and processing alternatives.

The development of MFH that can cope with the challenges of using existing and future Fiber to the Home PON infrastructures, such as NG-PON2, will have a strong impact on the emerging C-RAN architectures for future 5G inbuilding and outbuilding communication systems. The use of WDM transmission in combination with SCM presents an attractive solution to increase the number of supported RRHs. In addition, by increasing the order of the modulation, higher spectral efficiency will be achieved, which turns these transceivers into an appealing solution in terms of spectral efficiency and also cost, mainly due to the use of relatively standard DACs and optoelectronic devices.

9.7 Acknowledgments

This work was supported by the Instituto de Telecomunicações UID/EEA/50008/2013 within the project Hydra-RoF Ref:P01230, by FEDER, through COMPETE/QREN, under the project no. 38901 and DCT project ADIN Ref: PDTC/EEI-TEL/2990/2012.

References

1. Cisco Systems, Cisco visual networking index: Global mobile data traffic forecast update, 2014–2019, *Digital Publication*, 2015.
2. T. Pötsch, S. N. K. Marwat, Y. Zaki, and C. Goerg, Influence of future M2M communication on the LTE system, in Wireless and Mobile Networking Conference, Dubai, UAE, 2013.
3. G. Fetweiss, The tactile Internet: Applications and challenges, *Vehicular Technology Magazine, IEEE* 9(1): 64–70, 2014.
4. F. Boccardi, R. W. Heath, A. Lozano, and T. L. Marzetta, Five disruptive technology directions for 5G, *IEEE Communications Magazine* 52(2): 74–80, 2014.
5. C. Chen, C-RAN: The road towards green radio access network. Available from: http://ss-mcsp.riit.tsinghua.edu.cn/cran/C-RAN%20ChinaCOM-2012-Aug-v4.pdf, 2012.
6. C-RAN: Virtualizing the Radio Access Network. European 2014 Conference. Available from: http://www.uppersideconferences.com/cloudran2014/cloudran2014intro.html.
7. A. Saleh and R. R. Roman, Distributed antennas for indoor radio communications, *IEEE Transactions on Communications* 35(12): 1024–1035, 1987.
8. A. Pizzinat, P. Chanclou, F. Saliou, and T. Diallo, Things you should know about fronthaul, *Journal of Lightwave Technology* 33(5): 1077–1083, 2015.
9. A. de La Oliva, X. C. Perez, A. Azcorra, A. D. Giglio, F. Cavaliere, D. Tiegelbekkers, J. Lessmann, T. Haustein, A. Mourad, and P. Iovanna, *Xhaul: Towards an Integrated Fronthaul/Backhaul Architecture in 5G Networks*, Adava Publications, available from: http://eprints.networks.imdea.org/1059/, 2015.
10. J. Hoydis, M. Kobayashi, and M. Debbah, Green small-cell networks, *IEEE Vehicular Technology Magazine* 6(1): 38–43, 2011.

11. P. T. Hoa and T. Yamada, Cooperative control of connected micro-cells for a virtual single cell for fast handover, 7th International Symposium on Communication Systems Networks and Digital Signal Processing (CSNDSP), Newcastle upon Tyne, IEEE, pp. 852–856, 2010.

12. Common Public Radio Interface (CPRI), *CPRI Specification V6.1 Interface Specification*, available from: http://www.cpri.info/downloads/CPRI_v_6_1_2014-07-01.pdf, 2014.

13. OBSAI, *OBSAI Reference Point 3 Specification Version 4.2 OBSAI*, available from: http://www.obsai.com, 2010.

14. NGMN Alliance, *Project RAN Evolution, Backhaul and Fronthaul Evolution. Version 1.01*, available from: https://www.ngmn.org/uploads/media/NGMN_RANEV_D_BH_FH_Evolution_V1.01.pdf, 2015.

15. ETSI, Open Radio Equipment Interface Webpage, available from: http://www.etsi.org/technologies-clusters/technologies/ori.

16. NGMN Alliance, *Fronthaul Requirements for C-RAN*, available from: https://www.ngmn.org/uploads/media/NGMN_RANEV_D1_C-RAN_Fronthaul_Requirements_v1.0.pdf, 2015.

17. F. Rusek, D. Persson, B. K. Lau, E. G. Larsson, T. L. Marzetta, O. Edfors, and F. Tufvesson, Scaling up MIMO: Opportunities and challenges with very large arrays, *Signal Processing Magazine, IEEE* 30(1): 40, 60, 2013.

18. T. S. Rappaport, S. Sun, R. Mayzus, Z. Hang, Y. Azar, K. Wang, G. N. Wong, J. K. M. Schulz, M. Samimi, and F. Gutierrez, Millimeter wave mobile communications for 5G cellular: It will work! *IEEE Access* 1: 335–349, 2013.

19. S. Park, O. Simeone, O. Sahin, and S. Shamai, Robust and efficient distributed compression for cloud radio access networks, *IEEE Transactions on Vehicular Technology* 62(2): 692–703, 2013.

20. S. Park, O. Simeone, O. Sahin, and S. Shamai, Joint decompression and decoding for cloud radio access networks, *IEEE Signal Processing Letters* 20(5): 503–506, 2013.

21. S. Park, O. Simeone, O. Sahin, and S. Shamai, Joint precoding and multivariate fronthaul compression for the downlink of cloud radio access networks, *IEEE Transactions on Signal Processing* 61(22): 5646–5658, 2013.

22. N. J. Gomes, P. Assimakopoulos, L. Vieira, and P. Sklika, Fiber link design considerations for cloud-radio access networks, IEEE International Conference on Communications (ICC 2014), Sydney, IEEE, pp. 10–14, 2014.

23. E. B. Budish, P. Cramton, and J. J. Shim, The high-frequency trading arms race: Frequent batch auctions as a market design response, Chicago Booth Research Paper No. 14–03, available from: http://ssrn.com/abstract=2388265 or http://dx.doi.org/10.2139/ssrn.2388265, 2015.

24. S. Julião, R. Nunes, D. Viana, P. Jesus, N. Silva, A. S. R. Oliveira, and P. Monteiro, High spectral efficient and flexible multicarrier D-RoF modem using up to 1024-QAM modulation format, 41st European Conference in Optical Communications, Valencia, ECOC, 2015.

25. G. Anjos, D. Riscado, J. Santos, A. S. R. Oliveira, P. Monteiro, N. V. Silva, and P. Jesus, Implementation and evaluation of a low latency and resource efficient compression method for digital radio transport of OFDM signals, 4th International Workshop on Emerging Technologies for 5G Wireless Cellular Networks - IEEE GLOBECOM 2015, San Diego, CA, December 2015.

26. N. J. Gomes, P. P. Monteiro, and A. Gameiro, Eds., *Next Generation Wireless Communications Using Radio over Fiber*, New York: John Wiley, 2012.

27. P. P. Monteiro and A. Gameiro, Hybrid fiber infrastructures for cloud radio access networks, 16th International Conference on Transparent Optical Networks (ICTON), Graz, IEEE, 2014.

28. S. Pato, F. Ferreira, P. Monteiro, and H. Silva, On supporting multiple radio channels over a SCM-based distributed antenna system: A feasibility assessment, Proceedings of ICTON 12th International Conference on Transparent Optical Networks, Munich, IEEE, 2010.

29. L. F. Henning, P. P. Monteiro, and A. P. Pohl, Temperature and bias current behavior of uncooled light sources for application in passive optical networks, 17th International Conference on Transparent Optical Networks (ICTON), Budapest, IEEE, 2015.

30. A. Hekkala and M. Lasanen, Performance of adaptive algorithms for compensation of radio over fiber links, Wireless Telecommunications Symposium, 2009, Prague, IEEE, 2009.

31. A. S. Karar, Y. Jiang, J. C. Cartledge, J. Harley, D. J. Krause, and K. Roberts, Electronic precompensation of the nonlinear distortion in a 10 Gb/s 4-ary ASK directly modulated laser, in *Optical Communication (ECOC), 2010 36th European Conference and Exhibition on*, Torino, IEEE, pp. 1–3, 2010.

32. Y. Pei, K. Xu, A. Zhang, Y. Dai, Y. Ji, and J. Lin, Complexity-reduced digital predistortion for sub carrier multiplexed radio over fiber systems transmitting sparse multiband RF signals, *Optics Express* 21(3): 3708–3714, 2013.

33. L. Vieira, Digital Baseband Modelling and Predistortion of Radio over Fiber Links, PhD dissertation, Kent University, Kent, 2012.

34. R. Costa, P. P. Monteiro, and M. C. R. Medeiros, Nonlinearities mitigation in a RoF link based on OFDM signal transmission by a direct modulation of a VCSEL, to be presented at CONFTEL 2015.

35. M. C. R. Medeiros, R. Costa, H. A. Silva, P. Laurêncio, and P. P. Monteiro, Cost effective hybrid dynamic radio access supported by radio over fiber, Transparent Optical Networks (ICTON), 2015 17th International Conference on, Budapest, IEEE, 2015.

36. A. Ghosh and R. Ratasuk, *Essentials of LTE and LTE-A*. Cambridge: Cambridge University Press, 2011.

37. M. Hajduczenia, H. J. A. da Silva, and P. P. Monteiro, EPON versus APON and GPON: A detailed performance comparison, *Journal of Optical Networking* 5: 298–319, 2006.

38. CommScope Solutions Marketing, GPON–EPON Comparison, White Paper, available from: www.commscope.com/Docs/GPON_EPON_Comparison_WP-107286.pdf, 2013.

39. Recommendation ITU-T G.984.2, Gigabit-capable Passive Optical Networks (GPON): Physical Media Dependent (PMD) layer specification, available from: https://www.itu.int/rec/T-REC-G.984.2-200303-I/en, 2003.

40. IEEE Std. 802.3ah-2004 (Amendment to IEEE Std. 802.3-2002), Media access control parameters, physical layers, and management parameters for subscriber access networks, 2004.

41. Recommendation ITU-T G.987.2, 10-Gigabit-capable passive optical networks (XG-PON): Physical media dependent (PMD) layer specification, 2010.

42. IEEE Std. 802.3av-2009 (Amendment to IEEE Std. 802.3-2008), Physical layer specifications and management parameters for 10 Gb/s passive optical networks, 2009.
43. Recommendation G.984.5 (05/14); Gigabit-capable passive optical networks (G-PON): Enhancement band, available from: https://www.itu.int/rec/T-REC-G.984.5-201405-I/en, 2014.
44. D. Nesset, NG-PON2 technology and standards, *Journal of Lightwave Technology* 33(5): 1136–1143, 2015.
45. J. Salgado, R. Zhao, and N. Monteiro, New FTTH-based technologies and applications, White Paper by the Deployment & Operations Committee, available from: http://www.ftthcouncil.eu/documents/Publications/DandO_White_Paper_2014.pdf, 2014.
46. P. P. Monteiro, D. Viana, J. da Silva, D. Riscado, M. Drummond, A. S. R. Oliveira, N. Silva, and P. Jesus, Mobile fronthaul RoF transceivers for C-RAN applications, Transparent Optical Networks (ICTON), 2015 17th International Conference on, Budapest, IEEE, 2015.

Chapter 10

Power Control in Heterogeneous Networks Using Modulation and Coding Classification

Anestis Tsakmalis, Symeon Chatzinotas, and Björn Ottersten

Contents

Cognitive communications, a promising Dynamic Spectrum Management (DSM) technology [1], enables flexible spectrum usage and is considered to be a fifth-generation (5G) enabling mechanism. In this chapter, a heterogeneous network (HN) application is presented within the DSM framework that achieves an efficient spectral coexistence of a legacy network user and an adaptive, smart, and flexible cognitive radio (CR) network [2]. The wireless networks described as *legacy systems* are networks of outdated technology, still occupying frequency bands, which will not be replaced soon. One drawback of their operation is the inefficient use of the spectrum portion to which they are assigned. One way to better exploit the capacity of this frequency band is to consider a coexistence scenario using an underlay CR technique, whereby secondary users (SUs), the CR network, may transmit in the frequency band of the primary user (PU), the legacy network user, as long as the interference induced on the PU is under a certain limit and thus does not harmfully affect the legacy system's operability. In the examined case study, a centralized power control (PC) scheme and an interference channel gain learning algorithm are jointly tackled to allow the CR network to access the frequency band of a legacy network user operating based on an adaptive coding and modulation (ACM) protocol.

The enabler of the learning part is a cooperative modulation and coding classification (MCC) technique, which provides estimates of the modulation and coding scheme (MCS) of the PU [3]. Due to lack of cooperation between the PU and the SUs, the CR network exploits this multilevel MCC sensing feedback as an implicit channel state information (CSI) of the PU link to constantly observe the impact of the aggregated interference it causes. Statistical signal processing and a powerful machine learning (ML) tool, the Support Vector Machine (SVM) [4], are employed for each SU to determine the PU MCS and later forward these estimates to a cognitive base station (CBS) to cooperatively decide the PU MCS.

The goal is to develop an algorithm to maximize the total CR throughput, the PC optimization objective, and simultaneously learn how to mitigate PU interference, the constraint of the optimization problem, by using only the MCC information. Ideal candidate learning approaches for this problem setting with high convergence rates are the cutting plane methods (CPMs) [5]. Here, we focus on the analytic center cutting plane method (ACCPM) and the center of gravity cutting plane method (CGCPM), and the effectiveness of the proposed techniques is demonstrated through numerical simulations.

10.1 Introduction

5G wireless systems encounter the challenge of radically increasing user data rates. This demanding goal can be achieved by combining technologies under unifying frameworks such as spectrum management. Measurement studies in spectrum usage indicate that some frequency bands are already congested, but the vast majority of them are being underused [6]. The main reason for this uneven spectrum usage is the static nature of this architecture, which assigns particular services to specific frequency bands. This limits accessibility and creates the false perception of spectrum saturation.

A breakthrough idea proposed by the research community and gaining more and more ground is the DSM concept, suggesting that radio spectrum should be distributed to users and services based not on rigid regulations [1] but on flexible restraints that take into consideration overall spectrum availability, access needs, service priorities, market perspectives, network optimization objectives, and quality-of-service (QoS) requirements.

The DSM approach can be implemented by exploiting techniques developed within the CR scheme. This technology, which recommends the development of intelligent radio devices and networks, was first introduced by J. Mitola, who actually borrowed the term *cognitive* from the computer science world [2]. Cognition signifies a radio that is capable of sensing, understanding, adapting to, and interacting with the radio environment.

In the CR regime, there are three main application categories, which are also characterized as DSM examples: the underlay, the overlay, and the interweave mechanisms [7]. In the underlay methods, a CR network (CRN) may transmit in a PU-licensed frequency band as long as the CRN-generated interference to the PUs is under a certain threshold. As far as the overlay methods are concerned, PUs share knowledge of their codebooks and possibly messages with the CRs so as to reduce the interference on the PU receiver side, or even relay the PU message to enhance the PU communication link. In the interweave approach, CRs identify holes in space, time, or frequency from which PUs are absent, and they transmit only in the case of these vacancies. Here, we focus on underlay CRN paradigms, which enable a simultaneous coexistence in frequency of PUs and CRs, also called SUs.

To accomplish this DSM approach, the new smart radio must have some essential functions. These abilities are usually classified into two major groups: spectrum sensing (SS) and decision-making (DM) [2]. The first applies advanced signal processing and other methods and enables the CR to "observe" the spectrum. The second performs an adaptive configuration of the radio operational parameters, such as transmit power, modulation, coding, and frequency, to interact with the environment and consequently reach some goals, such as system throughput or signal-to-interference-plus-noise ratio (SINR) maximization and interference mitigation. The DM process is a broad scientific area and usually employs optimization tools or other mathematical mechanisms to enhance spectrum use. Here, we focus solely on DM that concerns the CR transmit power level, often called the *PC policy*. A new trend in CR technology, and closely related to its origins, is the application of ML to both SS and PC.

10.2 Spectrum Sensing: A Machine Learning Approach

The main enabler of the CRs is the sensing part. SS has thus become an important aspect of this idea, to give "eyes" and "ears" to the "body" of this intelligent radio. Throughout most of the existing literature, SS focuses on testing one hypothesis: the existence or absence of a PU in a frequency band. Another way of making the CR aware of its environment is to detect the type of PU signal. This new radio must be able to identify all kinds of signals, and a simple approach may be the recognition of their MCSs. This SS technique concerning MCS detection, MCC, has been realized by extracting features of the signal and either classifying it based on these features or applying likelihood-based classifiers [8].

10.2.1 Features

Communication signals contain many characteristics that give us information about their nature. Some of them are extracted in a straightforward way and others in a more complex way. The simplest features related to modulation classification were investigated in [9,10] with successful results, and they are the first to be explained thoroughly here. Now, let us consider that the PU signal sample sensed by the CR receiver is

$$r_{SU}[i] = h_S * s_{PU}[i] + n_{SU}[i] \tag{10.1}$$

where:

$\quad h_S \quad$ is the sensing channel gain
$\quad s_{PU}[i] \quad$ is the transmitted symbol from the PU
$n_{SU} \sim \mathcal{N}(0, N_{SU})$ is the additive white Gaussian noise (AWGN)

If we assume that $A(i)$ is each sample's amplitude and N_s is the number of samples, the first feature that can be derived is the maximum value of the spectral power density of the normalized centered instantaneous amplitude γ_{max}, and it is defined as [9]

$$\gamma_{max} = \frac{\max \left| \text{FFT}\left[A_{cn}(i) \right] \right|^2}{N_s} \tag{10.2}$$

where:

FFT[.] is the fast Fourier transform operator

A_{cn} is the zero centered unitary instantaneous amplitude, which is defined as

$$A_{cn}(i) = \frac{A(i)}{E\{A(i)\}} - 1 \qquad (10.3)$$

This parameter is considered to be efficient for recognizing different amplitude-modulated signals. Additionally, another characteristic of the PU sensed signal is the standard deviation of the absolute value of the nonlinear component of the normalized instantaneous amplitude σ_{aa}, which is expressed as [9]

$$\sigma_{aa} = \sqrt{E\{A_{cn}^2(i)\} - E^2\{|A_{cn}(i)|\}} \qquad (10.4)$$

and is used to distinguish M-ary amplitude shift keying (M-ASK), M-ary quadrature amplitude modulation (MQAM), and other amplitude-modulated signals. Furthermore, a third easy-to-extract PU signal feature is the standard deviation of the absolute value of the nonlinear component of the sample phase in the nonweak signal samples, σ_{ap}, which is calculated as [9]

$$\sigma_{ap} = \sqrt{E\{\phi_{NN}^2(i)\} - E^2\{|\phi_{NN}(i)|\}} \qquad (10.5)$$

where ϕ_{NN} is expressed as

$$\phi_{NN}(i) = \phi_N(i) - E\{\phi_N(i)\} \qquad (10.6)$$

and $\phi_N(i)$ corresponds only to the phase of the nonweak signal samples. These samples are the ones whose amplitude is above a certain amplitude threshold. Very similar to the previous feature, but also nevertheless useful, is the standard deviation of the nonlinear component of the sample phase in the nonweak signal samples σ_{dp}, computed as [9]

$$\sigma_{dp} = \sqrt{E\{\phi_{NN}^2(i)\} - E^2\{\phi_{NN}(i)\}} \qquad (10.7)$$

A fifth practical feature that is useful for M-ary frequency shift keying (M-FSK) signal classification is the standard deviation of the absolute value of the nonlinear component of the sample frequency in the nonweak signal samples, σ_{af} estimated as [9]

$$\sigma_{\alpha f} = \sqrt{E\left\{f_{NN}^2(i)\right\} - E^2\left\{f_{NN}(i)\right\}} \qquad (10.8)$$

where:

f_{NN} is expressed as

$$f_{NN}(i) = \frac{f_N(i) - E\left\{f_N(i)\right\}}{R_s} \qquad (10.9)$$

$\phi_N(i)$ corresponds only to the frequency sample of the nonweak signal samples
R_s is the PU signal symbol rate

Another group of signal characteristics more indicative of the modulation scheme are the cyclostationary (CS) ones. These are derived from signals whose statistical properties vary periodically with time. This sophisticated and very interesting signal processing approach was first introduced by Gardner [11] and is capable of detecting underlying periodicities. Most researchers who focus on modulation recognition consider signals that exhibit cyclostationarity in second-order statistics, such as the autocorrelation function. However, CS signal processing of second-order statistics is not adequate for distinguishing QAM and PSK schemes among themselves, whereas CS processing of higher-order statistics is capable of this. Notable studies that incorporate second-order CS processing to detect modulation schemes are [12–16].

Since signal samples are taken into account, all the exhibited CS functions of the signal $r_{SU}[i]$ are determined in discrete time. The first is the cyclic autocorrelation function (CAF), $R_r^\alpha(l)$. If we denote with $\langle . \rangle$ the time-averaging operation as

$$\langle \rangle = \lim_{N \to \infty} \frac{1}{2N+1} \sum_{m=-N}^{N} (.) \qquad (10.10)$$

then $R_r^\alpha(l)$ is defined as [11]

$$R_r^\alpha(l) = \langle r_{SU}[i] r_{SU}^*[i-l] e^{-j2\pi\alpha i} \rangle e^{-j\pi\alpha l} \qquad (10.11)$$

where r_{SU}^* is the complex conjugate signal. A closer look at Equation 10.11 can help us understand that the CAF measures the correlation of different frequency-shifted versions of a signal. This second-order underlying periodicity becomes clear when $R_r^\alpha(l)$ is different from zero for some nonzero frequency α. The frequency α is called the *cyclic frequency*, and the set of cyclic frequencies α for which $R_r^\alpha(l) \neq 0$ is called the *cyclic spectrum*. Now, to detect the cyclic frequencies more easily, there must be a transfer from the time domain to the frequency domain. For this reason, another useful function in CS signal processing is the Fourier transform of CAF. This is called the *spectral correlation density* (SCD) function and is given by [11]

$$S_r^\alpha(f) = \sum_{l=-\infty}^{\infty} R_r^\alpha(l) e^{-j2\pi fl} \tag{10.12}$$

Furthermore, it is beneficial to measure the degree of local spectral redundancy derived from spectral correlation. A metric to compute this is the spectral coherence function, $C_r^\alpha(f)$, a normalized version of $S_r^\alpha(f)$, which is determined as [11]

$$C_r^\alpha(f) = \frac{S_r^\alpha(f)}{\sqrt{S_r^0(f+\alpha/2)S_r^0(f-\alpha/2)}} \tag{10.13}$$

The final step of the computationally expensive CS processing is acquiring the α-domain profile or cyclic domain profile (CDP). This is calculated as [11]

$$I(\alpha) = \max_f |C_r^\alpha(f)| \tag{10.14}$$

and exhibits the peak values of the spectral coherence function, which are more convenient to handle for modulation classification. The use of cyclic spectral analysis has been very efficient in signal classification, especially when combined with robust detection tools. In [13,14], the set of binary phase-shift keying (BPSK), quadrature phase-shift keying (QPSK), frequency shift keying (FSK), minimum shift keying (MSK), and amplitude modulation (AM) signals was categorized with a probability of detection $P_{DE} = 100\%$ at a signal-to-noise ratio (SNR) $= -7$ dB.

Other features exploited for modulation classification are the sensed signal cumulants, which have distinctive theoretical values among different modulation schemes, and even though they demand a large number of samples, they are easy to calculate. The second-, fourth-, sixth-, and eighth-order mixed cumulants of r_{PU}, $C_{2,0}^r$, $C_{2,1}^r$, $C_{4,0}^r$, $C_{4,1}^r$, $C_{4,2}^r$, $C_{6,0}^r$, $C_{6,1}^r$, $C_{6,2}^r$, $C_{6,3}^r$, $C_{8,0}^r$, $C_{8,1}^r$, $C_{8,2}^r$, $C_{8,3}^r$, and $C_{8,4}^r$ have been used successfully by the research community [12–18] and have delivered results with a high probability of detection.

Cumulants are best expressed in terms of raw moments. A generic formula for the joint cumulants of several random variables $X_1, ..., X_n$ is

$$C_{X_1,...,X_n} = \sum_{\pi} (|\pi|-1)!(-1)^{|\pi|-1} \prod_{B\in\pi} E\left\{\prod_{i\in\pi} X_i\right\} \tag{10.15}$$

where:

π runs through the list of all partitions of 1,..., n

B runs through the list of all blocks of the partition π

$|\pi|$ is the number of parts in the partition

For example,

$$C_{X_1,X_2,X_3} = E\{X_1X_2X_3\} - E\{X_1X_2\}E\{X_3\}$$
$$-E\{X_1X_3\}E\{X_2\} - E\{X_2X_3\}E\{X_1\}$$
$$+2E\{X_1\}E\{X_2\}E\{X_3\} \tag{10.16}$$

Consequently, the p-order mixed cumulant $C_{p,q}^r$ of the complex received signal can be derived from the joint cumulant formula in Equation 10.15 as

$$C_{p,q}^r = C_{\underbrace{r,...,r}_{(p-q)\ \text{times}}\ ,\ \underbrace{r^*,...,r^*}_{(q)\ \text{times}}} \tag{10.17}$$

Because of the symmetry of the considered signal constellations, pth-order mixed cumulants for p odd are equal to zero, and also it can be easily proven that for p even, $C_{p,q}^r = C_{p,p-q}^r$.

10.2.2 Classifiers

As far as the classification methods used in the literature are concerned, there are two general categories of classifiers. The first is feature based, in which learning machines collect some of the features described in Section 10.2.1, process them, and classify a group of signal samples as noise or any other kind of signal. The second is the likelihood-based classifier group, which distinguishes the different signal classes by comparing likelihood ratios.

In the feature-based classifier literature, there has been significant progress using ML to identify correctly the modulation scheme of the signal. These learning algorithms are divided into two categories: supervised and unsupervised classifiers. The first requires training with labeled data, while the second does not. Assuming that one of the feature-extraction procedures mentioned in Section 10.2.1 is chosen by the CR to process the sensed signal samples, and the feature vectors are $x_i \in X$, then the supervised methods also need the corresponding label $y_i \in Y$, in our case the modulation scheme, to be trained, whereas the unsupervised methods can learn without y_i by clustering similar feature vectors.

A popular supervised classifying option is Artificial Neural Networks (ANNs). ANNs have been successfully used for modulation detection, and they were one of the early applications of ML in this field [9,14]. They are biologically inspired computational machines that imitate the function of a set of interconnected neurons, the basic processing units of the brain. They have been proven able to store experiential knowledge through a learning process, and thus, they have become a powerful classification technique. In the ANN model, each of these neurons has a transformation function, which it performs on the weighted sum of its inputs to produce an output. A number of these units can create a sequence of neuron layers, the first of which is the input layer and the last is the output layer. Between these two layers, there are the intermediate or hidden layers, which are not directly connected to the outside world. The kinds of ANNs used in this research area are the Multilayer Feedforward Neural Networks [9] and the Multilayer Linear Perceptron Network [14], which were able to identify signals of 2-ASK, 4-ASK, 2-FSK, BPSK, QPSK, and AM.

Another powerful supervised classification tool used in modulation classification literature [10,16] is SVMs. Their mathematical foundation is statistical learning theory, and they were developed by Vapnik [4]. A major drawback of SVMs is that initially they require many computations to train themselves offline, but they can become very accurate. SVMs operate by finding a hyperplane in a high-dimensional space that divides the training samples into two classes. This hyperplane is chosen so that the distance from it to the nearest data points on each side is maximized, as shown in Figure 10.1. This is called the maximum-margin hyperplane. But the most interesting contribution of the SVMs is in the nonlinear separation of data. This machinery, with some small adaptations and using the so-called kernel trick, can be used to map input feature vectors indirectly into a high-dimensional space in which they become linearly separable [4]. The impressive

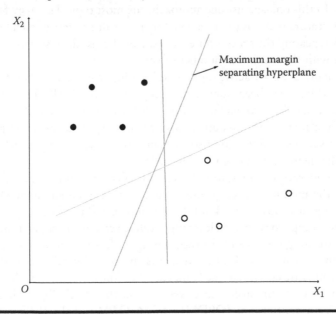

Figure 10.1   Maximum margin separating hyperplane.

part of this high-dimensional approach is that it happens without any extra computational effort. The reason why this nonlinear mapping Θ does not add any extra computational burden lies in the way the SVM operates. For a simple linear separation in the initial feature space, the SVM training has to solve a quadratic programming problem, which considers only the dot products of the training feature vectors. Extending this idea to a higher-dimension space in which the feature vector "images" are linearly separable, the SVM again needs to know only the dot products of the dimensional expansions of the training feature vectors. This enables us to overcome the obstacle of not knowing this nonlinear mapping Θ and just calculate the dot products of the training feature vector mappings. This is the point where the kernel trick is used. Given two vectors from the training feature space x_i and x_j, the dot product of their mappings in some high-dimensional feature space is

$$K(x_i, x_j) = \Theta(x_i) \cdot \Theta(x_j) \qquad (10.18)$$

where $K(x_i, x_j)$ denotes the kernel function. In most classification applications, the polynomial function kernel and the Gaussian radial basis function (GRBF) kernel are used. In previous work, the most commonly used kernel is the GRBF, which is actually a polynomial kernel of infinite degree. Originally, the SVM was a binary classifier, but it can also be used for multiclass classification into one of the available classes, here modulation schemes, if we consider a combination of binary classifiers to find which class the feature vector most likely belongs to compared with every other class. In this one-against-one approach, the most typical strategy for labeling a test signal feature vector is to cast a vote for the resulting class of each binary classifier. After repeating the process for every pair of classes, the test signal is assigned to the class with the maximum number of votes.

Additionally, a less attractive supervised learning machine, the Hidden Markov Models (HMMs), has been applied to recognize AM, BPSK, FSK, MSK, and QPSK sensed signals in combination with CS features [13]. Even though HMMs are powerful classifiers, they require a huge memory space to store a large number of past observations, and they are also computationally very complex, which makes them unsuitable for embedding in a CR device.

The second category of feature-based classifiers, the unsupervised category, is tackling the modulation recognition problem in a more autonomous manner. Since learning methods of this kind do not need labeling of the training feature vectors, CRs equipped with such learning modules can detect signal types without someone indicating the class of the training signal features to the CR. One of the initial attempts at using such mechanisms is the work introduced in [19], where three algorithms are employed, the K-means, the X-means, and the self-organizing maps (SOMs), to distinguish 8-level vestigial sideband (8VSB), orthogonal frequency-division multiplexing (OFDM), and 16-QAM signals. The first technique is clustering the observed feature vectors into a certain given number of classes so as to minimize the sum of the squared distances of all samples from their class centroids. The second is a variation of the K-means method capable of training without any knowledge of the class number. The SOMs are a special kind of ANN that represent training samples in a low-dimensional space in which the ANN's neurons organize themselves through a neuron weight updating process.

Furthermore, nonparametric unsupervised learning procedures have been used to classify signals, with notable results. The authors of [15,20] recommended the Dirichlet Process Mixture Model (DPMM)-based classifier and used the Gibbs sampling to sample from the posterior distribution and to update the DPMM hyperparameters. Also, they proposed a simplified and a sequential DPMM classifier to reduce the computationally demanding DPMM classifier by exploiting the Chinese Restaurant Process property of the Dirichlet process and improve the selection strategy of the Gibbs sampler. Applying this classifier has been successful for distinguishing Wi-Fi and Bluetooth signals with great accuracy [20].

As far as the likelihood-based classifiers are concerned, contributions have been made in both modulation and coding classification. An overview of these techniques concerning the modulation scheme detection can be found in [8], where three variations of this likelihood approach are demonstrated: the average likelihood ratio test (ALRT), the generalized likelihood ratio test (GLRT), and the hybrid likelihood ratio test (HLRT). Nevertheless, because the modulation scheme classification has already been extensively presented, the coding recognition will be further discussed in this section. The main reason why coding scheme identification is classed in the likelihood-based classifier category and not the feature-based one is because extracting features from coded bit sequences is, in general, dependent on the type of coding, and even though noteworthy work has been done in this direction, the likelihood approach is still the most popular one. All previous researchers in this area have taken advantage of the log-likelihood ratios (LLR) to identify codes such as space-time block codes (STBCs) [21], low-density parity-check (LDPC) code rates, and other coding schemes with parity-check relations [22,23].

Of particular interest, mostly because of their practicality, are the classification methods based on the unique parity-check matrix of each code [22,23]. Assuming that a CR intends to recognize the encoder of a PU transmitter and has a priori information of all the possible PU encoders, a reasonable piece of knowledge about the PU system, then the CR must detect the most likely encoder being used. Each candidate encoder θ' has an exclusive parity-check matrix $\mathbf{H}_{\theta'} \in \mathbb{Z}_2^{N_{\theta'} \times N_c}$, where $N_{\theta'}$ is the number of parity-check relations of the candidate encoder and N_c is the length of the codeword produced by the encoder θ'. Given a codeword $\mathbf{c}_\theta \in \mathbb{Z}_2^{N_c \times 1}$ from encoder θ, in a noiseless environment,

$$\mathbf{H}_{\theta'}\mathbf{c}_\theta = \mathbf{0} \tag{10.19}$$

holds over the Galois field $\mathbb{GF}(2)$ if and only if $\theta' = \theta$. Due to noise in the codeword, though, some errors occur in Equation 10.19 even when the correct encoder θ is chosen. These errors are called *code syndromes*, e^k, and for a candidate encoder θ' in vector form, they are defined as

$$\mathbf{e}_{\theta'} = \mathbf{H}_{\theta'}\mathbf{c}_{\theta} \qquad (10.20)$$

where $\mathbf{e}_{\theta'} \in \mathbb{Z}_2^{N_{\theta'} \times 1}$, and every line represents a parity-check relation.

To use the code syndromes $\mathbf{e}_{\theta'}$ for code identification, one needs to calculate the LLR of each bit of the codeword $\mathbf{c}_{\theta}$, which after some processes in the log-likelihood domain is obtained as

$$\mathrm{LLR}\left(c[m] \mid r_{\mathrm{SU}}[n]\right) = \mathrm{LLR}\left(r_{\mathrm{SU}}[n] \mid c[m]\right) \qquad (10.21)$$

where:

$c[m]$ is the considered bit
$r_{\mathrm{SU}}[n]$ is the corresponding received symbol sample

This is the result of the log-likelihood soft decision demodulation. Subsequently, if $e_{\theta'}^k$ is the syndrome derived from the kth parity-check relation of the candidate encoder θ'

$$e_{\theta'}^k = c[k_1] \oplus c[k_2] \oplus \cdots \oplus c[k_{N_k}] \qquad (10.22)$$

where N_k is the number of codeword bits taking part in the XOR operations of the parity-check relation, then the LLR of $e_{\theta'}^k$ is given by

$$\mathrm{LLR}\left(e_{\theta'}^k\right) = 2\tanh^{-1}\left(\frac{\prod\limits_{q=1}^{N_k} \tanh\left(\mathrm{LLR}\left(c[k_q]\right)\right)}{2}\right) \qquad (10.23)$$

a log-likelihood property of the XOR operation shown in [24]. Based on these LLRs of the code syndromes, two different approaches have been proposed to identify the right encoder. The first suggests a GLRT test that assumes a priori information on the distributions of the GLRT nuisance parameters [23], and the second proposes as a soft decision metric the average LLR of the code syndromes. This is calculated as

$$\Gamma_{\theta'} = \frac{\sum\limits_{k=1}^{N_{\theta'}} \mathrm{LLR}(e_{\theta'}^k)}{N_{\theta'}} \qquad (10.24)$$

Once the average syndrome LLRs of all the candidate encoders are calculated, the estimated encoder can be identified as

$$\hat{\theta} = \arg\max_{\theta' \in \Theta} \Gamma_{\theta'} \qquad (10.25)$$

where Θ is the set of the encoder candidates.

10.2.3 Modulation and Coding Classification

In this section, a combination of the most effective techniques mentioned in the chapter is proposed to be capable of performing MCC. The idea is to equip an SU with this MCC module to detect the MCS of the sensed PU signal. The methods chosen from Section 10.2.2 to execute the MCC are a feature-based classifier for the modulation scheme detection and a likelihood-based classifier for the coding detection [3]. The former estimates the signal cumulants and feeds them into a highly sophisticated supervised classifier, the SVM, which identifies modulation schemes with better accuracy than other ML methods, and the latter uses the average syndrome LLRs to detect the PU encoder. The presented MCC module is tested in the SNR range of [−11, 14] for the sensed signal of a PU that uses MCSs of QPSK 1/2, QPSK 3/4, 16-QAM 1/2, 16-QAM 3/4, 64-QAM 2/3, 64-QAM 3/4, and 64-QAM 5/6 with LDPC coding. Also, the number of symbol samples considered to be sensed in the simulations is $Ns = 64,800$ to obtain all the necessary statistical features employed by the aforementioned techniques. The metric used to measure the detection performance of the MCC method for a class j is the probability of correct classification (P_{cc}). In Figure 10.2, the P_{cc} of the simulations is shown. Initially, an obvious remark is that the higher the SNR of the test signal, the higher the P_{cc}. Furthermore, one may note that the lower the order of the constellation or the code rate to be classified, the easier it is to recognize it. Another conclusion is that the P_{cc} curves are very steep, mostly due to the performance of the coding classifier.

10.3 Power Control in Cognitive Radio Networks

The PC subject in CR communications is a vast research area, which has attracted much attention over the past years. Previous work in the field has considered a great variety of assumptions, protocols, system models, optimization variables, objective functions, constraints, and other known or unknown parameters. The general form of the PC scenarios is the optimization of an SU system metric, such as total throughput, worst user throughput, or SINR of every SU, subject to QoS constraints for PUs, such as SINR, data rate, or outage probability. Moreover, the research community has combined these PC problems with beam-forming

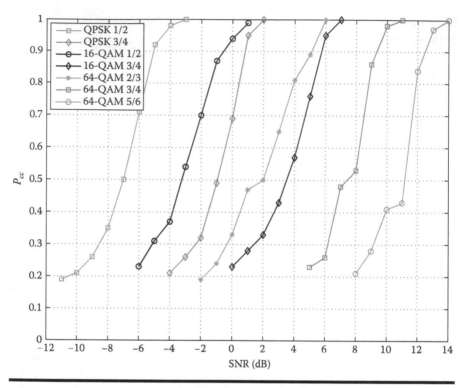

Figure 10.2 P_{cc} vs. SNR for $Ns = 64{,}800$ symbol samples.

patterns, base station assignment, bandwidth or channel allocation, and time schedules, leading to more complicated compound problems, but with the same basic form.

The main distinction between all the investigated schemes is based on the manner in which the PC policy is calculated and hence, the CRN structure. If the CR transmit power allocation is calculated by a central decision-maker, the CBS, the PC solution is characterized as centralized; if it is computed by each CR on its own, with or without any cooperation with the other CRs or the PU system, it is characterized as distributed. This means that the optimization problem solution is tackled in the first case with full knowledge and control over the variables of the problem (the CR transmit power levels), and in the second case, with partial or no knowledge or control of them. In the centralized schemes, all the scenarios are formulated as an optimization problem dealt with using appropriate techniques depending on the nature of the scheme. In the distributed scenarios, the optimization problem is decomposed into a set of coupled optimization problems and tackled under a game theoretic framework or by using a distributed optimization algorithm. In Sections 10.3.1 through 10.3.3, we will demonstrate some of the most notable research work produced in these categories.

10.3.1 Distributed Techniques Based on Game Theory

In Game Theory (GT), the decomposition of the overall optimization problem delivers, as mentioned in Section 10.3, a set of coupled optimization problems, which deal with the maximization of each decision-maker's utility function. The concept of the utility function is very common in games, and it reflects the satisfaction level of each player, or in the CR case, of the SU. Usually, in communication games, this function is related to an SU QoS metric, such as the throughput or the SINR. Another concept in GT is the Nash Equilibrium (NE), which concerns the solution of the GT problem. At an NE, given the power levels of other users, no player can improve its utility level by making individual changes in its power.

One of the earliest GT approaches in wireless communications, which is often referred to by recent researchers, is the distributed code-division multiple-access (CDMA) uplink power control problem, formulated as a noncooperative PC game based on pricing to obtain a socially better power allocation solution [25]. Practically, social welfare means the sum of the utility functions, another interpretation of the overall optimization problem objective. The proposed utility function is the efficiency function defined as the user rate per transmitted energy unit. Later on, many other contributions were made under the general framework of GT in wireless communications, and this encouraged researchers to apply it in the CR regime. One of these attempts was the work presented in [26], which achieves efficient and fair power allocation policies using an iterative punishment scheme in a repeated game. Furthermore, repeated Stackelberg game formulations were used to describe energy-efficient distributed PC problems [27], and bargaining solutions were developed to obtain a CRN spectrum efficiency increase and compliance with Nash theorem axioms subject to PU interference and SU minimum SINR constraints [28].

A very interesting subgroup of this category is the PC game that considers spectrum market models in which SUs reward the PU system for spectrum usage and pursue PU revenue maximization. Initially, a competitive pricing scheme was developed in which PUs, modeled as an oligopoly, offered spectrum access opportunities to SUs subject to PU QoS constraints [29]. The problem modeling was based on the Bertrand game, and additionally a punishment mechanism was implemented to impose collusion among PUs. Furthermore, PU monopoly markets were investigated using a suboptimal pricing scheme in [30], and a dynamic spectrum leasing approach in which the PU dynamically manages the acceptable PU interference threshold according to the obtained reward and SUs simultaneously aim to energy-efficient transmissions was outlined in [31].

The last subcategory of this section incorporates learning solutions in PC games. A general learning framework for stochastic games, which are repeated games with probabilistic action transitions, was given in [32], where the authors identified three possible learning options: myopic adaptation, reinforcement learning, and action-based learning. Further research in the latter two categories has been conducted using a Bush–Mosteller reinforcement learning procedure [33] and a no-regret learning technique [34].

10.3.2 Other Distributed Techniques

In this category, we include all the distributed solutions proposed in the literature that are not based on GT. The development of distributed solutions outside the GT framework is usually substantiated using distributed algorithms of proven convergence, decomposition methods, and the mathematical theory of variational inequalities (VI), or is simply based on simulation results. The first category includes the earliest contributions in distributed wireless communication scenarios. The authors of [35] proposed a fixed-step PC algorithm using 1-bit control commands with proven stability and convergence properties for power consumption minimization under QoS constraints. Other previous work in this field included the simultaneous iterative water-filling algorithm (SIWFA) [36] and the asynchronous iterative water-filling algorithm (AIWFA) [37], which belong, respectively, to the algorithmic classes of general Jacobi strategies and totally asynchronous schemes and solve the sum-rate maximization problem subject to SU power and PU interference constraints.

In the class of decomposition method–based distributed PC solutions, the optimization objective is the maximization of the CRN throughput. All problems in this category are converted into geometric programming problems and solved in a distributed way by decomposing them to simpler maximization problems for each SU power variable. The scenarios solved in this case differ only in the constraints. In [38], minimum and maximum SINR constraints for PUs and SUs are introduced; in [39], a PU outage probability constraint is handled; and [40] handles a robust form of the optimization problem by maintaining the PU interference below a given threshold.

Furthermore, the mathematical theory of VI is referred to as a promising general framework for developing distributed algorithms in wireless network applications [41]. A noteworthy VI application in PC for CR is the work presented in [40], where a robust AIWFA is suggested. In the final group, a 1-bit feedback gradient-based PC is established to address a satellite uplink and terrestrial fixed-service coexistence scenario in [42], and a model-free reinforcement learning technique is recommended in [43], known as Q-learning, to learn the interference channel gains and manage the PU interference. Their effectiveness is shown through numerical simulations.

10.3.3 Centralized Techniques

This group of PC centralized solutions contains some of the earlier work in CR, since the centralized approaches are usually the first to be investigated in any optimization problem. The authors of [44] provide us with a unified algorithmic framework for different types of water-filling solutions, which correspond to several constrained optimization problems in wireless communications. These complicated solutions include multiple water levels and multiple constraints, and are often applicable to many CR scenarios. The next research work presented in this section is a joint beamforming and power allocation problem, which tackles cases of multiple PUs and SUs with PU interference and SU peak power constraints [45]. Specifically, the SU sum-rate maximization and the SU SINR balancing problems are addressed, and solutions are given by decoupling the original optimization problems.

Robust formulations of centralized PC problems have also been formulated and discussed by the research community. In [46], a network utility maximization problem was tackled subject to PU interference constraints satisfied within certain limits and based on the statistics of the SU-to-PU channels. Additionally, a joint rate and power allocation scenario was considered, in which the optimization objective is a data rate proportional fair solution with outage and interference constraints met within specific limits [47].

Another joint rate and power allocation case was studied in [48], aided by admission control under high network load conditions. The targets are proportional and max–min fairness subject to PU interference and SU QoS constraints. The last work presented here is a stochastic approach of resource management optimization using CSI limited-rate feedback [49]. The authors' main contribution is maximizing a generic function of user average rates by adapting modulation, coding, and power modes and subject to PU and SU rate and power constraints when channel statistics are unknown. To accomplish this, a stochastic dual algorithm was implemented to learn channel statistics online and converge in probability to the optimal solution with known channel statistics.

10.4 Power Control Using Modulation and Coding Classification

Besides the centralized and decentralized categorization described in Sections 10.3.1 through 10.3.3, all the aforementioned PC techniques can also be distinguished based on another very fundamental criterion: knowledge of the parameters of the optimization problems formed in every scenario. This piece of information is assumed known, following certain statistics, or totally unknown. Although the first presumption helped to devise sophisticated optimization problems, it is not applicable in most cases. In the second hypothesis, certain statistical parameters are also required, and obtaining them is a further issue. The most realistic assumption is to take into account no prior knowledge of these parameters. This premises that a learning mechanism is implemented by a central decision-maker or each SU individually. A necessary condition for the learning process is the availability of feedback that can be acquired by the PU, the CBS, or each SU.

In the techniques described in Sections 10.3.1 through 10.3.3 that incorporate a learning process, depending on the type of cooperation between the CRN and the PU system, this feedback can be a price value or a punishment factor propagated by the PU or the CBS; a constraint violation indicator, such as the one used in water-filling processes; or a piece of information obtained by the CRN itself. As mentioned before, this information concerns the optimization problem parameters, and especially those describing the connection of the CRN operation with the PU system operation, such as the PU interference threshold or the interference channel gains. A very common feedback that is useful for learning optimal PC policies is the binary ACK/NACK packet captured from the PU feedback channel. This packet is an implicit CSI feedback of the PU link and can be obtained by eavesdropping on the PU feedback message and decoding it.

This idea, though, involves difficulties and obstacles that make it impractical. The first disadvantage is that decoding the PU feedback message requires the full implementation of a PU system receiver on the CRN side and raises a CR hardware complexity issue. Second, to successfully decode this PU message, the CR or the CRN must sense the PU signal through an adequately high SINR sensing channel, which cannot be guaranteed. Finally, even if these conditions are fulfilled, there is still a security complication, since decoding any message from a PU system can be considered malicious and harmful. Another, less complicated idea proposed in [3] is to use the MCC process described in Section 10.2.3 so that the can detect the PU MCS, an implicit CSI indicator for the PU link. The MCC tackles all three obstacles described before, since it is less complex to implement on the hardware level and more robust in low-SINR conditions and does not pose any security issue from the actual decoding of the PU message. In Section 10.4.3, the use of this SS technique, the MCC, in PC scenarios will be presented.

10.4.1 State of the Art

The main issue about the PC methods described in Sections 10.3.1 through 10.3.3 that made use of a feedback to reach the optimal PC policies of the defined problems is that they depend on parameter-updating mechanisms that are very slow and passively receive every feedback without introducing any intelligent probing process to the PU. An interesting idea was proposed in [50], called *proactive SS*, in which the SU strategically probes the PU and senses its effect from the PU power fluctuation. Also, the exploitation of the MCC feedback, used in our work, is suggested briefly by the authors of [50] in a footnote. Primarily, though, the most common piece of information being used to estimate optimization problem parameters, and specifically constraint parameters such as the interference channel gains, is the binary feedback, the ACK/NACK packet.

To address the issue of intelligent probing, which makes the learning process faster and more efficient, in this section we will describe rapid state-of-the-art learning methods applied in wireless communications. In addition, a practical approach for the CRN would be for the SUs to be coordinated by a CBS using a dedicated control channel, which signifies a centralized PC scheme [51]. Specifically, a central decision-maker, the CBS, must learn the interference channel gains, elaborate an intelligent selection of the SU transmit power levels, and communicate it to the SUs. Notably, the fastest and most sophisticated methods suitable for the CBS to learn the interference channel gains of multiple SUs with the use of feedback come from multiple antenna underlay cognitive scenarios. At this point, we need to explain how channel learning in beam-forming problems can easily be translated into channel learning in centralized PC problems. If we assume that each of the multiple antennas corresponds to an SU in a CRN, then coordinating the beam-forming vectors to estimate the CR-to-PU channel gains is similar to a CBS coordinating the transmit powers of a CRN for the same purpose. In fact, designing the transmit powers is actually much simpler than composing each antenna's complex coefficient in the beam-forming scenarios, since in PC, no phase parameters are incorporated.

Previous researchers in this field have exploited slow stochastic approximation algorithms [52,53], the one-bit null space learning algorithm (OBNSLA), and an ACCPM-based learning algorithm [54,55]. The last two approaches were introduced as channel correlation matrix learning methods, with the ACCPM-based technique outperforming the OBNSLA. All these learning techniques are based on a simple iterative scheme of probing the PU system and obtaining feedback indicating how the PU operation is changed. Another thing that this work has in common is the discrimination of the channel learning phase and the transmission phase that is optimum to an objective, such as the maximum total throughput or maximum SINR transmission. Thus, the optimization objective is achieved only after the learning process is terminated. Nonetheless, the ideal would be to tackle them jointly and learn the interference channel gains while at the same time pursuing the optimization objective without this affecting the learning convergence time. On this rationale, the authors of [56] proposed an ACCPM-based learning algorithm in which probing the PU system targets both learning channel correlation matrices and maximizing the SNR at the SU receiver side.

10.4.2 System Model

In this case study, we exploit the aforementioned idea in the underlay and centralized PC problem by using the MCC sensing feedback instead of the binary ACK/NACK packet captured from the PU feedback channel. For this problem formulation, learning the interference channel gains from each SU to one PU receiver is performed concurrently with maximizing the total SU throughput under an interference constraint that depends on these channel gains. Additionally, this method is discussed, enhancements are introduced, and its results are compared with a benchmark learning technique [57].

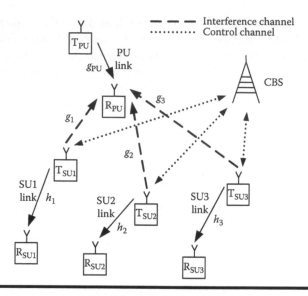

Figure 10.3 Centralized CRN interfering in the PU link.

Now, let us consider a PU link and N SU links existing in the same frequency band, as shown in Figure 10.3. Furthermore, SUs do not interfere with each other using a multiple access method. As far as the interference to the PU link is concerned, this is caused by the transmitter part of each SU link to the receiver of the PU link. The examined scenario considers the PU channel gain and the unknown interference channel gains to be static. Here, we focus on channel power gains g, which in general are defined as $g = \|c\|^2$, where c is the channel gain. From this point, we will refer to channel power gains as channel gains. Taking into account that the SU links transmit solely in the PU frequency band, the aggregated interference on the PU side is defined as

$$I_{PU} = \mathbf{g}^{\mathsf{T}} \mathbf{p} \tag{10.26}$$

where:

 $\mathbf{g}$ is the interference channel gain vector $[g_1,..., g_N]$ with g_i being the SU$_i$-to-PU interference channel gain
 $\mathbf{p}$ is the SU power vector $[p_1,..., p_N]$ with p_i being the SU$_i$ transmit power

Additionally, the SINR of the PU is defined as

$$SINR_{PU} = 10 \log\left(\frac{g_{PU} p_{PU}}{I_{PU} + N_{PU}}\right) dB \tag{10.27}$$

where:

g_{PU} is the PU link channel gain
p_{PU} is the PU transmit power
N_{PU} is the PU receiver noise power

The problem being addressed is that of total SU throughput $\left(U_{SU}^{tot}\right)$ maximization without causing harmful interference to the PU system, which can be written as

$$\underset{\mathbf{p}}{\text{maximize}} \quad U_{SU}^{tot}(\mathbf{p}) = \sum_{i=1}^{N} \log\left(1 + \frac{h_i\, p_i}{N_i}\right) \qquad (10.28\text{a})$$

$$\text{subject to} \quad \mathbf{g}^{\mathsf{T}}\mathbf{p} \le I_{th} \qquad (10.28\text{b})$$

$$0 \le \mathbf{p} \le \mathbf{p}_{max} \qquad (10.28\text{c})$$

where:

$\mathbf{p}_{max} = \left[p_{max_1}, \dots, p_{max_N}\right]$ with p_{max_i} being the maximum transmit power level of the SU$_i$ transmitter
h_i is the channel gain of the SU$_i$ link
N_i is the noise power level of the SU$_i$ receiver

The channel gain parameters h_i and the noise power levels N_i are considered to be known to the CR network and not changing in time. An observation necessary for tackling this problem is that the g_i gains normalized to I_{th} are adequate for defining the interference constraint. Therefore, the new version of Equation 10.28b will be

$$\tilde{\mathbf{g}}^{\mathsf{T}}\mathbf{p} \le 1 \qquad (10.29)$$

where $\tilde{\mathbf{g}} = \mathbf{g}/I_{th}$.

This optimization problem is convex, and using the Karush–Kuhn–Tucker (KKT) approach, a capped multilevel water-filling (CMP) solution is obtained for each SU$_i$ of the closed form [45]:

$$p_i^* = \begin{cases} p_{max_i} & \text{if } \dfrac{1}{\lambda \tilde{g}_i} - \dfrac{N_i}{h_i} \ge p_{max_i} \\[2ex] 0 & \text{if } \dfrac{1}{\lambda \tilde{g}_i} - \dfrac{N_i}{h_i} \le 0 \qquad i = 1,\dots,N \\[2ex] \dfrac{1}{\lambda \tilde{g}_i} - \dfrac{N_i}{h_i} & \text{otherwise} \end{cases} \qquad (10.30)$$

where λ is the KKT multiplier of the interference constraint (Equation 10.29), which can be determined as presented in [45].

10.4.3 Modulation and Coding Classification Feedback

Even though this problem setting is well known and already investigated in Section 10.4.4, we will demonstrate how to tackle it without knowing the interference constraint (Equation 10.29). But first, let us examine the enabler of the interference constraint learning, which is defined by the unknown $\tilde{g}_i$ parameters. Initially, the outputs of the MCC procedure have to be defined. Taking into consideration the centralized structure of our PC scenario and the existence of a cognitive control channel, in our previous work [57], a cooperative MCC method is described in which all the SUs are equipped with a secondary omnidirectional antenna only for sensing the PU signal and an MCC module that enables them to identify the MCS of the PU. Specifically, each SU collects PU signal samples, estimates the current MCS, and forwards it through the control channel to the CBS, and finally, the CBS, using a hard decision fusion rule, combines all this information to get to a decision based on a plurality voting system. After casting every vote, the CBS identifies the PU MCS.

Strong interference links may have a severe effect on the MCS chosen by the PU link, which changes to more robust modulation constellations and coding rates depending on the level of the $SINR_{PU}$. Let $\{MCS_1,..,MCS_J\}$ denote the set of the MCS candidates of the ACM protocol and $\{\gamma_1,...,\gamma_J\}$ the corresponding minimum required $SINR_{PU}$ values, at any violation of which the PU adjusts its MCS. Furthermore, consider these sets arranged such that γs appear in ascending order. Assuming that N_{PU} and the received power remain the same at the PU receiver side, the $\{\gamma_1,...,\gamma_J\}$ values correspond to particular maximum allowed I_{PU} values, designated as $\left\{ I_{th_1}, I_{th_J} \right\}$. Hence, whenever the PU is active, for every MCS_j, there is an interference threshold, I_{th_j}, above which the PU is obliged to change its transmission scheme to a lower-order modulation constellation or a lower code rate, and whose level is unknown to the CRN. Here, it has to be pointed out that it is reasonable to assume that the CRN has some basic information about the legacy PU system whose frequency band it is attempting to exploit, and therefore, it is reasonable to assume that the CRN has a priori knowledge of the PU system's ACM protocol and of its γ_j values.

Now, we shall explain how to take advantage of the MCC feedback to transform the MCS deterioration into a multilevel piece of information instead of a binary one. Taking as reference the PU MCS when the SU system is not transmitting at all, $MCS_{ref} = MCS_k$, and the corresponding $\gamma_{ref} = \gamma_k$, where $k \in \{1,..., J\}$, the following γ ratios can be defined:

$$c_j = \frac{\gamma_j}{\gamma_{ref}} \tag{10.31}$$

where $j \neq k$ and $j \in \{1,\dots,J\}$. Supposing a high SNR_{PU} regime, $g_{\text{PU}} p_{\text{PU}} \gg N_{\text{PU}}$, the I_{th_j} ratios can also be determined as

$$\frac{I_{\text{th}_j}}{I_{\text{th}_{ref}}} = \frac{\gamma_{\text{ref}}}{\gamma_j} = \frac{1}{c_j} \qquad (10.32)$$

where $I_{\text{th}_{ref}}$ is the interference threshold of MCS_{ref}.

Knowing these ratios is very important when any probing procedure is applied to the PU system to estimate the induced interference. In this case study, it is necessary to consider not only that the aggregated interference defined by Equation 10.26 exceeds $I_{\text{th}_{ref}}$, but also approximately how close to this limit it is. Therefore, with the acquisition of an MCC feedback, the SU system, and specifically the CBS, can compare this with the initial MCS, where no interference is caused, and derive two inequalities depending on the MCS deterioration. Let MCS_{ref} be the sensed MCS when the CRN is silent and no probing occurs, $\mathbf{p} = \mathbf{0}$, and MCS_j be the deteriorated MCS after the SU system has probed the PU using an arbitrary SU power vector $\mathbf{p}$. The information gained by the CBS is that

$$I_{\text{th}_{j+1}} < \mathbf{g}^{\mathsf{T}} \mathbf{p} \leq I_{\text{th}_j} \qquad (10.33)$$

These inequalities can be rewritten using the I_{th} ratios as

$$\frac{I_{\text{th}_{ref}}}{c_{j+1}} < \mathbf{g}^{\mathsf{T}} \mathbf{p} \leq \frac{I_{\text{th}_{ref}}}{c_j} \Leftrightarrow \frac{1}{c_{j+1}} < \tilde{\mathbf{g}}^{\mathsf{T}} \mathbf{p} \leq \frac{1}{c_j} \qquad (10.34)$$

where $\mathbf{g}$ is normalized as in Equation 10.29, with $I_{\text{th}} = I_{\text{th}_{ref}}$ as $\tilde{\mathbf{g}} = \mathbf{g} / I_{\text{th}_{ref}}$.

Thus, the MCC feedback allows us to detect more accurately where the probing SU power vector lies within the feasible region without uselessly searching the power vector feasible region. The inequalities in Equation 10.34 can also be formulated in a normalized version:

$$\tilde{\mathbf{g}}^{\mathsf{T}} \tilde{\mathbf{p}}_{\mathbf{u}} > 1$$

$$\tilde{\mathbf{g}}^{\mathsf{T}} \tilde{\mathbf{p}}_{\mathbf{l}} \leq 1 \qquad (10.35)$$

where $\tilde{\mathbf{p}}_{\mathbf{l}} = c_{j+1} \mathbf{p}$ and $\tilde{\mathbf{p}}_{\mathbf{u}} = c_j \mathbf{p}$.

10.4.4 A Novel Algorithm for Simultaneous Power Control and Interference Channel Learning

The advantage of using the multilevel MCC feedback instead of a simple binary indicator, such as the ACK/NACK packet of the PU link, will be employed by a CPM-based learning technique to estimate the unknown interference channel gain vector, $\tilde{\mathbf{g}}$, and concurrently achieve the optimization objective defined by Equation 10.30 using the SU system probing power vectors as training samples. In this probing procedure, the SU system may intelligently choose the training samples to learn and not just receive them from a teaching process. In ML, this kind of learning is called *Active Learning*: the learner chooses training samples that are more informative so that it can reach the learning solution faster, with fewer training samples and less processing. The learning speed, and thus the number of probing power vectors, is an essential part of the suggested idea, because the SU system must learn the interference constraint quickly so that it will not interfere with the PU for a long time and reduce the PU QoS. Effective Active Learning methods for this task are the CPMs, newly introduced to this field, which have attractive convergence properties because of their geometric characteristics.

Besides the fast convergence, the main advantage of CPMs is that the training sample, $\mathbf{p}$ in this case, can be chosen based on any rationale without affecting the decrease of the uncertainty region in the parameter $\tilde{\mathbf{g}}$ space. In our problem, this rationale can be the solution of the optimization problem defined in Equation 10.30. Hence, approaching the actual values of the parameter vector $\tilde{\mathbf{g}}$ can happen in parallel with maximizing the SU system throughout, the optimization objective. More specifically, at each learning step, the CPM only dictates the center of the $\tilde{\mathbf{g}}$ uncertainty set, an estimation of $\tilde{\mathbf{g}}$, and the hyperplane/cutting plane passing through this center, which is actually determined by $\mathbf{p}$, can be the solution of Equation 10.28. Since the chosen cutting plane passes through it, the SU system power allocation vector is considered to satisfy the equality of the interference constraint estimated so far.

Here, we examine the CGCPM and the ACCPM and their corresponding centers, the center of gravity and the analytic center, as being the two types of center points "deeper" in a convex set and therefore efficient in dissecting the uncertainty set more evenly, a necessary condition for rapidly reaching the point that is sought. Now, consider that the initial sensing MCC feedback by the CRN before any probing occurs, $\mathbf{p}(0) = \mathbf{0}$, is MCS_{ref}. Following t probing attempts, the CBS has collected t MCC pieces of feedback, which correspond to t pairs of inequalities:

$$\tilde{\mathbf{g}}^{\mathsf{T}}\tilde{\mathbf{p}}_{\mathbf{u}}(k) > 1$$
$$, k = 1,\dots,t \tag{10.36}$$
$$\tilde{\mathbf{g}}^{\mathsf{T}}\tilde{\mathbf{p}}_{\mathrm{l}}(k) \le 1$$

Equation 10.36 inequalities are derived as described in Section 10.4.3 in the form of Equation 10.35 and additionally consider inequalities coming from probing power vectors that do not cause MCS deterioration. An additional constraint for the $\tilde{g}_i$ parameters is that $\tilde{g}_i$ s have to be positive as channel gains:

$$\tilde{g}_i \geq 0, \quad i = 1,\ldots,N \tag{10.37}$$

The inequalities in Equations 10.36 and 10.37 define a convex polyhedron $\mathcal{P}_t$, the uncertainty set of the search problem:

$$\mathcal{P}_t = \left\{ \tilde{\mathbf{g}} \mid \tilde{\mathbf{g}} \geq \mathbf{0}, \tilde{\mathbf{g}}^\mathsf{T} \tilde{\mathbf{p}}_u (k) > 1, \tilde{\mathbf{g}}^\mathsf{T} \tilde{\mathbf{p}}_l (k) \leq 1, k = 1,\ldots,t \right\} \tag{10.38}$$

In the CGCPM, the center of gravity CG of the convex polyhedron $\mathcal{P}_t$ is calculated in vector form as

$$\tilde{\mathbf{g}}_{\mathrm{CG}}(t) = \frac{\displaystyle\int_{\mathcal{P}_t} \tilde{\mathbf{g}}\, dV_{\tilde{g}}}{\displaystyle\int_{\mathcal{P}_t} dV_{\tilde{g}}} \tag{10.39}$$

where $V_{\tilde{g}}$ represents volume in the parameter $\tilde{\mathbf{g}}$ space. The advantages of the CGCPM are that its convergence to the targeted point is guaranteed and that the number of uncertainty set cuts or inequalities needed are of $\mathcal{O}(N \log_2(N))$ complexity [5]. This convergence rate is ensured by the fact that any cutting plane passing through the CG reduces the polyhedron by at least 37% at each step.

In the ACCPM, the analytic center AC of the convex polyhedron Π_t is calculated in vector form as

$$\tilde{\mathbf{g}}_{\mathrm{AC}}(t) = \arg\min_{\tilde{\mathbf{g}}} \left(-\sum_{k=1}^{t} \log\left(\tilde{\mathbf{g}}^\mathsf{T} \tilde{\mathbf{p}}_u (k) - 1\right) - \sum_{k=1}^{t} \log\left(1 - \tilde{\mathbf{g}}^\mathsf{T} \tilde{\mathbf{p}}_u (k) - \sum_{i=1}^{N} \log(\tilde{g}_i)\right) \right) \tag{10.40}$$

and can be calculated efficiently using interior point methods. Furthermore, an upper bound for the number of inequalities needed to approach the point sought has been evaluated to prove the convergence of the ACCPM, which is of $\mathcal{O}(N^2)$ complexity, also referred to as iteration complexity.

Even though this framework seems ideal for learning the interference constraint and at the same time pursuing the optimization objective, there is still a problem. The optimization part, which is responsible for choosing the training power vectors, focuses on cutting planes of specific direction. These training power vectors basically adhere to the power level ratios that maximize $U_{SU}^{tot}(\mathbf{p})$ and are subjected to the initial interference hyperplane estimation. Thus, they contribute only to reducing uncertainty in these directions. This indicates that choosing the training power vectors based solely on the optimization problem is not a good strategy. Instead, the SU system should start probing the PU system in an exploratory manner by initially diversifying the training power vectors and gradually, when enough knowledge of the interference constraint is obtained, shift to an exploitive behavior that allocates power levels to the SUs specified by the optimization problem solution (Equation 10.30).

The authors of [56] proposed to make this shift from *exploration* to *exploitation* by mixing the optimization objective, the maximization of the SNR received by the SUs, with a similarity metric of the beam-forming vectors. The impact of this similarity metric in the design of these probing vectors was determined to be a decreasing function of time, so that the desirable transition could happen. This is a combination of two tactics known in the ML community as the ϵ-decreasing and contextual-ϵ-greedy strategies, according to which the choice of the training samples is performed using an exploration or randomization factor, ϵ. In these strategies, this factor decreases as time passes or depending on the similarity of the training samples, resulting in explorative behavior at the beginning and exploitative behavior at the end. Nevertheless, not only does this logic require tuning of the exploration factor time dependency according to performance results, but it also does not guarantee that enough diversification has occurred to reach the learning goal, which in the case of [56] is the channel correlation matrix, since time on its own cannot be an indicator of approaching the exact values of the parameters sought.

The enhancement introduced here is to relate the exploration factor, ϵ, to the geometry of $\mathcal{P}_r$. According to this, the smaller the uncertainty region becomes, the closer the learning algorithm gets to the exact value $\tilde{\mathbf{g}}$, and thus exploration should occur less and the training power vectors should be more relative to the optimization problem solution (Equation 10.30).

10.4.5 Results

Finally, we present the simulation results of the two CPM methods compared with the benchmark method developed in [57] and also considering a binary feedback to show the multilevel MCC feedback's effectiveness. The following diagrams correspond to a static interference channel scenario with $N=5$ and a PU system operating with MCSs of QPSK 1/2, QPSK 3/4, 16-QAM 1/2, 16-QAM 3/4, 64-QAM 2/3, 64-QAM 3/4, and 64-QAM 5/6 with LDPC coding. The first figure shows the channel estimation error diagrams for the benchmark, ACCPM-based, and CGCPM-based methods, depending on the number of time flops, where each time flop is the time period necessary to coordinate the CRN, probe the PU system, sense the MCC feedback, and collectively decide the PU MCS. It

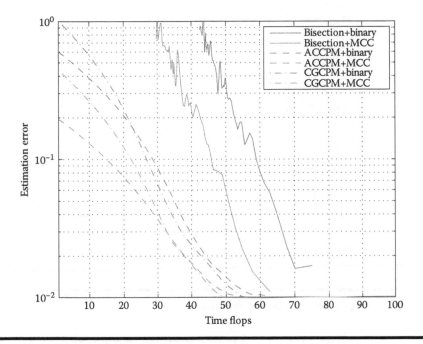

Figure 10.4   Channel estimation error in time.

can be clearly seen in Figure 10.4 that the CPM-based methods outperform the learning benchmark method. For an estimation error of approximately 1%, the gain of the necessary number of feedbacks to converge is 18 time flops for the binary feedback and 12 time flops for the MCC feedback. This proves that using the bisection method in the parameter space, as performed by the CPMs, is much faster than using bisection in the power vector space. Another outcome is that the use of the MCC feedback instead of the binary ACK/NACK packet reduces the convergence time significantly in the benchmark method and noticeably in the CPM-based learning methods. Specifically, for an estimation error of 1%, in the benchmark technique, this gain of time flops is almost 13, and in the CPM-based technique, it is nearly six. Even though the convergence time reduction is small in the CPM case, it is regarded as a marked enhancement, considering that CPM-based techniques are already fast enough. The final conclusion derived from this figure concerns the comparison of the two CPM-based learning mechanisms. Despite the clear precedence of the ACCPM-based approach, it is observed that the CGCPM-based scheme manages to surpass the former below certain estimation error boundaries. In particular, it is noticed that for an estimation error of 1%, the CGCPM-based procedure outperforms the ACCPM-based one in the binary feedback case by seven time flops and in the MCC feedback case by five time flops.

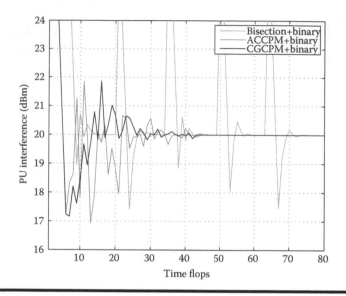

Figure 10.5 I_{PU} **in time for binary feedback.**

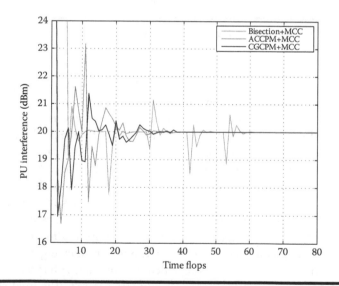

Figure 10.6 I_{PU} **in time for MCC feedback.**

The next diagrams show the aggregated interference caused to the PU during the simultaneous learning and CRN capacity maximization method for binary feedback in Figure 10.5 and MCC feedback in Figure 10.6. First of all, it is clear that taking advantage of MCC feedback rather than binary feedback causes smaller and fewer interference peaks and achieves faster convergence. Secondly, it is observed

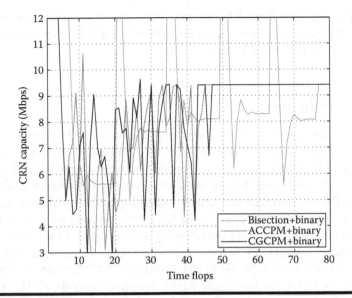

Figure 10.7   CRN capacity in time for binary feedback.

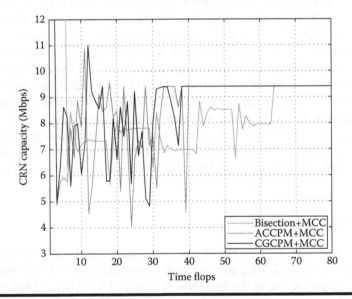

Figure 10.8   CRN capacity in time for MCC feedback.

that the CGCPM-based scheme converges to the PU interference threshold limit with fewer variations and much more smoothly than the ACCPM-based scheme.

The last diagrams depict the CRN capacity progress in time for binary feedback in Figure 10.7 and MCC feedback in Figure 10.8. The results for the CRN capacity in Figures 10.7 and 10.8 initially show, as stated before, the benefit of using MCC feedback. The CRN capacity variations, which are mostly interpreted as CRN throughput degradation, are fewer than in the binary feedback scenarios, and especially in the CGCPM case.

10.5 Conclusions

In this chapter, a simultaneous centralized PC and interference channel gain learning algorithm allows a CRN to access the frequency band of a PU operating based on an ACM protocol. Since no cooperation between the PU and the SUs is established, the CRN uses the multilevel MCC sensing information as an implicit CSI of the PU link and therefore as a multilevel interference violation indicator to learn the interference channel gains. To implement all this, an attempt has been made to demonstrate the trend toward using ML in signal classification as an SS problem and in underlay PC scenarios. Highly sophisticated supervised learning machines, SVMs, were employed to distinguish the PU MCS from even low SNR sensed signals, and newly introduced Active Learning methods, CPMs, were used to develop the simultaneous interference channel gain learning and CRN throughput maximization technique. The exploitation of advances in ML has been proved to be essential to CRs, not only because ML is competent to find solutions to a great variety of problems but also because it makes CR actually "cognitive." The demonstrated underlay scenario using MCC and intelligent centralized PC was shown to be a promising DSM technology whereby a CRN can coexist with a PU without causing significant interference to the PU and exchanging any information between the two systems. This HN application within the CR framework can increase the flexibility of the spectrum usage and enable 5G, the next major evolutionary step of telecommunication systems, to attain greater spectrum usage efficiency and speed. The 5G challenge will push for even further progress in all kinds of enabling technologies, and it is our belief that one of them will be ML.

Acknowledgment

This work was supported by the National Research Fund, Luxembourg, under the CORE projects "SeMIGod: SpEctrum Management and Interference mitiGation in cOgnitive raDio satellite networks" and "SATSENT: SATellite SEnsor NeTworks for spectrum monitoring."

References

1. Q. Zhao and B. Sadler, A survey of dynamic spectrum access, *IEEE Signal Processing Magazine* 24(3): 79–89, 2007.

2. J. Mitola, Cognitive radio an integrated agent architecture for software defined radio, *PhD dissertation*, KTH Royal Institute of Technology Stockholm, Stockholm, Sweden, IEEE, 2000.
3. S. Chatzinotas, A. Tsakmalis, and B. Ottersten, Modulation and coding classification for adaptive power control in 5G cognitive communications, in *IEEE International Workshop on Signal Processing Advances in Wireless Communications (SPAWC)*, Toronto, ON, IEEE, pp. 234–238, 2014.
4. V. N. Vapnik. *The Nature of Statistical Learning Theory*. New York: Springer, 1999.
5. S. Boyd and L. Vandenberghe, *Localization and Cutting-Plane Methods. EE364b Lecture Notes*. Stanford, CA: Stanford University, 2008.
6. Federal Communications Commission, Spectrum policy task force report. *ET Docket 02-155*, 2002.
7. A. Goldsmith, S. A. Jafar, I. Maric, and S. Srinivasa, Breaking spectrum gridlock with cognitive radios: An information theoretic perspective, *Proceedings of the IEEE* 97(5): 894–914, 2009.
8. O. A. Dobre, A. Abdi, Y. Bar-Ness, and W. Su, Survey of automatic modulation classification techniques: Classical approaches and new trends, *IET Communications* 1(2): 137–156, 2007.
9. J. J. Popoola and R. van Olst, A novel modulation-sensing method, *IEEE Vehicular Technology Magazine* 6(3): 60–69, 2011.
10. M. Petrova, P. Mahonen, and A. Osuna, Multi-class classification of analog and digital signals in cognitive radios using support vector machines, in *7th International Symposium on Wireless Communication Systems (ISWCS)*, York, IEEE, pp. 986–990, 2010.
11. W. Gardner, A. Napolitano, and L. Paura, Cyclostationarity: Half a century of research, *Signal Processing (Elsevier)* 86(4): 639–697, 2006.
12. B. Ramkumar, Automatic modulation classification for cognitive radios using cyclic feature detection, *IEEE Circuits and Systems Magazine* 9(2): 27–45, 2009.
13. K. Kim, I. A. Akbar, K. K. Bae, J. Um, C. M. Spooner, and J. H. Reed, Cyclostationary approaches to signal detection and classification in cognitive radio, in *2nd IEEE International Symposium on New Frontiers in Dynamic Spectrum Access Networks*, Dublin, IEEE, pp. 212–215, 2007.
14. A. Fehske, J. Gaeddert, and J. H. Reed, A new approach to signal classification using spectral correlation and neural networks, in *1st IEEE International Symposium on New Frontiers in Dynamic Spectrum Access Networks*, Baltimore, MD, IEEE, pp. 144–150, 2005.
15. M. Bkassiny, S. K. Jayaweera, Y. Li, and K. A. Avery. Wideband spectrum sensing and non-parametric signal classification for autonomous self-learning cognitive radios, *IEEE Transactions on Wireless Communications* 11(7): 2596–2605, 2012.
16. H. Hu, J. Song, and Y. Wang, Signal classification based on spectral correlation analysis and SVM in cognitive radio, in *22nd International Conference on Advanced Information Networking and Applications (AINA)*, Okinawa, IEEE, pp. 883–887, 2008.
17. H.-C. Wu, M. Saquib, and Z. Yun, Novel automatic modulation classification using cumulant features for communications via multipath channels, *IEEE Transactions on Wireless Communications* 7(8): 3098–3105, 2008.
18. O. A. Dobre, Y. Bar-Ness, and W. Su, Higher-order cyclic cumulants for high order modulation classification, IEEE *Military Communications Conference (MILCOM)* 1: 112–117, 2003.

19. T. C. Clancy, A. Khawar, and T. R. Newman, Robust signal classification using unsupervised learning, *IEEE Transactions on Wireless Communications* 10(4): 1289–1299, 2011.
20. M. Bkassiny, S. K. Jayaweera, and Y. Li, Multidimensional Dirichlet process-based non-parametric signal classification for autonomous self-learning cognitive radios, *IEEE Transactions on Wireless Communications* 12(11): 5413–5423, 2013.
21. V. Choquese, M. Marazin, L. Collin, K. C. Yao, and G. Burel, Blind recognition of linear space-time block codes: A likelihood-based approach, *IEEE Transactions on Signal Processing* 58(3): 1290–1299, 2010.
22. T. Xia and H. C. Wu, Novel blind identification of LDPC codes using average LLR of syndrome a posteriori probability, *IEEE Transactions on Signal Processing* 62(3): 632–640, 2014.
23. R. Moosavi and E. G. Larsson, A fast scheme for blind identification of channel codes, in *IEEE Global Telecommunications Conference (GLOBECOM)*, Houston, TX, IEEE, pp. 1–5, 2011.
24. J. Hagenauer, E. Offer, and L. Papke, Iterative decoding of binary block and convolutional codes, *IEEE Transactions on Information Theory* 42(2): 429–445, 1996.
25. C. U. Saraydar, N. B. Mandayam, and D. J. Goodman, Efficient power control via pricing in wireless data networks, *IEEE Transactions on Communications* 50(2): 291–303, 2002.
26. R. Etkin, A. Parekh, and D. Tse, Spectrum sharing for unlicensed bands, *IEEE Journal on Selected Areas on Communications* 25(3): 517–528, 2007.
27. M. Le Treust and S. Lasaulce, A repeated game formulation of energy-efficient decentralized power control, *IEEE Transactions on Wireless Communications* 9(9): 2860–2869, 2010.
28. C. G. Yang, J. D. Li, and Z. Tian, Optimal power control for cognitive radio networks under coupled interference constraints: A cooperative game-theoretic perspective, *IEEE Transactions on Vehicular Technology* 59(4): 1696–1706, 2010.
29. D. Niyato and E. Hossain, Competitive pricing for spectrum sharing in cognitive radio networks: Dynamic game, inefficiency of Nash Equilibrium, and collusion. *IEEE Journal on Selected Areas on Communications* 26(1): 192–202, 2008.
30. H. Yu, L. Gao, Z. Li, X. Wang, and E. Hossain, Pricing for uplink power control in cognitive radio networks, *IEEE Transactions on Vehicular Technology* 59(4): 1769–1778, 2010.
31. S. K. Jayaweera and T. Li, Dynamic spectrum leasing in cognitive radio networks via primary-secondary user power control games, *IEEE Transactions on Wireless Communications* 8(6): 3300–3310, 2009.
32. M. van der Schaar and F. Fu, Spectrum access games and strategic learning in cognitive radio networks for delay-critical applications, *Proceedings of the IEEE* 97(4): 720–740, 2009.
33. P. Zhou, Y. Chang, and J. A. Copeland, Reinforcement learning for repeated power control game in cognitive radio networks, *IEEE Journal on Selected Areas on Communications* 30(1): 54–69, 2012.
34. M. Maskery, V. Krishnamurthy, and Q. Zhao, Decentralized dynamic spectrum access for cognitive radios: Cooperative design of a non-cooperative game, *IEEE Transactions on Communications* 57(2): 459–469, 2009.
35. J. D. Herdtner and E. K. P. Chong, Analysis of a class of distributed asynchronous power control algorithms for cellular wireless systems, *IEEE Journal on Selected Areas on Communications* 18(3): 436–446, 2002.

36. G. Scutari, D. P. Palomar, and S. Barbarossa, Simultaneous iterative water-filling for Gaussian frequency-selective interference channels, *IEEE International Symposium on Information Theory*, Seattle, IEEE, pp. 600–604, 2006.
37. G. Scutari, D. P. Palomar, and S. Barbarossa, Asynchronous iterative water-filling for Gaussian frequency-selective interference channels, in *IEEE Transactions on Information Theory*, La Jolla, CA, IEEE, pp. 2868–2878, 2008.
38. N. Gatsis, A. G. Marques, and G. B. Giannakis, Power control for cooperative dynamic spectrum access networks with diverse QoS constraints, *IEEE Transactions on Communications* 58(3): 933–944, 2010.
39. S. Huang, X. Liu, and Z. Ding, Decentralized cognitive radio control based on inference from primary link control information, *IEEE Journal on Selected Areas in Communications* 29(2): 394–406, 2011.
40. P. Setoodeh and S. Haykin, Robust transmit power control for cognitive radio, *Proceedings of the IEEE* 97(5): 915–939, 2009.
41. G. Scutari, D. P. Palomar, F. Facchinei, and J. S. Pang, Convex optimization, game theory, and variational inequality theory, *IEEE Signal Processing Magazine* 27(3): 35–49, 2010.
42. E. Lagunas, S. K. Sharma, S. Maleki, S. Chatzinotas, and B. Ottersten, Power control for satellite uplink and terrestrial fixed-service co-existence in Ka-band. In *IEEE Vehicular Technology Conference (VTC)*, IEEE, 2015.
43. A. Galindo-Serrano and L. Giupponi, Power control and channel allocation in cognitive radio networks with primary users' cooperation, *IEEE Transactions on Vehicular Technology* 9(3): 1823–1834, 2010.
44. D. P. Palomar and J. R. Fonollosa, Practical algorithms for a family of waterfilling solutions, *IEEE Transactions on Signal Processing* 53(2): 686–695, 2005.
45. L. Zhang, Y. C. Liang, and Y. Xin, Joint beamforming and power allocation for multiple access channels in cognitive radio networks, *IEEE Journal on Selected Areas in Communications* 26(1): 38–51, 2008.
46. E. Dall'Anese, S. J. Kim, G. B. Giannakis, and S. Pupolin, Power control for cognitive radio networks under channel uncertainty, *IEEE Transactions on Wireless Communications* 10(10): 3541–3551, 2011.
47. D. I. Kim, L. B. Le, and E. Hossain, Joint rate and power allocation for cognitive radios in dynamic spectrum access environment, *IEEE Transactions on Wireless Communications* 7(12): 5517–5527, 2008.
48. L. B. Le and E. Hossain, Resource allocation for spectrum underlay in cognitive radio networks, *IEEE Transactions on Wireless Communications* 7(12): 5306–5315, 2008.
49. G. Marques, X. Wang, and G. B. Giannakis, Dynamic resource management for cognitive radios using limited-rate feedback, *IEEE Transactions on Signal Processing* 57(9): 3651–3666, 2009.
50. G. Zhao, G. Y. Li, and C. Yang, Proactive detection of spectrum opportunities in primary systems with power control, *IEEE Transactions on Wireless Communications* 8(9): 4815–4823, 2009.
51. F. Akyildiz, W.-Y. Lee, M. C. Vuran, and S. Mohanty, NeXt generation/dynamic spectrum access/cognitive radio wireless networks: A survey, *Computer Networks Journal (Elsevier)* 50(13): 2127–2159, 2006.
52. B. C. Banister and J. R. Zeidler, A simple gradient sign algorithm for transmit antenna weight adaptation with feedback, *IEEE Transactions on Signal Processing* 51(5): 1156–1171, 2003.

53. R. Mudumbai, J. Hespanha, U. Madhow, and G. Barriac, Distributed transmit beamforming using feedback control, *IEEE Transactions on Information Theory* 56(1): 411–426, 2010.
54. Y. Noam and A. J. Goldsmith, The one-bit null space learning algorithm and its convergence, *IEEE Transactions on Signal Processing* 61(24): 6135–6149, 2013.
55. J. Xu and R. Zhang, Energy beamforming with one-bit feedback, *IEEE Transactions on Signal Processing* 62(20): 5370–5381, 2014.
56. B. Gopalakrishnan and N. D. Sidiropoulos, Cognitive transmit beamforming from binary CSIT, *IEEE Transactions on Wireless Communications* 14(2): 895–906, 2014.
57. A. Tsakmalis, S. Chatzinotas, and B. Ottersten, Power control in cognitive radio networks using cooperative modulation and coding classification, in *10th International Conference on Cognitive Radio Oriented Wireless Networks and Communications*, Doha, Qatar, IEEE, 358–369, 2015.

Chapter 11

On the Energy Efficiency–Spectral Efficiency Trade-Off in 5G Cellular Networks

Oluwakayode Onireti and Muhammad Ali Imran

Contents

In the context of energy saving and operational cost reduction, energy efficiency (EE) has emerged as an important performance metric in cellular networks. According to the famous Shannon capacity theorem, maximizing the EE and maximizing the spectral efficiency (SE) are conflicting objectives; hence, both metrics can be jointly studied via the EE–SE trade-off. In this context, the aim of this chapter is to investigate the fundamental trade-off between EE and SE in the futuristic fifth-generation (5G) cellular networks in which a distributed multiple-input multiple-output (D-MIMO) scheme is used for meeting their high data rate requirement. More specifically, it presents a framework for comprehensively analyzing the D-MIMO system from both an EE and an SE perspective by means of an accurate closed-form approximation (CFA) of its EE–SE trade-off. The chapter first introduces the EE–SE trade-off concept in a generic fashion. Next, it provides the relevant background information regarding D-MIMO's system model, capacity expression, realistic power consumption model (PCM), and EE–SE trade-off formulation. Subsequently, the generic CFA of the D-MIMO EE–SE trade-off is presented and then analyzed for specific scenarios with one or a larger number of active radio access units (RAUs), respectively, and with practical antenna settings. Next, the D-MIMO EE–SE trade-off CFA is compared with the nearly exact approach (based on Monte Carlo simulation and a linear search algorithm), and the great accuracy of the former is established over a wide range of SE values for both the uplink and downlink channels as well as for both idealistic and realistic PCMs.

Since the D-MIMO EE–SE trade-off CFA becomes more complicated to formulate when the number of active RAUs is greater than two, to get insights into the EE of D-MIMO for a large number of RAUs, one can rely on lower and upper bounds instead of the full CFA. Hence, this chapter also presents the low- and high-SE approximations of the D-MIMO EE–SE trade-off in both the uplink and downlink channels. As an application for the D-MIMO EE–SE trade-off CFAs, this chapter also presents the achievable EE gain of the D-MIMO system over the traditional colocated MIMO (C-MIMO) system, that is, the one-RAU scenario, in the downlink channel and for both the idealistic and realistic PCMs.

11.1 EE–SE Trade-Off

Maximizing the EE, or equivalently minimizing the consumed energy per bit, and maximizing the SE are conflicting objectives, which implies the existence of a trade-off [1]. The EE–SE trade-off concept was first introduced for power-limited systems and accurately defined for the low-power/low-SE regime by Verdú in [2]. The EE–SE trade-off has become the metric of choice for efficiently designing future communication systems, since it incorporates the EE metric, which recently emerged as a key system design criterion, alongside the well-established SE. With the current shift in EE research from power-limited to power-unlimited applications, the concept of EE–SE trade-off must be generalized for power-unlimited systems and accurately defined for a wider range of SE regimes, as initiated by the work in [3].

In general, the existing works on the EE–SE trade-off can be categorized into two types: the first type aims at maximizing the EE based on a given SE requirement [4–8]; the second category looks at expressing the EE as a function of the SE [2,3,9–14]. The main limitation of the former approach is that maximizing EE while making the SE a constraint puts a limit on both EE and SE performances. With the latter approach, which is the focus of this chapter, the operator can select the operating point that best suits the system requirements.

11.1.1 EE–SE Trade-Off Concept

The EE–SE trade-off concept can be simply described as how to express the EE in terms of SE for a given available bandwidth. According to the famous Shannon capacity theorem [15], the maximum achievable SE or, equivalently, the channel capacity per unit bandwidth C (bit/s/Hz) is a function of the signal-to-noise ratio (SNR), γ, such that

$$C = f(\gamma) \tag{11.1}$$

where:

$\gamma \quad = P/N$ is the ratio between the transmit power P and the noise power N

$N \quad = N_0 W$, with N_0 (J) being the noise power spectral density

In the general case, $f(\gamma)$ can be described as an increasing function of γ mapping SNR values in $[0,+\infty)$ to capacity per unit bandwidth values in $[0,+\infty)$. As long as $f(\gamma)$ is a bijective function, $f(\gamma)$ would be invertible, such that

$$\gamma = f^{-1}(C) \tag{11.2}$$

where $f^{-1}:C \in [0,+\infty) \mapsto \gamma \in [0,+\infty)$ is the inverse function of f. For instance, over the additive white Gaussian noise (AWGN) channel, $f(\gamma)$ and $f^{-1}(C)$ are simply given in [10,15] as

$$f(\gamma) = \log_2(1+\gamma) \quad \text{and} \quad f^{-1}(C) = 2^C - 1 \tag{11.3}$$

respectively. As has been explained in [10], the transmit power P can be expressed as RE_b, and hence the SNR, γ, can be reexpressed as a function of both the achievable SE, S, and EE, C_J, such that

$$\gamma = \frac{P}{N_0 W} = \frac{R}{W} \frac{E_b}{N_0} = \frac{S}{N_0 C_J} \tag{11.4}$$

Inserting Equation 11.4 into Equation 11.2, the EE–SE trade-off expression in the general case can simply be formulated as

$$C_J = \frac{W}{N} \frac{S}{f^{-1}(C)} \tag{11.5}$$

Equation 11.5 describes the EE–SE trade-off for the case of $P_T = P$, that is, the idealistic PCM; however, for more generic PCM, such that $P_T = g(P)$, the EE–SE trade-off can be reformulated as

$$C_J = W \frac{S}{g\left(Nf^{-1}(C)\right)} \tag{11.6}$$

To provide some insights into the EE–SE trade-off, Figure 11.1 plots the EE–SE trade-off expression in Equations 11.5 and 11.6 for $g(Nf^{-1}(C)) = Nf^{-1}(C) + P_0$ by considering $f^{-1}(C)$ as given in Equation 11.3, $S = C$, and $W = N = 1$. The results indicate that maximizing the EE and maximizing the SE are conflicting objectives, and hence, an EE–SE trade-off exists between these two metrics [10]. Indeed, in the case of $P_0 = 0$ W, the maximum EE is achieved when $C = 0$ bit/s/Hz, and conversely, the maximum SE on the graph, $C = 15$ bit/s/Hz, is achieved for very low EE. Consequently, the most energy-efficient policy over the AWGN channel is not to transmit anything at all when $P_0 = 0$ W. However, when the total consumed power P_T is not restricted to the transmit power P, but an overhead power P_0 is consumed, the existence of an optimal EE operation point becomes apparent, which is circled in Figure 11.1 for different values of P_0. The existence of such a point illustrates the growing importance of the EE–SE trade-off as a system design criterion.

11.2 Distributed MIMO System

The D-MIMO system was originally proposed to simply cover dead spots in indoor wireless communications [16]. However, the increasing demand for a high data rate and the limitation on power resources in wireless networks has led to the development of novel applications for the D-MIMO system, since the D-MIMO scheme

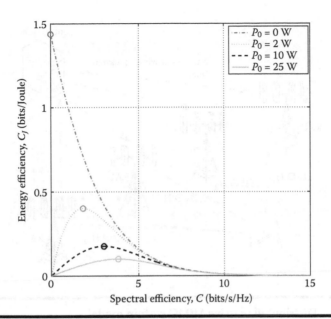

Figure 11.1   EE–SE trade-off over the AWGN channel for different values of overhead power.

can improve the capacity performance by shortening the transmission distance [17–19]. In the D-MIMO system, the antenna units, referred to as RAUs, are geographically distributed, which is in contrast to the traditional C-MIMO system, in which the antennas are just a few wavelengths apart. The capacity gain and improved power efficiency of the D-MIMO system over the C-MIMO scheme results from its ability to exploit both macro- and microdiversities [17–22].

In the D-MIMO system, the main processing unit is located at a centralized location referred to as the central unit (CU), which itself is connected to the RAUs via a high-speed delay-less error-free channel such as radio-frequency (RF) or optical fiber links, as illustrated in Figure 11.2. The RAUs and the CU exchange signaling information, and they are assumed to be perfectly synchronized. In the uplink of the D-MIMO system, the user terminals (UTs) simultaneously communicate with a group of geographically dispersed antennas (RAUs), which are connected to the CU, where the signals are jointly processed. In the downlink of the D-MIMO system, signal preprocessing is performed at the CU, and the processed signal is sent through the backhaul to a group of geographically distributed antennas (RAUs), which then transmit the message to the UT.

The channel capacity of a single-cell D-MIMO system has been presented for the uplink in [18,20,22] and the downlink channel in [17,18,22–25], while its expression for the multicell environment is given in [26] and [27,28], for the uplink and downlink channels, respectively. The distributed massive MIMO (DM-MIMO) system is made up of hundreds (or even thousands) of geographically distributed

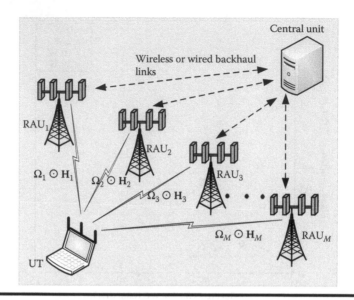

Figure 11.2 Distributed massive MIMO system model.

base station (BS) antennas, contrary to the D-MIMO system, in which the number of geographically distributed BS antennas is in the tens at most. The large number of BS antennas in the DM-MIMO can be in combinations ranging from a large number of geographically distributed RAUs, each with few antennas, to a small number of geographically distributed RAUs, each with a large number of antennas. Consequently, an asymptotic approximation can be obtained for their capacity expressions. The asymptotic closed-form expressions of the DM-MIMO capacity, which are applicable to both the uplink and downlink channels, can be found in [29–31], while DM-MIMO high-SE/SNR approximations have been derived in [32] for the generic case and in [21] for the DM-MIMO uplink channel by using the large random matrix theory [33]. As far as the single-cell DM-MIMO EE–SE trade-off is concerned, its lower and upper bounds at high SE/SNR have been obtained for some limited antenna configurations in the uplink channel [34].

11.2.1 D-MIMO Channel Model

A D-MIMO communication system consisting of M RAUs, each equipped with p antennas, and a user terminal, equipped with q antennas, is considered here, as illustrated in Figure 11.2. As a result of the large distance separating each RAU from the UT, each corresponding channel matrix is formed of independent microscopic and macroscopic fading components. The matrices Ω_i and $\mathbf{H}_i$ represent the deterministic distance-dependent path loss/shadowing (macroscopic fading component) and the MIMO Rayleigh fading channel (microscopic fading component), respectively, between the ith RAU and the UT, $i \in \{1,...,M\}$.

The channel model of the D-MIMO system can then be defined as $\tilde{\mathbf{H}} = \mathbf{\Omega}_V \odot \mathbf{H}_V$, where $\mathbf{H}_V = [\mathbf{H}_1^\dagger, \mathbf{H}_2^\dagger, \ldots, \mathbf{H}_M^\dagger]^\dagger$, $()^\dagger$ is the complex conjugate transpose, $\odot$ denotes the Hadamard product, $\tilde{\mathbf{H}} \in \mathbb{C}^{N_r \times N_t}$, $\mathbf{H}_V \in \mathbb{C}^{N_r \times N_t}$, and $\mathbf{\Omega}_V \in \mathbb{R}_+^{N_r \times N_t}$ with $R_+ = \{x \in \mathbb{R} \mid x \geq 0\}$. Moreover, considering the multiple antennas at the UT and the RAU, $\mathbf{\Omega}_V = \Lambda \triangleq \alpha \otimes \mathbf{J}$ and $\mathbf{\Omega}_V = \Lambda^\dagger$ in the uplink and downlink cases, respectively, where $\otimes$ denotes the Kronecker product, $\mathbf{J}$ is a $p \times q$ matrix with all its elements equal to one, $\alpha \triangleq [\alpha_1, \ldots, \alpha_M]^\dagger$ represents the average channel gain vector, and α_i represents the average channel gain between the UT and the ith RAU. Furthermore, the total number of transmit and receive antennas of the D-MIMO system is defined as N_t and N_r, respectively. Note that $N_t = n = q$ and $N_r = Mp$ in the uplink case, and $N_t = Mp$, $N_r = q$, and $n = p$ in the downlink case, where n is the number of transmit antennas per node. The received signal $\mathbf{y} \in \mathbb{C}^{N_r \times 1}$ can be expressed as

$$\mathbf{y} = \tilde{\mathbf{H}}\mathbf{s} + \mathbf{z} \tag{11.7}$$

where:

$\mathbf{s} \in \mathbb{C}^{N_t \times 1}$ is the transmit signal vector with average transmit power P

$\mathbf{z} \in \mathbb{C}^{N_r \times 1}$ is the noise vector with average noise power N

Moreover, $\mathbf{H}_V$ is assumed to be a random matrix having independent and identically distributed (i.i.d.) complex circular Gaussian entries with zero mean and unit variance.

11.2.2 D-MIMO Ergodic Capacity Review

In the case that the channel state information (CSI) is unknown at the transmitting node and perfectly known at the receiver, equal power allocation is adopted at the transmitter. Thus, the ergodic channel capacity per unit bandwidth of the D-MIMO system in both the uplink and downlink channels can then be expressed as [18,21]

$$C = f(\gamma) = \mathbb{E}_{\tilde{H}}\left\{\log_2\left|\mathbf{I}_{N_r} + \frac{\gamma}{n}\tilde{\mathbf{H}}\tilde{\mathbf{H}}^\dagger\right|\right\} \tag{11.8}$$

where:

$\mathbf{I}_{N_r}$ is an $N_r \times N_r$ identity matrix

$\gamma \triangleq P / N_0 W$ is the average SNR

W (Hz) is the bandwidth

N_0 is the noise power spectral density

$\mathbb{E}\{.\}$ and $|.|$ stand for the expectation and determinant operator, respectively

11.2.3 Asymptotic Approximation of D-MIMO System Capacity

It has been shown in [30,31] that the mutual information $I(\tilde{\mathbf{H}})$ of the D-MIMO system is asymptotically equivalent to a Gaussian random variable, such that both the uplink and downlink ergodic capacity per unit bandwidth can be approximated as

$$f(\gamma) \approx \tilde{f}(\gamma) = \frac{n}{2\ln(2)}\left[\kappa\sum_{i=1}^{M}\left(-1+2\ln(1+d_i)+\frac{1}{1+d_i}\right)\right.$$

$$\left.+\beta\left(-1+2\ln(1+g)+\frac{1}{1+g}\right)\right](\text{bit/s/Hz}) \qquad (11.9)$$

for large values of N_t and N_r, where

$$d_i = d_0\alpha_i^2/\rho, \quad g = \kappa\sum_{i=1}^{M}\alpha_i^2\left(\rho^2+d_0\alpha_i^2\rho\right)^{-1}, \quad \rho=1/\sqrt{\gamma}, \quad \kappa=(p/q), \text{ and } \beta=1$$

in the uplink while $\kappa=1$ and $\beta=q/p$ in the downlink. In addition, d_0 is the unique positive root of the $(M+1)$th-degree polynomial equation

$$P_m(d) = (d\rho-\beta)\prod_{i=1}^{M}(d+\rho v_i) + d\kappa\sum_{i=1}^{M}\prod_{k=1,k\neq i}^{M}(d+\rho v_k) \qquad (11.10)$$

where $v_i = 1/\alpha_i^2$. Hence, the capacity per unit bandwidth of D-MIMO given in Equation 11.9 can be reexpressed as

$$C \approx \tilde{f}(\gamma) = \frac{1}{\ln(2)}\left(S_q + \sum_{i=1}^{M}S_{p_i}\right) \qquad (11.11)$$

where S_q and S_{p_i} are given by

$$S_q = \frac{q}{2}\left(-1+\frac{1}{1+g}+2\ln(1+g)\right) \quad \text{and}$$

$$S_{p_i} = \frac{p}{2}\left(-1+\frac{1}{1+d_i}+2\ln(1+d_i)\right), \quad \forall i \in \{1,...,M\} \qquad (11.12)$$

respectively, for both the uplink and the downlink channels.

11.2.4 D-MIMO Power Model

The EE of a communication system is closely related to its total power consumption. In a realistic D-MIMO setting, power components such as signal processing, direct current (dc)-dc/alternating current (ac)-dc converter, backhaul powers and power losses from cooling, main supply, and amplifier inefficiency must be taken into account in addition to the transmit power when evaluating the actual EE of such a system. To model the power consumption of the D-MIMO system, each RAU is assumed to be a remote radio head (RRH), such that the power amplifier (PA) and RF units are mounted at the same physical location as the RAU, while the baseband processing unit is located at the CU, as illustrated in Figure 11.3. In comparison with the usual BS transceiver, an RRH transceiver does not require feeder cables, such that feeder loss is mitigated; furthermore, its PAs are naturally cooled by air circulation, and hence, a cooling unit is not necessary. By using the realistic PCM for RRH in [35], the total consumed powers in the uplink and the downlink of the D-MIMO system are given by

$$P_{T_u} = \frac{P}{\mu_{UT}} + qP_{ct} + M(pP_{0_u} + P_{bh}) \tag{11.13}$$

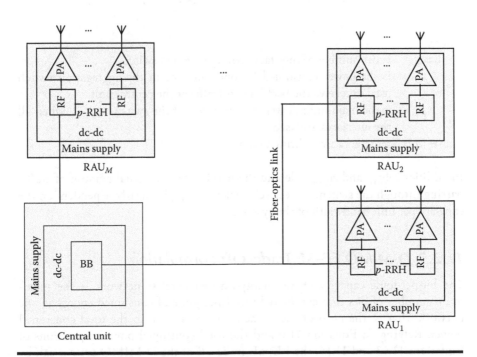

Figure 11.3 Distributed MIMO power model.

and

$$P_{T_d} = M\left(\Gamma P + pP_{0_d} + P_{bh}\right) + qP_{cr} \qquad (11.14)$$

respectively, where:

P_{ct} and P_{cr} are the UT's transmit and receive circuit power, respectively

μ_{UT} denotes the UT amplifier efficiency

P_{bh} is the backhauling induced power

Furthermore, P_{0_u} and P_{0_d} are the uplink and downlink components, respectively, of the power consumption at the minimum nonzero output power, which is defined in [35]. In addition, $P \in [0, P_{max}]$, with P_{max} being the maximum transmit power. The backhauling architecture is assumed to be based on fiber optic, and all switches and interfaces are identical, as in [36]. Moreover, an optical small form factor pluggable (SFP) interface is used to transmit data from each RAU over the backhauling fiber, and it has a power consumption of c watts; hence, the total backhauling induced power P_{bh} per RAU can be expressed as [36]

$$P_{bh}(\mathcal{C}) = \left(\frac{1}{\max_{dl}}\left(\phi\ p_b + (1-\phi)\frac{Ag_{switch}(\mathcal{C})}{Ag_{max}}\ p_b\right) + p_{dl} + c\right) \qquad (11.15)$$

where:

$\max_{dl}$ is the number of interfaces per aggregation switch

p_{dl} is the power consumed by one interface in the aggregation switch used to receive the backhauled traffic at the central unit

p_b is the maximum power consumption of the switch, that is, when all the interfaces are used

$\phi \in [0,1]$ represents a weighting factor

In addition, Ag_{max} and Ag_{switch} denote the maximum and actual amount of traffic passing through the switch, respectively, where Ag_{switch} is linearly dependent on the capacity per unit bandwidth of the system.

11.2.5 D-MIMO EE–SE Trade-Off Formulation

The bit-per-joule capacity of an energy-limited wireless network is the maximum number of bits that can be delivered per joule of consumed energy in the network, that is, the ratio of the capacity of the system to the total consumed power. Relying on Equation 11.6 and the total consumed power expressions of Equations 11.13 and 11.14, the EE–SE trade-off of the D-MIMO system can be formulated as

$$C_{J_u} = \frac{S}{N_0} \left[\frac{f^{-1}(C)}{\mu_{UT}} + \frac{qP_{ct} + M\left(pP_{0_u} + P_{bh}(C)\right)}{N} \right]^{-1} \qquad (11.16)$$

and

$$C_{J_d} = \frac{S}{MN_0} \left[\Gamma f^{-1}(C) + \frac{pP_{0_d} + P_{bh}(C) + \dfrac{q}{M} P_{cr}}{N} \right]^{-1} \qquad (11.17)$$

in the uplink and downlink channels, respectively, where $N = N_0 W$ is the noise power. Note that Equations 11.16 and 11.17 revert to Equation 11.5 when $P_{0_d} = P_{0_u} = P_{ct} = P_{cr} = P_{bh} = 0$ and $\Gamma = \mu_{UT} = 1$, that is, in the idealistic case.

11.3 Closed-Form Approximation of the EE–SE Trade-Off

The EE–SE trade-off expressions of the D-MIMO system given in Equations 11.16 and 11.17 require the knowledge of $f^{-1}(C)$; however, obtaining an explicit solution of $f^{-1}(C)$ from the ergodic capacity expression of Equation 11.8 is not feasible. However, $\tilde{f}(\gamma)$ in Equation 11.11 can be inverted, and since $\tilde{f}(\gamma)$ is an accurate approximation of $f(\gamma)$, it is expected that $\tilde{f}^{-1}(C)$ would be an accurate approximation of $f^{-1}(C)$.

Based on Equation 11.9, it has been recently proved in [13,14] that the EE–SE trade-off for the ergodic D-MIMO Rayleigh fading channel can be explicitly formulated by means of an accurate CFA as

$$\tilde{f}^{-1}(C) = \frac{-\left[1 + W_0(g_q(S_q))^{-1}\right]\left[\displaystyle\sum_{i=1}^{M}\left(\frac{-\left[1 + W_0(g_p(S_{p_i}))^{-1}\right]}{\Delta_i} - \frac{1}{\Delta_i} + 1\right)\right] - M}{2M\left(\kappa \displaystyle\sum_{i=1}^{M} \alpha_i^2 x_i + \alpha_i^2 \beta\right)}$$

$$(11.18)$$

where:

$$g_q(S_q) = -\exp\left(-\left(\frac{S_q}{q} + \frac{1}{2} + \ln(2)\right)\right) \quad g_p(S_{p_i}) = -\exp\left(-\left(\frac{S_{p_i}}{p} + \frac{1}{2} + \ln(2)\right)\right)$$

$$x_i = \frac{W_0\left(g_p\left(S_{p_i}\right)\right)}{W_0\left(g_p\left(S_{p_1}\right)\right)}$$

$$\Delta_i = \frac{\alpha_i^2}{\alpha_1^2}$$

M is the number of RAUs

$\kappa = p/q$ and $\beta = 1$ in the uplink, while $\kappa = 1$ and $\beta = q/p$ in the downlink

In addition, $W_0(x)$ is the real branch of the Lambert function. The Lambert W function is the inverse function of $f(w) = w\exp(w)$ and is such that $W(z)e^{W(z)} = z$, where w, $z \in \mathbb{C}$ [37].

11.4 Use Case Scenarios

Note that S_q and S_{p_i} in Equation 11.12 are functions of γ. Thus, to use the D-MIMO EE–SE trade-off expression of Equation 11.18, S_q and S_{p_i} first need to be expressed as a function of $\mathcal{C}$. This section starts by providing expressions of S_q and S_{p_i} as a function of $\mathcal{C}$ for some specific scenarios, such as when $M = 1$ RAU, and then for $M \geq 2$ RAUs.

11.4.1 One Radio Access Unit

The case in which only one RAU is active, that is, the 1-RAU D-MIMO case, is a very important scenario, since it is equivalent to the point-to-point MIMO channel, whose EE–SE trade-off CFA has recently been given in [3,9]. It can easily be shown that the inverse function, $\tilde{f}^{-1}(\mathcal{C})$, in Equation 11.18 simplifies into

$$\tilde{f}^{-1}(\mathcal{C}) = \frac{-1 + \left[1 + \dfrac{1}{W_0\left(g_q\left(S_q\right)\right)}\right]\left[1 + \dfrac{1}{W_0\left(g_p\left(S_{p_1}\right)\right)}\right]}{2\alpha_1^2(\kappa + \beta)} \tag{11.19}$$

in both the uplink and downlink channels, since $x_1 = \Delta_1 = 1$ when only one RAU is active, that is, when $M = 1$. Note that Equation 11.19 is equivalent to equation (12) in [3], which is used in obtaining the EE–SE trade-off closed-form expression of the point-to-point MIMO over the Rayleigh fading channel. Also, for the massive MIMO system, $p \gg q$. When the number of antennas at the RAU is greater than the number of antennas at the UT, the problem of defining a closed-form expression for the EE–SE trade-off is equivalent to expressing both S_q and S_{p1} as a function of C in Equation 11.19. Indeed, for the 1-RAU scenario, the system capacity per unit bandwidth can be expressed as $C\ln(2) \approx S_q + S_{p1}$ in Equation 11.11; therefore, a parametric function $\Phi_{p,q}(C)$ can be defined, such that $\Phi_{p,q}(C) \approx S_q + S_{p1}$; then, S_q and S_{p1} are obtained as a function of solely p, q, and C by solving two simple linear equations. The difference $S_q - S_{p1}$ can be reexpressed as

$$\Phi_{p,q}(C) \approx S_q - S_{p1} = \ln\left(\frac{(1+g)^q}{(1+d_1)^p}\right) \tag{11.20}$$

since $q(-1 + (1/1 + g)) - p(-1 + (1/1 + d_1)) = 0$. Using the curve-fitting method proposed in [38], a parametric function $\phi_{p,q}$ that tightly fits $e^{S_q - S_{p1}/q}$ is designed for $p > q$ and C values between 0 and 50 bit/s/Hz. According to Figure 11.4, in which $e^{S_q - S_{p1}/q}$ has been plotted in a logarithm scale as a function of C and for various p

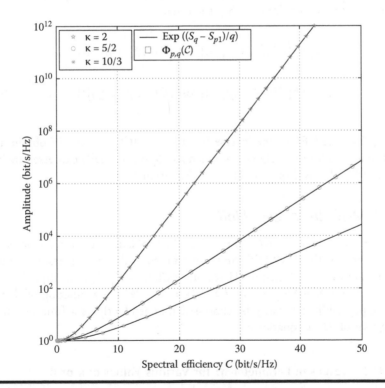

Figure 11.4 **Comparison of $e^{S_q - S_{p1}/q}$ with the parametric function $\phi_{p,\,q}(C)$ obtained from Equation 11.21 as a function of the spectral efficiency for various receive/transmit antenna ratios with $\kappa = 1/\beta > 1$.**

and q such that $p > q$, it can be seen that $S_q - S_{p1}/q$ is monotonically increasing in a logarithmic way at low C and in a linear way at high C. Moreover, this function is equal to zero at $C = 0$. In an effort to obtain the function that best fits these curves, the curve-fitting method leads to the parametric function

$$\phi_{p,q}(C) = \cosh\left(C\ln(2)\right)/\left(q\eta_1\right)^{\eta_1} \tag{11.21}$$

which provides a satisfying approximation for any of the $e^{S_q - S_{p1}/q}$ curves, as shown in Figure 11.4 for $\kappa = p/q = 2$, $5/2$, $10/3$ and $\eta_1 = 2.55$, 2.247, 1.988. Consequently,

$$\Phi_{p,q}(C) = q\eta_1 \ln\left(\cosh\left(C\ln(2)/\left(q\eta_1\right)\right)\right) \tag{11.22}$$

provides an accurate approximation for $S_q - S_{p1}$ as a function of C when $p \geq 2q$. The values of the parameter η_1 are given in Table 11.1 for various antenna configurations.

Finally, S_q and S_{p1} are expressed solely as a function of p, q, and C by using Equation 11.22 with $C\ln(2) \approx S_q + S_{p1}$, such that

$$S_q \approx 0.5\left(C\ln(2) + q\eta_1 \ln\left(\cosh\left(C\ln(2)/\left(q\eta_1\right)\right)\right)\right) \text{ and}$$

$$S_{p1} \approx 0.5\left(C\ln(2) - q\eta_1 \ln\left(\cosh\left(C\ln(2)/\left(q\eta_1\right)\right)\right)\right) \tag{11.23}$$

when $p \geq 2q$. The CFA of the inverse function, $f^{-1}(C)$, required in obtaining the EE–SE trade-off for the 1-RAU case when $p \geq 2q$ is eventually expressed by inserting S_q and S_{p1}, that is, Equation 11.23, in Equation 1.19.

11.4.2 M-Radio Access Unit

Whenever more than one RAU are active, the problem of defining a closed-form expression for the D-MIMO EE–SE trade-off is equivalent to expressing S_q and S_{pi} as a function of C in Equation 11.18. Since $C\ln(2) \approx S_q + \Sigma S_{pi}$ in Equation 11.11, then by defining $S_q - \Sigma S_{pi}$ and $S_{pi}/S_{p_1}, \forall i \in \{1,\dots,M\}$, as a function of C, S_q and $S_{pi}, \forall i \in \{1,\dots,M\}$, can easily be expressed independently as a function of C by solving a set of $M + 1$ equations.

Table 11.1 Values of Parameter η1 for Various Values of κ or β

$\kappa = p/q$	9	8	7	6	5	9/2	4	7/2	10/3	3	2
$\eta 1$	1.616	1.640	1.671	1.713	1.777	1.820	1.877	1.955	1.987	2.067	2.558

Approximation of $S_q - \sum S_{pi}$

In the following, the parametric function $\Phi_{p,q}$, which accurately approximates $S_q - \sum S_{pi}$ by means of a heuristic curve-fitting method, is proposed, such that [3]

$$\Phi_{p,q}(\mathcal{C}) \approx S_q - \sum S_{pi} = \ln\left(\frac{(1+g)^q}{\prod_{i=1}^{M}(1+d_i)^p}\right) \tag{11.24}$$

since it can be proved by direct substitution that $p(-1+(1/1+g))-q(-M+\sum_{i=1}^{M}(1/1+d_i))=0$ in Equation 11.12, as in the 1-RAU case. Similarly to the 1-RAU case, a curve-fitting method is used to design a parametric function $\phi_{p,q}(\mathcal{C})=e^{\Phi_{p,q}(\mathcal{C})/q}$ that tightly fits $e^{S_q - \sum S_{pi}/q}$ for $p > q$. Then, $e^{S_q - \sum S_{pi}/q}$ is numerically evaluated as a function of $\mathcal{C}$ for a fixed channel gain offset, that is, $\Delta_i = \alpha_i^2 / \alpha_1^2, \forall i \in \{1,\ldots,M\}$, and also for various values of M, p, and q, as shown in Figure 11.5. Similarly to the 1-RAU case, $e^{S_q - \sum S_{pi}/q}$ presents the feature of a logarithmic function at low $\mathcal{C}$ and

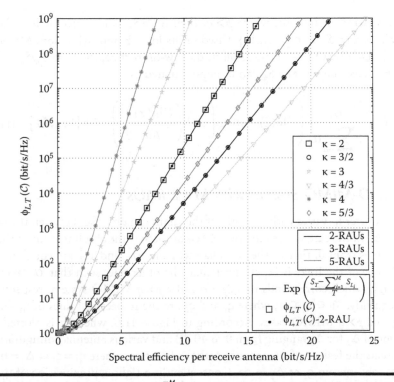

Figure 11.5 Comparison of $e^{S_q \sum_{i=1}^{M} S_{pi}/q}$ with the parametric function $\Phi_{p,q}(\mathcal{C})$ obtained from Equations 11.22 and 1.28 as a function of the spectral efficiency per RAU -receive antenna for various numbers of RAUs, receive/transmit antenna ratios with $\kappa = 1/\beta > 1$ and $\Delta_i = 10(i-1)$ dB.

a linear function at high C. The function is also monotonic, and its value at $C=0$ is zero. The parametric function given in Equation 11.22 for $S_q - S_{p1}$ in the 1-RAU case also provides a satisfying approximation for $S_q - \sum_{i=1}^{M} S_{pi}$ when $p > q$, as shown in Figure 11.5. Then, S_q is obtained as

$$S_q \approx 0.5 \Big(C \ln(2) + q \eta_1 \ln \big(\cosh \big(C \ln(2) / (q \eta_1) \big) \big) \Big) \tag{11.25}$$

by solving Equations 11.22 and $C \ln(2) \approx S_q + \sum_{i=1}^{M} S_{pi}$.

Approximation of S_{pi}

Furthermore, by solving Equations 11.22 and $C \ln(2) \approx S_q + \sum_{i=1}^{M} S_{pi}$, $\sum_{i=1}^{M} S_{pi}$ is also obtained as

$$\sum_{i=1}^{M} S_{pi} \approx 0.5 \Big(C \ln(2) - q \eta \ln \big(\cosh \big(C \ln(2) / (q \eta) \big) \big) \Big) \tag{11.26}$$

In the scenario of interest, where $p > q$, then $S_q \gg S_{pi}$, and hence, it is sufficient to evaluate any $S_{pi} \forall i \in \{1,\dots,M\}$ based on its low-SE approximation. Moreover, from the D-MIMO EE–SE trade-off in the low-SE regime, the ratio $S_{pi}/S_{p1} \approx \Delta_i$ at low SE. Consequently, any S_{pi} can be approximated as

$$S_{pi} \approx \frac{\alpha_i^2}{2 \sum_{i=1}^{M} \alpha_i^2} \left(C \ln(2) - q \eta_1 \ln \left(\cosh \left(\frac{C \ln(2)}{q \eta_1} \right) \right) \right), \quad i \in \{1,\dots,M\} \tag{11.27}$$

11.4.3 D-MIMO System with $M=2$ RAUs

The parameter η_1 varies with the ratio of the channel gain offset between the links, that is, Δ_i. In the case where only two RAUs are active, the absolute value of the log of Δ_2 varies from 0, that is, the two channel gains α_1^2 and α_2^2 are equal, to $+\infty$, that is, one of the links is far stronger than the other one, such that $\alpha_2^2 \gg \alpha_1^2$ or $\alpha_1^2 \gg \alpha_2^2$, which corresponds to a $2p \times q$ and a $p \times q$ MIMO system, respectively. Consequently, $\eta_1 \in [\varsigma_1, \varsigma_2]$, where ς_1 and ς_2 are the respective values of η_1 for the $2p \times q$ and $p \times q$ MIMO cases. According to Figure 11.6, where η_1 is plotted as a function of Δ_2 for Δ_2 ranging from 0 to 40 dB and various antenna configurations, η_1 presents the feature of a tangent hyperbolic function, where $\eta_1 = \varsigma_1$ at $\Delta_2 = 0$ dB, and η_1 converges to ς_2 as $\Delta_2 \to \infty$. Consequently, a tight approximation of η_1 can be defined by means of a curve-fitting method as

$$\eta_1 \approx \varsigma_1 + (\varsigma_2 - \varsigma_1) \tanh \left(10 \log_{10}(\Delta_2) \lambda_1 \right)^{\lambda_2} \tag{11.28}$$

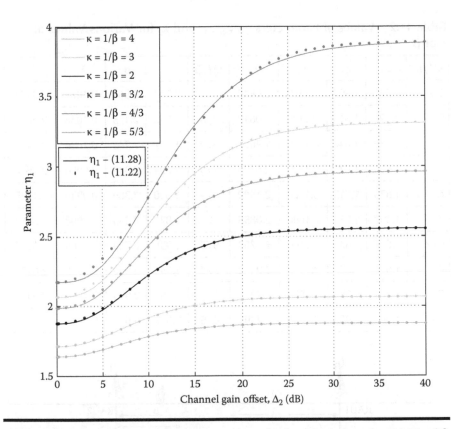

Figure 11.6 Comparison of the parameter η_1 obtained from Equation 11.22 with η_1 obtained via our interpolation approach of Equation 11.28 as a function of the channel gain offset Δ_2 in dB.

where the tightness of this approximation is shown in Figure 11.6. An accurate approximation of $S_q - (S_{p1} + S_{p2})$ is then obtained via $\Phi_{p,q}(C)$ by inserting Equation 11.28 into Equation 11.22, as illustrated in Figure 11.5. Note that the parameters σ_1, σ_2, λ_1, and λ_2 are given in Table 11.2 for some selected antenna settings.

Inserting η_1 in Equation 11.28 into Equations 11.25 and 11.27, S_q, $S_{pi} \forall i + \in \{1,2\}$ and $x_i \forall i \in \{1,2\}$ are easily obtained solely as a function of the variable C and for various parameters. Finally, the CFA of the EE–SE trade-off for the uplink and downlink of the 2-RAU D-MIMO system is then formulated by substituting S_q, $S_{pi} \forall I + \in \{1,2\}$, and $x_i \forall i \in \{1,2\}$ into $\tilde{f}^{-1}(C)$ in Equation 11.18 and inserting $\tilde{f}^{-1}(C) \approx \tilde{f}^{-1}(C)$ in Equations 11.16 and 11.17, respectively.

11.4.4 Accuracy of the CFAs: Numerical Results

This section verifies the accuracy of the CFA of the D-MIMO EE–SE trade-off for the 2-RAU scenario in both the uplink and downlink channels over the Rayleigh

Table 11.2 Values of Parameters ς_1, ς_2, λ_1, and λ_2 for Various Values of κ or β

$\kappa\lvert 1/\beta$	ς_1	ς_2	λ_1	λ_2	$\kappa\lvert 1/\beta$	σ_1	σ_2	λ_1	λ_2
10	1.517	1.597	0.1132	2.0640	3	1.713	2.067	0.1072	2.3121
9	1.526	1.616	0.1120	2.0619	8/3	1.752	2.175	0.1036	2.2831
8	1.536	1.640	0.1080	1.9627	5/2	1.777	2.243	0.1036	2.3447
7	1.551	1.671	0.1100	2.0683	7/3	1.804	2.330	0.0996	2.2676
6	1.570	1.713	0.1128	2.1882	9/4	1.820	2.389	0.0936	2.1345
5	1.597	1.777	0.1080	2.0855	2	1.877	2.558	0.1008	2.5063
9/2	1.616	1.820	0.1100	2.1763	9/5	1.938	2.769	0.0988	2.5974
4	1.640	1.877	0.1100	2.2472	7/4	1.955	2.836	0.0980	2.6178
7/2	1.671	1.955	0.1076	2.2272	5/3	1.987	2.964	0.0964	2.6490
10/3	1.683	1.987	0.1084	2.2805	8/5	2.017	3.086	0.0944	2.6490

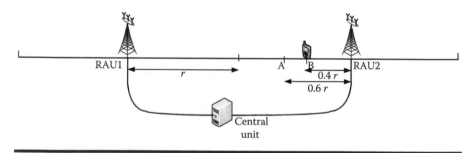

Figure 11.7 2-RAU D-MIMO linear layout.

fading channel by comparing it with the Monte Carlo simulation–based nearly-exact approach for various antenna configurations. A linear 2-RAU D-MIMO system in which the two RAUs are positioned as illustrated in Figure 11.7 is considered. It should be noted that the results and insights drawn from the linear architecture can be applied to any 2-RAU D-MIMO architecture. To present practical results, the average channel gain between the UT and the ith RAU, that is, α_i, is determined by the following path loss model:

$$\alpha_i = \sqrt{L_0\left(1+\frac{D_i}{D_0}\right)^{-\eta}} \tag{11.29}$$

Table 11.3   RAU System Parameters

Parameter	Symbol	Value
RAU cell radius (m)	r	100
Reference distance (m)	D_0	1
Reference path loss value (dB)	L_0	34.5
Path loss exponent	η	3.5
Max. UT transmit power (dBm)	P_{max}	27
Max. RAU transmit power (dBm)	P_{max}	46
Thermal noise density (dBm/Hz)	N_0	−169
Channel bandwidth (MHz)	W	10

where:

D_i is the distance between the UT and the ith RAU
η is the path loss exponent
L_0 is the power loss at a reference distance D_0

The values of the parameters η and L_0 are set according to those defined for the path loss model in the urban macro scenario and are given in Table 11.3 along with other system parameters [39].

Figure 11.8 compares the CFA of the D-MIMO EE–SE trade-off with the nearly exact EE in the downlink for an idealistic PCM, UT positioned at points A and B and for some specific values of $\kappa=1/\beta$, that is, $\kappa=\{2,1.5,1.6,1.5\}$, which corresponds to the antenna configurations $p\times q=\{2\times1,3\times2,5\times3,6\times4\}$. Results clearly show the tight fitness of the CFA with the nearly exact EE; hence, it is a graphical illustration of the accuracy of the CFA for the downlink channel. In addition, these results reveal that the most energy-efficient point occurs at $C\to0$ when an idealistic PCM is assumed, since from Equation 11.5, the maximum idealistic EE is obtained at $C\to0$. Note also that moving the UT from point A to point B, that is, closer to its serving RAU (RAU2), obviously improves the EE. Figure 11.9 compares the CFA of the D-MIMO EE–SE trade-off with the nearly exact EE in the downlink for the realistic PCM and some specific antenna configurations. The results obtained here are very different from those obtained in the idealistic setting. In the latter, increasing the number of antennas at either the RAU or the UT results in an improvement in both the SE and the EE. However, in the realistic PCM, increasing the number of antennas at the RAU results in an improvement in SE but not necessarily in EE, as a result of the additional power that is consumed by both the baseband processing and the RF unit. The power consumed by an additional unit, that is, the PA, increases linearly with p in the downlink.

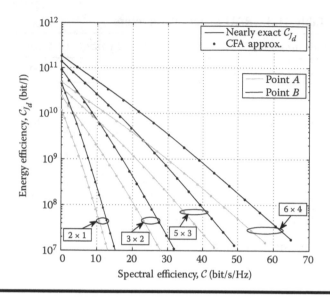

Figure 11.8   Comparison of the EE–SE trade-off for the downlink of a 2RAU D-MIMO system obtained via the nearly exact approach and by CFA based on the idealistic PCM.

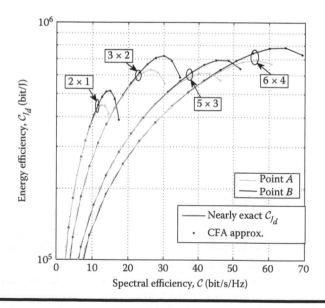

Figure 11.9   Comparison of the EE–SE trade-off for the downlink of a 2RAU D-MIMO system obtained via the nearly exact approach and by CFA based on the realistic PCM.

11.5 Low-SE Approximation of the D-MIMO EE–SE Trade-Off

The results that have been obtained so far in this chapter clearly indicate that the low-SE regime is the energy-efficient regime in the D-MIMO system when considering the idealistic PCM. It is known from Equation 11.30 that in the low-SE regime, $C \to 0$, and hence, the D-MIMO EE–SE trade-off expression in this regime can be expressed as [2]

$$f^{-1}(C)_{C \to 0} = \frac{N_t C \ln(2)}{\mathbb{E}\left(\text{tr}\left[\widetilde{\mathbf{H}}^\dagger \widetilde{\mathbf{H}}\right]\right)} \tag{11.30}$$

where N_t is the total number of transmit antennas, such that $N_t = p$ and $N_t = Mp$ in the uplink and downlink cases, respectively. This implies that $\mathbb{E}(\text{tr}[\widetilde{\mathbf{H}}\,\widetilde{\mathbf{H}}])$ has to be evaluated; however, a direct evaluation of this term is tedious. In order to circumvent this, one can resort to evaluating the D-MIMO EE–SE trade-off CFA of Equation 11.18 at $C \to 0$, as in the following proposition.

In the low-SE regime, $C \to 0$, such that Equation 11.18 can be simplified, and hence, the low-SE approximation of the inverse function $f^{-1}(C)$, which is used in characterizing the D-MIMO EE–SE trade-off over the Rayleigh fading channel, is given by

$$\widetilde{f}_l^{-1}(C) = \frac{C \ln(2)}{p\beta \sum_{i=1}^{M} \alpha_i^2} \tag{11.31}$$

where:
$\beta = 1$ and $\beta = q/p$ in the uplink and downlink scenarios, respectively
C is the capacity per unit bandwidth (SE)
M is the number of RAUs
α_i is the average channel gain between the UT and the ith RAU

It can be observed from Equation 11.31 that the low-SE approximation of the idealistic D-MIMO EE–SE trade-off is independent of the number of transmit antennas, which is in line with the results in [2,40] for the point-to-point MIMO Rayleigh fading channel. In addition, increasing the number of receive antennas increases the EE as a result of an improved diversity gain.

To present some numerical results, the D-MIMO architecture depicted in Figure 11.10 is considered, in which seven RAUs communicate with the UT. The system parameters are given in Table 11.3, while the average channel gain between the UT and the ith RAU, that is, α_i, is defined by the path loss model given in Equation 11.29. Table 11.4 shows RAU realistic power model parameters, which should be considered in practical path-loss calculations. Figures 11.11 and 11.12

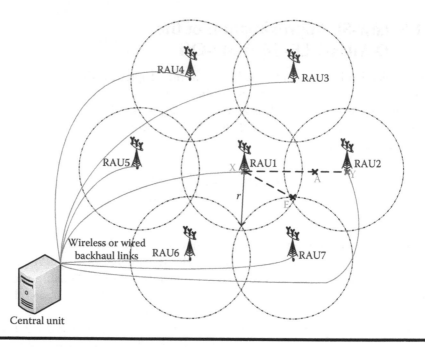

Figure 11.10 D-MIMO system model.

Table 11.4 RAU Realistic Power Model Parameters

Parameter	Symbol	Value
Receive part of RRH P_0 (W)	P_{0_u}	24.8
Transmit part of RRH P_0 (W)	P_{0_d}	59.2
RRH load-dependent PCM slope	Γ	2.8
UT receive circuit power (W)	P_{cr}	0.1
UT transmit circuit power (W)	P_{ct}	0.1
UT power amplifier efficiency (%)	μ_{UT}	100%
Weighting factor	ϕ	0.5
No. of interfaces per aggregation switch	max_{dl}	24
Power consumed by one interface (W)	p_{dl}	1
Power consumed optical SFP (W)	C	1
Max. power consumed by switch (W)	p_b	300
Max. traffic through switch (Gb/s)	Ag_{max}	24

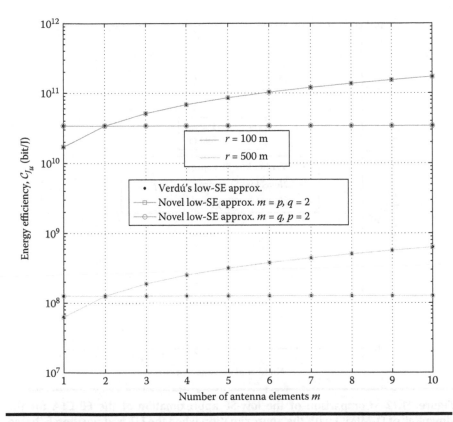

Figure 11.11 **Comparison of the low-SE approximation of the EE CFA for the uplink of D-MIMO with the approximation when the UT is at position E, based on the idealistic PCM. (From Verdú, S.,** *IEEE Transactions on Information Theory,* **48, 1319–1343, 2002.)**

present some numerical results on the low-SE approximations of the EE of the D-MIMO system for both the uplink and downlink scenarios, respectively, at $r = 100$ and 500 m. Assuming that the UT is located at point E, the novel EE approximation at low SE in Equation 11.31 is compared with Verdú's low-SE approximation, which is given in Equation 11.30 for both the uplink and downlink channels [2]. The results show a tight match between the novel low-SE approximation of the D-MIMO EE in Equation 11.31 and Verdú's low-SE approximation. In line with the insights drawn in Section 11.5 on the low-SE regime approximation, Figures 11.11 and 11.12 show that the EE of the D-MIMO system is independent of the number of transmit antennas, that is, q in the uplink channel and p in the downlink channel, while increasing the number of receive antennas, that is, p and q in the uplink and downlink channel, respectively, improves the EE when an idealistic PCM is considered. Furthermore, the EE of the system increases when the channel quality is improved as a result of the shorter cell radius.

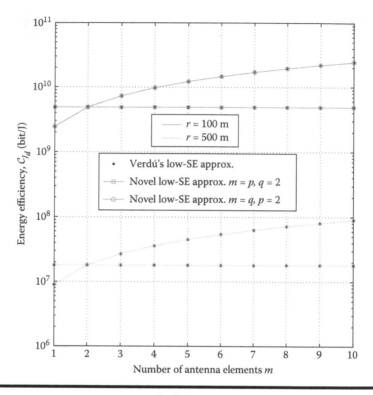

Figure 11.12 **Comparison of the low-SE approximation of the EE CFA for the downlink of D-MIMO with the approximation when the UT is at position E, based on the idealistic PCM. (From Verdú, S., *IEEE Transactions on Information Theory*, 48, 1319–1343, 2002.)**

11.6 High-SE Approximation of the D-MIMO EE–SE Trade-Off

In general, high-SNR/SE approximations are of practical interest for accurately assessing the SE or EE of communication networks that operate in the mid- to high-SNR/SE regime [41]. The high-SE approximation of the D-MIMO EE–SE trade-off can be obtained via the high-SNR approximation of the unique real positive root of the $(M+1)$th-degree polynomial given in Equation 11.10, that is, d_0. According to [32], the formulation of the asymptotic approximation of this root is dependent on the relationship between the total number of antennas at the RAUs and the total number of antennas at the UT, that is, $Mp > q$, $Mp = q$, or $Mp < q$. Here, we focus on DM-MIMO where $Mp > q$. Based on [32], it has recently been proved in [14] that the high-SE approximation of the DM-MIMO EE–SE trade-off, that is, $\tilde{f}_b^{-1}(C)$, is formulated by means of an accurate CFA as

$$\tilde{f}_h^{-1}(C) = -\frac{1}{V}\left[1 + \left(2W_0\left(-2^{-((C/q)+1)}e^{\left(\sum_{i=1}^{M} S_{p_i}^{\infty}/q - (1/2)\right)}\right)\right)^{-1}\right] \tag{11.32}$$

where

$$V = \kappa\left[\sum_{i=1}^{M}\frac{\alpha_i^2\left(\kappa\sum_{k=1}^{M}\frac{\alpha_k^2}{\alpha_1^2}x_k^{\infty} - \beta\right)}{\kappa\sum_{k=1}^{M}\frac{\alpha_k^2}{\alpha_1^2}x_k^{\infty} - \beta\left(1 - \frac{\alpha_i^2}{\alpha_1^2}\right)}\right] \tag{11.33}$$

In addition, x_i^{∞} and $\sum_{i=1}^{M} S_{p_i}^{\infty}$, which are independent of C, are given, respectively, as

$$x_i^{\infty} = \frac{u_i}{u_1} = \frac{\kappa\sum_{k=1}^{M}\Delta_k x_k^{\infty} - \beta\Delta_i x_i^{\infty}}{\kappa\sum_{k=1}^{M}\Delta_k x_k^{\infty} - \beta}, \quad i \in \{1,\ldots,M\} \tag{11.34}$$

and

$$S_{p_i}^{\infty} = \frac{p}{2}\left[-1 + \frac{\kappa\sum_{k=1}^{M}\Delta_k x_k^{\infty} - \beta}{\kappa\sum_{k=1}^{M}\Delta_k x_k^{\infty} - \beta(1-\Delta_i)} + 2\ln\frac{\kappa\sum_{k=1}^{M}\Delta_k x_k^{\infty} - \beta(1-\Delta_i)}{\kappa\sum_{k=1}^{M}\Delta_k x_k^{\infty} - \beta}\right]$$

$$\tag{11.35}$$

where:

C	is the capacity per unit bandwidth (SE)
M	is the number of RAUs
$\Delta_i = \alpha_i^2/\alpha_1^2$	is the average channel gain between the UT and the ith RAU
$W_0(x)$	is the real branch of the Lambert function [42]
$\kappa = p/q, \beta = 1$	in the uplink channel
$\kappa = 1, \beta = q/p$	in the downlink channel

The main advantage of the high-SE approximation of the D-MIMO EE–SE trade-off over its exact closed form is that it can be easily evaluated, as it does not require any parameter lookup table for its evaluation. It can be seen from Equation 11.32 that the high-SE approximation of the EE–SE trade-off is dependent on the capacity C, the number of antennas at the UT and the RAU, that is, q and p, respectively, the channel gain between the UT and the RAU, that is, α_i^2, and the number of active RAUs. Furthermore, by using the properties of the Lambert function, $\tilde{f}_b^{-1}(\mathcal{C})$ increases linearly (on a log scale) with a linear increase in $\mathcal{C}$ when all other variables are fixed in Equation 11.32. Consequently, the idealistic EE decreases linearly (on a log scale) as $\mathcal{C}$ increases. In addition, it can be observed in Equation 11.32 that, while $Mp > q$, increasing q leads to a greater improvement in the diversity gain than increasing either M or p. This improvement in the diversity gain results in a reduction in $\tilde{f}_b^{-1}(\mathcal{C})$ and consequently an improvement in the idealistic EE, as depicted in Figure 11.13. To present some numerical results on the high-SE approximation of the D-MIMO EE–SE trade-off, the scenario in which only RAU1, RAU2, and RAU7 are active in the D-MIMO architecture of Figure 11.10 is considered, that is, $M = 3$, and the UT is assumed to be positioned at point A, which is $0.6r$ from RAU2 with $r = 50$ m. In Figure 11.13, the antenna configurations $p \times q = \{1 \times 1, 2 \times 2, 2 \times 4, 4 \times 2\}$, which ensure that $Mp > q$, are also used to demonstrate the accuracy of high-SE approximation in the downlink channel. The high-SE approximation is valid, and the results in Figure 11.13

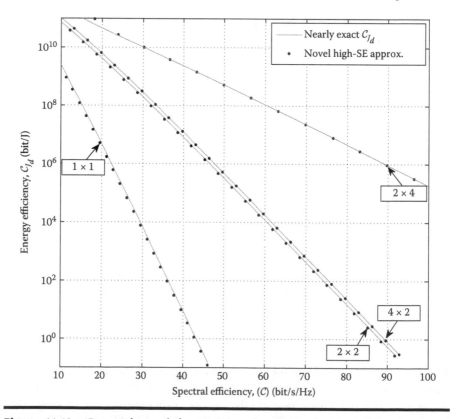

Figure 11.13 Comparison of the EE–SE trade-off for the downlink of a 3RAU D-MIMO system obtained via the nearly exact approach with its high-SE approximations when the UT is at A, based on the idealistic PCM.

indicate its great accuracy for any antenna settings, since it tightly matches the nearly exact EE results in any situation. Moreover, not only are the novel high-SE approximations very accurate at high SE, but they are also accurate at mid-SE, that is, for $C > 10$ bit/s/Hz, in Figure 11.13. In Figure 11.13, the impact of increasing the number of antenna elements at the UT, that is, q, and each RAU, that is, p, on the idealistic EE are also investigated. For the case where $Mp > q$, that is, $p \times q = \{2 \times 2, 4 \times 2, 2 \times 4\}$, increasing p has a less significant impact than increasing q on improving the idealistic EE. As can be seen in Equation 11.32, q is dividing C, such that it directly affects the diversity and modifies the slope of the trade-off curves, as is clearly depicted in Figure 11.13, whereas p acts as an EE multiplicative gain, since curves with different p values are parallel to each other. These results are in line with the insights previously drawn from Equation 11.32.

11.7 EE Gain of D-MIMO over C-MIMO

The power efficiency gain of the D-MIMO system over the C-MIMO system in which all the antenna elements are separated by a few wavelengths has been demonstrated while considering the idealistic PCM. In this section [17], the idealistic and realistic EE gains of the D-MIMO system over the C-MIMO system are demonstrated in the downlink channel. To evaluate how the D-MIMO system compares with the C-MIMO system in terms of EE, the EE gain of D-MIMO over C-MIMO can be expressed according to Equation 11.17 and the realistic PCM of [43] as

$$G_E = G_{Id,SE} \frac{\Gamma f_1^{-1}(C_C) + (MpP_{0_d} / N)}{M\left(\Gamma f_M^{-1}(C) + \bar{P_0} / N\right)} \tag{11.36}$$

where:

$G_{Id,SE} = C/C_C$ is the idealistic SE gain

C_C and C are the capacities of the C-MIMO and D-MIMO systems, respectively

$f_1^{-1}(C_C)$ and $f_M^{-1}(C)$ are approximated in Equations 11.19 and 1.18, respectively

From a PCM perspective, the C-MIMO is considered to use an RRH. This EE gain can result from D-MIMO transmit power reduction capability when both systems are required to achieve the same SE, that is, $C = C_C$. The idealistic EE gain due to power reduction, which is denoted by $G_{Id,PR}$, is then simply $G_{Id,PR} = f_1^{-1}(C_C) / M f_M^{-1}(C_C)$, since $G_{Id,SE} = 1$, while its realistic value, $G_{Re,PR}$, is simply a ratio of the total consumed power in the two systems, as observed from Equation 11.36.

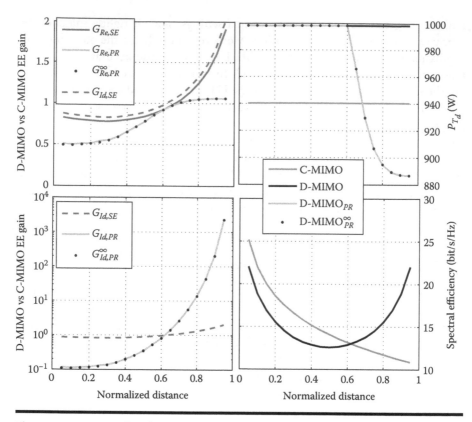

Figure 11.14 **EE gain of D-MIMO ($M=7$) over C-MIMO when the UT is moving from point X to point Y, based on both the idealistic and realistic PCMs in the downlink channel.**

The EE gain of D-MIMO over C-MIMO can also be approached via its SE improvement capability when a fixed total transmit power is assumed for both systems, that is, $Mf_M^{-1}(\mathcal{C}) = f_1^{-1}(\mathcal{C}_C)$. Hence, this EE gain is simply equivalent to the SE gain, that is, $G_{Id,SE}$, in the idealistic setting. Note that the D-MIMO system incorporates an additional backhauling induced power in comparison with the C-MIMO system; hence, the realistic EE gain as a result of its SE improvement capability, denoted as $G_{Re,SE}$, is always lower than $G_{Id,SE}$, as can be seen in Figure 11.14.

Figure 11.14 compares the D-MIMO with the C-MIMO system in terms of EE for various normalized UT positions and for the antenna configuration $p \times q = \{2 \times 1\}$. Here, it is considered that all RAUs are active in the D-MIMO system, that is, $M=7$, and that all RAUs are colocated at RAU1 in the C-MIMO case. In the lower right of the graph, the SEs of both the D-MIMO and the C-MIMO system are plotted when the total transmit power is set to 46 dBm. As expected, the C-MIMO system has a higher SE than D-MIMO when the UT is within its range, since the C-MIMO has Mp colocated antennas giving microdiversity gain, while in the D-MIMO system, each distributed RAU is equipped with p antennas, and the combination of macro- and microdiversity gains results in a higher SE when the UT is close to the cell edge (RAU2). In the upper left graph, the idealistic and realistic EE gains of the D-MIMO system over the C-MIMO system, that is, $G_{Id,SE}$ and $G_{Re,SE}$, which are obtained from Equation 11.36, are plotted. It can be observed that the SE improvement capability of D-MIMO when the UT is in close proximity with RAU2 results in EE gain. In line with earlier analysis in Section 11.7, the idealistic EE gain as a result of the SE improvement ability of D-MIMO, that is, $G_{Id,SE}$, is always greater than the realistic EE gain, that is, $G_{Re,SE}$. Furthermore, to demonstrate the EE gain of D-MIMO over C-MIMO as a result of power reduction, the novel high-SE approximation is used to plot this EE gain for both the idealistic and realistic PCMs, that is, $G_{Id,PR}^{\infty}$ and $G_{Re,PR}^{\infty}$, respectively, when the total transmit power of the C-MIMO is fixed to 46 dBm and the D-MIMO system achieves the same SE as the C-MIMO system. The EE gains, $G_{Id,PR}$ and $G_{Re,PR}$, are also plotted based on a numerical search approach to further demonstrate the accuracy of the novel high-SE approximations. In the upper right graph, the total power consumption of the C-MIMO and D-MIMO systems when the total transmit power is fixed at 46 dBm is plotted. In addition, the total power consumption of the D-MIMO system as a result of the EE gains $G_{Re,PR}$ and $G_{Re,PR}^{\infty}$, that is, D-MIMO$_{PR}$ and D-MIMO$_{PR}^{\infty}$, respectively, is also plotted. The results indicate that a reduction in the total power consumption in the D-MIMO system can be achieved by sacrificing the SE gain of D-MIMO while transmitting at a lower power for cell-edge users.

11.8 Summary

In this chapter, the fundamental trade-off between the EE and the SE of the D-MIMO system, which is a candidate architecture for the future 5G deployment, was presented. To this end, a generic accurate closed-form expression of the EE–SE trade-off for both the uplink and the downlink of the D-MIMO Rayleigh fading channel was derived by considering the idealistic and realistic PCMs. The D-MIMO EE–SE trade-off CFA was then shown to simplify into the MIMO expression for the case when only one RAU is active. Next, details on how the parameters for the trade-off expression are generated for the case of practical antenna settings ($p > q$) were provided by using a heuristic curve-fitting method. The accuracy of the CFA was shown graphically for various practical antenna configurations and for a wide range of SEs. The CFA was then used to show that in an idealistic PCM, increasing the number of antennas at either the RAU or the UT results in an improvement in both the SE and the EE, while in a realistic PCM, increasing the number of antennas at the RAU results in an improvement in SE but not necessarily in EE.

The low- and high-SE approximations of the EE–SE trade-off for the generic *M*-RAUs D-MIMO system were also presented. The accuracy of these approximations was verified with the Monte Carlo simulation–based nearly exact approach. The low-SE approximation was shown graphically to be accurate in the low-SE regime, while on the other hand, the high-SE approximation was shown to be accurate in both the mid- and high-SE regimes. Next, the EE gain of D-MIMO over C-MIMO was formulated in the downlink channel for both the idealistic and realistic PCMs. Furthermore, low- and high-SE approximations of the D-MIMO EE–SE trade-off were used in evaluating the EE gains. In both PCMs, D-MIMO was found to be more energy efficient than C-MIMO for cell-edge UTs.

References

1. Y. Chen, S. Zhang, S. Xu, and G. Y. Li, Fundamental trade-offs on green wireless networks, *IEEE Communications Magazine* 49(6): 30–37, 2011.
2. S. Verdú, Spectral efficiency in the wideband regime, *IEEE Transactions on Information Theory* 48(6): 1319–1343, 2002.
3. F. Héliot, O. Onireti, and M. A. Imran, An accurate closed-form approximation of the energy efficiency-spectral efficiency trade-off over the MIMO Rayleigh fading channel, in *Proceedings of the IEEE International Communications Workshops (ICC) Conference*, Kyoto, IEEE, pp. 1–6, 2011.
4. C. He, B. Sheng, P. Zhu, and X. You, Energy efficiency and spectral efficiency trade-eoff in downlink distributed antenna systems, *IEEE Wireless Communications Letters* 1(3): 153–156, 2012.
5. C. Isheden and G. P. Fettweis, Energy-efficient multi-carrier link adaptation with sum rate-dependent circuit power, in *Proceedings of the IEEE Globecom*, Miami, FL, IEEE, pp. 1–6, 2010.
6. G. Miao, N. Himayat, and G. Y. Li, Energy-efficient link adaptation in frequency-selective channels, *IEEE Transactions on Communications* 58(2): 545–554, 2010.
7. G. Miao, N. Himayat, G. Y. Li, and S. Talwar, Distributed interference-aware energy-efficient power optimization, *IEEE Transactions on Wireless Communications* 10(4): 1323–1333, 2011.
8. C. Xiong, G. Y. Li, S. Zhang, Y. Chen, and S. Xu, Energy- and spectral-efficiency trade-eoff in downlink OFDMA networks, *IEEE Transactions on Wireless Communications* 10(11): 3874–3886, 2011.
9. F. Héliot, M. A. Imran, and R. Tafazolli, On the energy efficiency-spectral efficiency trade-off over the MIMO Rayleigh fading channel, *IEEE Transactions on Communications* 60(5): 1345–1356, 2012.
10. H. Kwon and T. Birdsall, Channel capacity in bits per joule, *IEEE Journal of Oceanic Engineering* 11(1): 97–99, 1986.
11. O. Onireti, F. Héliot, and M. Imran, On the energy efficiency-spectral efficiency trade-off in the uplink of CoMP system, *IEEE Transactions on Wireless Communications* 11(2): 556–561, 2012.

12. O. Onireti, F. Héliot, and M. A. Imran, Trade-off between energy efficiency and spectral efficiency in the uplink of a linear cellular system with uniformly distributed user terminals, in *Proceedings of the IEEE 22nd International Personal Indoor and Mobile Radio Communications (PIMRC) Symposium*, Toronto, ON, IEEE, pp. 2407–2411, 2011.

13. O. Onireti, F. Héliot, and M. A. Imran, On the energy efficiency-spectral efficiency trade-off of the 2BS-DMIMO system, in *Proceedings of the IEEE 76th Vehicular Technology Conference (VTC Fall)*, Canada, IEEE, pp. 1–5, 2012.

14. O. Onireti, F. Heliot, and M. A. Imran, On the energy efficiency-spectral efficiency trade-off of distributed MIMO systems, *IEEE Transactions on Communications* 61(9): 3741–3753, 2013.

15. C. Shannon, A mathematical theory of communication, *Bell System Technical Journal* 27: 379–423, 623–656, 1948.

16. A. A. M. Saleh, A. Rustako, and R. Roman, Distributed antennas for indoor radio communications, *IEEE Transactions on Communications* 35(12): 1245–1251, 1987.

17. D. Castanheira and A. Gameiro, Distributed antenna system capacity scaling, *IEEE Wireless Communications Magazine* 17(3): 68–75, 2010.

18. W. Roh and A. Paulraj, MIMO channel capacity for the distributed antenna, in *Proceedings of the VTC 2002-Fall Vehicular Technology Conference 2002 IEEE 56th*, Canada, IEEE, vol. 2, pp. 706–709, 2002.

19. W. Roh and A. Paulraj, Outage performance of the distributed antenna systems in a composite fading channel, in *Proceedings of the IEEE 56th Vehicular Technology Conference (VTC Fall)*, Canada, IEEE, vol. 3, pp. 1520–1524, 2002.

20. L. Dai, A comparative study on uplink sum capacity with co-located and distributed antenna, *IEEE Journal on Selected Areas in Communications* 29(6): 1200–1213, 2011.

21. D. Wang, X. You, J. Wang, Y. Wang, and X. Hou, Spectral efficiency of distributed MIMO cellular systems in a composite fading channel, in *Proceedings of the IEEE International Conference on Communications ICC'08*, Beijing, IEEE, pp. 1259–1264, 2008.

22. X. You, D. Wang, B. Sheng, X. Gao, X. Zhao, and M. Chen, Cooperative distributed antenna systems for mobile communications, *IEEE Wireless Communications Magazine* 17(3): 35–43, 2010.

23. L. Xiao, L. Dai, H. Zhuang, S. Zhou, and Y. Yao, Information-theoretic capacity analysis in MIMO distributed antenna systems, In *Proceedings of the VTC 2003-Spring Vehicular Technology Conference. The 57th IEEE Semiannual*, vol. 1, IEEE, pp. 779–782, 2003.

24. H. Zhang and H. Dai, On the capacity of distributed MIMO system, in *Proceedings of the 2004 Conference on Information Sciences and Systems (CISs)*, Ottawa, ON, pp. 1–5, 2004.

25. H. Zhuang, L. Dai, L. Xiao, and Y. Yao, Spectral efficiency of distributed antenna system with random antenna layout, *Electronics Letters* 39(6): 495–496, 2003.

26. L. Dai, S. Zhou, and Y. Yao, Capacity analysis in CDMA distributed antenna systems, *IEEE Transactions on Wireless Communications* 4(6): 2613–2620, 2005.

27. W. Choi and J. G. Andrews, Downlink performance and capacity of distributed antenna systems in a multicell environment, *IEEE Transactions on Wireless Communications* 6(1): 69–73, 2007.

28. W. Feng, Y. Li, S. Zhou, J. Wang, and M. Xia, Downlink capacity of distributed antenna systems in a multi-cell environment, in *Proceedings of IEEE Wireless Communications and Networking Conference WCNC 2009*, Budapest, IEEE, pp. 1–5, 2009.

29. D. Aktas, M. N. Bacha, J. S. Evans, and S. V. Hanly, Scaling results on the sum capacity of cellular networks with MIMO links, *IEEE Transactions on Information Theory* 52(7): 3264–3274, 2006.

30. F. Héliot, R. Hoshyar, and R. Tafazolli, An accurate closed-form approximation of the distributed MIMO outage probability, *IEEE Transactions on Wireless Communications* 10(1): 5–11, 2011.

31. F. Héliot, M. A. Imran, and R. Tafazoll, Energy efficiency analysis of idealized coordinated multi-point communication system, in *Proceedings of the IEEE 73rd Vehicular Technology Conference (VTC Spring)*, Yokohama, IEEE, pp. 1–5, 2011.

32. S. Lee, S. Moon, J. Kim, and I. Lee, Capacity analysis of distributed antenna systems in a composite fading channel, *IEEE Transactions on Wireless Communications* 11(3): 1076–1086, 2012.

33. A. M. Tulino and S. Verdú, *Random Matrix Theory and Wireless Communications*. Berkeley, CA: Now, 2004.

34. C. He, B. Sheng, D. Wang, P. Zhu, and X. You, Energy efficiency comparison between distributed MIMO and co-located MIMO systems, *International Journal of Communication Systems* 27(1): 1–14, 2012.

36. S. Tombaz, P. Monti, K. Wang, A. Vastberg, M. Forzati, and J. Zander, Impact of backhauling power consumption on the deployment of heterogeneous mobile networks, in *Proceedings of the IEEE Global Telecommunications Conference (GLOBECOM 2011)*, Houston, TX, IEEE, pp. 1–5, 2011.

37. R. M. Corless, G. H. Gonnet, D. E. G. Hare, D. J. Jeffrey, and D. E. Knuth, On the Lambert W Function, *Advances in Computational Mathematics* 5: 329–359, 1996.

38. N. C. Beaulieu and F. Rajwani, Highly accurate simple closed-form approximations to lognormal sum distributions and densities, *IEEE Communications Letters* 8(12): 709–711, 2004.

39. ETSI TR 125 996 V10.0.0, 3rd Generation Partnership Project. Technical Specification Group Radio Access Network; Spatial channel model for Multiple Input Multiple Output (MIMO) simulations Release 10 (3GPP TR 25.996 V10.0.0), Technical Report, 2011.

40. A. Lozano, A. M. Tulino, and S. Verdú, Multiple-antenna capacity in the low-power regime, *IEEE Transactions on Information Theory* 49(10): 2527–2544, 2003.

41. N. Jindal, High SNR analysis of MIMO broadcast channels, in *Proceedings of the International Symposium on Information Theory ISIT 2005*, Adelaide, SA, IEEE, pp. 2310–2314, 2005.

43. G. Auer, V. Giannini, C. Desset, I. Godor, P. Skillermark, M. Olsson, M. A. Imran, et al., How much energy is needed to run a wireless network? *IEEE Wireless Communications Magazine* 18(5): 40–49, 2011.

5G PHYSICAL LAYER

Chapter 12

Physical Layer Technologies in 5G

Pablo Ameigeiras, Francisco Javier Lorca
Hernando, and Jose Gabriel Martinez Martin

Contents

This chapter provides an overview of candidate physical layer technologies for fifth-generation (5G) systems. The chapter starts by describing new waveforms that may substitute orthogonal frequency-division multiplexing (OFDM) technology in 5G and discusses their advantages and disadvantages to satisfy 5G system requirements. Next, it introduces frequency and quadrature amplitude modulation (FQAM) as a modulation technique to improve the properties of intercell interference. Then, the chapter presents nonorthogonal multiple access (NOMA), in which multiple users transmit in the same frequency and time resources, but are multiplexed in the power domain. NOMA both with and without successive interference cancellation (SIC) is discussed. The chapter then describes faster than Nyquist signaling (FTN), in which the symbol rate is increased beyond the Nyquist rate. Finally, the chapter discusses full duplex radios and their application to wireless backhaul.

12.1 New Waveforms

OFDM, as well as its variant orthogonal frequency-division multiple access (OFDMA), has been widely adopted in the industry for both wireless and wireline communications. Systems such as Long-Term Evolution (LTE), IEEE 802.16 (WiMAX), several versions of IEEE 802.11 (Wi-Fi), Digital Video Broadcasting (DVB), and Asymmetric Digital Subscriber Line (ADSL) are some relevant examples of the pervasive use of OFDM and OFDMA. Among the greatest advantages of this multicarrier modulation are its capability to cope with variable channel bandwidths; its low complexity in signal processing by means of fast Fourier transforms (FFTs); its seamless integration with multiantenna systems; its ability to perform both time and frequency scheduling of users; and its inherent robustness to multipath. It is not a coincidence that these features make OFDM the ideal choice for wireless cellular systems such as LTE and LTE-Advanced.

Despite the benefits, there are a number of drawbacks, which become more apparent when moving away from the traditional horizontal mobile broadband applications (such as mobile video) toward exploring other vertical uses (such as the so-called Internet of Things):

- Frequency-synchronous operation puts stringent limits on the frequency offset and phase noise characteristics of transmitter and receiver oscillators.
- Time-synchronous operation forces devices to perform synchronization with the network up to the limits imposed by the cyclic prefix (CP). Specific techniques such as coordinated multipoint (CoMP) also require the network to be globally time synchronized [1].
- OFDM spectrum mask presents poor out-of-band (OOB) radiation behavior caused by sinc-like subcarriers, which motivate the introduction of large guard bands for the protection of spectrally adjacent systems.

- The presence of a CP for absorption of the maximum expected time dispersion of the channel introduces a loss in spectral efficiency, which can be especially relevant when the symbol length is very short.

These and other shortcomings motivate fundamental research on the evolution of OFDM with cyclic prefix (CP-OFDM) as the basic constituent waveform for 5G. Multicarrier waveforms continue to be the most interesting for 5G; therefore, most effort is being devoted toward applying suitable modifications to OFDM so as to overcome these drawbacks. Currently, researchers worldwide are expending significant effort on seeking new waveforms for 5G that are robust against relaxed time and frequency synchronicity and that improve their spectral characteristics for application in narrow spectral regions (such as TV White Spaces [2]). The foreseeable explosion of inexpensive machine-type devices making use of the 5G network calls for a significant reduction in transceiver complexity, and operation in loose time/frequency synchronization is one way to alleviate the stability requirements of the oscillators, thereby reducing costs. The lack of available spectrum below 1 GHz also demands the ability to exploit very narrow frequency regions without impairing other incumbent services operating in adjacent bands. In Sections 12.1.1 through 12.1.4, we briefly describe some of the new waveform proposals, referring the reader to the available information included in the bibliography.

12.1.1 Filterbank Multicarrier (FBMC)

FBMC is recognized as one of the most promising waveforms for 5G. Widely studied by Saltzberg [3], the basic idea comprises a bank of filters to be applied on the individual constituent subcarriers of the multicarrier signal. Filtering has the objective of reducing the large sidelobe levels of the sinc-shaped subcarriers in the frequency domain. As a result, FBMC can be suitably described by a "synthesis" filter bank at the transmitter and an "analysis" filter bank at the receiver, both performing appropriate filtering operations at the subcarrier level. The resulting shapes of the filtered subcarriers have deep implications for the overall scheme, among them being the choice of the modulation scheme to be applied on top of the subcarriers. Usually, this modulation will be offset quadrature amplitude modulation (OQAM) rather than quadrature phase-shift keying (QPSK) or multiple quadrature amplitude modulation (MQAM), to avoid undesired intersubcarrier interference effects (as will be shown in this section). The discrete-time baseband signal $x(t)$ at the output of an FBMC transmitter based on OQAM modulation can be expressed as [4]

$$x(t) = \sum_{k} \sum_{n=-\infty}^{\infty} s_{k,n} \theta_{k,n} \beta_{k,n} g(t - nN/2) e^{j(2\pi/N)kt}, \quad t = 0,1,\ldots,KN-1 \quad (12.1)$$

where:

    t is a discrete-time index

    $s_{k,n}$ is the transmitted sequence of user information at subcarrier k and symbol n

    $\theta_{k,n}$ $= j^{(k+n)}$

    $\beta_{k,n}$ $= (-1)^{kn)}\cdot\exp(-jk(KN-1)\pi/N)$

    K is an overlapping factor

    N is the number of subcarriers

    g represents the prototype filter impulse response

The first summation symbol runs over the subcarriers allocated to the user. Factor $\theta_{k,n}$ alternates real and imaginary between adjacent subcarriers and symbols. Besides the presence of $\theta_{k,n}$ and $\beta_{k,n}$ multiplicative factors, the waveform comprises the superposition of multiple signal outputs and can be described by a suitable bank of filters characterized by a prototype filter impulse response g.

Figure 12.1 illustrates a simplified diagram of an FBMC transceiver. In this figure, the prototype filter $g(t)$ (with real impulse response) operates independently on each of the subcarriers, and the real and imaginary parts of the baseband information symbols (respectively, $s_n^I(t)$, $s_n^Q(t)$) are separately processed with half a symbol delay between them (as expressed by the delayed impulse response $g(t-N/2)$). The receiver performs simple matched filter operations characterized by the time-reversed impulse responses $g(t)$ and $g(t+N/2)$, for the real and imaginary parts, respectively. The prototype filters g are designed to be half-Nyquist; that is, the squared magnitude of their frequency responses must satisfy the Nyquist criterion so as to avoid intersymbol interference (ISI).

To perform the matched filter operation corresponding to the designed filter bank, it is desirable to operate in the frequency domain, because in this case convolutions become simple multiplications. The overlapping factor K is the number of FBMC symbols that overlap in the time domain. Accordingly, factor K increases the resolution of the subcarriers in the frequency domain compared with OFDM at the cost of processing an increased number of samples (equal to $K\cdot N$). A factor $K=4$ is commonly used, because it presents good performance characteristics under reasonable complexity. Such implementation in the frequency domain is called *frequency spreading FBMC* (FS-FBMC); more details can be obtained in [5,6].

The resulting frequency response of the filtered subcarriers represents the essential difference of FBMC with respect to OFDM, as can be seen in Figure 12.2. OFDM exhibits large ripples in the frequency domain, although perfect orthogonality of the subcarriers can be ensured thanks to the properties of the sinc signal.*

---

* In OFDM, subcarriers are packed in such a way that the peak position at each subcarrier matches the null positions of all the other subcarriers. Although there is still some overlap between subcarriers, detection involves simple FFT processing of the central samples, and therefore no interference between subcarriers exists (under conditions of perfect frequency synchronization).

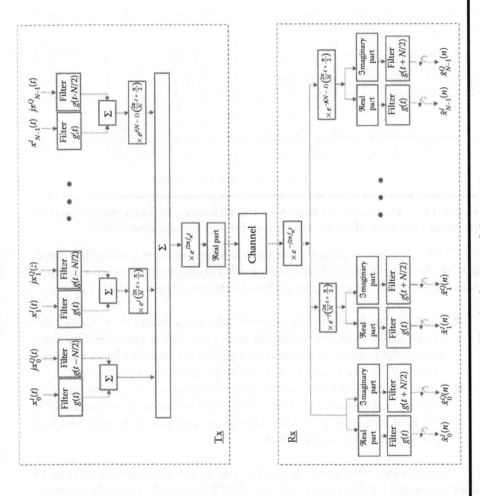

Figure 12.1 Schematic representation of the filter bank approach in FBMC.

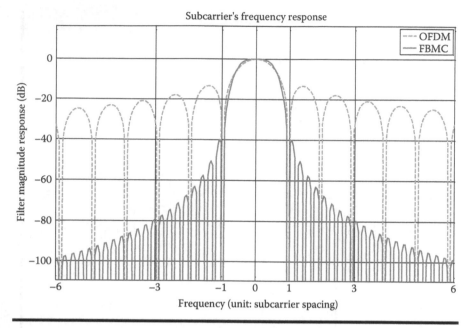

Figure 12.2 **Exemplary illustration of the difference between the subcarrier's frequency response of FBMC and OFDM, for a given subcarrier width.**

FBMC, in contrast, has negligible amplitude beyond the two subcarriers immediately adjacent to a given subcarrier, but perfect orthogonality between subcarriers is not maintained. The reasons for such nonorthogonality are twofold:

■ There is significant overlap between adjacent subcarriers in FBMC (see Figure 12.3). As opposed to OFDM, detection involves a matched filter operation at the receiver, which becomes significantly impaired by such overlap. This motivates the use of OQAM, as will be described in this section.

■ There is an additional (but small) spectral leakage beyond the two immediately adjacent subcarriers, when so-called nonperfect reconstruction (NPR) filter banks are employed.* NPR filter banks are easier to develop, and have thus become widely used in FBMC.

These two sources of nonorthogonality lead to the appearance of intercarrier interference (ICI). ICI induced by the use of NPR filter banks is usually negligible,

---

* PR filter banks are those for which the output signal is a delayed replica of the input signal. NPR filter banks, however, introduce some distortion error at its output. NPR filter banks, in the context of FBMC with OQAM, present some overlap of the associated frequency responses beyond the immediate neighbors, thus introducing a small amount of distortion that is usually negligible compared with channel-induced impairments. In contrast, PR filter banks do not present such distortion. See [7] for more information on PR and NPR filter banks.

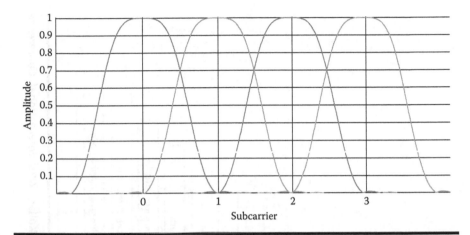

Figure 12.3 Schematic diagram showing the overlap between subcarriers in the frequency domain. Note that interference is significant only between each sub-carrier and the two immediately adjacent ones.

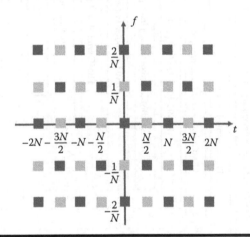

Figure 12.4 FBMC time–frequency lattice. (From B. Farhang-Boroujeny and C. H. Yuen, *EURASIP Journal on Advances in Signal Processing*, 1–17, 2010.) [8].

in the order of −65 dB for $K=4$ [9]. However, the overlap between immediately adjacent subcarriers is very significant and precludes the use of standard QPSK or MQAM modulations. A detailed analysis of the resulting ICI terms shows that interference between adjacent subcarriers is either purely real or purely imaginary, depending on the relative positions of the subcarriers in time and frequency [9], thereby motivating the use of OQAM. In OQAM, the real and imaginary parts of the complex baseband information symbols modulate the subcarriers in an inter-leaved way, as shown in Figure 12.4 (with black and gray squares meaning real and imaginary parts, respectively). Staggered mapping of the real and imaginary

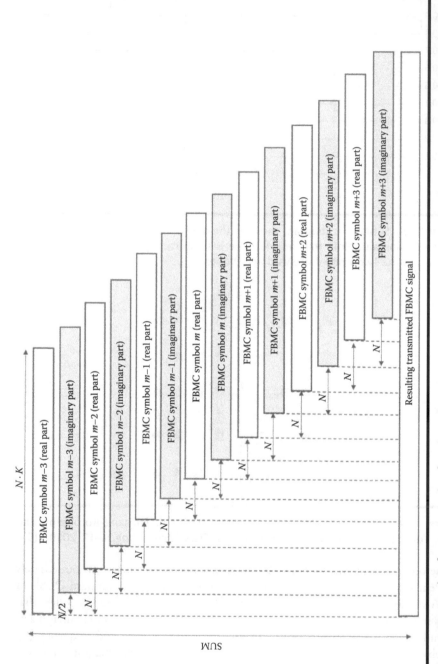

Figure 12.5 FBMC signal structure in the time domain. (From L. Vargas and Z. Kollar, *23rd International Conference Radioelektronika* (RADIOELEKTRONIKA), Budapest, IEEE, pp. 219–223, 2013.) [10].

components onto subcarriers in both time and frequency avoids any ICI between adjacent subcarriers, so that only ICI terms from the use of NPR filter banks remain; if perfect reconstruction (PR) filter banks are employed, no ICI will appear.

Under the scheme described in Figure 12.4, the resulting rate of the transmitted signal at each FBMC symbol is halved compared with OFDM, because subcarriers only carry half of the original information. This is therefore compensated by sending FBMC symbols at a doubled rate in the time domain, that is, with a time separation of $N/2$ samples. This is more clearly illustrated in Figure 12.5. Two sets of subsymbols are always sent in parallel, corresponding to the real and imaginary parts of the complex information. Note that there is significant overlap between symbols in the time domain (because each FBMC symbol comprises $K \cdot N$ samples), but the Nyquist criterion is obeyed by filter design, thus precluding any ISI in an additive white Gaussian noise (AWGN) channel without multipath.

However, in the presence of multipath, the Nyquist condition is not sufficient to prevent ISI, and some degradation in performance will occur. However, and in contrast to OFDM, no CP is necessary in this case to absorb the echoes from the previous symbols: the resulting degradation caused by ISI can be overcome by means of more sophisticated equalizers at the receive side, as opposed to OFDM, in which multipath destroys the signal's detectability in the absence of any CP. The reason for not requiring CP in FBMC is that the filter impulse response shape is designed to provide natural protection against multipath. Figure 12.6 shows a typical impulse response as used in the Physical Layer for Dynamic Access and Cognitive Radio (PHYDYAS) project [9]. The long ramps before and after the central region provide some robustness against multipath. Timing and frequency offset misalignments can also be absorbed at the receiver without strong degradation, as opposed to OFDM, in which tight synchronization is critical. The larger the K factor, the higher the protection against both multipath and time/frequency misalignments (at the cost of a longer symbol duration). The longer duration of the resulting symbols may, however, represent a drawback in machine-type applications in which very short bursts of information are to be processed.

There are also many other prototype filter designs, such as isotropic orthogonal transform algorithm (IOTA) pulses, which present good localization characteristics in time and frequency [11]. IOTA pulse shapes have been shown to alleviate ICI and ISI by inflicting zero-crossing at the other symbols' time–frequency positions located on multiples of the time and frequency dispersions, respectively [12].

With regard to transceiver implementation, in practice the most used architecture for FBMC is Polyphase Network FBMC (PPN-FBMC), schematically shown in Figure 12.7. In this figure, C2R and R2C represent complex-to-real and real-to-complex operations, respectively, selecting in each case the suitable real or imaginary parts of the information to be mapped on subcarriers. Polyphase filter structures are particularly efficient to implement in practical realizations [4,9].

Despite the multiple advantages of FBMC, it has been shown that multiple-in multiple-out (MIMO) extension requires excessive computational complexity

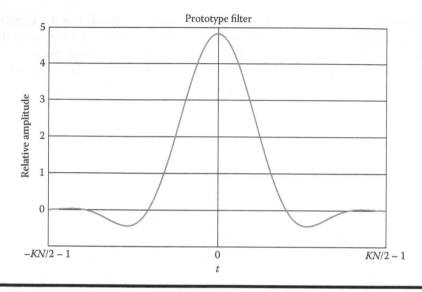

Figure 12.6 Schematic representation of the prototype filter impulse response. (From Bellanger, M., FBMC physical layer: A primer, *PHYDYAS—Physical Layer for Dynamic Access and Cognitive Radio*, 2010.)

when the flatness of the channel cannot be assumed at subcarrier level, due to long filter lengths for equalization and MIMO detection [11]. Further research is currently being conducted for the integration of MIMO techniques into FBMC systems.

In summary, FBMC has the following distinguishing features:

- No CP is needed.
- The transmission bandwidth can be exploited at full capacity using OQAM.
- Stringent OOB emission requirements can be easily satisfied with the choice of the proper prototype filter.
- Coexistence with other systems in the frequency domain can be easily achieved by leaving a single empty subcarrier at both edges of the intended transmission, thereby allowing almost full use of the available spectrum.
- Subcarriers can be grouped into independent blocks that can be digitally modulated at the baseband level.
- Operation is possible in cognitive radio applications involving spectrum sensing techniques and dynamic spectrum allocation.
- Reception is robust in the presence of time and frequency misalignments.
- Significantly higher symbol duration is expected, which must be taken into account in applications with very short bursts.

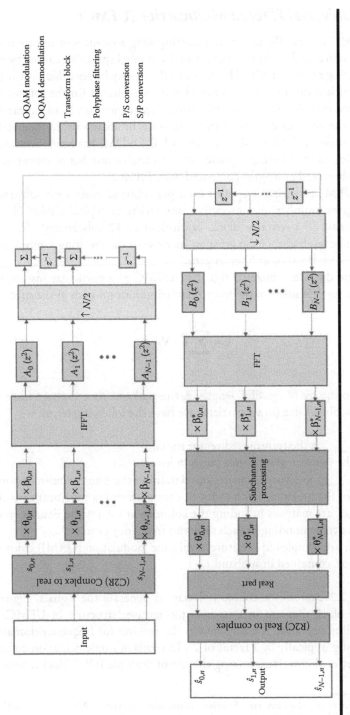

Figure 12.7 PPN-FBMC transceiver architecture.

12.1.2 Universal Filtered Multicarrier (UFMC)

While FBMC is very efficient in transmitting long sequences of information, it is not ideally suited to short bursts because of the symbol spreading caused by the sub-carrier filtering operation. CP-OFDM and FBMC may be regarded as two extreme cases in which transmissions are either band-wise filtered (in the former case, to meet spectral emission limits toward adjacent frequency bands) or subcarrier-wise filtered (in the latter case, to meet tight emission limits toward immediately adjacent transmissions). UFMC, also known as UF-OFDM, has been introduced as a generalization of the filtering approach over a variable number of subcarriers, thus leading to shorter filter lengths compared with FBMC.

With UFMC, filtering is applied on a per-subband basis, each subband comprising a given number of consecutive subcarriers (a typical choice in an LTE framework can be a resource block, equivalent to 12 subcarriers). This reduces OOB sidelobe levels without so great an increase in the resulting symbol length in accordance with the shorter filter lengths.

The time domain–generated signal in UFMC at a particular instant in time comprises a superposition of the filtered contributions for each of the subbands:

$$\mathbf{x} = \sum_{i=0}^{B-1} \mathbf{F}_i \cdot \mathbf{V}_i \cdot \mathbf{s_i} \tag{12.2}$$

where, denoting by N the FFT length, L the subband time-domain filter length, and n_i the subband size (in subcarriers), we have the following terms:

■ $[\mathbf{x}]_{(N+L-1)x1}$ is the transmitted vector signal
■ B is the number of subbands (with index i)
■ $[\mathbf{F}_i]_{(N+L-1)xN}$ are Toeplitz matrices containing the filter impulse responses for each of the subbands (with index i), thus performing the linear convolutions
■ $[\mathbf{V}_i]_{Nxn_i}$ are matrices including the columns of the inverse Fourier transform matrix corresponding to each subband frequency position*
■ $[\mathbf{s}_i]_{n_ix1}$ are complex (quadrature amplitude modulation [QAM]) constellation symbols contained in subband i

Equation 12.2 expresses a conceptual way to generate the signals. Figure 12.8 depicts the block diagram of a transmitter–receiver structure in UFMC, where an upsampling operation is considered at the receiver for frequency-domain symbol processing (typically by a factor of 2). In terms of implementation complexity, there are better alternatives avoiding the use of multiple IDFT blocks, because the

---

* The elements of the inverse Fourier transform matrix $[\mathbf{W}]_{(NxN)}$ are defined by $W_{m,n} = 1/\sqrt{N} \times (\exp-(j2\pi mn/N))_{m,n=0,1,...,N-1}$, where $j \equiv \sqrt{-1}$.

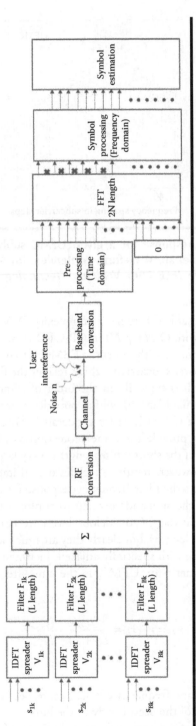

Figure 12.8 UFMC transceiver.

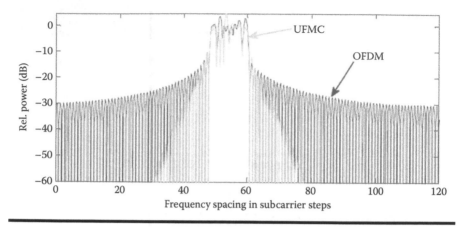

Figure 12.9 Frequency response for a given UFMC subband compared with OFDM. (From F. Schaich et al., Waveform contenders for 5G: latency transmissions, *Proceedings of the IEEE 79th Vehicular Technology Conference*, Seoul, Spring, 2014.)

"brute-force" approach described here has a complexity $O(N^2)$ real multiplications and additions compared with $O(N \log N)$ in CP-OFDM (see e.g., [13]).

The fundamental difference with respect to FBMC is that the filtering is applied over subbands instead of over subcarriers, which relaxes the filter impulse response length L. This filter length is typically in the order of standard LTE CP length (i.e., around 7% of the symbol length), which provides a "soft" protection against ISI without the need for a CP (at least for moderate delay spreads). Filter ramp-up and ramp-down areas provide the same protection as FBMC, although at a much lower level because of the shorter filter lengths. Very large delay spread values or timing offsets may, however, require specialized multitap equalizers to combat ISI. Figure 12.9 depicts the filter frequency response for a given subband, and Figure 12.10 shows how the subbands overlap to conform to the combined frequency response. Note that the OOB sidelobe levels are not as low as in FBMC because of the shorter filter lengths, but clearly they are much lower than in OFDM, thus improving coexistence with spectrally adjacent incumbent systems.

A suitable figure of merit for UFMC is the time–frequency efficiency r_{TF}, defined as follows [14]:

$$r_{TF} = r_T \cdot r_F = \frac{L_D}{L_D + L_T} \cdot \frac{N_u}{N'} \tag{12.3}$$

where:

r_T and r_F are the time and frequency efficiencies, respectively

L_D is the length of the useful body of the burst

L_T is the length of the burst tail not including useful information

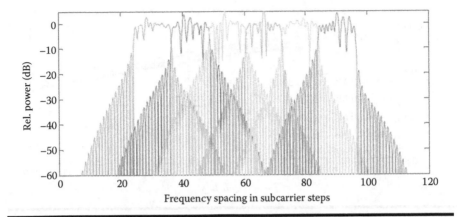

Figure 12.10 Combined frequency response of six UFMC subbands. (From F. Schaich et al., Waveform contenders for 5G: Suitability for short packet and low latency transmissions, *Proceedings of the IEEE 79th Vehicular Technology Conference*, Seoul, Spring, 2014.)

N_u is the number of usable subcarriers (excluding guards)
N' is the overall number of subcarriers

In CP-OFDM, the burst tail corresponds to the CP, while in UFMC, it is equal to the filter length. The number of usable subcarriers, N_u, depends in UFMC on the filter length and the sidelobe level α_{SLA}, and is always very close to the overall number of subcarriers N'. Figure 12.11 shows the time–frequency efficiency in UFMC using LTE as reference and assuming a transmission bandwidth of 10 MHz with subcarrier spacing 15 kHz (with normal CP). It is apparent that UFMC outperforms CP-OFDM in all cases by about 10%.

In short burst communications (for very low latency or small packet transmissions), UFMC offers advantages over FBMC and CP-OFDM, while keeping improved spectral properties over CP-OFDM, including robustness to carrier frequency offset [14,15].

12.1.3 Generalized Frequency-Division Multiplexing (GFDM)

GFDM is a flexible multicarrier modulation scheme [16] that is also being investigated for 5G systems. GFDM is a generalization of OFDM that modulates the data in a two-dimensional time–frequency block structure, in which each block consists of a number of subcarriers and subsymbols. The subcarriers are filtered with a flexible pulse-shaping filter that is circularly shifted in both time and frequency domains. A single CP for the entire block is inserted, which can be used to improve spectral efficiency.

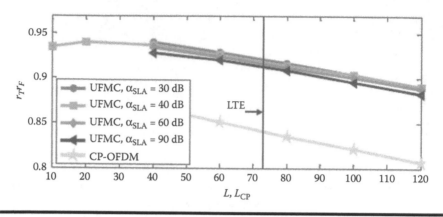

Figure 12.11 Time–frequency efficiency of UFMC and CP-OFDM for different sidelobe attenuation factors α_{SLA}. L is the tail burst length in UFMC, and L_{CP} is the cyclic prefix length in CP-OFDM. (From F. Schaich et al., Waveform contenders for 5G: Suitability for short packet and low latency transmissions, *Proceedings of the IEEE 79th Vehicular Technology Conference*, Seoul, Spring, 2014.)

GFDM can provide a reduced OOB emission compared with OFDM, as it applies the pulse-shaping filters per subcarrier. Different filter impulse responses can be used to filter the subcarriers, and this choice affects the OOB emissions. The usage of pulse-shaping filters eliminates orthogonality and introduces ICI and ISI. Hence, the use of receiving techniques, such as iterative interference cancellation, is required to mitigate this interference. However, impairments due to imperfect synchronism also affect the performance of multiple access scenarios in LTE systems in a similar way. So, by overcoming this problem, GFDM aims to relax the current requisites of oscillator accuracy of 0.1 ppm in LTE up to 10–100 times (1–10 ppm) and allow the design of simpler transmitters, leaving out complex synchronization procedures and reducing signaling overhead [17]. The circular structure of GFDM allows zero-forcing channel equalization, as efficiently used in OFDM, to be employed in the frequency domain.

Let $\vec{s}$ denote a symbols data block that contains $N = K \cdot M$ elements, which are organized into K subcarriers with M subsymbols each:

$$\vec{s} = (s_{0,0}, s_{1,0}, \ldots, s_{K-1,0}, s_{0,1}, s_{1,1}, \ldots, s_{K-1,1}, \ldots, s_{0,M-1}, s_{1,M-1}, \ldots, s_{K-1,M-1}) \qquad (12.4)$$

where the element $s_{k,m}$ represents the symbol transmitted in the kth subcarrier and mth subsymbol of the block. Each symbol $s_{k,m}$ is pulse shaped with the filter

$$g_{k,m}[n] = g\left[(n - m \cdot K) \bmod N\right] \cdot \exp\left(j2\pi \frac{k}{K} n\right) \qquad (12.5)$$

where n denotes the sampling index. The filter $g_{k,m}[n]$ represents a time-shifted version of a prototype filter $g[n]$, where the modulus operation makes the time shifting a circular operation. The factor $\exp(j2\pi(k/K)n)$ performs the filter shifting in the frequency domain. The transmitted signal for one block is the result of the superposition of all shifted impulse responses weighted by the corresponding information symbols:

$$x[n] = \sum_{m=0}^{M-1}\sum_{k=0}^{K-1} s_{k,m} \cdot g_{k,m}[n] \qquad n = 0,1,...,N-1 \qquad (12.6)$$

Figure 12.12 shows the structure of the GFDM modulator that implements Equation 12.6. The fact that GFDM uses a circular convolution in the filtering process is referred to as *tail biting* [18].

Let us define $\vec{g}_{k,m} = \left(g_{k,m}[n]\right)^{T}$ and then the matrix **A** as

$$\mathbf{A} = \begin{bmatrix} \vec{g}_{0,0} & \cdots & \vec{g}_{K-1,0} & \vec{g}_{0,1} & \cdots & \vec{g}_{K-1,M-1} \end{bmatrix} \qquad (12.7)$$

The transmitted signal for one block can equivalently be expressed as

$$\vec{x} = \mathbf{A} \cdot \vec{s} \qquad (12.8)$$

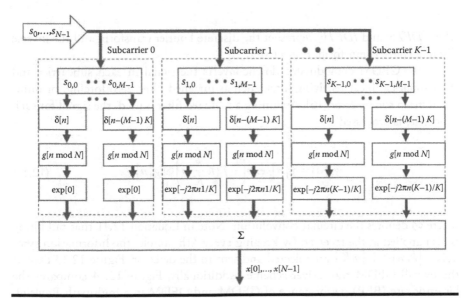

Figure 12.12   GFDM modulator.

At the transmitter side, after GFDM modulation, a CP is added to $x[n]$, which yields $\tilde{x}[n]$. Finally, $\tilde{x}[n]$ is sent to the radio channel. To prevent interference between subsequent data blocks, the duration of the CP should be $T_{cp} = T_g + T_h + T_g$, where T_g denotes the duration of the prototype filter and T_h denotes the length of the channel impulse response $h[n]$. Note that T_{cp} accounts for both transmit and receive filters. To reduce the overhead introduced by the CP, T_{cp} should be kept small. On the other hand, large values of T_g can improve the frequency localization of the filter. However, the usage of the tail-biting procedure allows the filtering part in T_{cp} to be neglected, which could not be achieved with linear convolution instead of circular. The usage of tail biting keeps the length of the CP independent from T_g, thereby reducing the length of the CP without cutting short the pulse-shaping filter length.

The received signal at the receiver is the convolution of the transmitted signal and the channel impulse response $h[n]$ plus AWGN noise:

$$\tilde{y}[n] = h[n] * \tilde{x}[n] + w[n] \tag{12.9}$$

At the receiver, the CP is removed, which yields $y[n]$, and then zero-forcing channel equalization is performed:

$$\bar{y}[n] = IDFT\left(\frac{DFT(y[n])}{DFT(h[n])}\right) \tag{12.10}$$

where $DFT(\cdot)$ and $IDFT(\cdot)$ represent the discrete Fourier transform and the inverse discrete Fourier transform, respectively.

Next, the GFDM demodulator downconverts the signal in each subcarrier and then applies a linear receiving filter, such as matched filter, zero forcing, or minimum mean square error [16]. Assuming a matched filter is used, the signal for each subcarrier is processed as

$$\bar{y}_k[n] = \bar{y}[n] \cdot \exp\left(-j2\pi \frac{k}{K} n\right) \otimes g[n] \tag{12.11}$$

where $\otimes$ denotes the circular convolution. Note in Equation 12.11 that tail biting is also applied at the receiver. By keeping every Kth sample, the information symbols $\bar{s}[k,m] = \bar{y}_k[mK]$ are selected and sent to the detector. Figure 12.13 depicts the overall GFDM transceiver scheme. Additionally, Figure 12.14 compares the bit error rate (BER) performance of GFDM and OFDM in a multipath Rayleigh channel.

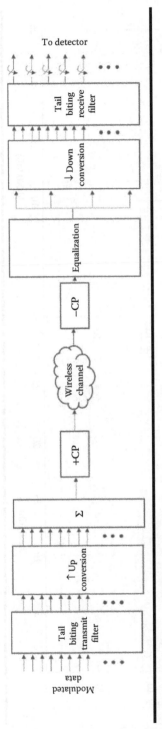

Figure 12.13 GFDM transceiver scheme.

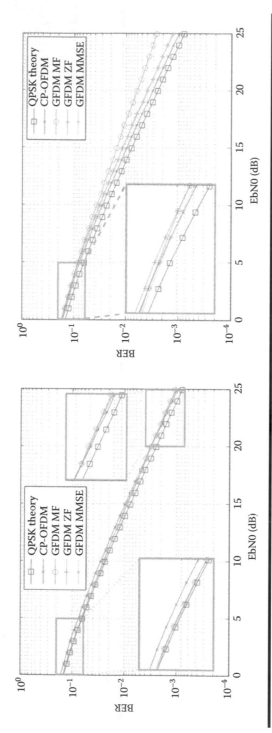

Figure 12.14 OFDM and GFDM BER performance for uncoded QPSK transmission in multipath Rayleigh channel. Matched filter (MF), zero-forcing (ZF), and minimum mean square error (MMSE) receivers applied for GFDM. (a) Root raised cosine filter with roll-off factor 0.1, and (b) root raised cosine filter with roll-off factor 0.5. (From Michailow, N. et al., Bit error rate performance of generalized frequency-division multiplexing, *Vehicular Technology Conference (VTC Fall), 2012 IEEE*, Quebec City, QC, IEEE, pp. 1–5, 2012.) [19].

12.1.4 Filtered OFDM (f-OFDM)

Slightly more restrictive than UFMC, f-OFDM retains the subband-wise spectrum shaping as a characteristic of the system. However, in this case, spectral shaping is performed over the whole system bandwidth by means of digital filters that keep OOB leakage radiation to the minimum, while keeping the remaining properties of OFDM unchanged (including the CP) [20].

Filtering allows proper coexistence between fragmental spectrum chunks, as shown in Figure 12.15. Subcarrier spacing can be made dependent on the actual system bandwidth, thus leading to a flexible time–frequency lattice for the coexistence of very dissimilar time–frequency granularities. Filtering also allows more robustness to time and frequency misalignments compared with OFDM. A rough comparison of the spectral characteristics of f-OFDM as compared with OFDM and UFMC is shown in Figure 12.16, where UFMC is referred to as universal filter OFDM (UF-OFDM) [21]. Note that in this figure, the spectral characteristics of OFDMA are obtained considering one single user (hence leading to the same spectral mask as OFDM), and the frequency axis is negative because of the particular simulation setup assumed in [21].

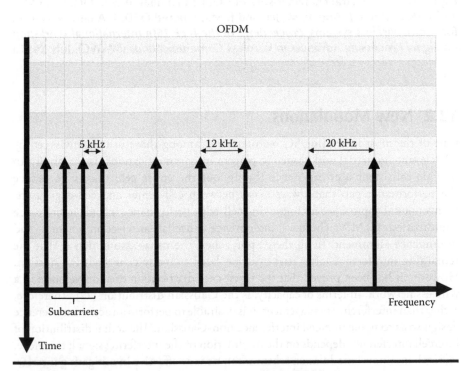

Figure 12.15   Flexible subcarrier spacing with f-OFDM.

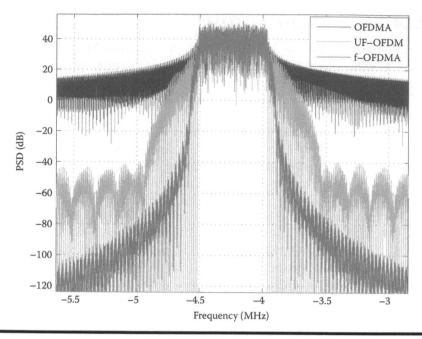

Figure 12.16 Spectral characteristics of f-OFDM compared with OFDM, FBMC, and UFMC. (From J. Abdoli, M. Jia, and J. Ma, Filtered OFDM: A new waveform for future wireless systems, *Proceedings of the IEEE 16th International Workshop on Signal Processing Advances in Wireless Communications* **(SPAWC), July 2015.)**

12.2 New Modulations

One of the most interesting 5G requirements, among those usually envisaged by all relevant industrial and academic players in communications, is the ability to provide consistent user experience throughout the entire cell: that is, to reduce the performance gap that always exists between cell-center and cell-edge users. Conventional approaches for improving cell-edge performance include interference coordination, CoMP techniques, interference cancellation/rejection schemes, and interference alignment. In all these approaches, the usual assumption is that the remaining interference (after part of it has been properly cancelled) is Gaussian. However, it has been proved that the worst-case distribution of additive noise in a wireless network, in terms of capacity, is the Gaussian distribution [22]. Therefore, rather than interference management, it is desirable to perform an active interference design so as to make intercell interference non-Gaussian. The actual distribution of intercell interference depends on the modulation of the interfering signal; therefore, research into novel modulation schemes can be very effective in mitigating interference. One such example is FQAM.

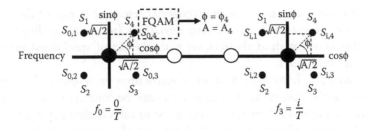

Figure 12.17 16-ary FQAM signal constellation.

12.2.1 Frequency and Quadrature Amplitude Modulation

This modulation is a combination of frequency shift keying (FSK) and QAM [11]. The performance of FQAM in uncoded systems and trellis-coded modulation (TCM) systems has been extensively analyzed [23,24]. The performance of Reed–Solomon-coded OFDM using FQAM was analyzed in [25] and shown to be superior to that in QAM at a given bit rate, even using noncoherent detection.

Figure 12.17 illustrates an example of 16-ary FQAM modulation as a combination of 4-ary FSK and 4-ary QAM.

Research results for single-cell scenarios (thus precluding intercell interference) show that the bandwidth efficiency of FQAM is dramatically enhanced compared with FSK systems, while the error performance is kept similar. Moreover, FQAM approaches the channel capacity limit when combined with coded modulation (CM) systems* in low signal-to-noise (SNR) conditions, in which channel capacity cannot be achieved with either QAM or FSK modulation [11].

Moreover, performance in the downlink of a multicell OFDMA network shows that intercell interference has a heavier tail than the Gaussian distribution in FQAM. As a result, the performance of cell-edge users significantly improves compared with QAM [26].

12.3 Nonorthogonal Multiple Access

One of the concepts being actively revisited in 5G is the need for maintaining strict orthogonality between users in multiple access schemes. Intracell orthogonality has been a cornerstone of cellular access since the second generation (2G), and even if it is rarely encountered in practice (mostly due to intercell effects or channel impairments, such as multipath), systems have been traditionally

---

* CM systems are obtained from concatenation of an error correction code and a signal constellation. Examples of CM systems are TCM, block-coded modulation (BCM), and turbo-coded modulation.

designed so that users are allocated orthogonal resources in any of the time, frequency, code, or spatial domains.* Any deviations from ideal orthogonality are thus further coped with at the receiver with the aid of specialized interference cancellation techniques.

This paradigm, however, puts a lot of restrictions on resource allocation, and most importantly, requires all devices to be under the control of the network prior to any transmission of information. In contrast to this classical approach, there are research initiatives whereby a controlled amount of nonorthogonality is allowed by design. One such example is NOMA. This concept is really an umbrella under which a number of multiple access techniques are being researched, all of them having in common a more or less explicit assumption of nonorthogonality in the access.

In one of its forms (heavily promoted by NTT Docomo), NOMA superposes multiple users in the power domain under certain conditions and adopts a SIC receiver as the baseline receiver for multiple access [27]. An alternative receiver design is also feasible where signals from multiple users are jointly detected, thus avoiding SIC operation [28]. In both cases, the underlying resource allocation mechanism is denoted as multiuser power allocation (MUPA), whereby users are grouped and scheduled according to their path loss levels.

NOMA in its basic form can increase the total system throughput compared with OFDMA systems, at the cost of added complexity at the transmitter side (from MUPA) and at the receiver side (from SIC or multiuser detection). There are other variants of NOMA that further exploit the nonorthogonality assumption, such as sparse-code multiple access (SCMA)† and multiuser shared access (MUSA),‡ mostly addressing the specific use case of massive machine-type communication (MTC). The basic NOMA concept based on power allocation is suitable for traditional mobile broadband applications, in which an extra capacity boost is required to address the ever-growing traffic demand. The 3rd Generation Partnership Project (3GPP) recently initiated a Study Item in March 2015 on multiuser superposition transmission (MUST), which will specifically analyze the eventual performance benefits of NOMA beyond Release 13 [29].

In Sections 12.3.1 and 12.3.2, we briefly describe basic NOMA based on power allocation, both with and without SIC operation at the receiver side.

---

* There are multiple access schemes, such as code-division multiple access (CDMA), in which orthogonality can only be achieved with perfect power control in the absence of multipath. Hence, some intercell interference will always exist, even in isolated cells, but orthogonality can be asymptotically approached under certain channel conditions.
† SCMA is heavily promoted by Huawei as a candidate NOMA scheme for 5G, relying on sparse codes with nonlinear detection at the receiver [30].
‡ MUSA is heavily promoted by ZTE as a candidate NOMA scheme for 5G, based on multicarrier, nonorthogonal code spreading with linear detection and SIC at the receiver [31].

12.3.1 Basic NOMA with SIC

12.3.1.1 Single-Antenna Base Station (BS)

For simplicity, let us assume that two users are to be allocated resources in NOMA (Figure 12.18), and that the BS does not exploit the presence of multiple transmitter antennas. User UE_1 (in black) is assumed to be near the cell center, while UE_2 (in gray) is located at the cell edge. Given that the BS wants to transmit a signal s_i for UE_i ($i = 1, 2$), with transmission power P_i, both signals are superposition coded in the same time–frequency resources, thus yielding

$$x = \sqrt{P_1} s_1 + \sqrt{P_2} s_2 \qquad (12.12)$$

where it is assumed that $E[|s_i|^2] = 1$ and $P_1 + P_2 \leq P$ (the maximum transmission power). It is apparent from Figure 12.18 that the power allocated to the cell-edge user UE_2 (in gray) is stronger than that of the user at the cell center UE_1 (in black) to overcome the increased path loss.

Assuming flat channel conditions (as in OFDM at subcarrier level), the received signal at UE_i is represented as

$$y_i = h_i x + w_i \qquad (12.13)$$

where:

h_i is the complex channel coefficient between UE_i and the BS
w_i is a complex noise plus interference term

The SIC process at the receiver tries to estimate the signals x_i from the received signal y_i by getting rid of the interuser interference. As can be seen in Figure 12.18, user UE_1 (with good signal-to-interference-plus-noise ratio [SINR] conditions) first

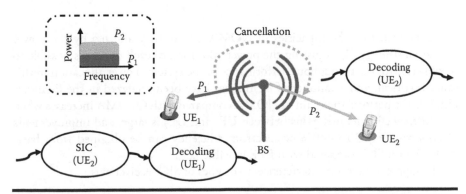

Figure 12.18   **Basic NOMA with SIC applied to two users, UE_1 and UE_2, in the downlink.**

cancels the interference caused by the strong signal transmitted to UE$_2$ (with a higher transmit power, as shown by the dotted arrow). SIC receivers can cope with strong interference components, thereby being able to separate the signals intended for UE$_1$ and UE$_2$ and further detect their own signal. In parallel, user UE$_2$ (under bad SINR conditions) does not need to perform SIC, as its own signal is much higher than the residual interference from the signal intended for UE$_1$.

The optimal order for decoding is in the order of increasing channel gain ($|h_i|^2$) normalized by the noise and intercell interference power ($n_{0,i}$), $|h_i|^2 / n_{0,i}$. Assuming error-free detection of x_2 at UE$_1$ (perfect SIC), the capacities of UE$_1$ and UE$_2$ can be written as

$$C_1^{\text{NOMA}} = B \log_2 \left(1 + \frac{P_1 |h_1|^2}{n_{0,1}} \right)$$

$$C_2^{\text{NOMA}} = B \log_2 \left(1 + \frac{P_2 |h_2|^2}{P_1 |h_2|^2 + n_{0,2}} \right) \tag{12.14}$$

where B is the signal bandwidth. The corresponding capacities for OFDMA, in which a fraction $0 < \beta < 1$ of the signal bandwidth is allocated to UE$_1$ and $(1 - \beta)$ to UE$_2$, are

$$C_1^{\text{OFDMA}} = B\beta \log_2 \left(1 + \frac{P_1 |h_1|^2}{\beta N_{0,1}} \right)$$

$$C_2^{\text{OFDMA}} = B(1-\beta) \log_2 \left(1 + \frac{P_2 |h_2|^2}{(1-\beta) N_{0,2}} \right) \tag{12.15}$$

It can be seen that the capacities in NOMA are strongly affected by the power allocation strategy. By adjusting the power allocation of each user, it is possible to flexibly control the throughput performance of the system. Power allocation in this sense is critical, and a number of techniques have been analyzed in the literature [32,33]. The performance gain of NOMA compared with OFDMA increases when the difference in channel gains between UE$_1$ and UE$_2$ is large, and improvements of approximately 30%–40% of spectrum efficiency can be obtained from basic NOMA with SIC compared with LTE baseline [27].

To suppress interuser interference, two types of SIC receiver exist:

■ Symbol-level SIC (SLIC), in which interfering modulation symbols are detected without decoding

■ Codeword-level SIC (CWIC), in which interfering data is detected and decoded prior to being cancelled out

The CWIC receiver has better performance than the SLIC receiver at the cost of higher complexity [34]. Practical receiver designs present error propagation effects, whereby imperfect interference cancellation from the cell-edge user impacts the decoding of the cell-center user (especially in SLIC receivers). To this end, link-level performances of real receivers are also studied in [35]. This fact partly motivates the introduction of NOMA receivers without SIC, as will be presented in Section 12.3.2.

The described procedure can be extended to the use of multiple receiver antennas at the user side by employing maximum ratio combining (MRC) so as to boost the received SINR. More interestingly, the BS can also exploit the presence of multiple transmitter antennas through advanced MIMO techniques. The coexistence of NOMA with MIMO is thus exemplified in Section 12.3.1.2.

12.3.1.2 Multiantenna Base Station

The NOMA concept can coexist with multiantenna techniques (Figure 12.19). The transmitter at the BS can generate multiple beams in multiuser MIMO (MU-MIMO) and superposes multiple users within each beam by means of superposition coding (in Figure 12.19, UE_1–UE_2 for the upper beam and UE_3–UE_4 for the lower beam). Each beam will be characterized by different precoding weights at the transmitter.

Two interference cancellation schemes must be present at the receiver side to address the resulting interuser interference:

■ Interference rejection combining (IRC) for interbeam user multiplexing, that is, interference suppression among the groups applying different precoding weights at the transmitter. Multiple receiver antennas are needed in this case,

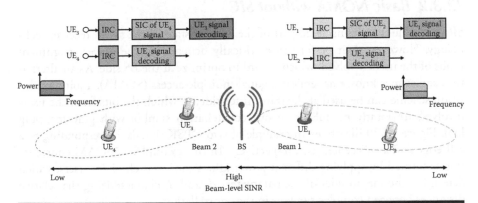

Figure 12.19 NOMA/MIMO scheme applied to four users with combined IRC/ SIC operation.

but performance degrades when the spatial correlation between own and interfering signal is high. This spatial correlation can be very significant in current devices, especially at lower frequencies (below 1 GHz). Careful radio-frequency (RF) design must take into account the radiation characteristics of the whole printed circuit board, including the receiver antennas.

■ SIC for intrabeam user multiplexing, that is, interference cancellation among the users belonging to a group applying the same precoding weights. Operation in this case would be similar to that in single-antenna BSs.

So far, the description has been limited to only two simultaneous users. The maximum number of simultaneous users in the common resources, usually known as the *overload factor*, could in principle be unlimited with ideal superposition coding and SIC operation. However, increasing the number of users will not always provide a performance gain, as the transmit power per layer will decrease and the amount of interuser interference will become more relevant [11]. Finding the optimal number of multiplexed users remains a future research item. In practical terms, this number is often limited to fewer than three users.

It is apparent that many new challenges appear when one tries to apply NOMA with both single and multiple antennas under realistic conditions. Power allocation, adaptive modulation and coding, beam multiplexing, multiuser scheduling, and practical overload factors are examples of highly challenging issues that demand new design paradigms compared with those in OFDMA. System-level performance under several downlink environments has been analyzed in [36]. Resource allocation in downlink NOMA has been studied in [37]. Finally, extension to multiple sites can also be investigated for a scenario with several remote radio heads and a common baseband processing unit, as in [38].

12.3.2 Basic NOMA without SIC

SIC techniques are considered part of the current state of the art in receiver technology. However, their performance critically depends on the power assignment ratios of the users involved, which is hard to optimize at the BS side. As an alternative, sometimes known as semiorthogonal multiple access (SOMA), a joint modulation scheme can be used at the transmitter side, in which the bits from the users involved are jointly modulated in one constellation symbol with gray mapping [35]. Figure 12.20 illustrates an example of two QPSK signals, corresponding to a cell-edge and a cell-center user, respectively. In this example, a 16QAM constellation is formed by applying different power allocations to each of the users, which determine the magnitudes of the parameters d_1 and d_2 characterizing the relative distances between complex symbols in the constellation.

Figure 12.21 illustrates a comparison between suitable NOMA transmitter structures both with SIC and without SIC; in the latter case, joint constellation

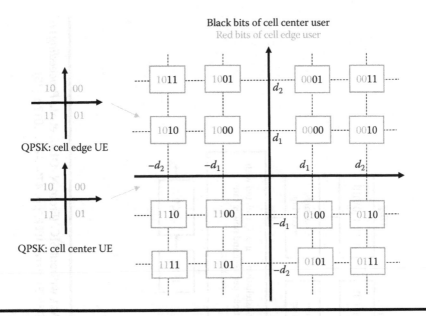

Figure 12.20 Constellation of jointly modulated signal at the transmitter.

mapping is performed under different power allocations. The difference in performance when jointly detecting both signals at the receiver is compensated by means of different modulation and coding schemes (MCSs) on transmission, to be determined as a function of the allocated transmit powers and the channel state information from each user.* Independent channel encoding is thus performed prior to joint modulation.

The receivers of both the cell-center and cell-edge users must jointly demodulate the symbols as if one larger constellation were transmitted. However, only the soft output estimates of the bits intended for each user may be obtained, disregarding the others (e.g., only the first and second bits in Figure 12.20 for the cell-edge user). Proper channel decoding can be further performed for the intended information stream without the error propagation effects caused by SIC processing. Complexity is therefore halved for the cell-center user compared with NOMA with SIC receivers. This approach is, however, more problematic when higher-order modulations are used: for example, two 16QAM modulations yield a combined 256QAM constellation, which poses more challenges for reception at the devices.

---

* This difference in performance comes from the different Euclidean distances between symbols containing a "0" and a "1" for each of the constituent bits. Referring to Figure 12.20, the first two bits (corresponding to the cell-edge user) have better protection than the other two bits (for the cell-center user) in accordance with their higher transmit power.

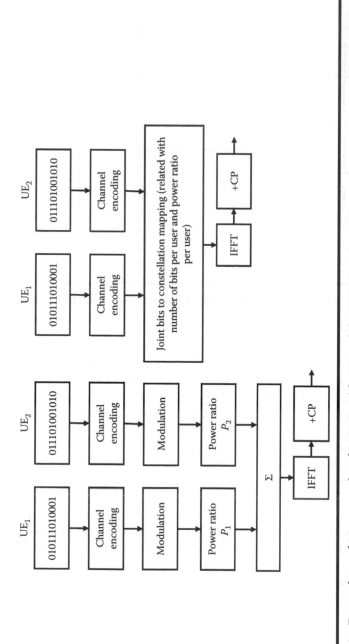

Figure 12.21 Transmitter of (a) conventional NOMA with SIC and (b) NOMA without SIC. (From N. Otao, et al., *Proceedings of the IEEE International Symposium on Wireless Communications Systems (ISWCS)*, Paris, IEEE, pp. 476–480, 2012.) [35].

12.4 Faster than Nyquist Communications

Similar to the nonorthogonality assumption in the multiple access scheme that leads to the NOMA concept, there is also growing interest in relaxing the assumption of ISI-free transmission for a single user. Thus, instead of designing the waveforms so as to obey the Nyquist criterion for ISI avoidance in a band-limited channel, a certain degree of ISI is allowed even in ideal AWGN channels, by increasing the data rate beyond the Nyquist rate. The resulting ISI must be compensated at the receiver by means of complex equalization.

The continuous-time Nyquist rate transmission over a band-limited channel with bandwidth W can be expressed as

$$x(t) = \sum_{n=0}^{N-1} s_n g(t - nT) \tag{12.16}$$

where:

N	is the packet size
$g(t)$	is the continuous-time modulation waveform
$T = 1/(2W)$	is the symbol period
s_n	is the modulated symbol for time instant $n = 0, ..., N-1$

This is the usual assumption in wireless communications, even if in practice, radio channels exhibit significant multipath, thus leading to ISI.

We can, however, relax this assumption and increase the rate beyond the Nyquist limit by means of a nonorthogonal transmission scheme. The continuous-time FTN signaling transmission can be expressed as [39]

$$x(t) = \frac{1}{\sqrt{D}} \sum_{n=0}^{ND-1} s_n g\left(t - n\frac{T}{D}\right) \tag{12.17}$$

where $D > 1$ is an FTN factor related to the time separation between consecutive data symbols, $\Delta t = T/D$, and $n = 0, 1, ..., ND-1$; D controls the relative increase of FTN signaling over the Nyquist rate and can also be expressed as the inverse $\tau = 1/D < 1$. The seminal work of Mazo in 1975 showed that for binary signaling, the symbol separation can be reduced by a factor 0.802 without asymptotic performance degradation, known as Mazo's limit, and the same happens with other pulse shapes [40]. Mazo showed that sending sinc pulses up to 25% faster does not increase the minimum Euclidean distance between symbols for an uncoded system using binary modulation [41]. The complexity, however, translates to the receiver, which must compensate the intentional ISI introduced at the transmitter. Receiver complexity especially increases in time-dispersive channels, in

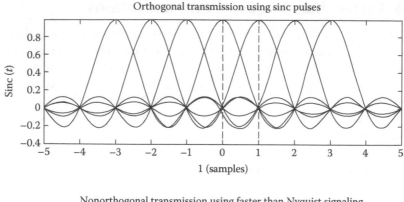

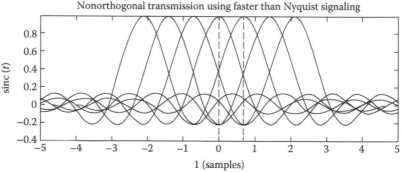

Figure 12.22 Illustration of the faster than Nyquist concept using sinc pulses. (From El Hefnawy, M. and Taoka, H., Overview of faster-than-Nyquist for future mobile communication systems, *Proceedings of the IEEE 77th Vehicular Technology Conference (VTC Spring),* **Dresden, IEEE, 2013.)**

which the ISI inherent in FTN signaling is combined with the ISI associated with multipath [42].

FTN signaling makes it possible to handle large amounts of ISI induced by this nonorthogonal transmission. Figure 12.22 illustrates FTN signaling in the time domain using sinc pulses. Raised cosine pulses are often used instead of sinc pulses, because the latter have an impulse response of infinite duration. Figure 12.23 shows both FTN and Nyquist signaling in the frequency domain using raised cosine pulses with an excess bandwidth $W = (1 + \alpha/2T)$, where $1/T$ is the Nyquist rate and α is the roll-off factor.

It is apparent that Nyquist signaling achieves the well-known criterion whereby pulses add up in frequency to a constant value, as given by the expression

$$\frac{1}{T} \sum_{k=-\infty}^{\infty} X\left(f + \frac{k}{T} \right) = 1 \text{ for all } f \qquad (12.18)$$

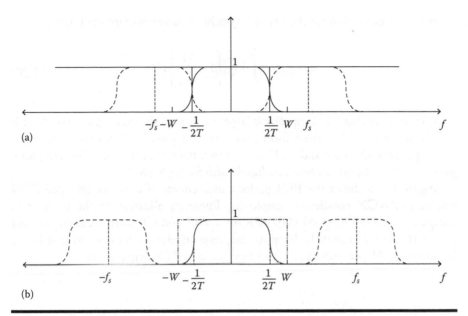

(a)

(b)

Figure 12.23 **(a) Nyquist and (b) FTN signaling in the frequency domain using raised cosine pulses. (From El Hefnawy, M. and Taoka, H., Overview of faster-than-Nyquist for future mobile communication systems, *Proceedings of the IEEE 77th Vehicular Technology Conference (VTC Spring)*, Dresden, IEEE, 2013.)**

where $X(f)$ is the Fourier transform of the transmitted signal $x(t)$. In FTN signaling, this condition does not hold, and the pulses at the frequency domain are shifted further apart. A low-pass filter with a bandwidth strictly higher than $1/2T$ must be applied at the receiver for reconstruction of the signal, whereas in the Nyquist case, a bandwidth $1/2T$ is sufficient for perfect reconstruction if an ideal rectangular filter is used (although practical implementations will employ realizable, nonrectangular filters with a higher bandwidth).

The larger bandwidth in FTN signaling is responsible for a higher capacity compared with the Nyquist case. Capacity can be given by the following expression, valid for a Gaussian alphabet in AWGN conditions with band-limited pulses to W Hz [43]:

$$C_{\text{FTN}} = \int_0^W \log_2 \left[1 + \frac{2P}{N_0} |H(f)|^2 \right] df \qquad (12.19)$$

where:
- W is the one-side bandwidth of the signal
- P is the average power
- N_0 is the white noise spectral density
- $H(f)$ is the frequency response of the signal

A similar expression for the case of Nyquist signaling is computed using

$$C_{\text{Nyquist}} = \int_0^{1/2T} \log_2\left[1 + \frac{2P}{N_0}\right] df \qquad (12.20)$$

It is apparent that FTN capacity is larger because of the excess pulse bandwidth *W*, which is governed by the roll-off parameter α. Figure 12.24 shows the normalized capacity for Nyquist and FTN cases with different values of α. The asymptotic gain tends to be equal to the excess bandwidth for high SNR.

Figure 12.25 shows the BER performance curves of a binary encoded FTN system in AWGN conditions, employing linear equalization at the receiver to compensate the resulting ISI [41]. BER for $\tau = 0.9$ and 0.8 (corresponding to data rates 1.11 and 1.25 times the Nyquist rate, respectively) is very close to that in the Nyquist case. Higher values of τ lead to significant BER degradation.

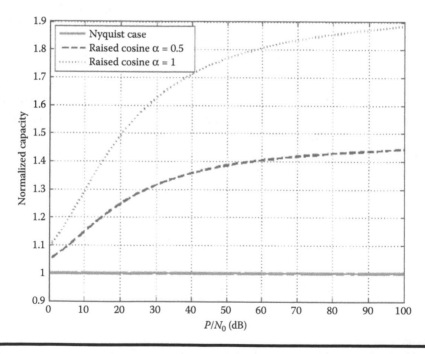

Figure 12.24 Normalized capacity for Nyquist and FTN cases, with different values of the roll-off factor. (From El Hefnawy, M. and Taoka, H., Overview of faster-than-Nyquist for future mobile communication systems, *Proceedings of the IEEE 77th Vehicular Technology Conference (VTC Spring)*, Dresden, IEEE, 2013.)

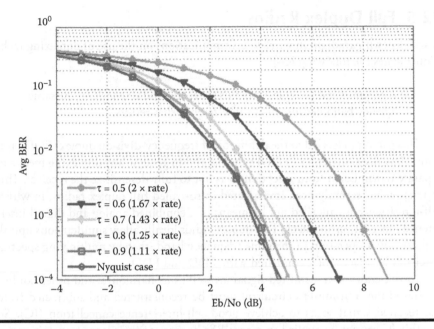

Figure 12.25 Average BER performance of Nyquist and FTN signaling for different rates. (From El Hefnawy, M. and Taoka, H., Overview of faster-than-Nyquist for future mobile communication systems, *Proceedings of the IEEE 77th Vehicular Technology Conference (VTC Spring)*, Dresden, IEEE, 2013.)

When combined with OFDM or single-carrier frequency-division multiplexing (SC-FDM),* frequency-domain equalization (FDE) can cope with ISI in frequency-selective channels, but complexity becomes prohibitively high for severely time-dispersive channels. Besides, some performance degradation occurs compared with Nyquist signaling, and the degradation increases with the symbol rate, as in the AWGN case. To cope with severe ISI, a combined iterative FDE scheme and a Hybrid Automatic Repeat Request (HARQ) mechanism can be applied [42].

Finally, an extension of the performance analysis of FTN to multiple-access channels, in which users compete to access the medium in synchronous and asynchronous operation, is analyzed in [39]. Multiuser demodulation and decoding can be exploited in this case, leading to capacity merit of FTN under different SNR ranges, at the cost of more complicated receiver designs.

---

* SC-FDM, also known as discrete Fourier transform-spread-OFDM (DFT-s-OFDM), is a variant of OFDM in which the signal modulated onto a given subcarrier is a linear combination of all the data symbols transmitted at the same instant (the linear relationship being expressed as a discrete Fourier transform). SC-FDM has a lower peak-to-average power ratio (PAPR) compared with OFDM, which makes it ideal for operation in the uplink, given the lower resulting battery consumption, as in LTE [1].

12.5 Full Duplex Radios

A well-known paradigm in wireless communications on the use of duplexing techniques states the following [44]:

> It is generally not possible for radios to receive and transmit on the same frequency band because of the interference that results.

This paradigm has led to the definition of frequency-division duplex (FDD) systems, time-division duplex (TDD) systems, and in general, any duplexing means to separate transmission from reception so as not to fully desensitize the receiver. This restriction has recently been investigated in so-called full duplex radios, in which self-interference is cancelled out by means of a combined analog and digital interference cancellation approach. Full duplex radios ideally allow simultaneous uplink and downlink operation in the same frequency band, thereby multiplying spectral efficiency by a factor of 2 compared with TDD and FDD.

To this end, the transmitted signal (as well as its harmonics and other nonlinearities of the transmitter's circuitry) must be reconstructed and subtracted from the received signal so as to achieve good self-interference cancellation (IC). So far, this IC cannot be applied to macro BSs because of their large transmit power (which can exceed 46 dBm), which for a noise floor level of around −104 dBm (in an example bandwidth of 10 MHz), leads to more than 140 dB IC requirement. A multistage approach can offer the best performance to date, with 110 dB IC [45]. More challenging than this requirement is the ability to achieve it over a whole frequency band while fitting it to the small form factors of today's devices.

Full duplex radios could in principle be applied to cellular communications between users and BSs, in such a way that uplink and downlink transmissions would share the same frequencies. However, in this case, severe interference issues would appear between users that could not be avoided with any self-IC strategy, and that would require some coordination between the transmission and reception instants, thereby rendering full duplex radios impractical for cellular use. Alternatively, there is a growing interest in applying full duplex radios for self-backhauled small cells (or relays), in which the same cellular frequencies are used for wireless backhaul. Figure 12.26 depicts such a scenario, in which an FDD small cell (acting as a repeater) integrates backhaul in a transparent way by reusing the same frequencies of the air interface. The backhaul link in the direction from small cell to macro cell is established by transmitting at the same frequency of the uplink radio access (in black), which does not harm the small cell receiver's sensitivity thanks to the self-IC technology. In the same way, the backhaul link in the direction from macro cell to small cell reuses the same frequency of the downlink radio access (in gray), which is not hampered by the small cell's transmitter after self-IC. A similar (and simpler) case in TDD would involve the same frequency for access and backhaul at the uplink/downlink time intervals.

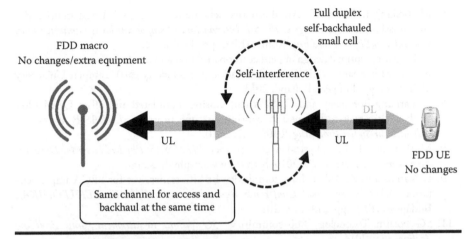

Full duplex
self-backhauled
small cell

FDD macro
No changes/extra equipment

Self-interference

DL

UL   UL

FDD UE
No changes

Same channel for access and
backhaul at the same time

Figure 12.26 Self-backhauled repeater using full duplex radios. (From Kumu Networks, *Full-duplex revolution*. Available from: http://comsocscv.org/docs/20140416-Kumu-Brand.pdf.) [46].

Device-to-device (D2D) communications would also naturally benefit from full duplex radios by establishing bidirectional links between nearby peers at a single frequency. Note that in this case, interuser interference would be limited by the relative proximity of the communicating devices, in conjunction with some smart power control and resource allocation strategies that minimize interferences to non-D2D devices (and to other D2D pairs).

In general terms, different application scenarios lead to different requirements of the self-IC performance in full duplex radios. Although they are highly promising, further improvements are needed for them to be effectively applied in scenarios other than self-backhauled small cells.

References

1. S. Sesia, I. Toufik, and M. Baker (eds), *LTE: The UMTS Long Term Evolution: From Theory to Practice*, 2nd edn, New York: Wiley, 2011.
2. FCC, Unlicensed operation in the TV broadcast bands, second report and order, FCC-08-260, 2008. Available from: https://apps.fcc.gov/edocs_public/attachmatch/FCC-08-260A1.pdf.
3. B. Saltzberg, Performance of an efficient parallel data transmission system, *IEEE Transactions on Communications Technology* 15(6): 805–816, 1967.
4. M. Tanda, T. Fusco, M. Renfors, J. Louveaux, and M. Bellanger, Data-aided synchronization and initialization (single antenna), PHYDYAS: Physical layer for dynamic access and cognitive radio, *Deliverable D2.1*, 2008.
5. M. Bellanger, FS-FBMC: A flexible robust scheme for efficient multicarrier broadband wireless access, *IEEE Globecom Workshops*, Anaheim, CA, IEEE, pp. 192–196, 2012.

6. M. Bellanger, FS-FBMC: An alternative scheme for filter bank based multicarrier transmission, in *Proceedings of the 5th International Symposium on Communications, Control and Signal Processing*, Rome, IEEE, pp. 1–4, 2012.
7. R. Bregovic, Optimal design of perfect-reconstruction and nearly perfect-reconstruction multirate filter banks, PhD Thesis dissertation (introductory part), Tampere University of Technology, Tampere, Finland, 2013.
8. B. Farhang-Boroujeny and C. H. Yuen, Cosine modulated and offset QAM filter bank multicarrier techniques: A continuous-time prospect, *EURASIP Journal on Advances in Signal Processing* 2010: 1–17, 2010.
9. M. Bellanger, FBMC physical layer: A primer, *PHYDYAS: Physical Layer for Dynamic Access and Cognitive Radio*, 2010. http://www.ict-phydyas.org/.
10. L. Vargas and Z. Kollar, Low complexity FBMC transceiver for FPGA implementation, *23rd International Conference Radioelektronika (RADIOELEKTRONIKA)*, Budapest, IEEE, pp. 219–223, 2013.
11. 5G Forum Technology Sub-committee, *5G Vision, Requirements, and Enabling Technologies*, 2014.
12. B. Le Floch, M. Alard, and C. Berrou, Coded orthogonal frequency division multiplex, *Proceedings of the IEEE* 83(6): 982–996, 1995.
13. T. Wild and F. Schaich, A reduced complexity transceiver for UF-OFDM, *Proceedings of the IEEE 81st Vehicular Technology Conference*, Glasgow, Spring, IEEE, pp. 1–6, 2015.
14. F. Schaich, T. Wild, and Y. Chen, Waveform contenders for 5G: Suitability for short packet and low latency transmissions, *Proceedings of the IEEE 79th Vehicular Technology Conference*, Seoul, Spring, IEEE, pp. 1–4, 2014.
15. T. Wild, G. Wunder, F. Schaich, Y. Chen, M. Kasparick, and M. Dryjanski, *5GNOW: Intermediate Transceiver and Frame Structure Concepts and Results*, Bologna, 2014. Available from: https://mns.ifn.et.tu-dresden.de/Lists/nPublications/Attachments/1007/GC-WS-2014-5GNOW_final.pdf.
16. N. Michailow, M. Matthe, I. S. Gaspar, A. N. Caldevilla, L. L. Mendes, A. Festag, and G. Fettweis, Generalized frequency division multiplexing for 5th generation cellular networks, *Communications, IEEE Transactions on* 62(9): 3045, 3061, 2014.
17. 5G NOW Project, *5G Waveform Candidate Selection D3.1 Deliverable*, 2013.
18. G. Fettweis, M. Krondorf, and S. Bittner, GFDM: Generalized frequency division multiplexing, *Vehicular Technology Conference, 2009. VTC Spring 2009. IEEE 69th*, Barcelona, IEEE, pp. 1–4, 2009.
19. N. Michailow, S. Krone, M. Lentmaier, and G. Fettweis, Bit error rate performance of generalized frequency division multiplexing, *Vehicular Technology Conference (VTC Fall), 2012 IEEE*, Quebec City, QC, IEEE, pp. 1–5, 2012.
20. P. Zhu, 5G enabling technologies: A unified adaptive software defined air interface, *IEEE PIMRC Huawei Keynote Presentation*, 2014. Available from: http://www.ieee-pimrc.org/2014/2014-09-03%205G%20Enabling%20Technologies%20PMIRC%20Huawei_Final.pdf.
21. J. Abdoli, M. Jia, and J. Ma, Filtered OFDM: A new waveform for future wireless systems, *Proceedings of the IEEE 16th International Workshop on Signal Processing Advances in Wireless Communications* (SPAWC), July 2015.
22. I. Shomorony and A. S. Avestimehr, Worst-case additive noise in wireless networks, *IEEE Transactions on Information Theory* 59(6), 3833–3847, 2013.

23. R. Padovani and J. Wolf, Coded phase/frequency modulation, *IEEE Transactions on Communications* 34(5), 446–453, 1986.
24. S. S. Periyalwar and S. M. Fleisher, Multiple trellis coded frequency and phase modulation, *IEEE Transactions on Communications* 40(6), 1038–1046, 1992.
25. A. Latif and N. D. Gohar, BER performance evaluation and PSD analysis of noncoherent hybrid MQAM-LFSK OFDM transmission system, in *Proceedings of the IEEE 2nd International Conference on Emerging Technologies*, Peshawar, IEEE, pp. 53–59, 2006.
26. A. Latif and N. D. Gohar, Performance of Hybrid MQAM-LFSK (HQFM)OFDM transceiver in Rayleigh fading channels, in *Proceedings of the IEEE Multi Topic Conference*, Islamabad, IEEE, pp. 52–55, 2006.
27. Y. Saito, Y. Kishiyama, A. Benjebbour, T. Nakamura, A. Li, and K. Higuchi, Nonorthogonal multiple access (NOMA) for cellular future radio access, *Proceedings of the IEEE 77th Vehicular Technology Conference (VTC Spring)*, Dresden, IEEE, pp. 1–5, 2013.
28. C. Yan, A. Harada, A. Benjebbour, Y. Lan, A. Li, and H. Jiang, Receiver design for downlink non-orthogonal multiple access (NOMA), *Proceedings of the IEEE 81st Vehicular Technology Conference (VTC Spring)*, Glasgow, IEEE, pp. 1–6, 2015.
29. 3GPP TSG RAN Meeting #66, *RP-142315 New Study Item Proposal: Enhanced Multiuser Transmissions and Network Assisted Interference Cancellation for LTE*, Maui.
30. H. Nikopour, E. Yi, A. Bayesteh, K. Au, M. Hawryluck, H. Baligh, and J. Ma, SCMA for downlink multiple access of 5G wireless networks, *Proceedings of the IEEE Global Communications Conference* (Globecom), pp. 3940–3945, 2014.
31. Future Mobile Communications Forum, 5G White Paper, 2015. Available from: http://euchina-ict.eu/wp-content/uploads/2015/01/5G-SIG-white-paper-first-version.pdf.
32. A. Li, A. Harada, and H. Kayama, Investigation on low complexity power assignment method and performance gain of non-orthogonal multiple access systems, submitted to *IEICE Transactions on Communications*.
33. N. Otao, Y. Kishiyama, and K. Higuchi, Performance of nonorthogonal access with SIC in cellular downlink using proportional fair-based resource allocation, *Proceedings of the IEEE International Symposium on Wireless Communications Systems (ISWCS)*, Paris, IEEE, pp. 476–480, 2012.
34. K. Saito, A. Benjebbour, A. Harada, Y. Kishiyama, and T. Nakamura, Link-level performance of downlink NOMA with SIC receiver considering error vector magnitude, *Proceedings of the IEEE 81st Vehicular Technology Conference (VTC Spring)*, Japan, IEEE, pp. 1–5, 2015.
35. C. Yan, A. Harada, A. Benjebbour, Y. Lan, A. Li, and H. Jiang, Receiver designs for downlink non-orthogonal multiple access (NOMA), *Proceedings of the IEEE 81st Vehicular Technology Conference* (VTC Spring), May 2015.
36. Y. Saito, A. Benjebbour, Y. Kishiyama, and T. Nakamura, System-level performance of downlink non-orthogonal multiple access (NOMA) under various environments, *Proceedings of the IEEE 81st Vehicular Technology Conference* (VTC Spring), May 2015.
37. M. Hojeij, J. Farah, C. A. Nour, and C. Douillard, Resource allocation in downlink non-orthogonal multiple access (NOMA) for future radio access, *Proceedings of the IEEE 81st Vehicular Technology Conference* (VTC Spring), May 2015.

38. S. Khattak, W. Rave, and G. Fettweis, Distributed iterative multiuser detection through base station cooperation, *EURASIP Journal on Wireless Communication and Networking* 2008(17): 1–15, 2008.

39. Y. Feng and J. Bajcsy, Improving throughput of faster-than-Nyquist signaling over multiple-access channels, *Proceedings of the IEEE 81st Vehicular Technology Conference* (VTC Spring), May 2015.

40. J. E. Mazo, Faster-than-Nyquist signaling, *The Bell System Technical Journal* 54: 1451–1462, 1975.

41. M. El Hefnawy and H. Taoka, Overview of faster-than-Nyquist for future mobile communication systems, *Proceedings of the IEEE 77th Vehicular Technology Conference* (VTC Spring), June 2013.

42. R. Dinis, B. Cunha, F. Ganhão, L. Bernardo, R. Oliveira, and P. Pinto, A hybrid ARQ scheme for faster than Nyquist signaling with iterative frequency-domain detection, *Proceedings of the IEEE 81st Vehicular Technology Conference* (VTC Spring), May 2015.

43. F. Rusek and J. Anderson, Constrained capacities for faster-than-Nyquist signaling, *IEEE Transactions on Information Theory* 55(2): 764–775, 2009.

44. A. Goldsmith, *Wireless Communications*, New York: Cambridge University Press, 2005.

45. D. Bharadia, E. McMilin, and S. Katti, Full duplex radios, *Proceedings of ACM SIGCOMM* 2013, pp. 375–386, 2013.

46. Kumu Networks, *Full-duplex revolution*. Available from: http://comsocscv.org/docs/20140416-Kumu-Brand.pdf.

Chapter 13

GFDM: Providing Flexibility for the 5G Physical Layer

Luciano Mendes, Nicola Michailow,
Maximilian Matthé, Ivan Gaspar,
Dan Zhang, and Gerhard Fettweis

Contents

Mobile networks have revolutionized the way that society communicates, and today, cellular and smartphones are part of our everyday life. Since the first generation (1G) of mobile networks, deployed during the 1980s, cellular systems have evolved toward higher capacity to cover the ever-growing number of subscribers. With the introduction of digital services, already in the second generation (2G), the demand for higher throughput has driven the development of new standards. Hence, the third generation (3G) and the fourth generation (4G) were designed to address higher capacity and data rates. Now, the fifth generation (5G) is being discussed by the research community, and it is clear that the new services and scenarios foreseen for the future mobile communication systems will impose requirements and constraints that go beyond throughput and capacity. Certainly, bitpipe communication will require data rates several times higher than Long-Term Evolution–Advanced (LTE-A) can support. Also, capacity in terms of the number of users is a huge challenge when a massive number of small and power-limited devices must connect to the network in an Internet of Things (IoT) scenario. New services are challenging several parameters of the mobile networks that have not evolved during the last generations. Fragmented and dynamic spectrum access will be necessary for the efficient provision of data rates and capacity to accommodate more users, which demands a waveform with very low out-of-band (OOB) emission. Tactile Internet must have very low latency to provide a good quality of experience (QoE) and avoid issues such as cybersickness and lack of responsiveness [1]. The current target for 5G is to achieve 1 ms of end-to-end latency, at least one order of magnitude below the LTE-A latency [2]. Coverage of low-populated areas is also an issue that cannot be properly addressed by wired or wireless technologies today. IEEE 802.22 aims to solve this problem by using cognitive radio (CR) technologies on vacant TV channels. However, the use of orthogonal frequency-division multiplexing (OFDM) is a huge obstacle in terms of low OOB and spectrum flexibility.

The requirements imposed on 5G are challenging and require a flexible waveform that can be optimized for the different scenarios. In this chapter, we will explore how generalized frequency-division multiplexing (GFDM) [3], a block-based multicarrier filtered modulation scheme, addresses the foreseen challenges for the physical layer (PHY) of future mobile networks. In this flexible scheme, each of the K subcarriers transmits up to M data symbols in different time slots, defined as subsymbols. GFDM employs circular convolution to filter the subcarriers individually, meaning that the overall GFDM frame is self-contained in $N = MK$ samples, implying that cyclic prefix (CP) can be efficiently used to mitigate the multipath channel, and its block structure allows most of the techniques developed for OFDM to be reused. A single prototype filter is circularly shifted in time and frequency to provide all the necessary impulse responses and, unlike other filtered multicarrier modulations, GFDM can easily benefit from transmit diversity to achieve robustness over mobile channels. The GFDM waveform can be shaped to address low latency, and, combined with the Walsh–Hadamard transform (WHT), this scheme can deliver high performance in applications that require single-shot transmission. Actually, by reviewing the signal processing chain, it is clear that GFDM can be combined with precoding to pursue other benefits such as a low peak-to-average power ratio (PAPR) or frequency diversity. WHT–GFDM is just one special case of a much broader flexible system. If nonorthogonal filters are employed to generate the GFDM signal, self-interference will rise due to intersymbol interference (ISI) and intercarrier interference (ICI). A matched filter (MF) receiver can maximize the signal-to-noise ratio (SNR), but successive interference cancellation (SIC) is necessary to achieve a symbol error rate (SER) performance equivalent to OFDM systems. A zero-forcing (ZF) receiver can remove the self-interference, simplifying the demodulation process, but at the cost of noise enhancement. The performance loss might be negligible for high SNRs but becomes significant for low SNRs. The minimum mean square error (MMSE) receiver is of interest because it minimizes the influence of noise (as the MF) at low SNRs and mitigates the self-interference at high SNRs (as ZF), at the cost of estimating the noise statistics. Hence, the MMSE receiver achieves a good balance between noise and interference. Another approach to avoid self-generated interference is to combine GFDM with offset quadrature amplitude modulation (OQAM) mapping.

At first glance, implementation complexity seems to be an issue for GFDM modulators and demodulators. However, filters with a sparse frequency response allow efficient implementation in the frequency domain, resulting in slightly higher complexity when compared with OFDM. Also, a reinterpretation of the data flow allows the GFDM signal to be generated as a time-domain element-wise multiplication of the discrete Fourier transform (DFT) of the data symbols by the filter's impulse response, which is equivalent to a frequency-domain circular convolution. This approach is very efficient, even for filters that do not have sparse frequency responses. GFDM has been proved flexible, and it can be configured to cover several other waveforms. OFDM and single-carrier frequency-domain equalization (SC-FDE) can be identified as corner cases of GFDM, which can also cover burst transmission of the filterbank multicarrier (FBMC). Slight modifications in the definition of the subcarrier and subsymbol spacing broaden the flexibility of GFDM to include faster than Nyquist (FTN) signaling as a special case as well. The unprecedented degrees of freedom provided by GFDM make this flexible waveform a strong candidate for the 5G PHY. This chapter will explore the fundamentals of this modulation scheme, showing how it can be employed and combined with other techniques to overcome the challenges imposed by 5G networks.

13.1 5G Scenarios and Motivation for Flexible Waveforms

From the current research regarding 5G networks, it is clear that the main point that must be addressed is flexibility [4]. Several new applications are being proposed to provide different services. These new applications are being organized in various scenarios, some of which are depicted in Figure 13.1. Clearly, the requirements of the different scenarios cannot be simultaneously addressed, and understanding the new services and the associated needs is fundamental for designing waveforms for the next generation of mobile networks. These scenarios and their requirements are described in this section.

13.1.1 Bitpipe Communication

Video on demand is becoming the main service for media consumption nowadays, and mobile devices are the preferable player for most users. Also, high-resolution cameras embedded in smartphones generate large images and videos that users want to share instantaneously through social media. Being able to upload and access dense content from anywhere is one key feature that is being imposed on

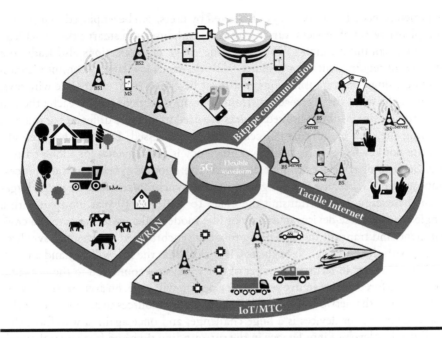

Figure 13.1   Four of the main scenarios envisioned for 5G networks.

5G networks. Estimates suggest that the next mobile network standards must be able to deal with throughput 100 times larger than the data rate available in the current generation [2]. This requirement can only be addressed if the capabilities of the 5G PHY are expanded beyond conventional orthogonal approaches. To achieve such high capacity, the density of the cells will have to increase, leading to heterogeneous networks. The 5G waveform must reduce the OOB emission to minimize interference in surrounding cells and also allow cooperative multipoint (CoMP) algorithms [5] to be more efficient. Access to large bandwidths will also be essential to provide data rates beyond 1 Gbps. Spectrum aggregation and dynamic spectrum access play an important role in this requirement. In this case, it is mandatory that the waveform has excellent spectral localization, so that the flexibility of accessing noncontinuous chunks of spectrum will not be hindered by the necessity for radio-frequency (RF) filtering. Spectrum efficiency is another key feature of the 5G waveform, and clearly, the overhead to combat the multipath channel must be reduced without introducing any performance loss. Finally, the 5G waveform must be compatible with multiple-input multiple-output (MIMO) techniques to achieve the necessary robustness and throughput in the mobile environment.

13.1.2 Internet of Things

The IoT is being highlighted as one of the main breakthroughs for 5G, although the commercial models are not clearly defined yet. Devices measuring different parameters related to users, but not triggered by users, can be employed to provide a complete new set of services, varying from smart houses and smart cities to vehicle-to-vehicle communication (V2V). The wide range of applications also leads to a wide set of requirements. For instance, V2V demands low-latency communication based on single-shot transmissions that must be conveyed in short time windows, for example, when cars approaching from opposite directions pass by each other. In smart cities, simple power-limited devices are used to measure different parameters, sending the data to information centers to be processed. Latency and through-put are not key issues here, but power efficiency is fundamental for a long battery life. Such limited devices cannot deal with the rigorous synchronization process imposed by 4G, because the energy required to synchronize would be much larger than the energy used to transmit the data. To achieve a 10 year life span with a single battery, these devices must stay in idle mode, wake up to measure the environment, and transmit the data being roughly synchronized (or even not synchronized) with the network. Nevertheless, it is clear that the IoT will demand a much higher scalability in terms of number of users (devices) connected to the network. Today, it is forecast that in five years, every person will have, on average, 6.5 devices connected to the network [6]. Scheduling the PHY resources to accommodate this massive number of devices is a huge challenge, and once again, a waveform that produces low levels of interference in the surrounding time and frequency channels is mandatory for achieving an efficient system.

13.1.3 Tactile Internet

Tactile Internet is a new concept [1] in which the interaction between users and devices is taken to the next level. In Tactile Internet applications, the feedback between an action triggered by the user and the response from a system must happen within a time window of 1 ms. This latency is at least one order of magnitude below the delay provided by LTE-A. A 5G network capable of achieving such small delays can provide new mobile services and bring new revenue for operators. Low latency is particularly important for online action gaming and wearable devices, such as smart glasses and smart gloves, to avoid cybersickness. Controlling real and virtual objects using mobile terminals is also an application in which low latency is mandatory for a good quality of service (QoS) and precision. Several areas of the economy will benefit from the Tactile Internet, and it could be the biggest breakthrough in mobile communication since the advent of short message service (SMS).

13.1.4 Wireless Regional Area Network

Several communication media can provide good Internet access for inhabitants of densely populated areas: coaxial cable (data over cable service interface specification), twisted pair (digital subscriber line), fiber optics (fiber to the home),

and wireless (LTE-A, WiMAX, and Wi-Fi). On the other hand, sparsely populated areas have poor Internet services, or no service at all, because the current
technology cannot provide proper coverage. Satellite communication is an interesting solution, but current prices for the service are prohibitive for most users.
Mobile networks are the most promising solution to provide digital services in
remote and rural areas, but the small coverage of the current generation, typically
around 10 km with a single base station (BS), hinders this due to economic issues.
Enlarging the BS coverage to tens of kilometers will allow an operator to deploy an
economically feasible wireless regional area network (WRAN) to provide access
in rural areas. IEEE 802.22 has tried to address this issue by using CR technology to explore vacant ultra-high-frequency (UHF) channels. However, the use of
OFDM in the PHY makes it challenging to access different channels dynamically
once the high OOB of OFDM waveforms demands RF filtering to address the
emission masks imposed by regulatory agencies. WRAN channels also present
long channel delay profiles, which demand longer CP. Since OFDM needs one
CP per symbol, this can result in a prohibitive overhead that drastically reduces
the overall system capacity. A 5G network, with a low OOB emission and a highly
efficient waveform, can operate in a mode that covers large regions, bringing high-
quality digital service to rural areas. This application will surpass the last frontier
for mobile networks. Services such as cattle monitoring and rural automation can
improve the quality of life of a large number of people and bring agricultural business efficiency to a new level.

13.2 GFDM Principles and Performance

The main characteristic that makes GFDM one of the most promising waveforms
for 5G networks is its flexibility. This section describes its principles, including
GFDM signal transmission and detection, and assesses its performance regarding
OOB emission and SER. An important limitation in the GFDM parametrization
is discussed, and an efficient procedure to design the receiver filter with the help of
Zak transform is presented [7]. The description presented in this section will make
all the advantages and potentials of GFDM clear to readers.

13.2.1 GFDM Waveform

As can be seen in Figures 13.2 and 13.3, GFDM [3] is a multicarrier modulation
scheme in which each of the K subcarriers transmit M complex-valued data symbols
$d_{k,m}$ in different time slots, defined as subsymbols. The overall payload of a GFDM
block consists of $N=MK$ complex symbols. In this sense, the data symbols are
organized in a time–frequency grid, as illustrated in Figure 13.3. Each subcarrier is
pulse-shaped by a transmit filter $g_{k,m}[n]$, which is generated by circularly shifting a
prototype filter $g[n]$ in both time and frequency, leading to

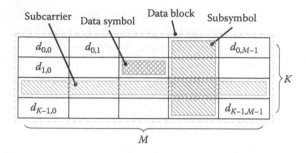

Figure 13.2 **Block structure of GFDM and corresponding terminology.**

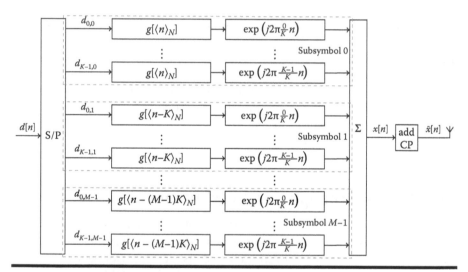

Figure 13.3 **Block diagram of GFDM modulator.**

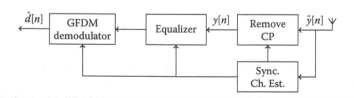

Figure 13.4 **Block diagram of GFDM receiver.**

$$g_{k,m}[n] = g\left[\langle n - mK \rangle_N\right] \exp\left(j2\pi \frac{k}{K} n\right) \qquad (13.1)$$

where:

$k = 0, 1, \ldots, K-1$ is the subcarrier index
$m = 0, 1, \ldots, M-1$ is the subsymbol index
$n = 0, 1, \ldots, N-1$ is the sample index

Circular convolution is used to modulate the transmit filters with the data symbols, meaning that the GFDM block is self-contained in N samples, and it is given by

$$x[n] = \sum_{k=0}^{K-1} \sum_{m=0}^{M-1} d_{k,m} \delta\left[\langle n - mK \rangle_N\right] \circledast g[n]\exp\left(j2\pi \frac{k}{K} n\right) = \sum_{k=0}^{K-1} \sum_{m=0}^{M-1} d_{k,m} g_{k,m}[n]$$

$$(13.2)$$

One CP can be added to protect the M subsymbols from interframe interference (IFI) introduced by multipath channels. This leads to a lower overhead when compared with OFDM, in which one CP is needed for each symbol. Notice that when $M = 1$, GFDM simplifies to OFDM. Let $\tilde{x}[n]$ be the GFDM signal with an appended CP of $N_{CP} < N$. Assume a multipath time-invariant channel with impulse response $h[n]$ of length L, meaning that $N_{CP} \geq L$. Under these circumstances, the signal at the GFDM receiver, depicted in Figure 13.4, is given by

$$\tilde{y}[n] = h[n] * \tilde{x}[n] + \tilde{w}[n] \qquad (13.3)$$

where $\tilde{w}[n]$ is the additive white Gaussian noise (AWGN) vector with length $\tilde{N} = N + N_{CP}$. After removal of the CP, the linear convolution becomes a circular convolution, and the received signal without CP is written as

$$y[n] = x[n] \circledast h[n] + w[n] \qquad (13.4)$$

where $w[n]$ is the AWGN sequence without the samples corresponding to the CP. Note that the circular convolution in Equation 13.4 is evaluated within a period of N samples. Because of circular convolution, GFDM can also take advantage of simple equalization in the frequency domain. Therefore, the equalized receive signal can be written as

$$y_{eq}[n] = \mathcal{F}_N^{-1}\left(\frac{\mathcal{F}_N(y[n])}{\mathcal{F}_N(h[n])}\right) \qquad (13.5)$$

where $\mathcal{F}_N$ and $\mathcal{F}_N^{-1}$ represent the N-point DFT and inverse discrete Fourier transform (IDFT), respectively.

After equalization, the data symbols can be recovered by using a set of receive filters, whose design will be detailed in Sections 13.2.2 and 13.2.4. For now, assume that a prototype receive filter $\gamma[n]$ is used to demodulate the data transmitted with $g[n]$. In this case, the estimated data symbols are given by

$$\hat{d}_{k,m} = \left(\gamma^* \left[\langle -n \rangle_N \right] \circledast y_{\text{eq}} \left[n \right] \exp \left(-j2\pi \frac{k}{K} n \right) \right) \Bigg|_{n=mK} \tag{13.6}$$

A slicer is used to recover the data bits and deliver them to the data sink. The pulse shape used to filter the subcarriers can be designed to achieve different goals, such as low self-interference and good spectral localization. In general, $g[n]$ leads to a family of nonorthogonal filters that introduces ISI and ICI, which means that the demodulator must be designed to deal with self-generated interference. Section 13.2.2 presents a matrix-vector representation of the GFDM signal, bringing insights into how a different set of linear GFDM demodulators can be designed.

13.2.2 Matrix Notation for GFDM

The GFDM modulation and demodulation processes can be represented by matrix operations, which is useful to design the receive filters. Especially, the data symbols of the GFDM block are organized in a $(K \times M)$ data matrix, given by

$$\mathbf{D} = \begin{bmatrix} d_{0,0} & d_{0,1} & \cdots & d_{0,M-1} \\ d_{1,0} & d_{1,1} & \cdots & d_{1,M-1} \\ \vdots & \vdots & \ddots & \vdots \\ d_{K-1,0} & d_{K-1,1} & \cdots & d_{K-1,M-1} \end{bmatrix} = [\mathbf{d}_{c,0} \quad \mathbf{d}_{c,1} \quad \cdots \quad \mathbf{d}_{c,M-1}] = \begin{bmatrix} \mathbf{d}_{r,0} \\ \mathbf{d}_{r,1} \\ \vdots \\ \mathbf{d}_{r,K-1} \end{bmatrix} \tag{13.7}$$

The columns of $\mathbf{D}$ are $(K \times 1)$ vectors $\mathbf{d}_{c,m}$ with the data symbols transmitted in the mth subsymbol, while the rows of $\mathbf{D}$ are $(M \times 1)$ vectors $\mathbf{d}_{r,k}$ with the data symbols transmitted in the kth subcarrier. Hence, $\mathbf{D}$ represents the time–frequency resource grid of the GFDM block.

Let $\mathbf{g}_{k,m}$ be an $(N \times 1)$ vector with the samples of the transmit filter impulse response $g_{k,m}[n]$. These vectors can be organized in a modulation matrix, given by

$$\mathbf{A} = \begin{bmatrix} \mathbf{g}_{0,0} & \mathbf{g}_{1,0} & \cdots & \mathbf{g}_{K-1,0} & \mathbf{g}_{0,1} & \cdots & \mathbf{g}_{K-1,M-1} \end{bmatrix} \tag{13.8}$$

Figure 13.5 shows the structure of the modulation matrix for a given prototype filter. The GFDM transmit vector is then given by

$$\mathbf{x} = \mathbf{Ad} \tag{13.9}$$

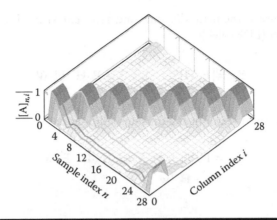

Figure 13.5   Structure of modulation matrix A with $M=7$, $K=4$, and raised cosine pulse with roll-off equals 0.4 as prototype filter.

where $\mathbf{d}=\mathrm{vec}(\mathbf{D})$ is the data vector obtained by stacking the columns of $\mathbf{D}$ one after another, that is,

$$\mathbf{d} = \begin{bmatrix} \mathbf{d}_{c,0}^{\mathrm{T}} & \mathbf{d}_{c,1}^{\mathrm{T}} & \cdots & \mathbf{d}_{c,M-1}^{\mathrm{T}} \end{bmatrix}^{\mathrm{T}} \tag{13.10}$$

with $(\cdot)^{\mathrm{T}}$ representing the transpose operation.

A CP can be added to the GFDM vector by copying the last N_{CP} samples of $\mathbf{x}$ to its beginning, leading to the transmit vector $\tilde{\mathbf{x}}$.

Let $\mathbf{h}$ be an $(N \times 1)$ vector where the N_{ch} first elements represent the channel impulse response and the last $N-N_{\mathrm{ch}}$ elements are zeroed. Hence, the received signal after removing the CP is given by

$$\mathbf{y} = \mathbf{H}\mathbf{x} + \mathbf{w} \tag{13.11}$$

where $\mathbf{H}$ is an $(N \times N)$ circulant matrix based on $\mathbf{h}$, and $\mathbf{w}$ is an $(N \times 1)$ vector with AWGN samples. The received signal in the frequency domain can be obtained with the help of the $(N \times N)$ size Fourier matrix $\mathbf{F}_N$, leading to

$$\begin{aligned} \mathbf{Y} = \mathbf{F}_N \mathbf{y} &= \mathbf{F}_N \mathbf{H}\mathbf{x} + \mathbf{F}_N \mathbf{w} \\ &= \mathbf{F}_N \mathbf{H}\mathbf{F}_N^{\mathrm{H}} \mathbf{F}_N \mathbf{x} + \mathbf{W} \\ &= \mathbf{F}_N \mathbf{H}\mathbf{F}_N^{\mathrm{H}} \mathbf{X} + \mathbf{W} \end{aligned} \tag{13.12}$$

where $\mathbf{X}$ and $\mathbf{W}$ are the transmit and AWGN vectors in the frequency domain, respectively. Note that $\mathbf{F}_N \mathbf{H}\mathbf{F}_N^{\mathrm{H}}$ is a diagonal matrix containing the channel frequency response as the main diagonal and zeros elsewhere. Using a frequency-domain equalizer (FDE) yields

$$\mathbf{Y}_{eq} = \left(\mathbf{F}_N \mathbf{H} \mathbf{F}_N^H\right)^{-1} \mathbf{Y} = \mathbf{X} + \mathbf{F}_N \mathbf{H}^{-1} \mathbf{F}_N^H \mathbf{W} \qquad (13.13)$$

The equalized received vector in the time domain is given by

$$\mathbf{y}_{eq} = \mathbf{F}_N^H \mathbf{Y}_{eq} = \mathbf{F}_N^H \mathbf{X} + \mathbf{F}_N^H \mathbf{F}_N \mathbf{H}^{-1} \mathbf{F}_N^H \mathbf{W}$$
$$= \mathbf{x} + \mathbf{H}^{-1} \mathbf{w} \qquad (13.14)$$

A demodulation matrix **B** is used to recover the data symbols from the equalized received vector, that is,

$$\hat{\mathbf{d}} = \mathbf{B}\mathbf{y}_{eq} = \mathbf{B}\mathbf{x} + \mathbf{B}\mathbf{H}^{-1}\mathbf{w}$$
$$= \mathbf{B}\mathbf{A}\mathbf{d} + \mathbf{B}\mathbf{H}^{-1}\mathbf{w} \qquad (13.15)$$

Different demodulation matrices can be used to recover the data symbols. The trivial solution for Equation 13.15 is to make $\mathbf{B}\mathbf{A}=\mathbf{I}_N$, which leads to the ZF demodulation matrix

$$\mathbf{B}_{ZF} = \mathbf{A}^{-1} \qquad (13.16)$$

The ZF demodulator eliminates the self-generated interference introduced by the nonorthogonal transmit filters. The drawback of this approach is that the frequency response of the ZF filter for one specific subcarrier spreads to surrounding subcarriers, meaning that noise from undesired bands is collected, as can be seen in Figure 13.6, leading to noise enhancement and performance degradation that depends on the chosen prototype filter.

The MF demodulation matrix, given by

$$\mathbf{B}_{MF} = \mathbf{A}^H \qquad (13.17)$$

where $(\cdot)^H$ is the Hermitian operator, can maximize the SNR for each detected symbol, but it suffers from self-generated interference. Figure 13.6 shows that the impulse response of the MF demodulator is constrained to the desired bandwidth, and no noise enhancement is added to the demodulated symbols. However, the MF demodulator is unable to deal with the self-interference introduced on the transmit side, which means that an error floor is expected due to ISI and ICI among the demodulated data. Figure 13.7 shows the absolute value of the ambiguity matrix

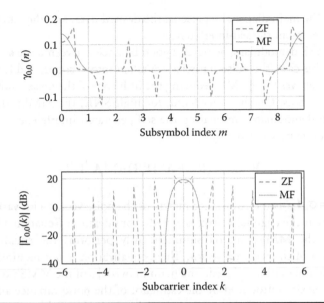

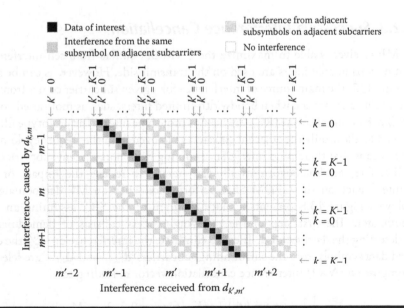

Figure 13.6 Impulse and frequency response of MF and ZF prototype receive filters.

Figure 13.7 Example of ambiguity matrix for an MF demodulator assuming an RRC filter with roll-off equals 0.9, $M = 7$, and $K = 128$.

C=B$_{\mathrm{MF}}$**A.** The main diagonal represents the desired data, and the remaining non-zero values represent the self-interference.

The MMSE demodulator achieves a good balance between noise and self-interference. In this case, the noise statistics are taken into account, and for low SNR, the MMSE behaves like an MF, reducing the influence of the noise on the demodulated data symbols. On the other hand, for high SNR, the MMSE demodulator acts as a ZF demodulator, mitigating the self-generated interference. The MMSE demodulation matrix is given by

$$\mathbf{B}_{\mathrm{MMSE}} = (\mathbf{R}_w + \mathbf{A}^{\mathrm{H}}\mathbf{H}^{\mathrm{H}}\mathbf{HA})^{-1}\mathbf{A}^{\mathrm{H}}\mathbf{H}^{\mathrm{H}} \tag{13.18}$$

where $\mathbf{R}_w = \sigma_w^2 \mathbf{I}_N$ is the covariance matrix of the noise vector with variance σ_w^2.

Notice that the channel matrix is already considered in Equation 13.18, which means that the MMSE matrix simultaneously performs the equalization and demodulation of the received vector. Thus, equalization is not required when the MMSE demodulator is employed. The main drawbacks of the MMSE demodulator are the complexity introduced by the estimator of the noise variance and the need to reevaluate the demodulation matrix when the channel impulse response changes.

It is important to note that the first row of **B** contains the receive prototype filter γ for MF and ZF demodulators, and therefore, the receive prototype γ[*n*] to be used in Equation 13.6 can be readily obtained from **B**. This statement also holds for the MMSE demodulator under the assumption of AWGN channels.

13.2.3 Successive Interference Cancellation

The MF receiver is able to maximize the SNR, but suffers from self-interference when nonorthogonal filters are used on the transmit side. However, as can be seen in Figure 13.7, the main source of interference for a given subcarrier comes from the surrounding subcarriers, when bandwidth-limited filters, such as root raised cosine (RRC) MF or raised cosine (RC), are used as prototype filters. The prototype filter is designed to allow only the adjacent subcarriers to overlap in the frequency domain, which means that there is no interference coming from nonadjacent subcarriers.

Therefore, an SIC algorithm [8] can iteratively reduce the impact of the self-interference on the GFDM SER performance when an MF demodulator is employed. Figure 13.8 presents a block diagram of the SIC, assuming an MF demodulator. The first step of the receiving process consists of demodulating and detecting the data vector $\hat{\mathbf{d}}$. Then, to eliminate the interference on $\mathbf{d}_{r,k}^{(i)}$, the estimated data symbols from the surrounding subcarriers $\hat{\mathbf{d}}_{r,k-1}$ and $\hat{\mathbf{d}}_{r,k+1}$ are selected to compose an ($N \times 1$) interference cancellation vector **o**, leading to

$$[\mathbf{o}]_\iota = \begin{cases} \hat{d}_{r,k\pm1,m} & \iota = k\pm1+mK \quad \text{for } m=0,1,2,\ldots,M-1 \\ 0 & \text{otherwise} \end{cases} \tag{13.19}$$

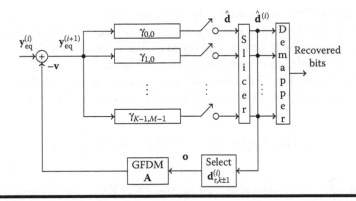

Figure 13.8 Block diagram of SIC receiver.

The interference cancellation vector is applied to the GFDM modulator to generate the error signal

$$\mathbf{v} = \mathbf{Ao} \tag{13.20}$$

which is subtracted from the equalized received signal to produce the signal without ICI on the kth subcarrier, given by

$$\mathbf{y}_{eq}^{(i+1)} = \mathbf{y}_{eq}^{(i)} - \mathbf{v} \tag{13.21}$$

Once all subcarriers are cleaned, the ICI-free signal is demodulated. This process can be repeated for multiple iterations to achieve a better SER performance.

At the cost of complexity on the receiver side, the SIC can achieve a SER performance comparable with orthogonal systems [9]. However, error propagation can be a problem when dense constellations are used in low SNR. One solution for this problem is to use soft decision instead of hard decision, whereby the constellation symbols fed back to the SIC are weighted by their specific reliability (or log-likelihood ratios) delivered by the decoder [10].

13.2.4 Design of the Receive Filter using Zak Transform

The matrix notation is useful to design the receive filter based on the transmit matrix. However, it is a challenge to obtain the ZF and MMSE demodulators when M and K are large because of the required matrix inversions. The challenge is even higher for MMSE demodulators, because the matrix inversion must be performed every time the channel impulse response or the noise variance changes. Also, there are cases in which $\mathbf{A}$ is singular, meaning that the ZF demodulator is not available. Hence, a procedure to investigate the reasons that make $\mathbf{A}$ singular is essential to define proper GFDM parameters. Also, it is important to define a solution that allows the receiver filter to be designed based only on the transmit prototype pulse, avoiding costly matrix inversion computations. The Gabor theory [7] is an interesting tool that provides an efficient solution for this issue.

Assume a set of $T = TF$ discrete and finite sequences $u_{f,t}[n]$ with N samples that are obtained by circularly shifting an elementary sequence $u[n]$, that is,

$$u_{f,t}[n] = u\left[(n - t\Delta_T) \bmod N\right] \exp\left(j2\pi \frac{f\Delta_F}{N} n \right) \qquad (13.22)$$

The length of the sequences is defined by $N = T\Delta_T = F\Delta_F$. The time and frequency shifts are defined by $\Delta_T = N/T$ and $\Delta_F = N/F$, respectively.

Consider a periodic sequence $r[n]$ with length N that is expanded from the $u_{f,t}$ sequences as

$$r[n] = \sum_{f=0}^{F-1} \sum_{t=0}^{T-1} a_{f,t} u_{f,t}[n] \qquad (13.23)$$

In this sense, $a_{f,t}$ are called Gabor coefficients, and Equation 13.23 is a Gabor expansion. For $N = \Delta_F\Delta_T$, the Gabor frame is said to be *critically sampled*, and this is the condition that will be assumed from now on. In this case, the Gabor coefficients can be obtained with the help of a set of sequences $v_{f,t}[n]$ dual to $u_{f,t}[n]$, meaning that

$$\langle \mathbf{u}_{f',t'}, \mathbf{v}_{f,t} \rangle = \sum_{n=0}^{N-1} u_{f',t'}[n] v_{f,t}^*[n] = \begin{cases} A\delta[n] & \text{for } t' = t \text{ and } f' = f \\ 0 & \text{otherwise} \end{cases} \qquad (13.24)$$

where:

$v_{f,t}[n]$ as the analysis windows
$\langle \mathbf{a},\mathbf{b} \rangle$ is the inner product between two generic vectors $\mathbf{a}$ and $\mathbf{b}$
A is a constant value

Then, the Gabor transform can be defined as [7]

$$a_{f,t} = \sum_{n=0}^{N-1} r[n] v_{f,t}[n] \qquad (13.25)$$

When Equations 13.9 and 13.23 are compared, it becomes evident that GFDM modulation is a critically sampled Gabor expansion in which the data symbols are the Gabor coefficients $(a_{f,t} = d_{k,m})$, the prototype filter is the elementary sequence for the Gabor expansion $(u[n] = g[n])$, $\Delta_T = K$, $\Delta_F = M$, and $N = KM$. Comparing Equations 13.6 and 13.25 also leads to the conclusion that the GFDM demodulation is a Gabor transform of the received signal, in which the receive filter is the analysis window $(v[n] = \gamma[n])$. The main question is how to define an efficient approach to calculating the analysis window based on the elementary sequence of the Gabor frame. Because the Gabor frame is critically sampled, the Gabor expansion is unique, and there is only one analysis window for a given $g[n]$. Hence, the analysis window can be efficiently calculated from the elementary sequence with the help of the discrete Zak transform (DZT), defined as

$$\mathcal{G}[k,m] = \mathcal{Z}^{(K,M)}\{g[n]\} = \sum_{l=0}^{M-1} g[k+lK]\exp\left(-j2\pi\frac{m}{M}l\right) \qquad (13.26)$$

The DZT rearranges the samples of the input vector in a $(K \times M)$ matrix, in which the rows contain the lKth elements, that is,

$$\mathbf{G}_{\text{Zak}} = \begin{bmatrix} g[0] & g[K] & \cdots & g[(M-1)K] \\ g[1] & g[K+1] & \cdots & g[(M-1)K+1] \\ \vdots & \vdots & \ddots & \vdots \\ g[K-1] & g[2K-1] & \cdots & g[MK-1] \end{bmatrix} \qquad (13.27)$$

and the DFT is applied to each row of the resulting matrix, leading to

$$\mathcal{G} = \mathbf{G}_{\text{Zak}}\mathbf{F}_M^T \qquad (13.28)$$

which is the matrix form of Equation 13.26. The inverse DZT is obtained by applying the IDFT to each row of $\mathcal{G}$ and rearranging the resulting samples by stacking the columns, that is,

$$\mathbf{g} = \frac{1}{M}\text{vec}\left(\mathcal{G}\mathbf{F}_M^*\right) \qquad (13.29)$$

For a critically sampled time–frequency frame, such as GFDM, the DZT of the analysis window, $\Gamma[n,k]$, can be obtained from the DZT of the elementary sequence, $\mathcal{G}[n,k]$, as

$$\Gamma[k,m] = \frac{1}{K\mathcal{G}^*[k,m]} \qquad (13.30)$$

Applying the iDZT in Equation 13.30 leads to the desired analysis window $\gamma[n]$. Note that this procedure leads to the ZF receive filter [7]. The same principles can be used to obtain the MMSE receive filter under an AWGN channel. However, in this case, it is necessary to consider the following elementary sequence:

$$g_{\text{MMSE}}[n] = g[n] + \sigma_w^2 \gamma[n] \qquad (13.31)$$

The procedure to obtain the receive filter using the DZT gives important insights into the parameters that can be used to design GFDM filters. The Balian–Low theorem [7] states that sequences well localized in the time and frequency domains, such as the RC and RRC filters commonly used in the literature, cannot produce Gabor frames, and the continuous-time Zak transform of these sequences has at least one null value. If this null value is sampled in the DZT, the element-wise inversion of $\mathcal{G}$ cannot be calculated, and the analysis window does not exist. Figure 13.9 shows that the DZT of the prototype filter has a null value at $\mathcal{G}[k = K/2, \ m = M/2]$ when K and M are even. In this case, the modulation matrix is singular, no ZF exists, and GFDM performs poorly when other demodulators are employed. In fact, for a symmetric real-valued elementary sequence, the null value of the continuous-time Zak transform is always sampled when K and M are even, which means that this combination must be avoided when designing GFDM systems.

Another interesting observation from Figure 13.9 is that the minimal value of $\left|\mathcal{G}\left[k, m = \lfloor M/2 \rfloor\right]\right|$ decreases as α and M increase (assuming odd values for M). This means that the analysis window will have higher peak amplitudes, resulting in a higher noise enhancement for the ZF demodulator. Therefore, for a given number of subcarriers, small values of α and M will typically lead to GFDM schemes with better SER performance when the ZF demodulator is employed.

13.2.5 Solutions for Low OOB Emissions

Subcarrier filtering can reduce the GFDM OOB emissions when compared with OFDM, as can be seen in Figure 13.10. However, the abrupt transitions between the GFDM blocks, depicted in Figure 13.11, limit the overall reduction that can be achieved in OOB emission. Further reduction of undesired OOB emission can be obtained by using guard subsymbols or time windowing, as presented in Sections 13.2.5.1 and 13.2.5.2.

13.2.5.1 Guard Symbol-Generalized Frequency-Division Multiplexing

To reduce the OOB emission, it is necessary to smooth the transitions between the GFDM blocks. The circularity of the signal in the time domain allows a simple

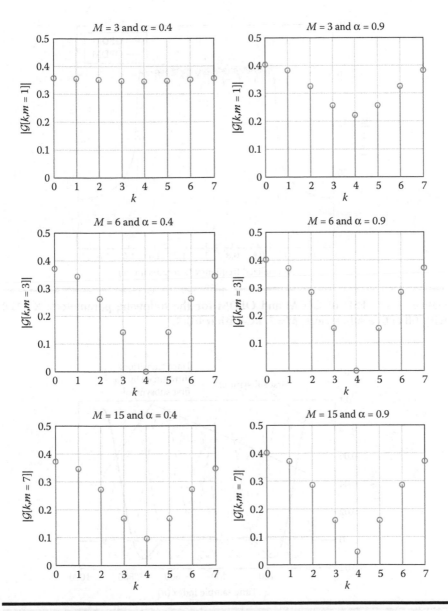

Figure 13.9   DZT of RC filters with different parameters: $K=8$; $M\in\{3,6,15\}$; roll-off factor $\alpha\in\{0.4,0.9\}$.

and elegant solution to reduce the abrupt change between GFDM blocks. As can be seen in Figure 13.11, the first subsymbol wraps around the edges of the block, introducing the abrupt amplitude discontinuities.

If the first subsymbol is erased, a GS is introduced between the GFDM blocks, and the edges of the signal fade out toward zero, making the transitions between

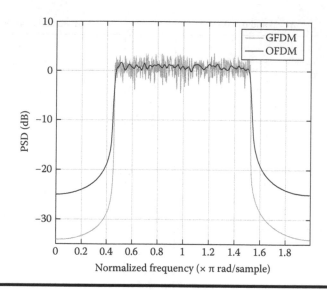

Figure 13.10 PSD of GFDM and OFDM for the following parameters: $K=128$ with 68 active subcarriers, $M=7$, RC with $\alpha=0.5$.

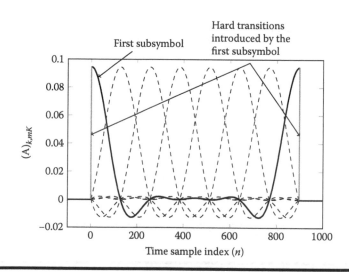

Figure 13.11 The first subsymbol of the GFDM block introduces abrupt variations in the time-domain signal, which leads to high OOB emissions.

blocks smooth. This technique is called *guard-symbol GFDM*, and Figure 13.12 shows the time-domain signal and the corresponding spectrum.

The addition of the CP would introduce hard transitions between blocks once again. One solution to avoid this problem is to also make the last subsymbol null

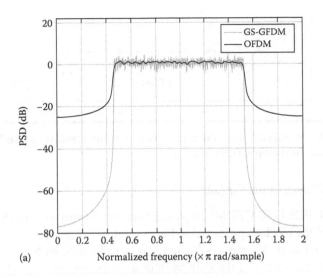

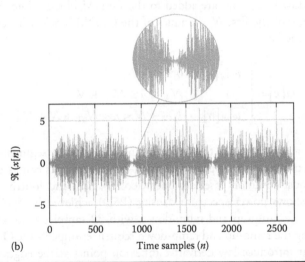

Figure 13.12 GS-GFDM signal. (a) Comparison of GS-GFDM and OFDM PSDs. (b) Real part of the GS-GFDM signal, highlighting the smooth transition between blocks.

and to make $N_{CP}=K$. The drawback of this approach is the throughput reduction, given by

$$R_{GS} = \frac{M-2}{M} \times \frac{KM}{KM+K} = \frac{M-2}{M+1} \tag{13.32}$$

From Equation 13.32, it is clear that this technique becomes interesting for scenarios where M is large.

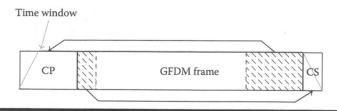

Figure 13.13 W-GFDM time-domain signal. A time window is used to smooth the transitions between GFDM blocks.

13.2.5.2 Windowed GFDM

Windowed GFDM (W-GFDM) employs a time window to smooth the transition between GFDM blocks, as depicted in Figure 13.13.

A CP with length $N_{CP} = N_{CH} + N_W$ and a cyclic suffix (CS) with length $N_{CS} = N_W$, where N_{CH} is the length of the channel impulse response and N_W is the length of the time window transition, are added to the GFDM block. Note that the CS is just the copying of the first N_{CS} samples of the GFDM block to its end. The time window is defined as

$$w[n] = \begin{cases} w_{\text{rise}}[n] & 0 \le n < N_W \\ 1 & N_W \le n \le N_{CP} + N \\ w_{\text{fall}}[n] & N_{CP} + N < n < N_{CP} + N + N_W \end{cases} \tag{13.33}$$

where $w_{\text{rise}}[n]$ and $w_{\text{fall}}[n]$ are the ramp-up and ramp-down segments of the time window, respectively. The ramp-up and ramp-down segments can assume different shapes. The most common cases are linear, cosine, RC, and fourth-order RC [3]. Figure 13.14 shows power spectrum density (PSD) achieved by W-GFDM when linear and cosine ramp-up and ramp-down, with 32 samples each, are employed. Clearly, varying the ramp-up and ramp-down sequences impacts the OOB emission. Sequences that introduce low derivative inflexion points at the edges of the signal provide lower OOB emission.

W-GFDM can be used to achieve low OOB emission and still keep high spectral efficiency, even when M is low. The rate loss introduced by the ramp-up and ramp-down edges is given by

$$R_W = \frac{N}{N + N_{CP} + 2N_W} \tag{13.34}$$

Since N_{CP} is defined as a function of the channel impulse response, it equally affects the rate loss of GS-GFDM and W-GFDM. The ramp-up and ramp-down sequences are much smaller than the GFDM block length, which means that R_W is typically higher than R_{GS}. From Figures 13.12 and 13.14, it is possible to

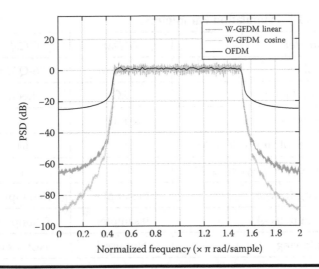

Figure 13.14 PSD of the W-GFDM using linear and cosine time windows compared with OFDM.

conclude that both GS-GFDM and W-GFDM can achieve similar OOB emission. Therefore, W-GFDM can be seen as a more promising solution for a highly efficient PHY with low OOB emissions.

13.2.6 Performance Analysis of the GFDM Symbol Error Rate

Nonorthogonal waveforms suffer from ICI and ISI, which must be considered on the receiver side. In this subsection, the GFDM SER performance is evaluated assuming different channel models for MF, ZF, and MMSE demodulators. Table 13.1 describes the parameters used for the simulations, while Table 13.2 shows the channel models. Three channel models will be considered: the AWGN channel; the frequency-selective channel (FSC), with 16 taps linearly varying from 0 dB to −10 dB; and the time-variant channel (TVC), consisting of one complex random tap with zero mean and unitary variance normal distribution.

13.2.6.1 SER Performance under AWGN Channel

The SER performance of GFDM depends on the demodulator employed to recover the data. A reference symbol error probability (SEP) equation can be easily obtained assuming a ZF demodulator. In this case, the self-interference is completely removed from the received signal, but because the receive filter can collect noise from undesired frequency bands, the noise is enhanced by a noise enhancement factor (NEF) defined as

$$\xi = \sum_{n=0}^{N-1} \left| \gamma_{ZF}[n] \right|^2 \tag{13.35}$$

Table 13.1   Simulation Parameters

Parameter	GFDM	OFDM
Mapping	16-QAM	16-QAM
Transmit filter	RC	Rect.
Roll-off (α)	0 or 0.9	0
Number of subcarriers (K)	64	64
Number of subsymbols (M)	9	1
CP length (N_{CP})	16 samples	16 samples
CS length (N_{CS})	0 samples	0 samples
Windowing	Not used	Not used

Table 13.2   Channel Models for Simulations

Channel Model	Impulse Response
Flat AWGN	$\mathbf{h}_{flat}=1$
FSC	$\mathbf{h}_{FSC} = \left(10^{\frac{-(2/3)i}{20}} \right)^{T}_{i=0,\dots,15}$
TVC	$\mathbf{h}_{TVC}=h,\ h \sim \mathcal{CN}(0,1)$

Hence, the SEP for GFDM with a ZF demodulator is given by

$$p_{AWGN}(e) = 2\left(\frac{\kappa-1}{\kappa}\right)\operatorname{erfc}\left(\sqrt{\varrho}\right)-\left(\frac{\kappa-1}{\kappa}\right)\operatorname{erfc}^2\left(\sqrt{\varrho}\right) \tag{13.36}$$

where $\kappa=\sqrt{2^\mu}$, μ is the number of bits per data symbol, and

$$\varrho = \frac{3R_T}{2(\kappa^2-1)}\cdot\frac{E_s}{\xi N_0} \tag{13.37}$$

where E_s is the average energy per data symbol, N_0 is the AWGN spectrum density, and

$$R_T = \frac{KM}{KM+N_{CP}+N_{CS}} \tag{13.38}$$

accounts for the SNR loss introduced by the CP and the CS.

The performance degradation introduced by the NEF depends on the prototype transmit filter. Typically, the higher the interference to be removed, the stronger the NEF. This means that when an RC filter is used, a high roll-off will lead to a high NEF, while small roll-offs result in negligible NEF. Figure 13.15 compares the performance of the three linear GFDM demodulators with Equation 13.36, while

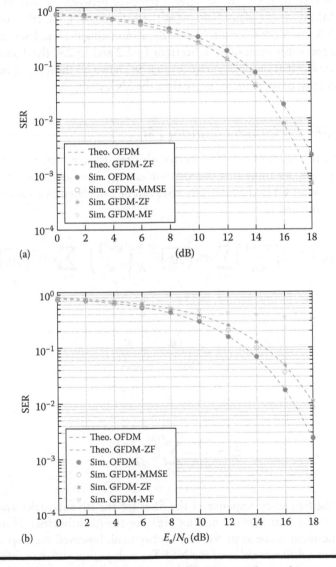

Figure 13.15 GFDM SER performance under AWGN channel. (a) RC filter with α = 0. (b) RC filter with α = 0.9.

OFDM SER is plotted as a reference. Figure 13.15a and b show the result for $\alpha = 0$ and $\alpha = 0.9$, respectively.

Figure 13.15a shows that GFDM outperforms OFDM due to the more efficient use of the CP. Also, all three demodulators have the same performance, because the RC filter with $\alpha = 0$ results in the Dirichlet pulse and makes GFDM orthogonal. In Figure 13.15b, the self-generated interference leads to a high error floor when an MF demodulator is used. The NEF reduces the performance of the ZF demodulator, while the MMSE demodulator achieves a trade-off between ZF and MF. For low SNR, the MMSE demodulator behaves as the MF, reducing the effect of the NEF. For high SNR, the MMSE approaches the ZF, eliminating the interference at the cost of NEF. As previously mentioned in Sections 13.2.2 and 13.2.4, the drawback of the MMSE is the fact that the noise variance must be known on the receiver side, and the receive filter must be recalculated every time the SNR changes.

13.2.6.2 SER Performance under FSC

Under FSC, the effect of the NEF varies with the receive filter and channel frequency responses, meaning that the SEP will change from subcarrier to subcarrier. Hence, the overall GFDM SEP under FSC can be calculated as an average, which can be approximated to

$$p_{\text{FSC}}(e) = 2\left(\frac{\kappa-1}{\kappa K}\right)\sum_{l=0}^{K-1}\text{erfc}\left(\sqrt{\varrho_l}\right) - \frac{1}{K}\left(\frac{\kappa-1}{\kappa}\right)^2\sum_{l=0}^{K-1}\text{erfc}^2\left(\sqrt{\varrho_l}\right) \quad (13.39)$$

where

$$\varrho_l = \frac{3R_T}{2(\kappa^2-1)} \cdot \frac{E_s}{\xi_l N_0} \quad (13.40)$$

and

$$\xi_l^2 = \frac{1}{MK}\sum_{k=0}^{MK-1}\left|\frac{G_{R_{l,0}}[-k]}{H[k]}\right|^2 \quad (13.41)$$

$G_{R_{l,0}}[k]$ is the frequency response of the filter for the lth subcarrier and the first subsymbol, and ξ_l is the corresponding NEF for the lth subcarrier. Note that the noise enhancement is the same for every subsymbol; however, the response of the FSC results in a different value of the NEF for each position of the prototype filter in the frequency domain. Figure 13.16 depicts the GFDM SER performance under the FSC presented in Table 13.2.

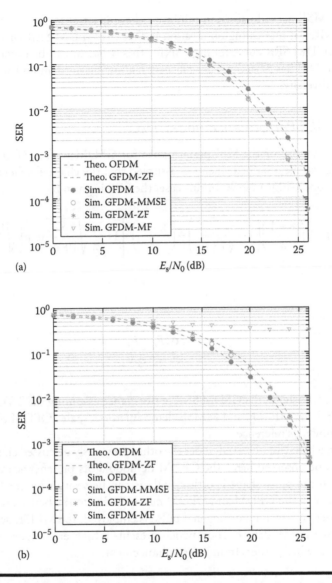

Figure 13.16 GFDM SER performance under FSC channel. (a) RC filter with α = 0. (b) RC filter with α = 0.9.

As can be seen from Figure 13.16a, the GFDM demodulators achieve the same performance when the transmit pulse is orthogonal. Again, GFDM outperforms OFDM because of the better use of the CP. Figure 13.16b shows that ISI and ICI severely impact the performance of the MF demodulator, leading to a high error floor. For the ZF demodulator, as expected, the noise enhancement introduces a performance loss when compared with the SER obtained with an orthogonal pulse, and the MMSE demodulator can minimize the impact of the noise enhancement for low SNR. An interesting observation is that the GFDM SER curve is steeper than the OFDM SER curve for high SNR. This behavior can be explained by the fact that GFDM has M samples per subcarrier, allowing the demodulator to explore frequency diversity.

13.2.6.3 SER Performance under TVC

An approximation for the SER performance for GFDM with a ZF demodulator under TVC can be obtained when it is assumed that the channel coherence time is larger than one GFDM block. In this case, the SEP is given by

$$p_{\text{TVC}}(e) = 2\left(\frac{\kappa - 1}{\kappa}\right)\left(1 - \sqrt{\frac{\varrho_r}{1 + \varrho_r}}\right) - \left(\frac{\kappa - 1}{\kappa}\right)^2\left[1 - \frac{4}{\pi}\sqrt{\frac{\varrho_r}{1 + \varrho_r}}\text{atan}\left(\sqrt{\frac{1 + \varrho_r}{\varrho_r}}\right)\right]$$

(13.42)

where

$$\varrho_r = \frac{3\sigma_r^2 R_T}{\kappa^2 - 1}\frac{E_s}{\xi N_0}$$

(13.43)

Figure 13.17 shows the GFDM SER performance assuming the TVC described in Table 13.2 for MF, ZF, and MMSE demodulators. Again, OFDM SER performance is shown as reference.

The same behavior observed under the other previously discussed channel models can also be observed here. The GFDM demodulators present the same performance when the pulse shape is orthogonal, as can be seen in Figure 13.17a, and the gap between the GFDM and OFDM curves is due to the better use of the CP by the former. When a nonorthogonal pulse is employed, the MMSE demodulator outperforms the MF demodulator, which presents a high error floor, and the ZF demodulator, which suffers from noise enhancement.

13.3 Offset QAM for GFDM

Good time–frequency localization is an important feature for 5G waveforms to deal with the requirements from the different scenarios. However, well-localized pulses in time and frequency cannot provide interference-free communication when signaling at the Nyquist rate [11]. Filtered multicarrier modulations overcome this limitation by employing OQAM mapping to transmit two real-valued sequences, achieving orthogonality and good time and frequency localization at the same

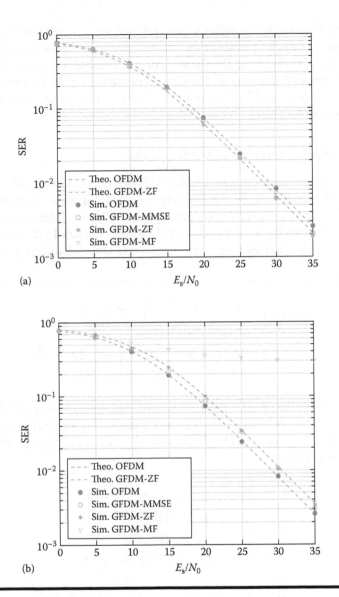

Figure 13.17 GFDM SER performance under TVC channel. (a) RC filter with α = 0. (b) RC filter with α = 0.9.

time. This principle can also be applied to GFDM, leading to a self-interference-free communication, meaning that performance loss can be avoided when using an MF receiver and the same SER observed by orthogonal system can be achieved with OQAM-GFDM. The duality between time and frequency allows two different approaches to combine OQAM with GFDM. The first is the conventional time-domain approach, in which the real and imaginary parts of the data symbols are transmitted using two modulation matrices, one time-shifted by half a subsymbol with respect to the other. The second approach consists of transmitting the data symbol components with two modified transmit matrices in the frequency domain, with a frequency shift of half a subcarrier relative to each other. Both approaches are presented in Sections 13.3.1 and 13.3.2.

13.3.1 Time-Domain OQAM-GFDM

The basic idea of time-domain OQAM-GFDM (TD-OQAM-GFDM) [12] is to avoid self-interference by transmitting two real sequences $i_{k,m} = \Re(d_{k,m})$ and $q_{k,m} = \Im(d_{k,m})$ using two modulation processes in which the pulse shape of the second one is time-shifted by half a subsymbol (or $K/2$ samples). In this case, two sets of pulse shapes are derived from a symmetric, real-valued, half-Nyquist prototype filter $g[n]$:

$$g_{k,m}^{(R)}[n] = g\left[\langle n - mK \rangle_N\right] \exp\left(j2\pi \frac{k}{K}n\right)$$

$$g_{k,m}^{(I)}[n] = jg\left[\left\langle n - \left(m + \frac{1}{2}\right)K \right\rangle_N\right] \exp\left(j2\pi \frac{k}{K}n\right) \qquad (13.44)$$

The goal is to design an orthogonal system, which means that an MF receiver is able to perfectly reconstruct the data sequences under an ideal noiseless channel. Consider the projection of $g_{k_2,m_2}^{(R)}[n]$ and $g_{k_2,m_2}^{(I)}[n]$ onto $g_{k_1,m_1}^{(R)}[n]$ and $g_{k_1,m_1}^{(I)}[n]$, respectively, given by

$$s_{k',m'}^{(RR)}[n] = g_{k_2,m_2}^{(R)}[n] \circledast g_{k_1,m_1}^{*(R)}[\langle -n \rangle_N]$$

$$s_{k',m'}^{(IR)}[n] = g_{k_2,m_2}^{(I)}[n] \circledast g_{k_1,m_1}^{*(R)}[\langle -n \rangle_N]$$

$$s_{k',m'}^{(RI)}[n] = g_{k_2,m_2}^{(R)}[n] \circledast g_{k_1,m_1}^{*(I)}[\langle -n \rangle_N]$$

$$s_{k',m'}^{(II)}[n] = g_{k_2,m_2}^{(I)}[n] \circledast g_{k_1,m_1}^{*(I)}[\langle -n \rangle_N], \qquad (13.45)$$

where $k' = k_2 - k_1$ and $m' = m_2 - m_1$. Figure 13.18 shows the behavior of these interferences for the cases of interest, assuming RRC, $K = 64$, and $M = 9$.

The signals $s_{0,0}^{(RR)}[n]$ and $s_{0,0}^{(II)}[n]$ show that the information transmitted on $g_{k,m}^{(R)}[n]$ and $g_{k,m}^{(I)}[n]$ can be recovered free of ISI due to the half-Nyquist pulse shape. The term $s_{0,0}^{(RI)}[n]$ shows that $g_{k,m}^{(R)}[n]$ does not introduce interference on the imaginary component. The terms $s_{1,0}^{(RR)}[n]$ and $s_{1,0}^{(IR)}[n]$ show that the adjacent subcarrier does not introduce interference on the real component of the desired subcarrier at $n = (m + 1/2)K$. Also, the imaginary component of the desired subcarrier is ICI free at $n = mK$. Finally, $s_{1,0}^{(II)}[n]$ shows that the imaginary part of the adjacent subcarrier

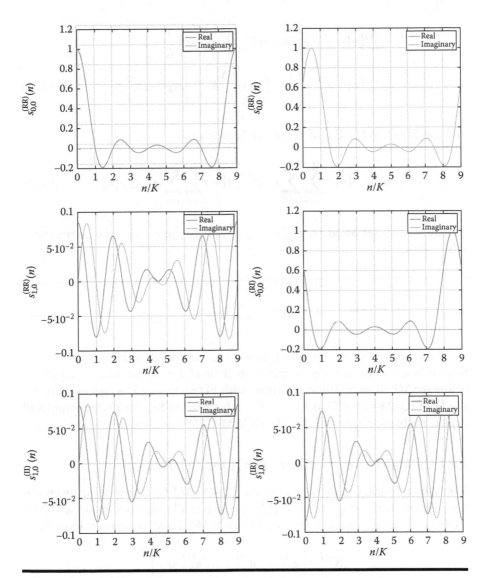

Figure 13.18   Behavior of interference in TD-OQAM-GFDM.

allows ICI-free reception at $n = mK$ for the real component and $n = (m + 1/2)K$ for the imaginary component.

Hence, it is clear that the adjacent subcarriers must have a phase rotation of $\pi/2$ radians to avoid ICI, which leads to the effect of interchanging the real and imaginary components. The pulse shapes used to transmit the real-valued data are redefined as

$$g_{k,m}^{(R)}[n] = j^{\langle k \rangle_2} g\left[\langle n - mK \rangle_N\right] \exp\left(j2\pi \frac{k}{K} n\right)$$

$$g_{k,m}^{(I)}[n] = j^{\langle k \rangle_2} g\left[\left\langle n - \left(m + \frac{1}{2}\right)K \right\rangle_N\right] \exp\left(j2\pi \frac{k}{K} n\right) \qquad (13.46)$$

and the TD-OQAM-GFDM signal can be written as

$$x_{\mathrm{OQAM}}[n] = \sum_{m=0}^{M-1}\sum_{k=0}^{K-1} i_{k,m} g_{k,m}^{(R)}[n] + j \sum_{m=0}^{M-1}\sum_{k=0}^{K-1} q_{k,m} g_{k,m}^{(I)}[n] \qquad (13.47)$$

An MF receiver based on $g_{k,m}^{(R)}[n]$ and $g_{k,m}^{(I)}[n]$ can be used to recover the transmitted data:

$$\hat{i}_{k,m} = \Re\left\{x[n] \circledast g_{k,m}^{*(R)}[\langle -n \rangle_N]\right\}_{n=mK}$$

$$\hat{q}_{k,m} = \Im\left\{x[n] \circledast -j\, g_{k,m}^{*(I)}[\langle -n \rangle_N]\right\}_{n=(m+1/2)K} \qquad (13.48)$$

Note that the MF receiver is only able to recover the information without interference in a flat channel. For a multipath channel, the signal must be equalized before the MF receiver. This can be done with FDE when a CP is inserted between the TD-OQAM-GFDM blocks, exactly as described in Section 13.2. Figure 13.19 depicts the block diagram of the TD-OQAM-GFDM transceiver.

13.3.1.1 Matrix Notation for TD-OQAM-GFDM

Matrix notation can be used to describe the TD-OQAM-GFDM by using two transmit matrices based on $g_{k,m}^{(R)}[n]$ and $g_{k,m}^{(I)}[n]$, which can be represented as the $(N \times 1)$ vectors $\mathbf{g}_{k,m}^{(R)}$ and $\mathbf{g}_{k,m}^{(I)}$, respectively. The TD-OQAM-GFDM matrices are given by

$$\mathbf{A}^{(R)} = \left[\mathbf{g}_{0,0}^{(R)} \; \cdots \; \mathbf{g}_{K-1,0}^{(R)} \; \cdots \; \mathbf{g}_{K-1,M-1}^{(R)}\right]$$

$$\mathbf{A}^{(I)} = \left[\mathbf{g}_{0,0}^{(I)} \; \cdots \; \mathbf{g}_{K-1,0}^{(I)} \; \cdots \; \mathbf{g}_{K-1,M-1}^{(I)}\right] \qquad (13.49)$$

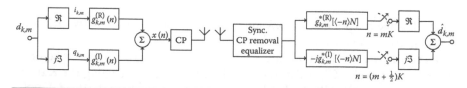

Figure 13.19 Block diagram of TD-OQAM-GFDM transceiver.

The TD-OQAM-GFDM transmit vector can be written as

$$\mathbf{x}_{OQAM} = \mathbf{A}^{(R)}\Re\{\mathbf{d}\} + j\mathbf{A}^{(I)}\Im\{\mathbf{d}\} \tag{13.50}$$

A CP is added to the TD-OQAM-GFDM vector before transmission over a multipath channel. The received signal, after synchronization, CP removal, and equalization, is given by

$$\mathbf{y}_{OQAM} = \mathbf{x}_{OQAM} + \mathbf{H}^{-1}\mathbf{w} \tag{13.51}$$

The data symbols are recovered using an MF receiver:

$$\hat{\mathbf{d}} = \Re\{\mathbf{B}^{(R)}\mathbf{y}_{OQAM}\} + j\Im\{-j\mathbf{B}^{(I)}\mathbf{y}_{OQAM}\} \tag{13.52}$$

where

$$\mathbf{B}^{(\cdot)} = \left(\mathbf{A}^{(\cdot)}\right)^{H} \tag{13.53}$$

with (·) standing for (R) or (I) to represent both modulation matrices.

Figure 13.20 depicts the block diagram of the TD-OQAM-GFDM chain based on the matrix model.

13.3.2 Frequency-Domain OQAM-GFDM

Assume a unitary transform matrix $\mathbf{U}_N$ applied to the modulation matrix $\mathbf{A}(\cdot)$. The MF filter regarding the transformed matrix can still be applied, leading to

$$\left(\mathbf{U}_N\mathbf{A}^{(\cdot)}\right)^{H}\mathbf{U}_N\mathbf{A}^{(\cdot)} = \left(\mathbf{A}^{(\cdot)}\right)^{H}\mathbf{U}_N^{H}\mathbf{U}_N\mathbf{A}^{(\cdot)}$$
$$= \left(\mathbf{A}^{(\cdot)}\right)^{H}\mathbf{A}^{(\cdot)} \tag{13.54}$$

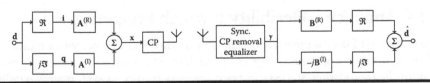

Figure 13.20 **Block diagram of TD-OQAM-GFDM transceiver based on the matrix model.**

Equation 13.54 shows that the real orthogonality is kept when the modulation matrices are unitary transformed. Assume now that the Fourier matrix $\mathbf{F}_N$ is used, leading to the frequency-domain modulation matrices given as

$$\mathcal{A}^{(\cdot)} = \mathbf{F}_N^{\mathrm{H}} \mathbf{A}^{(\cdot)} \qquad (13.55)$$

Equation 13.55 suggests that the data transmitted with $\mathcal{A}^{(\cdot)}$ are defined in the frequency domain, and the inverse Fourier transform conveys the time-domain signal. Thus, the frequency-domain OQAM-GFDM (FD-OQAM-GFDM) can be defined as [13]

$$\mathbf{x}_{\mathrm{OQAM}} = \mathcal{A}^{(\mathrm{R})} \Re\{\mathbf{d}\} + j\mathcal{A}^{(\mathrm{I})} \Im\{\mathbf{d}\} \qquad (13.56)$$

After synchronization, CP removal, and equalization, the received signal $\mathbf{y}_{\mathrm{OQAM}}$ can be used to recover the data symbols as

$$\hat{\mathbf{d}} = \Re\{\mathcal{B}^{(\mathrm{R})} \mathbf{y}_{\mathrm{OQAM}}\} + j\Im\{-j\mathcal{B}^{(\mathrm{I})} \mathbf{y}_{\mathrm{OQAM}}\} \qquad (13.57)$$

where

$$\mathcal{B}^{(\cdot)} = \left(\mathcal{A}^{(\cdot)}\right)^{\mathrm{H}} = \left(\mathbf{A}^{(\cdot)}\right)^{\mathrm{H}} \mathbf{F}_N \qquad (13.58)$$

FD-OQAM-GFDM is dual to TD-OQAM-GFDM, where the roles of subcarriers and subsymbols are exchanged. Moreover, the RRC filter defined in the frequency domain, which spreads over all subcarriers, will result in a well-localized pulse in the time domain, which will interact only with the two surrounding subsymbols. Hence, the half-Nyquist condition will ensure ICI-free communication, while the shift of $M/2$ samples (half a subcarrier) between the subcarriers will guarantee an ISI-free link. Clearly, in this case, the number of subsymbols must be even.

13.4 Enhancing Flexibility through Precoding

Up to now, GFDM has proved to be a flexible modulation scheme, with a multitude of configurations and several degrees of freedom. However, a different interpretation of the modulation process can open the opportunity to even more flexibility through precoding of the data symbols [14]. To explore this property, the modulation process must be analyzed from different points of view.

13.4.1 GFDM Processing per Subcarrier

Assume that

$$d_{r,k}[n] = \sum_{m=0}^{M-1} d_{k,m} \delta[n - mK] \tag{13.59}$$

is the data sequence to be transmitted on the kth subcarrier. Rewriting Equation 13.2 to expose the circular convolution between $d_k[n]$ and the pulse shape for every subcarrier leads to

$$x[n] = \sum_{k=0}^{K-1} d_{r,k}[n] \circledast g[n] \exp\left(j2\pi \frac{k}{K} n \right) \tag{13.60}$$

The circular convolution is implemented as a product in the frequency domain. Let us make

$$D_{r,k}[f] = \mathcal{F}_N\{d_{r,k}[n]\}$$

$$G_k[f] = \mathcal{F}_N\left\{ g[n] \exp\left(j2\pi \frac{kM}{N} n \right) \right\} = G[\langle f - kM \rangle_N] \tag{13.61}$$

where $G[f] = \mathcal{F}_N\{g[n]\}$. Since $D_{r,k}[f]$ is the N-point DFT of the M data symbols upsampled by K, $D_{r,k}[f]$ is equal to the K-fold repetition of $\mathcal{F}_M\{d_{r,k}[m]\}$. Now Equation 13.60 is written as

$$x[n] = \mathcal{F}_N^{-1}\left\{ \sum_{k=0}^{K-1} D_{r,k}[f] G_k[f] \right\} \tag{13.62}$$

Once again, matrix notation can be useful to describe the current approach of the modulation chain. The first step consists of taking the data symbols to be transmitted in the kth subcarrier $\mathbf{d}_{r,k} = [d_{k,0} d_{k,1} \dots d_{k,M-1}]^T$ to the frequency domain:

$$\mathbf{Y}_k^{(M)} = \mathbf{F}_M \mathbf{d}_{r,k} \tag{13.63}$$

Next, the data in the frequency domain is repeated K times to result in the N-point DFT. This can be achieved with the help of a repetition matrix defined as

$$\mathbf{R}^{(K,M)} = \mathbf{1}_{K,1} \otimes \mathbf{I}_M \tag{13.64}$$

where $\mathbf{1}_{i,j}$ is an $(i \times j)$ matrix of ones and $\otimes$ is the Kronecker product. The upsampled version of the data symbols in the frequency domain is

$$\mathbf{Y}_k^{(N)} = \mathbf{R}^{(K,M)} \mathbf{Y}_k^{(M)} \tag{13.65}$$

The next step consists of multiplying the upsampled data symbols in the frequency domain by the prototype filter frequency response, shifted to the center frequency of the subcarrier. Let

$$\mathbf{G} = \mathrm{diag}\left(\mathbf{F}_N \mathbf{g}\right) \tag{13.66}$$

where diag (**u**) returns a diagonal matrix with **u** as the main diagonal when **u** is a column vector or returns the main diagonal if **u** is a square matrix. Thus, **G** is a matrix containing the prototype filter frequency response as the main diagonal. To obtain the filter frequency response for the kth subcarrier, the rows of **G** must be properly shifted, leading to

$$\mathbf{G}_k = \mathbf{\Lambda}_k \mathbf{G} \tag{13.67}$$

where $\mathbf{\Lambda}_k$ is a shifting matrix given by

$$\mathbf{\Lambda}_k = \mathbf{\Psi}\left(\boldsymbol{\lambda}_K^{(k)}\right) \otimes \mathbf{I}_M \tag{13.68}$$

with $\mathbf{\Psi}(\cdot)$ being a function that returns the circulant matrix based on the input column vector and $\boldsymbol{\lambda}_K^{(k)}$ being a $(K \times 1)$ vector with 1 on the kth position and 0 elsewhere.
The therefore, the transmit vector is given by

$$\mathbf{x} = \mathbf{F}_N^{\mathrm{H}} \sum_{k=0}^{K-1} \mathbf{G}_k \mathbf{Y}_k^{(N)} \tag{13.69}$$

The demodulation process of the equalized receive vector $\mathbf{y}_{\mathrm{eq}}$ follows the opposite steps, leading to

$$\widehat{\mathbf{Y}}_k^{(M)} = \frac{1}{M}\left(\mathbf{R}^{(K,M)}\right)^{\mathrm{T}} \mathbf{\Gamma}\left(\mathbf{\Lambda}_k\right)^{\mathrm{T}} \mathbf{F}_N \mathbf{y}_{\mathrm{eq}} \tag{13.70}$$

where:
$\mathbf{\Gamma} \quad = \mathrm{diag}(\mathbf{F}_N \boldsymbol{\gamma})$
$\boldsymbol{\gamma} \quad =$ the receive filter (MF, ZF, or MMSE receiver)

The recovered data symbols are then given by

$$\hat{\mathbf{d}}_{r,k} = \mathbf{F}_M^H \hat{\mathbf{Y}}_k^{(M)} \tag{13.71}$$

The approach presented here can be used to obtain an implementation of the GFDM transceiver with affordable complexity [15,16].

13.4.2 GFDM Processing per Subsymbol

Instead of processing the subcarriers, the same GFDM transmit vector can also be generated by processing the subsymbols individually. This approach leads to a considerable reduction of the implementation complexity, as described in this section. Explicitly writing the circular convolution in Equation 13.60 leads to

$$x[n] = \sum_{m=0}^{M-1} g[\langle n - mK \rangle_N] \underbrace{\sum_{k=0}^{K-1} d_{k,m} \exp\left(j2\pi\frac{k}{K}n\right)}_{M \text{ copies of the IDFT of the data}} \tag{13.72}$$

The second summation in Equation 13.72 is the K-point IDFT of the data symbols transmitted in the mth subsymbol, but repeated M times because of the range of n and multiplied by K. Comparing Equation 13.62 with Equation 13.72 leads to the conclusion that the former is considerably more complex to implement, since it requires performing an M-point DFT K times and one N-point IDFT to generate the transmit sequence in the time domain, while for the latter, only M IDFTs of size K are necessary to obtain the transmit sequence.

Again, Equation 13.72 can be expressed with matrix notation. First, the K-point IDFT of the data symbols transmitted in the mth subsymbol is obtained by

$$\mathfrak{d}_m^{(K)} = \mathbf{F}_K^H \mathbf{d}_{c,m} \tag{13.73}$$

The repetition matrix defined in Equation 13.64 can also be used here to concatenate the IDFT of the data symbol as

$$\mathfrak{d}_m^{(N)} = \mathbf{R}^{(M,K)} \mathfrak{d}_m^{(K)} \tag{13.74}$$

and the GFDM transmit vector can be written as

$$\mathbf{x} = \sum_{m=0}^{M-1} \mathbf{\Lambda}_K^{(m)} \text{diag}(\mathbf{g}) \mathfrak{d}_m^{(N)} \tag{13.75}$$

The equalized received vector can be properly demodulated by

$$\hat{\mathfrak{d}}_m^{(K)} = \left(\mathbf{R}^{(M,K)}\right)^{\mathrm{T}} \operatorname{diag}(\gamma)\left(\mathbf{\Lambda}_K^{(m)}\right)^{\mathrm{T}} \mathbf{y}_{\mathrm{eq}} \qquad (13.76)$$

and the received data symbols are given by

$$\hat{\mathbf{d}}_{c,m} = \mathbf{F}_K \, \hat{\mathfrak{d}}_m^{(K)} \qquad (13.77)$$

Processing the GFDM signal per subsymbol considerably reduces the implementation complexity, and it is an interesting approach mainly for applications in which channel equalization is not necessary.

13.4.3 Precoding for GFDM

Equations 13.63 and 13.73 can be seen as precoding operations of the data symbols prior to transmission, while Equations 13.71 and 13.77 are the corresponding inverse operations on the receiver side. In the presented cases, the Fourier matrix has been used as precoding, but in general, any unitary transformation can be used to achieve specific goals.

To generalize the precoding, let us assume that GFDM is processed per subsymbol. Note, however, that the reasoning can be applied for GFDM processing per subcarrier as well. Consider now a $(j \times j)$ generic unitary transform matrix $\mathbf{\Delta}_j$ that satisfies the following condition:

$$\mathbf{\Delta}_j^{\mathrm{H}} \mathbf{\Delta}_j = \mathbf{I}_j \qquad (13.78)$$

In this case, the precoded samples to be transmitted are given by

$$\mathfrak{d}_m^{(K)} = \mathbf{\Delta}_K^{\mathrm{H}} \mathbf{d}_{c,m} \qquad (13.79)$$

and Equation 13.75 can be used to generate the transmit vector.

On the receiver side, after the estimated precoded samples are recovered using Equation 13.76, the estimated received symbols are obtained by

$$\hat{\mathbf{d}}_{c,m} = \mathbf{\Delta}_K \, \hat{\mathfrak{d}}_m^{(K)} \qquad (13.80)$$

Figure 13.21 shows the block diagram of the GFDM communication chain when precoding is used.

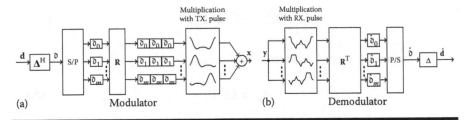

Figure 13.21 Block diagram of precoded GFDM communication chain.

Precoding opens the possibility of enhancing the flexibility and robustness of GFDM by employing or combining different transform matrices. A simple, yet powerful, example of how precoding can be used to benefit GFDM is presented next.

13.4.3.1 Walsh–Hadamard Transform (WHT)-GFDM

One challenge for low-latency scenarios is to achieve reliable communication with single-shot transmission over FSCs. In this case, relatively small packages must be received with a low probability of error, since the low latency requirement does not allow retransmissions of missed packages. Precoding using the WHT can be used efficiently to increase the GFDM robustness over multipath channels [17]. The main idea is to spread the data symbols over all subcarriers, so data symbols can be correctly detected on the receiver side even when a subset of subcarriers suffers from severe attenuation.

To achieve this goal, for each subsymbol, the data symbols $\mathbf{d}_{c,m}$ are linearly combined using the Walsh–Hadamard matrix

$$\boldsymbol{\Omega}_K = \frac{1}{\sqrt{2}}\begin{bmatrix} \boldsymbol{\Omega}_{K/2} & \boldsymbol{\Omega}_{K/2} \\ \boldsymbol{\Omega}_{K/2} & -\boldsymbol{\Omega}_{K/2} \end{bmatrix} \tag{13.81}$$

with $\boldsymbol{\Omega}_1 = 1$. Hence, the data coefficients transmitted in the mth subsymbol are given by

$$\mathbf{c}_m = \boldsymbol{\Omega}_K \mathbf{d}_{c,m} \tag{13.82}$$

On the receiver side, after the GFDM demodulation, the data symbols can be reconstructed as

$$\hat{\mathbf{d}}_{c,m} = \boldsymbol{\Omega}_K^H \hat{\mathbf{c}}_m = \boldsymbol{\Omega}_K \hat{\mathbf{c}}_m \tag{13.83}$$

The Walsh–Hadamard matrix can be combined with the conventional GFDM modulation chain using the precoding definition presented in Equation 13.73. Thus, the precoded samples are defined as

$$\mathfrak{d}_m^{(K)} = \mathbf{F}_K^{\mathrm{H}} \mathbf{\Omega}_K \mathbf{d}_{c,m} \tag{13.84}$$

Notice that the precoding matrix, in this case, is given by

$$\mathbf{\Delta}_K = \mathbf{\Omega}_K^{\mathrm{H}} \mathbf{F}_K \tag{13.85}$$

and that Equation 13.75 can be used to generate the transmit vector, while Equation 13.76 can be used to estimate the precoded samples. Finally, the data reconstruction is obtained by Equation 13.80.

The SER performance of the WHT-GFDM using a ZF demodulator can be estimated by Equation 13.39, but with the average SNR modified to

$$\varrho = \frac{3}{2} \frac{R_T \,|H_e|^2}{\left(\kappa^2 - 1\right)} \frac{E_s}{\xi_l N_0} \tag{13.86}$$

The change on the modulation matrix introduced by the WHT also affects the NEF, which has to be evaluated as

$$\xi_l = \sum_{i=0}^{N-1} \left| \left[\mathbf{\Psi} \mathbf{A}^{-1} \right]_{l,i} \right|^2 \tag{13.87}$$

where

$$\mathbf{\Psi} = \mathbf{I}_M \otimes \mathbf{\Omega}_K \tag{13.88}$$

Finally, the equivalent channel frequency response for every subcarrier is given by

$$H_e = \left(\frac{1}{K} \sum_{k=0}^{K-1} \frac{1}{|H[k]|^2} \right)^{-1/2} \tag{13.89}$$

where $H[k]$ is the channel frequency response.

Figure 13.22 shows the WHT-GFDM SER performance assuming the parameters presented in Table 13.3 and the channel delay profiles shown in Table 13.4. The figure shows that the gain introduced by the WHT highly depends on the channel delay profile. Higher gain is expected when the channel presents narrow and deep notches in the frequency response, while smaller gain is obtained under a channel with mild frequency responses. Therefore, WHT-GFDM is an

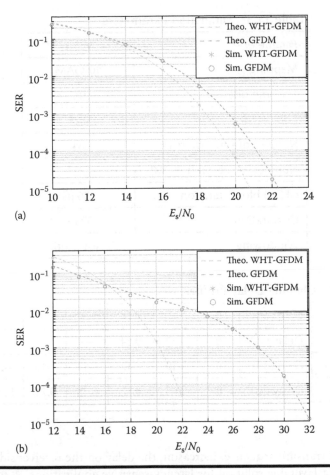

Figure 13.22 SER performance of WHT-GFDM over FSC. (a) SER performance over Channel A. (b) SER performance over Channel B.

interesting scheme to be used in scenarios where low OOB emission and robustness must be combined.

13.5 Transmit Diversity for GFDM

Robustness against the impairments introduced by the wireless channel is fundamental for 5G PHY, and MIMO plays an important role in enhancing the SER performance under time- and frequency-variant channels. The GFDM block structure can be explored to apply time-reverse space-time code (TR-STC) to the time-domain samples of the transmitted signal, leading to the same transmit diversity gain obtained with the traditional Alamouti scheme [18]. The drawback of this approach is that two GFDM frames are necessary to space-time

Table 13.3 Simulation Parameters

Parameter	Value
Mapping	16-QAM
Transmit filter	RC
Roll-off (α)	0.25
Number of subcarriers (K)	64
Number of subsymbols (M)	7
GFDM block duration	256 µs
CP duration	32 µs
Windowing	Not used

Table 13.4 Channel Delay Profiles

Channel A	Gain (dB)	0	−8	−14	—	—	—	—
	Delay (µs)	0	4.57	9.14	—	—	—	—
Channel B	Gain (dB)	0	−10	−12	−13	−16	−20	−22
	Delay (µs)	0	2.85	4.57	6.28	9.71	15.43	20

encode the transmit sequence, increasing the delay on the receiver side. Clearly, this is not a favorable solution for latency-sensitive applications. A solution for this problem consists in space-time encoding the data symbols using two adjacent subsymbols within one GFDM frame. However, the inherent ICI and ISI between subcarriers and subsymbols demand a joint demodulation, combining, and equalization using widely linear processing (WLP) [19] on the receiver side to harvest the diversity. Both approaches are described in Sections 13.5.1 and 13.5.2.

13.5.1 *Time-Reversal STC-GFDM*

TR-STC [20] has been proposed to allow the use of space-time code (STC) within a single-carrier (SC) system under FSC. The GFDM block structure allows TR-STC to be applied directly to the transmit sequence. Consider the two subsequent GFDM frames $\mathbf{x}_i$ and $\mathbf{x}_{i+1}$, where $(\mathbf{x}_i)_n = x_i[n]$ is the GFDM signal transmitted in the ith signaling window. The signals transmitted by both antennas for two subsequent time slots are given by

	Antenna 1	Antenna 2
Block i	$\mathbf{F}_N^H \mathbf{X}_i$	$-\mathbf{F}_N^H \mathbf{X}_{i+1}^*$
Block $i+1$	$\mathbf{F}_N^H \mathbf{X}_{i+1}$	$\mathbf{F}_N^H \mathbf{X}_i^*$

where $\mathbf{X}_i = \mathbf{F}_N \mathbf{x}_i$ and i is an even number. The name *time-reversal space-time coding* comes from the fact that

$$\left(\mathbf{F}_N^H \mathbf{X}_i^* \right)_n = x_i^* \left[\langle -n \rangle_N \right] \tag{13.90}$$

On the receiver side, after the CP is removed, the signals at the lth receiving antenna for the subsequent time windows in the frequency domain are given by

$$\mathbf{Y}_{i,l} = \mathbf{F}_N \mathbf{H}_{1,l} \mathbf{F}_N^H \mathbf{X}_i - \mathbf{F}_N \mathbf{H}_{2,l} \mathbf{F}_N^H \mathbf{X}_{i+1}^* + \mathbf{W}_{1,l}$$

$$\mathbf{Y}_{i+1,l} = \mathbf{F}_N \mathbf{H}_{1,l} \mathbf{F}_N^H \mathbf{X}_{i+1} + \mathbf{F}_N \mathbf{H}_{2,l} \mathbf{F}_N^H \mathbf{X}_i^* + \mathbf{W}_{2,l} \tag{13.91}$$

where $\mathbf{H}_{j,l}$ is the circulant matrix based on $\mathbf{h}_{j,l}$, the channel impulse response between the jth transmit and the lth receive antenna.

Let $\mathcal{H}_{j,l} = \mathbf{F}_N \mathbf{H}_{j,l} \mathbf{F}_N^H$, so Equation 13.91 can be rewritten as

$$\mathbf{Y}_{i,l} = \mathcal{H}_{1,l} \mathbf{X}_i - \mathcal{H}_{2,l} \mathbf{X}_{i+1}^* + \mathbf{W}_{1,l}$$

$$\mathbf{Y}_{i+1,l} = \mathcal{H}_{1,l} \mathbf{X}_{i+1} + \mathcal{H}_{2,l} \mathbf{X}_i^* + \mathbf{W}_{2,l} \tag{13.92}$$

Note that $\mathcal{H}_{j,l}$ is a diagonal matrix.

The signals can be combined as follows to achieve full diversity:

$$\widehat{\mathbf{X}}_i = \mathcal{H}_{eq}^{-1} \sum_{l=1}^{L} \left(\mathcal{H}_{1,l}^* \mathbf{Y}_{i,l} + \mathcal{H}_{2,l} \mathbf{Y}_{i+1,l}^* \right)$$

$$\widehat{\mathbf{X}}_{i+1} = \mathcal{H}_{eq}^{-1} \sum_{l=1}^{L} \left(\mathcal{H}_{1,l}^* \mathbf{Y}_{i+1,l} - \mathcal{H}_{2,l} \mathbf{Y}_{i,l}^* \right) \tag{13.93}$$

where

$$\mathcal{H}_{eq} = \sum_{l=1}^{L} \sum_{j=1}^{2} \mathcal{H}_{j,l} \odot \mathcal{H}_{j,l}^* \tag{13.94}$$

and $\odot$ is the Hadamard product (element-wise multiplication) of matrices.

The estimated data symbols can be obtained by demodulating the vectors from Equation 13.93 in the time domain:

$$\hat{\mathbf{d}}_i = \mathbf{B}\mathbf{F}_N^{\mathrm{H}}\hat{\mathbf{X}}_i \tag{13.95}$$

13.5.1.1 Multiuser Scenario

TR-STC-GFDM is an elegant solution to provide full transmit and receive diversity by exploring the GFDM block structure. This scheme can also be easily adapted to allow multiple access by sharing the subcarriers with multiple users. This solution is named TR-STC generalized frequency-division multiple access (TR-STC-GFDMA). Figure 13.23 shows the block diagram of the communication chain assuming the uplink channel of a mobile communication system.

In this scenario, U users share the K available subcarriers of the TR-STC-GFDM codeword. Hence, each user u employs two time slots to send the TR-STC-GFDM sequences based on the data matrices $\mathbf{D}_i^{(u)}$ that have nonzeroed symbols only in the subcarriers designated to the uth user. Because of the inherent nonorthogonality of GFDM, the users cannot have overlapping subcarriers. This implies that one subcarrier must be used as a guard band between users. Note that users with one or two transmit antennas can share the same TR-STC-GFDM block.

Assuming that every user has the same demand for QoS, two approaches can be used to allocate the subcarriers to the users. In the first approach, the channel state information (CSI) is not available for the transmitters, and each user receives $K_U = K/U - K_g$ adjacent subcarriers, where K_g is the number of guard subcarriers.

For the second approach, it is assumed that a central node has access to the CSI of all users, and the channel distribution is performed based on the channel quality, which is given by [21]

$$G_p^{(u)} = \sum_{k \in K_p} \left(\mathcal{H}_{\mathrm{eq}}^{(u)} \right)_k \tag{13.96}$$

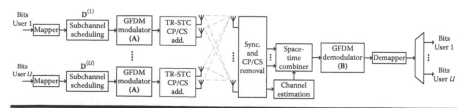

Figure 13.23   Block diagram of TR-STC-GFDMA.

where K_p is the set of subcarrier samples for the pth available subband, $p = 1, 2,..., U$. The users receive the best channel available, following a priority list. Several approaches can be used to define the priority list. In this context, two approaches will be considered: first, a random priority list; second, the user that has the best channel quality chooses first. Random channel sorting provides a fairer distribution of the channel resources among users, but it reduces the overall possible data rate once a subband can be assigned to a user that has a lower channel quality than other users. Channel sorting based on channel quality leads to better overall use of the channel, as the subbands will be occupied by the users that can best explore them. The downside of this approach is that users with a single transmit antenna are less likely to choose the channel first, leading to an average performance loss for these users when compared with the random priority list. On the other hand, the overall performance of the users with two transmit antennas is improved.

13.5.1.2 SER Performance Analysis for TR-STC-GFDMA

TR-STC-GFDMA achieves the same diversity gain as orthogonal STC schemes, but there is a penalty introduced by the noise enhancement when a ZF receiver and nonorthogonal filters are employed.

An approximation for the TR-STC-GFDM SER performance under a frequency-selective fading channel can be derived from the SEP for orthogonal schemes (chapter 13 in [11]) by considering the NEF of GFDM. The resulting approximation is given by

$$p_e \approx 4\zeta \sum_{i=0}^{JL-1} \binom{JL-1+i}{i}\left(\frac{1+\eta}{2}\right)^i \tag{13.97}$$

where J and L are the number of transmit and receive antennas, respectively,

$$\zeta = \left(\frac{\kappa-1}{\kappa}\right)\left(\frac{1-\eta}{2}\right)^{JL} \tag{13.98}$$

$$\eta = \sqrt{\frac{h_e^2 \varrho_r}{2 + h_e^2 \varrho_r}} \tag{13.99}$$

$$h_e^2 = \sum_n E[|h_n|^2] \tag{13.100}$$

The simulation results presented in this subsection are based on the parameters shown in Table 13.5.

Table 13.5 Simulation Parameters

Parameter	Value
Mapping	16-QAM
Transmit filter	RC
Roll-off (α)	0.25
Number of subcarriers (K)	64
Number of subsymbols (M)	9
CP length (N_{CP})	16 samples
CS length (N_{CS})	0 samples
Windowing	Not used
No. of TX. antennas (J)	1 or 2
No. of RX. antennas (L)	1
No. of users (U)	8
No. of guard subcarriers (K_g)	1

Table 13.6 Channel Power Delay Profile for the TR-STC-GFDMA Simulations

Tap (nth sample)	0	1	2	3	4	5	6
Tap gain $h(n)$ (dB)	0	−1	−2	−3	−8	−17.2	−20.8

The reference channel impulse response used in the simulations is given in Table 13.6; however, each tap is multiplied by a complex normal random variable $h_r \in \mathcal{CN}(0, 1)$.

Figure 13.24 shows the TR-STC-GFDMA SER performance for eight users, where three users have a single transmit antenna and five users have two transmit antennas [21]. No CSI is available on the transmit side; hence, a fixed channel assignment is employed. Perfect CSI is available on the receiver side.

Observing Figure 13.24, it is possible to conclude that the approximation presented in Equation 13.97 can be used to estimate the SER performance for TR-STC-GFDMA.

Also, the multiple access scheme allows users with one or two transmit antennas to share the TR-STC-GFDMA resources without introducing further interference. Clearly, the users with one transmit antenna do not benefit from the transmit diversity gain.

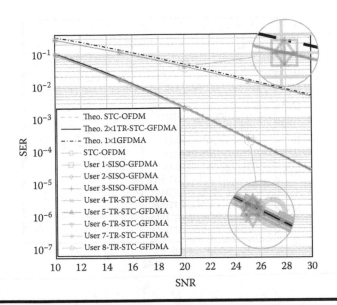

Figure 13.24   TR-STC-GFDMA SER performance with no CSI available on the transmit side. Fixed channel assignment is employed.

When CSI is available on the transmit side, the subbands can be more efficiently assigned to the users. Figure 13.25a shows the TR-STC-GFDMA SER performance when random priority between the users is employed, while Figure 13.25b presents the SER performance using the priority based on the channel quality [21].

From Figure 13.25a, it is possible to observe that the use of the CSI improves the SER performance for all users. Users with one transmit antenna benefit from an 8 dB gain when compared with the static subband distribution, while users with two transmit antennas collect around 4 dB gain. When the priority list is based on the maximum channel quality, the users with two transmit antennas benefit more from the subband distribution than the users with one transmit antenna. Figure 13.25b shows that users with two transmit antennas collect 5 dB gain compared with static subband distribution, while users with just one transmit antenna harvest 4 dB gain. Several other strategies can be used to distribute the subbands among users, for example, prioritizing users with poor channel condition or users with a single transmit antenna. The approaches presented here are examples to highlight the fact that the CSI can be used to improve the system performance when it is considered for the subband distribution among users.

13.5.2 Widely Linear Equalizer (WLE) STC-GFDM

The use of two GFDM blocks to build the STC codeword is not interesting for low-latency applications. In this case, the use of a single GFDM block, in which the

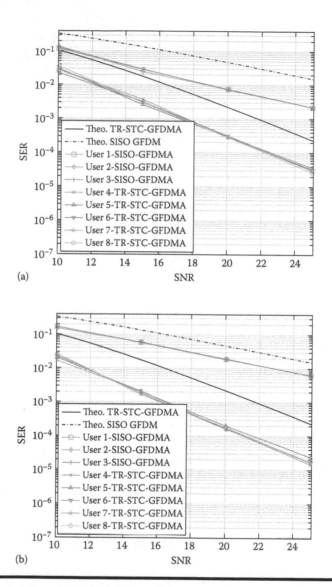

Figure 13.25 **TR-STC-GFDMA SER performance with CSI available on the transmit side. (a) Random priority list. (b) Priority list based on the best channel quality.**

STC codewords are built with data symbols, is more desirable. However, because of ISI and ICI, decoupling the subcarriers before combining will result in residual ISI, severely reducing the overall system performance [22]. The solution for this problem is to jointly demodulate, combine, and equalize the GFDM block using a WLE [23], as described in Section 13.5.2.1.

13.5.2.1 STC for GFDM Block

There are two basic approaches to build STC codewords within one GFDM block: (1) using the adjacent subsymbols from the same subcarrier and (2) using adjacent subcarriers from the same subsymbol. The advantage of the first approach is that the channel frequency response can be different for each subcarrier, which is suitable for low-latency applications. The disadvantage of the first approach is that GFDM needs an odd number of symbols to present a good performance, but the space-time coding demands an even number of subsymbols. Hence, one subsymbol must be left empty, reducing the throughput. However, the empty subsymbol can be used as a GS, leading to GS-GFDM, or to insert the pilots for synchronization and channel estimation. The STC applied to the adjacent subsymbols will be the approach considered here.

Let

$$\mathbf{D}_1^{(s)} = \begin{bmatrix} \mathbf{d}_{c,1} & \mathbf{d}_{c,2} & \cdots & \mathbf{d}_{c,M-2} & \mathbf{d}_{c,M-1} \end{bmatrix} = \mathbf{D}^{(s)} \tag{13.101}$$

be the data matrix to be transmitted by antenna 1. Notice that $\mathbf{D}^{(s)}$ is a $(K \times M - 1)$ matrix derived from $\mathbf{D}$ by removing the first subsymbol. The data matrix transmitted by the second antenna is

$$\mathbf{D}_2^{(s)} = \begin{bmatrix} -\mathbf{d}_{c,2}^* & \mathbf{d}_{c,1}^* & \cdots & -\mathbf{d}_{c,M-1}^* & \mathbf{d}_{c,M-2}^* \end{bmatrix} = \mathbf{D}^{(s)*} \mathbf{P}^{(s)} \tag{13.102}$$

where

$$\mathbf{P}^{(s)} = \mathbf{I}_{(M-1)/2} \otimes \begin{bmatrix} 0 & 1 \\ -1 & 0 \end{bmatrix} \tag{13.103}$$

A shortened version of the modulation matrix, $\mathbf{A}^{(s)}$, in which the first K rows related to the first subsymbol are discarded, is used to modulate the data vectors:

$$\mathbf{x}_j^{(s)} = \mathbf{A}^{(s)} \mathbf{d}_j^{(s)} \tag{13.104}$$

where $\mathbf{d}_j^{(s)} = \text{vec}\left(\mathbf{D}_j^{(s)}\right)$.

The signal at the lth receiving antenna, after removal of the CP, is given by

$$\mathbf{y}_l = \begin{bmatrix} \widehat{\mathbf{H}}_{1,l} & \widehat{\mathbf{H}}_{2,l}\mathbf{P} \end{bmatrix} \begin{bmatrix} \mathbf{d}^{(s)} \\ \mathbf{d}^{(s)*} \end{bmatrix} + \mathbf{w}_l \tag{13.105}$$

where:

$$\widehat{\mathbf{H}}_{j,l} = \mathbf{H}_{j,l}\mathbf{A}_s$$

$\mathbf{H}_{j,l}$ denotes the circulant channel matrix from the jth transmitting to the lth receiving antenna

$$\mathbf{P} = \mathbf{P}^{(s)^{\mathrm{T}}} \otimes \mathbf{I}_K$$

$\mathbf{w}_l$ is the AWGN at the lth receiving antenna

WLP can be used to jointly combine, demodulate, and equalize the received signals when these signals are improper processes. Figure 13.26 shows the block diagram of the WLE-STC-GFDM communication chain.

Assuming that the independent and identically distributed data symbols come from a rotationally invariant constellation with unitary symbol energy, then

$$\mathrm{E}\left[\mathbf{d}^{(s)}\mathbf{d}^{(s)\mathrm{H}}\right] = \mathbf{I}_{K(M-1)} \quad \mathrm{E}\left[\mathbf{d}^{(s)}\mathbf{d}^{(s)\mathrm{T}}\right] = \mathbf{0}_{K(M-1)} \tag{13.106}$$

where $\mathbf{0}_n$ is an $n \times n$ null matrix. The autocorrelation of the received signal is then given by

$$\mathbf{\Gamma}_l = \mathrm{E}\left[\mathbf{y}_l\mathbf{y}_l^{\mathrm{H}}\right] = \begin{bmatrix} \widehat{\mathbf{H}}_{1,l} & \widehat{\mathbf{H}}_{2,l}\mathbf{P} \end{bmatrix} \begin{bmatrix} \widehat{\mathbf{H}}_{1,l}^{\mathrm{H}} \\ \mathbf{P}^{\mathrm{H}}\widehat{\mathbf{H}}_{2,j}^{\mathrm{H}} \end{bmatrix} + \sigma_w^2\mathbf{I}_{MK} \tag{13.107}$$

while its pseudo-autocorrelation is given by

$$\mathbf{C}_l = \mathrm{E}\left[\mathbf{y}_l\mathbf{y}_l^{\mathrm{T}}\right] = \begin{bmatrix} \widehat{\mathbf{H}}_{1,l} & \widehat{\mathbf{H}}_{2,l}\mathbf{P} \end{bmatrix} \begin{bmatrix} \mathbf{P}^{\mathrm{T}}\widehat{\mathbf{H}}_{2,l}^{\mathrm{T}} \\ \widehat{\mathbf{H}}_{1,l}^{\mathrm{T}} \end{bmatrix} \tag{13.108}$$

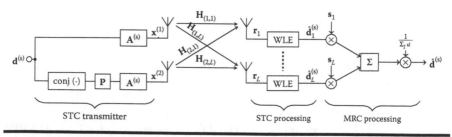

Figure 13.26   Block diagram of proposed WLE-STC-GFDM transceiver.

Since the pseudo-autocorrelation of the received signal is not null, the signal is improper, and WLP can be used to improve the receiver performance [22]. Unlike conventional linear processing, the WLE employs both the received signal and its conjugate to estimate the transmitted data symbols. Two filters are used to combine, demodulate, and equalize the signal received at the *l*th antenna:

$$\hat{\mathbf{d}}_j^{(s)} = \begin{bmatrix} \mathbf{U}_l \\ \mathbf{V}_l \end{bmatrix}^H \begin{bmatrix} \mathbf{y}_l \\ \mathbf{y}_l^* \end{bmatrix} \qquad (13.109)$$

The filters $\mathbf{U}_l$ and $\mathbf{V}_l$ are designed to minimize the mean-squared error (MSE) between $\hat{\mathbf{d}}_j^{(s)}$ and $\mathbf{d}^{(s)}$, and they are obtained by solving the following linear system:

$$\begin{bmatrix} \boldsymbol{\Gamma}_l & \mathbf{C}_l \\ \mathbf{C}_l^* & \boldsymbol{\Gamma}_l^* \end{bmatrix} \begin{bmatrix} \mathbf{U}_l \\ \mathbf{V}_l \end{bmatrix} = \begin{bmatrix} \boldsymbol{\Phi}_l \\ \boldsymbol{\Theta}_l^* \end{bmatrix} \qquad (13.110)$$

where

$$\boldsymbol{\Phi}_l = \mathrm{E}\left[\mathbf{y}_l \mathbf{d}^{(s)H} \right] = \hat{\mathbf{H}}_{1,l} \qquad (13.111)$$

and

$$\boldsymbol{\Theta}_l = \mathrm{E}\left[\mathbf{y}_l \mathbf{d}^{(s)T} \right] = \hat{\mathbf{H}}_{2,l} \mathbf{P} \qquad (13.112)$$

The solution of Equation 13.110 is given by

$$\mathbf{U}_l = \mathbf{S}_l^{-1}\left(\boldsymbol{\Phi}_l - \mathbf{C}_l \left(\boldsymbol{\Gamma}_l^{-1} \right)^* \boldsymbol{\Theta}_l^* \right)$$

$$\mathbf{V}_l = \left(\mathbf{S}_l^{-1} \right)^* \left(\boldsymbol{\Theta}_l^* - \mathbf{C}_l^* \boldsymbol{\Gamma}_l^{-1} \boldsymbol{\Phi}_l \right) \qquad (13.113)$$

where $\mathbf{S}_l = \boldsymbol{\Gamma}_l - \mathbf{C}_l \left(\boldsymbol{\Gamma}_l^{-1} \right)^* \mathbf{C}_l^*$.

Notice that because $\hat{\mathbf{H}}_{i,j}$ is a tall matrix, $\boldsymbol{\Gamma}_l$ is singular when the noise variance is null. Hence, a ZF estimation cannot be directly derived from Equation 13.113. However, the system model in Equation 13.105 can be rewritten to consider the WLP as the double-size linear system

$$
\underbrace{\begin{bmatrix} \mathbf{y}_l \\ \mathbf{y}_l^* \end{bmatrix}}_{\mathbf{y}_l^{(a)}} = \widehat{\mathbf{H}}_l^{(\mathrm{eq})} \underbrace{\begin{bmatrix} \mathbf{d}^{(s)} \\ \mathbf{d}^{(s)*} \end{bmatrix}}_{\mathbf{d}^{(a)}} + \underbrace{\begin{bmatrix} \mathbf{w}_l \\ \mathbf{w}_l^* \end{bmatrix}}_{\mathbf{w}_l^{(a)}}
\tag{13.114}
$$

where

$$
\widehat{\mathbf{H}}_l^{(\mathrm{eq})} = \begin{bmatrix} \widehat{\mathbf{H}}_{1,l} & \widehat{\mathbf{H}}_{2,l}\mathbf{P} \\ \widehat{\mathbf{H}}_{2,j}^*\mathbf{P} & \widehat{\mathbf{H}}_{1,l}^* \end{bmatrix}
\tag{13.115}
$$

The linear minimum mean square error (LMMSE) estimator for $\mathbf{d}^{(a)}$ in Equation 13.114 is given by

$$
\widehat{\mathbf{d}}_l^{(a,\mathrm{MMSE})} = \underbrace{\widehat{\mathbf{H}}_l^{(\mathrm{eq})\mathrm{H}}(\widehat{\mathbf{H}}_l^{(\mathrm{eq})}\widehat{\mathbf{H}}_l^{(\mathrm{eq})\mathrm{H}} + \sigma_w^2 \mathbf{I})^{-1}}_{\mathbf{B}_l^{(\mathrm{MMSE})}} \mathbf{y}_l^{(a)}
\tag{13.116}
$$

A direct calculation shows that

$$
\widehat{\mathbf{H}}_l^{(\mathrm{eq})}\widehat{\mathbf{H}}_l^{(\mathrm{eq})\mathrm{H}} + \sigma_w^2 \mathbf{I} = \begin{bmatrix} \boldsymbol{\Gamma}_l & \mathbf{C}_l \\ \mathbf{C}_l^* & \boldsymbol{\Gamma}_l^* \end{bmatrix}
\tag{13.117}
$$

Hence, Equation 13.116 is equivalent to Equations 13.109 and 13.110. Rewriting Equation 13.116 to its alternative form [24] leads to

$$
\widehat{\mathbf{d}}_l^{(a,\mathrm{MMSE})} = (\widehat{\mathbf{H}}_l^{(\mathrm{eq})\mathrm{H}}\widehat{\mathbf{H}}_l^{(\mathrm{eq})} + \sigma_w^2 \mathbf{I})^{-1}\widehat{\mathbf{H}}_l^{(\mathrm{eq})\mathrm{H}} \mathbf{y}_l^{(a)}
\tag{13.118}
$$

This reformulation allows us to derive the ZF estimator by assuming $\sigma_w = 0$, yielding

$$
\widehat{\mathbf{d}}_l^{(a,\mathrm{ZF})} = (\widehat{\mathbf{H}}_l^{(\mathrm{eq})\mathrm{H}}\widehat{\mathbf{H}}_l^{(\mathrm{eq})})^{-1}\widehat{\mathbf{H}}_l^{(\mathrm{eq})\mathrm{H}} \mathbf{y}_l^a = \mathbf{B}_l^{(\mathrm{ZF})}\mathbf{y}_l^{(a)}
\tag{13.119}
$$

where $\mathbf{B}_l^{(\mathrm{ZF})} = \widehat{\mathbf{H}}_l^{(\mathrm{eq})+}$ is the Moore–Penrose pseudo inverse of $\widehat{\mathbf{H}}_l^{(\mathrm{eq})}$.

Analogously, widely linear MMSE and ZF estimators can be derived when L receive antennas are jointly combined:

$$
\underbrace{\begin{bmatrix} \mathbf{y}_1^{(a)} \\ \mathbf{y}_2^{(a)} \\ \vdots \\ \mathbf{y}_L^{(a)} \end{bmatrix}}_{\mathbf{y}^{(a)}} = \underbrace{\begin{bmatrix} \widehat{\mathbf{H}}_1^{(eq)} \\ \widehat{\mathbf{H}}_2^{(eq)} \\ \vdots \\ \widehat{\mathbf{H}}_L^{(eq)} \end{bmatrix}}_{\widehat{\mathbf{H}}^{(eq)}} \mathbf{d}^{(a)} + \underbrace{\begin{bmatrix} \mathbf{w}_1^{(a)} \\ \mathbf{w}_2^{(a)} \\ \vdots \\ \mathbf{w}_L^{(a)} \end{bmatrix}}_{\mathbf{w}^{(a)}}
\tag{13.120}
$$

and widely linear MMSE and ZF estimators are given by

$$
\widehat{\mathbf{d}}^{(a,\text{MMSE})} = \widehat{\mathbf{H}}^{(eq)^{H}} \left(\widehat{\mathbf{H}}^{(eq)} \widehat{\mathbf{H}}^{(eq)^{H}} + \sigma_w^2 \mathbf{I} \right)^{-1} \mathbf{y}^{(a)}
$$

$$
\widehat{\mathbf{d}}^{(a,\text{ZF})} = \widehat{\mathbf{H}}^{(eq)^{+}} \mathbf{y}^{(a)}
\tag{13.121}
$$

This approach requires the inversion of a $2LMK \times 2LMK$ matrix, which becomes computationally expensive with increasing L. An alternative approach is to separately estimate $\widehat{\mathbf{d}}_l^{(s)}$ at every receiving antenna and then combine the $\widehat{\mathbf{d}}_l^{(s)}$ weighted by the quality of the channels.

Therefore, the ZF maximum ratio combiner (MRC) receiver can be easily derived as follows. Assuming the ZF receiver, we have

$$
\widehat{\mathbf{d}}_l^{(s,\text{ZF})} = \mathbf{d}^{(s)} + \mathbf{B}_l^{(s,\text{ZF})} \mathbf{w}_j^{(a)}
\tag{13.122}
$$

for every receive antenna, and thus the MSE of the estimated data equals

$$
\mathbf{e}_l = \text{diag}\left(\mathrm{E}\left[\left(\mathbf{d}_l^{(s,\text{ZF})} - \mathbf{d}^{(s)} \right)\left(\mathbf{d}_l^{(s,\text{ZF})} - \mathbf{d}^{(s)} \right)^{H} \right] \right)
\tag{13.123}
$$

$$
= \sigma_w^2 \text{diag}\left(\mathbf{B}_l^{(s,\text{ZF})} \mathbf{B}_l^{(s,\text{ZF})^{H}} \right)
\tag{13.124}
$$

The operator diag $(\cdot)$ returns the diagonal of a matrix argument and a diagonal matrix for a vector argument. The estimated data symbols from the J receiving antennas are now linearly combined, weighted by their inverse MSE:

$$
\mathbf{s}_l = \left[\text{diag}(\mathbf{e}_l) \right]^{-1}
\tag{13.125}
$$

according to

$$\hat{\mathbf{d}}^{(s)} = \left[\sum_{l=1}^{L} \mathbf{s}_l\right]^{-1} \left(\sum_{l=1}^{L} \mathbf{s}_l \hat{\mathbf{d}}_l^{(s)}\right) \tag{13.126}$$

Note that $\hat{\mathbf{d}}^{(s)} = \hat{\mathbf{d}}_1^{(s)}$ for $L=1$.

13.5.2.2 Performance Analysis of WLE-STC-GFDM

The SER performance for WLE-STC-GFDM can also be estimated by Equation 13.97. However, because only $M-1$ subsymbols are active, the throughput reduction factor given in Equation 13.43 is adjusted to

$$R_T = \frac{K}{K + N_{CP} + N_{CS}} \times \frac{M-1}{M} \tag{13.127}$$

Figure 13.27 depicts the WLE-STC-GFDM SER performance assuming the channel presented in Table 13.6 and the parameters shown in Table 13.7.

Figure 13.27 shows that the ISI is completely removed from the detected symbol by the WLP, and the WLE-STC-GFDM can achieve the same diversity gain obtained with TR-STC-GFDM under frequency-selective TVCs. For severe channel delay profile, the MSE per subcarrier presented by WLE-STC-GFDM can be uneven and may lead to a performance loss at high SNR [22]. Figure 13.27 also

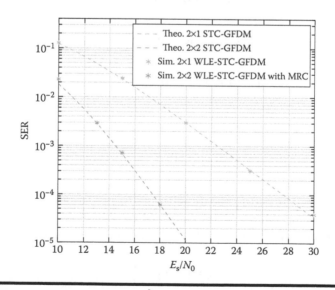

Figure 13.27 WLE-STC-GFDM SER performance over frequency-selective time-variant channels.

Table 13.7   Simulation Parameters

Parameter	Value
Mapping	16-QAM
Transmit filter	RC
Roll-off (α)	0.25
Number of subcarriers (K)	64
Number of subsymbols (M)	9
Number of active subsymbols	8
CP length (N_{CP})	16 samples
CS length (N_{CS})	0 samples
Windowing	Not used
# TX. antennas (J)	2
# RX. antennas (L)	1 or 2

shows the performance of the WLE-STC-GFDM combined with the MRC with two received antennas. Clearly, the proposal presented in Equation 13.126 achieves full diversity gain, and the simulation results follow the theoretical approximation presented in Equation 13.97. Hence, the proposed structure can reduce the complexity when multiple receive antennas are employed without reducing the overall SER performance of the system.

13.6 GFDM Parametrization for the LTE Resource Grid

The development of a new standard usually introduces disruptive and innovative technologies and features new services. However, it is also important to provide a certain level of compatibility with previous standards to allow a soft transition between subsequent generations. In mobile communication, the possibility of reusing the main reference clock is very important, because it simplifies the design of multistandard mobile units. For instance, the LTE master clock frequency is eight times higher than the one used in 3G networks. It will be advantageous for manufacturers and operators if 5G is based on a PHY that is able to reuse the LTE master clock. The next subsections show that GFDM can be parametrized to fit the LTE time–frequency grid when using the 30.72 MHz master clock. Here, two situations will be considered. First, the time–frequency resource grid based on the LTE clock will be used only by GFDM signals, and second, OFDM and GFDM signals will coexist in the same time–frequency grid [25].

Table 13.8   LTE Parameters for the FDD Mode

Parameter	Normal Mode	Extended Mode
Frame duration	10 ms or 307.200 samples	
Subframe duration	1 ms or 30.720 samples	
Slot duration	0.5 ms or 15.360 samples	
Subcarrier spacing	15 kHz	
Subcarrier bandwidth	15 kHz	
Sampling frequency (clock)	30.72 MHz	
No. of subcarriers	2048	
No. of active subcarriers	1200	
Resource block	12 subcarriers of one slot	
Number of OFDM per slot	7	6
CP length (samples)	First symbol: 160 Other symbols: 144	512

13.6.1 LTE Time–Frequency Resource Grid

Here, a 20 MHz frequency-division duplex (FDD) LTE system, with the main parameters presented in Table 13.8, is considered as a reference.

The LTE time–frequency grid is organized in resource blocks (RBs) with 12 subcarriers, leading to a total bandwidth of 180 kHz. The time duration of the RB is 0.5 ms, which consists of seven and six OFDM symbols for the normal and extended modes, respectively. A subframe composed by two RBs is the minimal resource allocation for one given user. Figure 13.28 depicts the LTE RB structure assuming normal operation mode.

13.6.2 GFDM Parametrization for the LTE Time–Frequency Grid

The main goal here is to configure GFDM to use the LTE time–frequency grid. This means that the GFDM block time must be 1 ms, and a set of GFDM subcarriers must fit in an integer multiple of 180 kHz. Table 13.9 presents one possible set of GFDM parameters for this scenario.

From Tables 13.8 and 13.9, it can be seen that the proposed GFDM approach has the same subframe duration as the LTE grid. Notice that three GFDM subcarriers occupy the bandwidth of four LTE RBs. This means that each GFDM subcarrier is 16 times wider than the LTE subcarriers. Because each GFDM subcarrier has $M = 15$ times more samples than an LTE subcarrier, the spectrum resolution

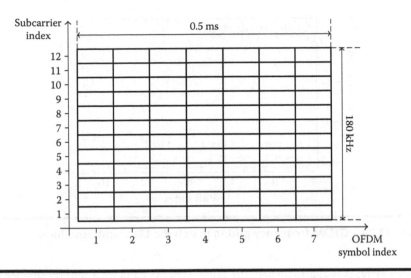

Figure 13.28 **Time–frequency structure of LTE resource block in normal operation mode.**

Table 13.9 GFDM Configuration Aligned with the LTE Grid

Parameter	Normal Mode
Subframe duration	1 ms or 30.720 samples
GFDM symbol duration	66.67 μs or 2048 samples
Subsymbol duration	4.17 μs or 128 samples
Subcarrier spacing	240 kHz
Subcarrier bandwidth	240 kHz
Sampling freq. (clock)	30.72 MHz
Subcarrier spacing factor N	128
Subsymbol spacing K	128
No. active subcarriers N_{on}	75
No. subsymbols per GFDM symbol M	15
No. of GFDM symbols per subframe	15
CP length	4.17 μs or 128 samples
Prototype filter	Dirichlet

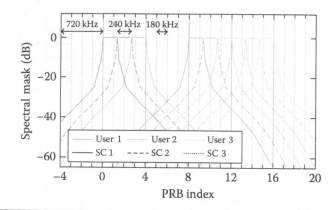

Figure 13.29   GFDM frequency grid to match the LTE frame structure.

of both systems is approximately the same. GFDM employs a slightly smaller CP length than LTE, and it does not require a larger CP for the first GFDM symbol. The Dirichlet pulse makes the system orthogonal, and because the roll-off factor in the frequency domain is zero, the GFDM subcarriers do not overlap with the surrounding subcarriers outside the used RBs. Therefore, the LTE time–frequency grid can be used to accommodate GFDM signaling from multiple synchronized users, as shown in Figure 13.29.

13.6.3 Coexistence of GFDM and LTE Signals

GFDM can also be configured to use two empty RBs, leaving a guard band to avoid interference in the surrounding RBs that are used to transmit the conventional LTE signal. In this case, the GFDM signal can be seen as a secondary signal that is used to explore vacant RBs for low-latency applications. Table 13.10 shows the GFDM parameters for this approach.

A new approach to generate the GFDM signal must be introduced here to keep the subcarrier spacing compatible with the LTE time–frequency grid. The subcarrier bandwidth is 320 kHz, while the subcarrier spacing must be a multiple of 180 kHz (the bandwidth of one RB). To achieve this frequency spacing, N must assume a value that differs from the subsymbol spacing K. Nevertheless, N must be carefully chosen to guarantee that an integer number of subcarrier periods is present within the duration of one GFDM frame; otherwise, there will be phase jumps between the CP and the GFDM signal, leading to a strong OOB emission. The parametrization presented in Table 13.10 achieves this goal.

From Tables 13.8 and 13.10, it can be seen that the GFDM subcarriers are 21.33 times larger than the LTE subcarriers in terms of bandwidth, while the GFDM symbol duration is 10 times smaller than the corresponding slot duration

Table 13.10 Parameters for Asynchronous GFDM Signaling

Parameter	Normal Mode
Subframe duration	1 ms or 30.720 samples
GFDM symbol duration	50 µs or 1536 samples
Subsymbol duration	3.125 µs or 96 samples
Subcarrier spacing	360 kHz
Subcarrier bandwidth	320 kHz
Sampling freq. (clock)	30.72 MHz
Subcarrier spacing factor	256/3
Subsymbol spacing	96 samples
No. of active subcarriers K_{on}	Half of available RBs
No. of subsymbols per GFDM symbol M	15
No. of GFDM symbols per subframe	20
CP length	3.125 µs or 96 samples
Prototype filter	Dirichlet

of the LTE system. Moreover, GFDM subcarrier resolution in the frequency domain is $M = 15$ times the resolution of the LTE subcarriers. The CP length has been shortened to two-thirds of the LTE CP length, which means that this approach is appropriate for a small cell size (typically for a diameter smaller than 4 km).

LTE equipment transmits system information periodically, even on empty RBs. Therefore, it is difficult to employ the configuration proposed in Table 13.10. One solution is to consider an LTE system operating in a spectrum hole that is larger than the system bandwidth: for example, a 5 MHz signal being transmitted in the center of a 10 MHz band. Since the LTE grid is the same for any bandwidth configuration, with the only difference being the number of active subcarriers, the GFDM signal presented in Table 13.10 can be appended on the edges of the LTE signal, as depicted in Figure 13.30.

Note that the low OOB emission of GFDM causes little interference in the LTE signal. However, the high OOB emissions of the LTE OFDM signal might be harmful for the GFDM signal [26]. The interaction of the LTE emissions with GFDM signals must be considered to specify the forward error control codes and other protective measures for the GFDM PHY layer.

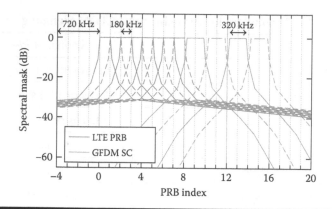

Figure 13.30 GFDM as a secondary signal in the LTE time–frequency grid.

13.7 GFDM as a Framework for Various Waveforms

At first glance, the requirements of each 5G scenario could not be addressed by a specific PHY. In fact, several waveforms are being proposed to address one specific scenario, but with disadvantages for other applications. For instance, given its low OOB emissions, FBMC [27] is being rediscovered for CR and dynamic spectrum allocation. On the other hand, the long impulse response of the filters, typically leading to the overlapping of at least four data symbols, prohibits its use for applications with sporadic traffic and tight latency constraints. FTN signaling [27] is another example. Taking advantage of the Mazo limit, FTN is a promising solution for high data rate scenarios. But the large complexity of the receiver makes it unsuitable for IoT. Also, OFDM and SC-FDE can still be explored in 5G networks, for example, to keep compatibility with legacy technology.

However, rather than a multitude of scenario-specific PHY, a better approach is to have a single PHY that can be used to cover all major waveforms proposed for 5G as corner cases. GFDM can achieve this goal if it is modified to have a subsymbol and subcarrier spacing different from the number of subcarriers and subsymbols. The classical definition of the GFDM transmit sequence from Equation 13.2 can be rewritten as

$$x[n] = \sum_{k=0}^{K-1} \sum_{m=0}^{M-1} d_{k,m} g[\langle n - mK \rangle_N] \exp\left(j2\pi \frac{kM}{N} n \right), \quad n = 0, 1, \dots, N-1 \quad (13.128)$$

Now consider that the prototype filter has N samples divided in P periods with S samples each, leading to $N = PS$. The subsymbols are K samples apart, and the space between subcarriers is M samples. In this case, Equation 13.128 can be expanded to

$$x[n] = \sum_{k=0}^{K-1} \sum_{m=0}^{M-1} d_{k,m} g\left[\langle n - m\mathcal{K} \rangle_N\right] \exp\left(j2\pi \frac{k\mathcal{M}}{\mathcal{N}} n \right), \quad n = 0,1,\ldots,\mathcal{N}-1 \tag{13.129}$$

At this point, it is useful to define the subsymbol and the subcarrier distance factors as

$$v_t = \frac{\mathcal{K}}{\mathcal{S}}$$

$$v_f = \frac{\mathcal{M}}{\mathcal{P}} \tag{13.130}$$

respectively. Using Equation 13.130 in Equation 13.129 leads to

$$x[n] = \sum_{k=0}^{K-1} \sum_{m=0}^{M-1} d_{k,m} g\left[\langle n - m v_t \mathcal{S} \rangle_N\right] \exp\left(j2\pi \frac{k v_f}{\mathcal{S}} n \right), \quad n = 0,1,\ldots,\mathcal{N}-1 \tag{13.131}$$

Table 13.11 summarizes the relationship among the GFDM parameters.

The density of data symbols per GFDM block samples, that is, N/$\mathcal{N}$, becomes larger than one when $v_t > 1$ or $v_f > 1$, which means that GFDM can now cover the

Table 13.11 Relationship among GFDM Parameters

Variable	Meaning
$\mathcal{S}$	Samples per period in the filter
$\mathcal{P}$	Periods in the filter
$\mathcal{N} = \mathcal{PS}$	Total number of samples in the signal
$\mathcal{K}$	Subsymbol spacing in time domain
$\mathcal{M}$	Subcarrier spacing in frequency domain
$v_t = \mathcal{K}/\mathcal{S}$	Subsymbols distance factor
$v_f = \mathcal{M}/\mathcal{P}$	Subcarriers distance factor
$K = \mathcal{PS}/\mathcal{M} = \mathcal{S}/v_f = \lfloor \mathcal{N}/\mathcal{M} \rfloor$	Subcarriers per block
$M = \mathcal{PS}/\mathcal{K} = \mathcal{P}/v_t = \lfloor \mathcal{N}/\mathcal{K} \rfloor$	Subsymbols per block
$N = KM$	No. of data symbols per block

FTN waveforms as well. Table 13.12 shows how GFDM must be parametrized to achieve the major 5G waveforms.

The different waveforms are characterized by two aspects. First, parameters related to the dimensions of the underlying resource grid are explored. These include the number of subcarriers K and subsymbols M in the system. The scaling factor in time v_t and frequency v_f can theoretically take values of any rational number larger than 0, while numbers close to 1 are meaningful because they relate to critically sampled Gabor frames. Additionally, the option to force specific data symbols in a block to carry the value 0, that is, so-called guard subsymbols [3], with M_s being a number between 0 and $M-2$, is relevant for some candidates. The second set of features is related to the properties of the signal. Here, the choice of the pulse-shaping filter is a significant attribute, and the presence or absence of circularity constitutes a characteristic feature. Moreover, the use of OQAM is needed for some waveforms, aiming to achieve higher flexibility. Further, some waveforms rely on a CP to allow transmission of a block-based frame structure in a time-dispersive channel, while others do not use CP to achieve higher spectrum efficiency.

The family of *classical waveforms* includes OFDM, block OFDM, SC-FDE, and single-carrier frequency-domain multiple access (SC-FDMA). Particularly OFDM and SC-FDMA have been relevant for the development of the 4G cellular standard LTE. All four waveforms in this category have in common that $v_f = 1$ and $v_t = 1$, which enables meeting the Nyquist criterion. Silent subsymbols are not employed, and the CP and regular QAM are used in the default configuration. OFDM and block OFDM are corner cases of GFDM, where a rectangular pulse is used. Additionally, OFDM is restricted to one subsymbol, while block OFDM constitutes the concatenation of multiple OFDM symbols in time to create a block with a single common CP. Similarly, SC-FDE and SC-FDMA can also be considered as corner cases of GFDM. However, here a Dirichlet pulse is used, and analogously, the number of subcarriers in SC-FDE is $K = 1$, while SC-FDMA is a concatenation in frequency of multiple SC-FDE signals. All waveforms in this category share the property of orthogonality.

The family of *filter bank waveforms* revolves around filtering the subcarriers in the system and still retaining orthogonality. As the names suggest, FBMC-OQAM [27] and its cyclic extension FBMC-COQAM [12] rely on offset modulation, while in FBMC-filtered multitone (FMT) and cyclic block-FMT (CB-FMT) [28], the spacing between the subcarriers is increased, such that they do not overlap, that is, $v_f > 1$. Also, a separation between cyclic and noncyclic prototype filters can be made. In this context, silent subsymbols become relevant. The best spectral efficiency is achieved with $M_s = 0$, while $M_s > 0$ helps to improve the spectral properties of the signal. Using a sufficiently large number of silent subsymbols at the beginning and the end of a block enables emulating noncyclic filters from a cyclic prototype filter response, to generate FBMC-OQAM and FBMC-FMT bursts. More precisely, M_p is the length of the prototype filter, and $M_s = M_p$. Lastly, the CP is only compatible with cyclic filters.

Table 13.12   GFDM Parameters for Virtualization of 5G Waveform Candidates

Design space	GFDM	OFDM	Block OFDM	SC-FDE	SC-FDMA	FBMC OQAM	FBMC FMT	FBMC COQAM	CB-FMT	FTN	SEFDM
Number of subcarriers	K	K	K	1	K	K	K	K	K	K	K
Number of subsymbols	M	1	M	M	M	M	M	M	M	M	1
Scaling frequency	v_f	1	1	1	1	1	>1	1	>1	1	<1
Scaling time	v_t	1	1	1	1	1	1	1	1	<1	1
Silent subsymbols	M_s	–	–	–	–	M_p	M_p	–	–	M_p	–
Filter imp. resp.	Cyclic	Rect	Rect	Dirichlet	Dirichlet	$\sqrt{}$Nyquist	$\sqrt{}$Nyquist	Cyclic	Cyclic	IOTA	Rect
Offset mod.	(yes)	(yes)	No	No	No	Yes	No	Yes	No	Yes	No
Cyclic prefix	Yes	Yes	Yes	Yes	Yes	No	No	Yes	Yes	No	Yes
Orthogonal	(yes)	Yes	Yes	Yes	Yes	Yes	Yes	Yes	Yes	No	No
Application scenarios	All	Legacy systems	Bitpipe	IoT/MTC	IoT/MTC	WRAN, bitpipe	WRAN	Tactile Internet	Tactile Internet	Bitpipe	Bitpipe
Beneficial features	Flex.	Orth.	Small CP overhead	Low PAPR	Low PAPR	Low OOB	Low OOB	No filter tail	No filter tail	Spectral eff.	Spectral eff.

Generally, the waveform can become nonorthogonal depending on the use of specific filters and for a given value of v_f and v_t. This is addressed in the final category, which consists of the *nonorthogonal multicarrier techniques* FTN [27] and spectrally efficient frequency-division multiplexing (SEFDM) [29]. The key property of FTN is $v_t < 1$, meaning that this waveform achieves higher spectrum efficiency. The isotropic orthogonal transform algorithm (IOTA) pulse, in combination with OQAM, has been proposed to avoid the need for a CP. Since the impulse response of the filter is not cyclic, M_p subsymbols are silent. Analogously, the idea of SEFDM is to increase the density of subcarriers in the available bandwidth, that is, $v_f < 1$. Here, $M = 1$, because each block consists of a single subsymbol that is filtered with a rectangular pulse, and a CP is prepended to combat multipath propagation. In this case, regular QAM is employed. The amount of squeezing that can be employed without severely impacting the error rate performance is limited. The Mazo limit states that this threshold is around 25% for both schemes.

13.8 Conclusions

The various scenarios and applications that must be covered by 5G networks will demand an unprecedented flexibility of the PHY layer. Although using a specific waveform to address the requirements of each scenario is one option, having a single waveform that can be shaped to deal with the challenges imposed by the different applications is more desirable. GFDM has proved to be a flexible multicarrier modulation that can be tuned to cover the major 4G and 5G waveforms. It can also reuse the LTE time–frequency grid and master clock, which means that compatibility with previous generations can be seamlessly achieved. The overall GFDM performance can be enhanced by transmit diversity, while OQAM can be applied in both time and frequency domains to obtain an orthogonal system. Precoding can be used to broaden the flexibility even more: for instance, precoding matrices can be used to achieve a higher performance over FSCs or reduce OOB emissions. Roughly synchronized devices can share the time–frequency grid by using a single subcarrier as the guard band, while an efficient use of the CP can deal with long channel delay profiles and time misalignments among users. All these features make GFDM a strong candidate for the 5G network PHY layer.

References

1. G. P. Fettweis, The Tactile Internet: Applications and challenges, *IEEE Vehicular Technology Magazine* 9(1): 64–70, 2014.
2. Next Generation Mobile Network 5G Initiative Team, NGMN 5G White Paper. In Final Report NGMN Project, 2015.

3. N. Michailow, M. Matthe, I. S. Gaspar, A. N. Caldevilla, L. L. Mendes, A. Festag, and G. Fettweis, Generalized frequency division multiplexing for 5th generation cellular networks, *IEEE Transactions on Communications* 62(9): 3045–3061, 2014.

4. G. Wunder, P. Jung, M. Kasparick, T. Wild, F. Schaich, Y. Chen, S. Brink, et al., 5GNOW: Non-orthogonal, asynchronous waveforms for future mobile applications, *IEEE Communications Magazine* 52(2): 97–105, 2014.

5. C. Yang, S. Han, X. Hou, and A. F. Molisch, How do we design CoMP to achieve its promised potential? *Wireless Communications, IEEE* 20(1): 67–74, 2013.

6. D. Evans, The Internet of Things: How the next evolution of the Internet is changing everything. In Cisco Internet Business Solutions Group (IBSG) White Paper, 2011.

7. M. Matthe, L. L. Mendes, and G. Fettweis, Generalized frequency division multiplexing in a Gabor transform setting, *IEEE Communications Letters* 18(8): 1379–1382, 2014.

8. R. Datta, N. Michailow, M. Lentmaier, and G. Fettweis, GFDM interference cancellation for flexible cognitive radio PHY design. In *IEEE Vehicular Technology Conference (VTC Fall)*, Quebec City, QC, IEEE, pp. 1–5, 2012.

9. N. Michailow, S. Krone, M. Lentmaier, and G. Fettweis, Bit error rate performance of generalized frequency division multiplexing. In *Proceedings 76th IEEE Vehicular Technology Conference (VTC Fall '12)*, Quebec City, QC, IEEE, pp. 1–5, 2012.

10. N. Michailow, Generalized frequency division multiplexing transceiver principles, PhD thesis, Technische Universität Dresden.

11. S. Benedetto and E. Biglieri, *Principles of Digital Transmission with Wireless Applications*. New York: Kluwer Academic/Plenum, 1999.

12. H. Lin and P. Siohan, Multi-carrier modulation analysis and WCP-COQAM proposal, *EURASIP Journal on Advances in Signal Processing*, 2014(79): 1–19, 2014.

13. I. Gaspar, L. Mendes, M. Matthé, N. Michailow, D. Zhang, A. Albertiy, and G. Fettweis, Frequency-domain offset-QAM for GFDM, *IEEE Communications Letters* 62(9): 3045–3061, 2015.

14. I. Gaspar, M. Matthé, N. Michailow, L. Mendes, D. Zhang, and G. Fettweis, GFDM transceiver using precoded data and low-complexity multiplication in time domain, *IEEE Communications Letters* 19(1): 106–109, 2015.

15. I. Gaspar, N. Michailow, A. Navarro, E. Ohlmer, S. Krone, and G. Fettweis, Low complexity GFDM receiver based on sparse frequency domain processing. In *IEEE 77th Vehicular Technology Conference (VTC Spring)*, Dresden, IEEE, pp. 1–6, 2013.

16. N. Michailow, I. Gaspar, S. Krone, M. Lentmaier, and G. Fettweis, Generalized frequency division multiplexing: Analysis of an alternative multi-carrier technique for next generation cellular systems. In *Wireless Communication Systems (ISWCS), 2012 International Symposium on*, Paris, IEEE, pp. 171–175, 2012.

17. N. Michailow, L. Mendes, M. Matthe, I. Gaspar, A. Festag, and G. Fettweis, Robust WHT-GFDM for the next generation of wireless networks, *IEEE Communications Letters* 19(1): 106–109, 2015.

18. S. M. Alamouti, A simple transmit diversity technique for wireless communications, *IEEE Journal on Selected Areas in Communications* 16(8): 1451–1458, 1998.

19. B. Picinbono and P. Chevalier, Widely linear estimation with complex data, *Signal Processing, IEEE Transactions on*, 43(8): 2030–2033, 1995.

20. N. Al-Dhahir, Single-carrier frequency-domain equalization for space-time block-coded transmissions over frequency-selective fading channels, *IEEE Communications Letters* 5(7): 304–306, 2001.

21. M. Matthe, L. L. Mendes, I. Gaspar, N. Michailow, D. Zhang, and G. Fettweis, Multi-user time-reversal STC-GFDMA for future wireless networks, *EURASIP Journal on Wireless Communications and Networking* 2015(1): 1–8, 2015.

22. M. Matthe, L. Mendes, N. Michailow, and G. Fettweis, Widely linear estimation for space-time-coded GFDM in low-latency applications, *IEEE Transactions on Communications* 62(9): 3045–3061, 2015.

23. B. Picinbono and P. Chevalier, Widely linear estimation with complex data, *Signal Processing, IEEE Transactions on* 43(8): 2030–2033, 1995.

24. S. M. Kay, *Fundamentals of Statistical Signal Processing: Estimation Theory*. Prentice Hall Signal Processing Series. Upper Saddle River, NJ: Prentice Hall, 1993.

25. I. Gaspar, L. Mendes, M. Matthe, N. Michailow, A. Festag, and G. Fettweis, LTE-compatible 5G PHY based on generalized frequency division multiplexing. In *11th International Symposium on Wireless Communications Systems (ISWCS)*, Barcelona, IEEE, pp. 209–213, 2014.

26. N. Michailow, M. Lentmaier, P. Rost, and G. Fettweis, Integration of a GFDM secondary system in an OFDM primary system. In *Proceedings of the Future Network & Mobile Summit*, Warsaw, IEEE, pp. 1–8, 2011.

27. P. Banelli, S. Buzzi, G. Colavolpe, A. Modenini, F. Rusek, and A. Ugolini, Modulation formats and waveforms for 5G networks: Who will be the heir of OFDM: An overview of alternative modulation schemes for improved spectral efficiency, *Signal Processing Magazine, IEEE* 31(6): 80–93, 2014.

28. A. M. Tonello, A novel multi-carrier scheme: Cyclic block filtered multitone modulation. In *Communications (ICC), 2013 IEEE International Conference on*, Budapest, IEEE, pp. 5263–5267, 2013.

29. I. Kanaras, A. Chorti, M. R. D. Rodrigues, and I. Darwazeh, Spectrally efficient FDM signals: Bandwidth gain at the expense of receiver complexity. In *IEEE International Conference on Communications*, Dresden, IEEE, pp. 1–6, 2009.

Chapter 14

A Novel Centimeter-Wave Concept for 5G Small Cells

Nurul H. Mahmood, Davide Catania, Mads Lauridsen, Gilberto Berardinelli, Preben Mogensen, Fernando M. L. Tavares, and Kari Pajukoski

Contents

The fifth generation (5G) of mobile radio access technologies is expected to be operating over millimeter-wave (mmWave) bands as well as centimeter-wave (cmWave) bands (below 30 GHz). A number of challenging design requirements have to be addressed to meet the demanding targets of the 5G radio access technology in terms of connectivity, latency, data rate, and energy efficiency. This chapter presents our vision for the physical layer and radio resource management layer aspects of a 5G cmWave concept for small cells. Fundamental technology components, such as optimized frame structure, dynamic scheduling of uplink/downlink transmission, interference suppression receivers, and rank adaptation, are discussed along with the design of a novel energy-saving enabler.

14.1 Introduction

The demand for mobile broadband services continues to increase at a staggering rate and is expected to increase 1000-fold over the next decade [1]. Existing radio technologies such as Long-Term Evolution–Advanced (LTE-A) and Wi-Fi have inherent design limitations that make their potential enhancements unable to cope with such a huge traffic demand. This suggests the need to design a completely novel fifth-generation (5G) radio access technology (RAT), targeting peak data rates in the order of 10 Gbps, minimum guaranteed rates of 100 Mbps, and a sub-millisecond over-the-air latency [2,3].

Increased traffic growth and heterogeneity, along with higher user expectations for faster services, are the main drivers of a 5G system. High-level requirements such as improved area spectral efficiency, better guaranteed cell-edge rates, higher peak data rates, lower latencies, faster setup times, lower energy consumption, and reduced device and service costs are some of the requirements that 5G is expected to satisfy [1,2,4]. Further key requirements include reducing the latency to provide better reliability for mission-critical applications and improving energy consumption and battery life. Due to the demand for satisfying such diverse and potentially conflicting requirements, a 5G system is envisioned to be a multitier architecture consisting of complementing access technologies handling different use cases [5,6].

Possible strategies for meeting the huge traffic demand of the future targeted by 5G include:

- Usage of a larger frequency spectrum
- Spectral efficiency enhancement
- Increase in the number of cells per area

Among these strategies, increasing the number of cells per area (by a factor of 50–1000 over the current density) is considered to be the most promising approach for capacity expansion, leading to a predicted massive deployment of small cells in areas with high traffic demand. Smaller cells enable reuse of the scarce spectrum and also entail a lower number of serviceable nodes per base station. Moreover, studies carried out in [7] show that most of the traffic is generated indoors, reinforcing the suitability of indoor small cells.

Introducing such a massive number of small cells involves a number of challenges that need to be addressed. The diversity of devices and traffic requirements implies the need for an ever-growing heterogeneous network. At the same time, multi-RAT association and mobility will be challenging tasks that need to be addressed in the presence of a large number of small cells [4]. Device cost also deserves appropriate attention. Furthermore, the random and massive deployment of small cells also entails added intercell interference, requiring the adoption of semi- or fully distributed interference coordination or suppression techniques. In addition, an appropriate flexible scheme for allocating the transmission direction between the uplink (UL) and downlink (DL) is necessary to reduce the latency and accommodate a fast response to the traffic variation.

Key technological components such as multiple transmit and receive antennas, dynamic UL/DL transmission, and full duplex communication are expected to play important roles in enhancing the spectral efficiency of the upcoming 5G system. For example, multiple antennas can be exploited to increase the system throughput through the joint impact of interference-aware transmission techniques [8] and advanced interference suppressing receivers [9].

Being a disruptive technology, 5G will be operating over frequency bands different from the ones used for existing RATs. Lower frequencies, especially in the sub-3 GHz region, are typically desirable in cellular networks due to their favorable propagation characteristics. However, the severe shortage of spectrum availability in this frequency region has led to the exploration of different spectrum ranges. In particular, the millimeter-wave (mmWave) bands in the 20–90 GHz region (including the E-band) have attracted major attention due to the availability of a large combined spectrum (around 10 GHz) [10]. However, further frequency bands are also available in the region below 6 GHz, that is, in the centimeter-wave (cmWave) region. In particular, it is anticipated that the 3.4–4.9 GHz bandwidth can be used for 5G small cells. The total amount of available spectrum is around 1 GHz, and the suggested frequency band per operator is 200–400 MHz.

Attractive from a spectrum availability point of view, the design of a 5G mmWave concept poses different challenges from those of a 5G cmWave concept due to the different propagation characteristics as well as the amount of usable spectrum. In this chapter, we focus on the cmWave concept for 5G. In particular, we address the main qualitative differences from the 5G mmWave design, and describe our vision of the key technology components for enabling energy-efficient multigigabits per second wireless transmission with low latency.

The rest of this chapter is structured as follows. Section 14.2 recalls the basic challenges of both mmWave and cmWave concepts. A detailed overview of our envisioned 5G cmWave system concept is introduced in Section 14.3, followed by an in-depth discussion of the novel flexible time-division duplex (TDD) concept in Section 14.4. Section 14.5 discusses the design of a throughput-enhancing interference-aware rank adaptation algorithm. Finally, energy-consumption issues are addressed in Section 14.6, followed by concluding remarks in Section 14.7.

14.2 mmWave and cmWave Challenges

Both the cmWave and the mmWave concepts aim at gigabit data rates and reduced latency. The radio propagation characteristics of mmWaves are still being explored with extensive measurement campaigns (e.g., [11]). To counteract the severe link budget due to paramount propagation losses and heavy attenuation of the potential obstructions, the presence of a robust line-of-sight (LOS) component is proved to be fundamental for establishing the communication link. The link budget can be significantly improved by using large antenna arrays steering highly directional beams at both ends of the communication link. In that respect, operating at extremely high frequency is beneficial for placing such a large antenna array in a small physical area. The beam forming will likely be performed in the analog radio-frequency domain, given the huge cost of analog-to-digital (ADC) and digital-to-analog (DAC) converters at mmWave frequencies [12]. The possibility of exploiting a large frequency spectrum alleviates the necessity of relying on the spatial multiplexing of data streams for conveying a large data rate, thus significantly reducing the computational complexity of the baseband processing. Further, in mmWaves, the intercell interference is not considered a major limiting factor, given the possibility of using highly directive beam-forming patterns.

Conversely, the cmWave concept is expected to be applied over bands whose propagation characteristics are quite similar to those of already known bands. Establishing the communication link at cmWave frequencies does not require LOS conditions. Given the lower amount of spectrum at cmWave frequencies, the use of multiple-in multiple-out (MIMO) spatial multiplexing becomes fundamental for achieving the targeted data rates. On the other hand, the maximum number of spatial data streams is limited in the cmWave concept, given the practical difficulties in placing a large number of antenna ports operating, for example, at below 6 GHz in a handheld device. Similar limitations may apply also to the small cell access point. We therefore believe 4×4 MIMO to be a realistic target for 2020 [1]. Further, the presence of a large number of significant scattered components besides the LOS component makes intercell interference the main limiting factor for the cmWave concept. Clearly, the mmWave concept does not fulfill all the requirements, such as low power consumption and extreme cost efficiency, required for machine-type communication (MTC).

A careful design for both cmWave and mmWave concepts should, then, tackle their respective issues in an agile and cost-effective manner. Further, significant research effort is also being devoted to the harmonization of both concepts from a numerology perspective. This would allow some of the baseband component to be used for both technologies, regardless of the different specific radio-frequency front-ends. The rest of the chapter will focus on our envisioned cmWave concept.

14.3 5G Small Cell cmWave System Overview

In this section, we will provide a detailed overview of our envisioned 5G cmWave system, [2] highlighting the most relevant concepts. A general overview of some of the key features introduced in the envisioned 5G small cell concept will be presented, followed by further details on the frame structure, MIMO and advanced receiver support, and the proposed TDD framework.

14.3.1 Key Features

The envisioned 5G small cell cmWave concept introduces several key technological components to provide increased peak throughputs, lower latencies, and higher robustness against intercell interference. The frame structure has been totally redesigned and shortened to 0.25 ms to accommodate lower latencies and a shortened round-trip time (RTT). It also includes support for advanced receivers via a special demodulation reference symbol (DMRS). Flexible resource allocation is supported by allowing fully flexible TDD on a per-frame basis, such that each frame can be independently allocated to the UL or DL direction. Such a feature is especially useful in small cells, in which the active number of users is typically much smaller than in a macro cell, and a newly arriving traffic burst can significantly shift the required balance of UL and DL resources.

Higher peak throughputs and increased robustness against intercell interference are achieved with MIMO systems in tandem with advanced receivers using an advanced interference rejection combining (IRC) receiver coupled with dynamic rank adaptation. In low-interference scenarios, the number of transmitted spatial streams can be increased, allowing higher peak throughputs. Conversely, when the interference levels are strong, the system can attempt to suppress this interference using its spatial degrees of freedom by reducing the transmission rank, thereby trying to guarantee a minimum throughput target.

Table 14.1   Summary of 5G Small Cell cmWave Requirements and Key Features Addressing Them

Requirement	Key Features
Higher peak throughputs	MIMO, larger bandwidths, flexible TDD
Guaranteed minimum throughput	Robustness to intercell interference via additional spatial streams and IRC, supported by appropriately designed frame structure
Lower latencies	Shortened frame structure
Fast setup time	Shortened and appropriate modifications to the frame structure
Flexibility	Full TDD flexibility on a per-frame basis

A summary of the requirements and the potential key technological components needed to address them is presented in Table 14.1.

14.3.2 Envisioned 5G Frame Structure

The frame structure is a critical element in the design of a novel RAT, since it significantly affects the latency and the required baseband processing for detecting the data. The millisecond latency target of 5G is justified by envisioned applications such as tactile Internet [13], which require extremely short RTTs. Further, the low latency reduces the necessity of storing large blocks of data, for example, for acknowledged transmissions. This allows the size of the buffers, which represent the most expensive component in the baseband chip, to be significantly reduced. In this section, we present a frame structure that comprises a control part located at the beginning and time separated from the data part [14], as shown in Figure 14.1.

The control part features in a TDD fashion both DL and UL control, while the data part can be assigned to only one transmission direction (i.e., either DL or UL) per frame. A short guard period (GP) is inserted at every switch of the transmission

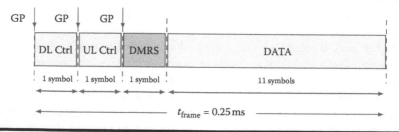

Figure 14.1   5G frame structure indicating symbols dedicated to control, DMRS, and data.

direction to allow the radio circuitry to power off/on. Multiple access points (APs) or user equipments (UEs) can simultaneously transmit their control information on the control part by using, for instance, orthogonal frequency blocks. Note that the first time symbol in the data part is dedicated to the transmission of DMRSs to enable channel estimation at the receiver for coherent detection.

The constraint of allowing a unique transmission direction per frame enables stabilization of the interference pattern within the frame itself. This is beneficial for the use of IRC receivers, as will be further elaborated in Section 14.3.3. The time separation between control and data allows a straightforward separation of their respective control planes. This enables cost-effective pipeline processing at the receiver, since the UE can process its dedicated control information while transmitting/receiving in the data part, thus reducing the latency. Note that this is different, for instance, from LTE, in which both physical UL control channels (PUCCHs) and physical DL control channels (PDCCHs) are multiplexed in the same time symbol and mapped over different frequency resources, forcing the UEs and the base station to detect both data and control for extracting the needed information [15].

UE-initiated data transmission requires three TDD cycles (scheduling request in the UL, scheduling grant in the DL, and data transmission in the UL) for a total of 0.75 ms. Similarly, the RTT of the Hybrid Automatic Repeat Request (HARQ) process (AP grant, AP transmission, and UE acknowledgement/negative-acknowledgement [ACK/NACK] transmission) requires 0.75 ms, including processing times. Unlike LTE-TDD, the HARQ RTT is here fixed and is independent from the UL/DL ratio; the control part in each radio frame offers at least one time symbol in each direction for the transmission of acknowledgments. The number of parallel HARQ processes is four, while in LTE-TDD it is up to 15; this allows a considerable reduction of the memory circuitry (buffers), which leads to significant cost savings in the baseband chip. The benefits in terms of energy consumption of our envisioned frame structure will be addressed further in Section 14.6.

The choice of TDD over frequency-division duplexing (FDD) was based on some important advantages that this technology has to offer. First of all, there is no need for paired spectrum, as in FDD. It also adapts well to unbalanced traffic scenarios via simple reconfiguration in LTE-TDD [16] and via full flexibility per slot in our envisioned 5G concept. This is particularly useful for small cell scenarios, in which the number of active users is typically much lower than in macro scenarios. Lower component cost and channel reciprocity between UL and DL are other features that make TDD attractive [17,18].

14.3.2.1 Differences from Long-Term Evolution–Time-Division Duplexing (LTE-TDD)

There are some noticeable differences between the envisioned 5G frame structure and the LTE-TDD frame structure, from both a numerology and a structural point of view. In contrast to LTE-TDD, in which orthogonal frequency-division multiple access (OFDMA) is used for DL and single-carrier frequency-domain multiple access (SC-FDMA) for UL, here it is assumed that orthogonal frequency-division multiplexing (OFDM) is used for both UL and DL. While there are several candidates for future potential 5G waveforms, OFDM was selected due to some inherent advantages related to its low implementation complexity when compared with competing emerging technologies such as filterbank multicarrier (FBMC). Moreover, OFDM lends itself easily to MIMO extension, an important key technology component in our 5G system concept. It is also robust to hardware impairments typical of low-end devices [19,20].

The frame size is also shortened to 0.25 ms as compared with LTE's 1 ms, making it more attractive from a latency and RTT minimization point of view. This shrinking of the frame size was done by increasing the subcarrier spacing from 15 to 60 kHz, hence obtaining a shorter symbol time. Shortening the symbol time increases the relative overhead of the cyclic prefix to the symbol duration, but since this frame structure was designed for small cell systems, lower delay spreads are expected, and therefore, compared with the LTE, the cyclic prefix can be shortened significantly without worrying too much about intersymbol interference. Within the UL–DL switching points of the frame itself, a GP is typically required to avoid power leakage between the transmitter and the receiver of a node. Given the lower transmit powers of small cell systems and hence the shorter associated rise and fall times, along with the evolution of technological components, it is assumed that the GP duration can also be shrunk considerably when compared with LTE.

A short summary of the main differences in numerology between the envisioned 5G and the LTE-TDD frame structure is provided in Table 14.2.

14.3.3 MIMO and Advanced Receiver Support

The availability of multiple antennas and MIMO allows the peak throughput performance to be increased under favorable conditions. On the other hand, the multiple antennas can be used to suppress parts of the received interference with the help of the IRC receiver in challenging interference conditions. Therefore, the additional antennas at our disposal give us the ability to configure the system to adapt to certain conditions, such that we can enjoy the best of both worlds, whether guaranteeing a minimum throughput performance or satisfying peak throughput requirements [2,21].

A key characteristic for effectively exploiting advanced receiver techniques employing interference suppression is linked to an appropriately designed frame structure. LTE was not built from the ground up to support such functionality, and the envisioned 5G frame structure described in Section 14.3.2 attempts to overcome this limitation. The design of the frame structure and proper support for advanced receiver techniques such as IRC becomes even more important in the light of fully flexible TDD and hence increased probability of cross-link interference. In such a situation, the interference channel conditions can easily change

Table 14.2 Numerology Differences between 5G Frame Structure and LTE-TDD

	LTE-TDD	*5G*
Number of symbols per frame	14	14
Subcarrier spacing (kHz)	15	60
Symbol time (μs)	66.67	16.67
Frame length (ms)	1	0.25
Subcarriers per PRB	12	165
PRB allocation (BW)	180 kHz	10 MHz
System bandwidth (MHz)	1.4–20	100 or 200
TDD flexibility	Set of TDD configurations (DL:UL, 2:3 to 9:1)	Full flexibility per frame

Source: Lähetkangas, E., et al. On the TDD subframe structure for beyond 4G radio access network. In *Future Network and Mobile Summit (FutureNetworkSummit), 2013*, Lisboa, IEEE, pp. 1–10, 2013.

from one frame to the next. Within the envisioned 5G concept, IRC receivers are expected to be operational at the AP as well as in the UEs.

To exploit the advantages of IRC, an updated estimation of the interference covariance matrix (ICM) needs to be available at the receiver, such that the receiver can adjust its weights accordingly to suppress interference [9]. The 5G frame structure was designed with this in mind, allowing the receiver to obtain this information. This is done in the DMRS symbol, whereby all the nodes scheduled in the data part will transmit simultaneously during this symbol, allowing the IRC-capable receivers to distinguish and identify the desired channel and accurately estimate the ICM, such that the appropriate weights can be applied to suppress this interference. The ICM estimation is independent of the link direction applied at each cell. Conducting a similar operation in LTE-TDD is much more challenging, especially for cross-link interference (AP–AP or UE–UE interference), since the DL and UL transmissions use different access technologies (ODFMA and SC-FDMA, respectively).

14.3.4 Flexible TDD Support

Flexible TDD is not a new concept, and limited support for this feature has already been introduced in other systems, such as LTE-TDD and WiMAX [22]. Its importance becomes even more crucial when one considers its application in small cell

Table 14.3 LTE-TDD Configurations

Configuration	Subframe Number										DL-UL Asymmetry
	0	1	2	3	4	5	6	7	8	9	
0	D	S	U	U	U	D	S	U	U	U	0.25–0.40
1	D	S	U	U	D	D	S	U	U	D	0.50–0.60
2	D	S	U	D	D	D	S	U	D	D	0.75–0.80
3	D	S	U	U	U	D	D	D	D	D	0.67–0.70
4	D	S	U	U	D	D	D	D	D	D	0.78–0.80
5	D	S	U	D	D	D	D	D	D	D	0.89–0.90
6	D	S	U	U	U	D	S	U	U	D	0.38–0.50

D = DL frame, U = UL frame, S = special frame.

systems, in which the number of users is typically low, and the traffic requirements between UL and DL can shift more frequently over time.

LTE-TDD introduces a limited set of options in allocating its resources. This is done by defining a set of seven configurations with different DL–UL traffic asymmetries. These configurations are shown in Table 14.3. The configurations can be reconfigured every x ms, and studies conducted in [23] have shown that there is additional benefit in lowering this reconfiguration time. The available frame configurations allow the DL to UL asymmetry to vary from 40% to 90%. Some of the subframes in Table 14.3 are aligned, while others are not, introducing the problem of cross-link interference, an aspect that introduces uncertainties, given LTE's radio access differences in UL and DL.

The frame structure presented in Section 14.3.2 gives us full flexibility to allocate each frame as UL or DL. This flexibility provides benefits in terms of traffic adaptation but will also undoubtedly create additional intercell interference variation, since a cell's neighbor could be shifting its allocation from UL to DL at its convenience to deal with its own traffic unbalance scenario. This issue presents a challenge resource allocation algorithm that relies on the short-term stability of the channel conditions, such as rank/link adaptation. This drawback, as we will see in Section 14.4, can be dealt with by the use of IRC receivers, for which our frame structure provides appropriate support to operate sufficiently well.

14.4 Flexible TDD

Fully flexible (dynamic) TDD allows the system to capture the instantaneous traffic conditions and react accordingly. Such a flexible system can avoid manual TDD configuration, requiring long-term statistical information related to the experienced DL and UL traffic profiles present in a system. The advantages of flexible TDD have already been acknowledged in existing systems such as LTE, which offer various options and degrees of flexibility. In this section, we will take a closer look at the introduction of fully flexible TDD as envisioned within the described 5G concept. We shall look at the problem from a traffic perspective and analytically assess the expected gains originating from flexible TDD when compared with a fixed static TDD system that can predict the long-term average traffic share between UL and DL.

While intuitively attractive, flexible TDD also poses some challenges that need to be overcome. In particular, flexible TDD introduces cross-link (AP–AP and UE–UE) interference and additional intercell interference variation, which hinder the full potential of flexible TDD. The threat of increased intercell interference variation can in part be mitigated by the use of advanced receivers such as IRC, which when supported by the appropriate frame structure, have the ability to suppress interference independently from the source of the interferer. As a further step, we therefore assess the suitability and impact of such receivers in relation to the problem mentioned in Section 14.3.3, considering a realistic multicell system to investigate whether the flexibility provided by flexible TDD outweighs the presented demerits of such a system.

14.4.1 Expected Gains from Flexible TDD

In this section, we will present an analytical approach to find the maximum achievable gain from fully dynamic TDD. Consider a single cell, consisting of an AP and a UE, each having an independent traffic arrival profile. Assume that the traffic profile of both link directions follows a bursty traffic model, as specified in the 3rd Generation Partnership Project (3GPP) FTP traffic model, with payload size K bits interpacket off period (t_{off}). This operation is illustrated in Figure 14.2.

The time to service a packet t_s in a system with time transmission interval (TTI) rate r can be represented as $t_s = K/R$, with R depending on the conditions of the current buffer sizes in DL and UL, K_{DL} and K_{UL}, respectively. The rate R varies based on whether instantaneously both DL and UL directions have data to be transmitted or not. For a scheduler allocating resources equally between UL and DL, and for packet sizes spanning multiple time slots, such that $t_s > t_{TTI}$, the rate of each link, R, is given as

$$R = \begin{cases} 0 & \text{if } K_{DL} = 0 \text{ and } K_{UL} = 0 \\ \dfrac{r}{2} & \text{if } K_{DL} > 0 \text{ and } K_{UL} > 0 \\ r & \text{if } K_{DL} > 0 \text{ and } K_{UL} = 0 \\ r & \text{if } K_{DL} = 0 \text{ and } K_{UL} > 0 \end{cases} \qquad (14.1)$$

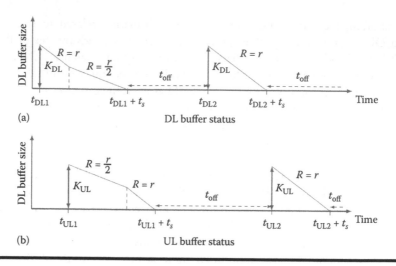

(a) DL buffer status

(b) UL buffer status

Figure 14.2 Buffer status with traffic arrivals in UL and DL.

Let us assume a fixed-slot 1:1 TDD system as a reference, with equal UL and DL arrivals. To assess the gain from flexible TDD, the probability that both links are active needs to be derived for different t_{off} periods. While the traffic arrivals in UL and DL are independent, the DL and UL channel occupation time and the individual probability of a link being active are dependent on whether there is any channel occupation from the opposing link, since this has the impact of halving the effective rate and hence doubling the service time of a packet.

We assume that t_{off} is negative exponentially distributed, such that $P(T = t_{off}) = \lambda e^{-\lambda t}$, where λ represents the arrival rate and $1/\lambda$ represents the mean off time. The probability of the arrival of a DL burst is simply given by $\int_0^\infty \lambda_{DL} e^{-\lambda_{DL} t_{DL}} dt_{DL}$, and the probability that a UL burst arrives before the termination of the previous DL burst arrival, denoted by $P_{UL \to DL}$, is given by

$$P_{UL \to DL} = \int_0^\infty \int_{t_{DL}}^{t_{DL} + (K/R)} \lambda_{DL} \lambda_{UL} e^{-\lambda_{DL} t_{DL}} e^{-\lambda_{UL} t_{UL}} dt_{UL} dt_{DL} = \frac{\lambda_{DL}}{\lambda_{DL} + \lambda_{UL}} \left(1 - e^{(-\lambda_{UL} K/r)}\right)$$

(14.2)

Similarly, for a UL burst, the probability that a DL burst arrives before its termination is given by

$$P_{DL \to UL} = \frac{\lambda_{UL}}{\lambda_{DL} + \lambda_{UL}} \left(1 - e^{(-\lambda_{DL} K/r)}\right)$$

The probability of DL and UL being active simultaneously is then given by

$$P_{\text{DL\&UL}} = \frac{\lambda_{\text{DL}}}{\lambda_{\text{DL}} + \lambda_{\text{UL}}}\left(1 - e^{(-\lambda_{\text{UL}}K/r)}\right) + \frac{\lambda_{\text{UL}}}{\lambda_{\text{DL}} + \lambda_{\text{UL}}}\left(1 - e^{(-\lambda_{\text{DL}}K/r)}\right)$$

Assuming that both UL and DL have identical traffic profiles, with identical payload sizes K and identical t_{off} distributions, $P_{\text{DL\&UL}}$ reduces to

$$P_{\text{DL\&UL}} = 1 - e^{(-\lambda K/r)} \tag{14.3}$$

For a fixed packet size K and for a set of rates per TTI r_{TTI}, the probability $P_{\text{DL\&UL}}$ varies according to the mean t_{off} and the rate r_{TTI}. The lower t_{off}, the higher the load in the network; and for a fixed service time t_s, the higher the probability of UL and DL being active simultaneously. Conversely, when t_{off} increases, the probability of UL and DL being active simultaneously diminishes. r_{TTI} dictates the service time t_s, so for a high rate, a lower service time is expected and hence a lower probability of concurrent UL and DL transmission. Figure 14.3 shows this probability as a function of the mean off time and the rate.

The expected session throughput of a single-cell system operating in dynamic TDD mode can be written as the sum consisting of the product of the full rate and the probability of UL and DL not being active simultaneously, and the product of

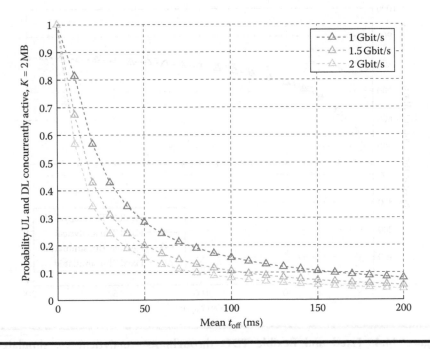

Figure 14.3 **Probability of concurrently active UL and DL, $K = 2$ MB.**

half the rate and the probability of UL and DL being active simultaneously; that is, dynamic TDD throughput = $r(1 - P_{DL\&UL}) + (r/2)P_{DL\&UL}$.

Therefore, given $P_{DL\&UL}$, the gain of flexible TDD over a fixed TDD scheme can be extracted. Let us denote the expected flexible TDD gain as ρ_{DL} and ρ_{UL} for the DL and UL gain, respectively. As a generalized solution, for a TDD pattern consisting of s slots, of which s_{UL} are UL slots and s_{DL} are DL slots, the gain ρ can be given as

$$\rho_{DL} = \left(\frac{r}{r\left(\frac{s_{DL}}{s_{DL} + s_{UL}}\right)}\right)(1 - P_{DL\&UL}), \quad \rho_{UL} = \left(\frac{r}{r\left(\frac{s_{UL}}{s_{DL} + s_{UL}}\right)}\right)(1 - P_{DL\&UL}) \quad (14.4)$$

For validation purposes, the result in Equation 14.4 is compared with simulation results evaluated in a system-level simulator with a single-cell system for a TDD system having equal UL and DL traffic shares and a fixed TDD slot configuration of 1:1, which consists of a DL and a UL slot, as presented in Figure 14.4. The session throughput for each payload, that is, the throughput experienced for each payload

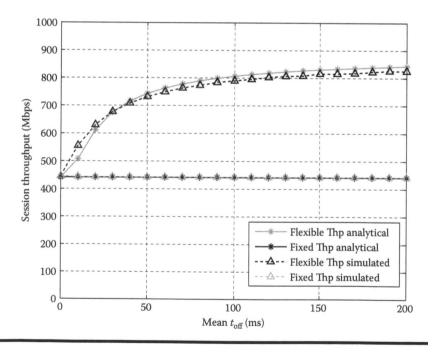

Figure 14.4 Fixed and flexible TDD throughputs—analytical vs. simulated approach for $r = 880$ Mbps, $K = 2$ MB, and $t_{off} = 0$–200 ms.

transmitted, is compared for the two schemes, and the relative gain of the flexible over the fixed scheme is compared with the expected gain from Equation 14.4.

The analysis can be extended for different UL to DL traffic asymmetries and other fixed DL:UL configurations. The probabilities can be derived accordingly from Equations 14.3 and 14.4 by updating the appropriate parameter values. For example, under low load and high traffic asymmetries, the flexible scheme will offer a small increase in performance in the traffic-heavy link direction and a large increase in performance in the lightly loaded link direction.

14.4.2 Demerits of Flexible TDD

From a radio perspective and in a realistic multicell system where intercell interference plays an important role, there are, however, some challenges to address when considering the use of flexible TDD. In this subsection, we will present some of these aspects, estimate their impact, and subsequently investigate potential tools that can mitigate the effect of these challenges. In a fully synchronized fixed TDD system, an AP transmission typically interferes with a UE's reception, and vice versa. Flexible TDD changes this paradigm, as it introduces cross-link interference, which is an additional interference level that can be experienced by a victim receiving node.

To understand better the behavioral differences between fixed and flexible TDD, we consider a small cell scenario consisting of 20 cells arranged in a 10×2 grid fashion, operating with a fixed TDD 1:1 scheme, two different flexible TDD schemes, and a random TDD scheme that randomly chooses the link direction. The first flexible TDD scheme, the delay fairness (DF)-based algorithm, consists of a simple algorithm that allocates time-slot resources to the direction having data. If data are present in both UL and DL buffers, the algorithm will converge, allocating the time-slot resources in an alternate fashion to UL and DL. Another flexible TDD scheme, the load fairness (LF)-based algorithm, is yet another flexible TDD traffic allocation algorithm, which considers the instantaneous traffic asymmetry between DL and UL and the previous slot allocations.

We offer an absolute load of 150 Mbps to each node, representing around 60% resource use, such that additional stress is placed on the intercell interference variation. In this investigation, it is assumed that all nodes are equipped with a 4×4 MIMO transceiver and operate with the conventional maximal ratio combining (MRC). A simple signal-to-interference-plus-noise ratio (SINR) trace for a particular node is illustrated in Figure 14.5, immediately capturing the added SINR variability introduced by the flexible TDD schemes.

The net effect of this added SINR variability causes link adaptation channel estimation errors, inducing increased HARQ retransmissions. Simulation results show that in this particular situation, in the fixed TDD case, 5% of the packets need to be retransmitted at least once, while in the flexible TDD cases, 14%–30% of the packets need to be retransmitted.

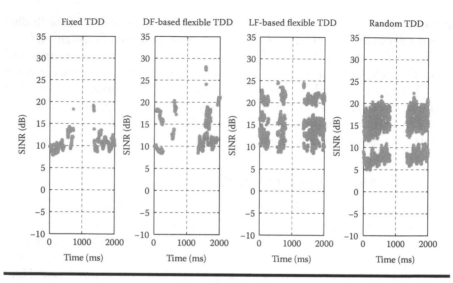

Figure 14.5   SINR time trace.

These additional retransmissions do affect the experienced end-user session goodput. The envisioned 5G concept discussed in Section 14.3 assumes the availability of MIMO and IRC receivers, which have been shown to effectively mitigate the impact of the introduced intercell interference variation in [9]. Given the availability of an appropriately designed frame structure such as the one explained in Section 14.3.2, IRC can suppress intercell interference independently from the source, hence also reducing the experienced intercell interference variation.

To further illustrate this finding, we reinspect the performance measure with the IRC receiver configuration. The experienced average SINR variation in the presence of IRC receivers is dramatically reduced, translating into a direct improvement of the number of induced HARQ retransmissions, as shown in Table 14.4, reducing the number of retransmissions from 14%–30% to 0.8%–2% for the flexible TDD schemes. In turn, this also improves the perceived session goodput, as shown in Figure 14.6, allowing us to exploit the full advantages of dynamic TDD without being affected by excessive retransmissions caused by the increased intercell interference variation introduced by this same feature.

Table 14.4   Number of HARQ Retransmissions

Receiver Type	MRC (%)	IRC (%)
Fixed TDD scheme	5	0.5
DF-based flexible TDD scheme	14	0.8
LF-based flexible TDD scheme	30	2
Random TDD scheme	36	3

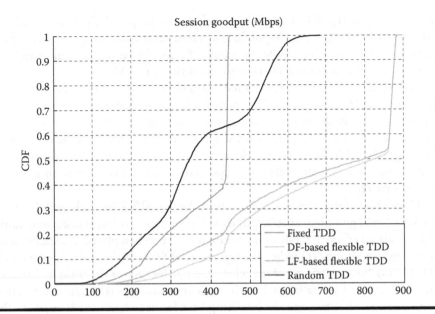

Figure 14.6 **Session goodput—IRC receiver.**

14.5 Rank Adaptation in 5G cmWave Small Cell Systems

The envisioned 5G concept presented here assumes a MIMO configuration. Due to the multiple available antennas, the system can be configured to transmit a single spatial stream, or with the maximum number of spatial streams. The former scheme provides a high degree of robustness to intercell interference variation in highly interfered scenarios [24], while the latter enables higher peak throughput under favorable channel conditions.

There is, therefore, a classical trade-off to be made between increased peak throughput and improved interference resilience. Ideally, the number of spatial streams used for transmission, hereinafter referred to as the transmission rank, needs to be low in highly interfered conditions. Conversely, under favorable channel conditions, a high transmission rank should be preferred. The dynamic adjustment of the transmission rank reflecting the interference conditions is known as *rank adaptation* and is the focus of this section. More specifically, a taxation-based interference-aware rank adaptation algorithm is proposed and evaluated in this section.

14.5.1 Proposed Taxation-Based Rank Adaptation Scheme

The proposed interference-aware taxation-based rank adaptation scheme (TB-RA) aims at choosing a rank k^* that maximizes a given utility function Π_k, that is, $k^* = \text{argmax}_k \; \Pi_k$. When choosing a rank k, the proposed algorithm considers the achievable rate and a corresponding taxation term based on a rank-dependent monotonically increasing weighting vector W_k, also based on the rank k and the rate impact of the prevailing interference conditions, $C(I/N)$.

On reception of a signal, the desired channel matrix H_D and interference covariance matrix H_I are extracted. The matrix H_D can be acquired over time and is expected to change slowly over time due to the assumed small cell scenario. On the other hand, the value of H_I can potentially vary from one frame to the next due to the usage of flexible TDD. Nonetheless, the presence of the DMRS symbol in our frame structure allows us to obtain an updated estimate of the interference covariance matrix H_I.

Once this information is retrieved, the effective SINR for each rank is calculated. This will give us an effective SINR value for each possible transmission rank. The calculated effective SINR values for each rank are then placed in a sliding window filter containing Q samples. The log-averaged SINR is then calculated for each of the ranks, and an estimate of the achievable rate for each of the ranks is obtained. Once the achievable rate for each of the rank transmissions is obtained, a taxation term based on the rank and the incoming interference conditions is applied. The rank k that maximizes utility function Π_k is then chosen. The considered utility function is represented mathematically by Equation 14.5:

$$\Pi_k = \underbrace{kC(\overline{\text{SINR}_{\text{effective}_k}})}_{\text{Estimated capacity for rank } k} - \underbrace{kW_kC\left(\frac{I}{N}\right)}_{\text{Taxation for rank } k} \qquad (14.5)$$

where $C(\overline{\text{SINR}_{\text{effective}_k}})$ is estimated as $C(\overline{\text{SINR}_{\text{effective}_k}}) = \log_2\left(1 + \overline{\text{SINR}_{\text{filtered}_k}}\right)$.

The taxation term $C(I/N)$ in Equation 14.5 is given by $C(I/N) = \log_2(1 + (tr(H_IH_I^H)/\sigma_n^2))$ and represents an estimate of the loss in throughput resulting from the incoming interference-over-noise ratio. When considering the taxation and specifically the interference term, one should ideally consider the outgoing rather than the incoming interference, since this represents the actual harm generated to the other nodes. Obtaining such an estimate in a fully distributed manner is, however, challenging given the available frame structure, and therefore, we assume that the incoming interference is equal to the outgoing interference, even if this is not always the case. Moreover, the incoming interference should give us an estimate of the currently perceived interference levels.

It is important to note that the rank transmission decision calculation is done at both the AP and the UE. The UE will simply decide a DL transmission rank based on the calculation of the utility function, considering its locally perceived interference conditions. This information is then fed back via the scheduling request (SR) message in the UL control channel, and the AP, the final decision-maker, will simply use this information when instructing which rank transmission to use in DL.

14.5.2 Performance Evaluation

In this section, we shall evaluate the performance of the TB-RA algorithm, showing how the rank adaptation algorithm adapts in different interference conditions.

The interference conditions will be controlled by varying the traffic load, showing whether the TB-RA algorithm can adapt in time to different interference conditions. In doing so, we will assume a flexible TDD slot allocation based on the DF-based flexible TDD scheme with an equal UL to DL traffic load. We will also benchmark the algorithm against a selfish scheme that applies no taxation, hereinafter denoted as SRA, and fixed rank 1 and 2 transmission schemes. The TB-RA algorithm will also be configured with two different W_k vector parametrizations representing a conservative and an aggressive rank transmission selection scheme. The W_k vectors are chosen to be $W_1 = [0, 0.5, 0.66, 0.75]$ and $W_2 = [0, 0.25, 0.66, 0.75]$, representing the conservative and aggressive configurations, respectively. These different configurations will be displayed as TB-RA conservative and TB-RA aggressive when showing the results.

The inspected key performance indicator (KPI) metric by which the performance of the individual schemes is evaluated is the average node session throughput at the application layer. This represents the average experienced session throughput over multiple sessions by a particular node during the course of a simulation.

14.5.2.1 Traffic Load

To assess the ability of the rank adaptation algorithm to adapt to different interference conditions, we load the system to use approximately 25% and 75% of resource use, corresponding to an offered load of 100 and 250 Mbps/node, respectively.

The cumulative distribution function (CDF) of the average node session throughput and the distribution of transmission ranks for the individual schemes at around 25% resource use are shown in Figure 14.7. In this case, the interference conditions are low, and the fixed rank 1 scheme's throughput is clipped. At low load conditions, the fixed rank 2 scheme offers a superior performance, even on the low end of the CDF curve. The result shown in Figure 14.7 stresses the importance of testing the performance of a scheme over different traffic loads and also proves the need for a rank adaptation scheme that can adapt to different interference and traffic conditions.

Figure 14.7 also shows that all the rank adaptation schemes can exploit temporary favorable channel conditions and hence make use of higher transmission ranks, allowing some nodes to experience higher average session throughputs. The provided result also shows that at the low end of the CDF curve, the rank adaptation schemes can almost achieve fixed rank 2 performance. From Figure 14.7, one can see that the conservative TB-RA approach tends to choose the transmission ranks conservatively, with the aggressively configured TB-RA scheme giving the best overall performance. The selfish scheme offers a satisfactory performance in

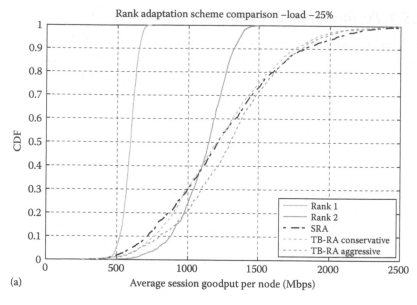

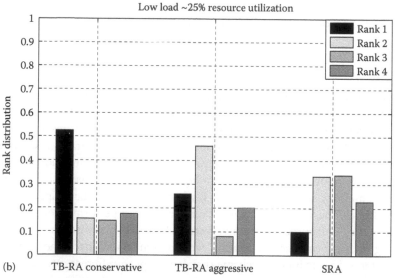

(a)

(b)

Figure 14.7 Average session goodput and rank distribution with TB-RA for RU ~25%.

this scheme but tends to choose the transmission ranks a bit too aggressively, leading to a slightly inferior performance at the low end of the CDF curve.

The performance results for the rank adaptation schemes at approximately 75% resource use are shown in Figure 14.8. While the fixed rank 1 scheme offers the best outage performance, it also limits the maximum achievable throughput. The

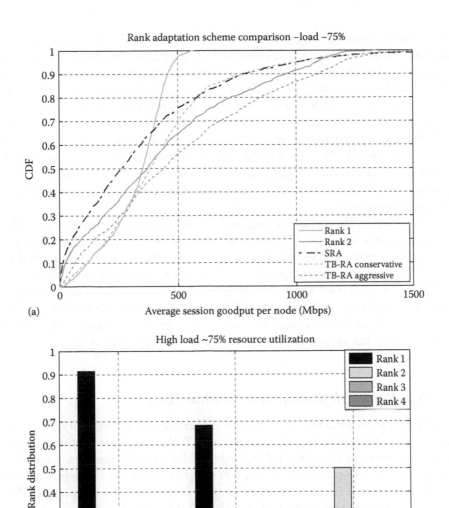

Figure 14.8 Average session goodput and rank distribution with TB-RA for RU ~75%.

fixed rank 2 scheme has an inferior outage performance in this case but can reach higher throughputs.

In this case, the different rank adaptation schemes offer a noticeable difference in performance. We once again collectively inspect the average node session throughput CDF along with the rank transmission distribution chosen by the schemes, shown in Figure 14.8. Here, we notice that all the rank adaptation schemes lower the chosen transmission ranks as a result of more severe interference conditions. The SRA scheme, however, still chooses the ranks aggressively, even if there is little gain in doing so. This happens because it is designed to maximize its own capacity in a selfish manner, leading to unsatisfactory outage performance, since the IRC receiver cannot suppress interference effectively. The conservative TB-RA scheme manages to match fixed rank 1 outage performance and also exploits the use of higher transmission ranks where possible. The aggressive TB-RA approach tends to choose slightly higher transmission ranks than the conservative approach in this case, resulting in higher peak throughputs at the cost of reduced outage performance.

These results show that the proposed TB-RA scheme outperforms the SRA scheme at both low and high loads and manages to exploit the usage of higher transmission ranks whenever it is possible to do so. The taxation term applied in the TB-RA approach limits the transmission rank to be used if high interference conditions are perceived and avoids choosing a higher transmission rank if there is little gain in doing so. The presented TB-RA algorithm can also be parametrized conservatively, to retain outage performance, or aggressively, to enjoy higher peak throughputs. The automatic parametrization of the algorithm is left as possible future work.

14.5.3 Rank Adaptation and Flexible TDD

In this section, we would like to investigate the performance of flexible TDD against fixed TDD when operating over the presented TB-RA scheme. In Section 14.4, the potential demerits of flexible TDD related to added intercell interference variation were outlined. It was, however, shown that this problem can be counteracted if the available degrees of freedom are used to suppress interference with the help of IRC receivers. The presented rank adaptation algorithm further allows the use of higher transmission ranks where possible, thus limiting the possibility of interference suppression independently from the source. Moreover, it increases the intercell interference variation due to the introduced liberty of choosing a transmission rank in an adaptive and varying manner.

The goal of this section is to inspect whether flexible TDD still offers noticeable gains in such conditions. To verify this, we run system-level simulations at high load scenarios with the proposed TB-RA algorithm for fixed TDD and flexible TDD configurations, as presented in Figure 14.9.

There is a difference in performance between the conservative and the aggressive approach. For the flexible TDD case, there is a clear benefit in being conservative when choosing the W_k parametrization, especially if outage performance needs to be improved. For the aggressive TB-RA parametrization, there is similar behavior to the previous results for both fixed and flexible TDD schemes. With this parametrization, higher transmission ranks are favored, allowing the system to

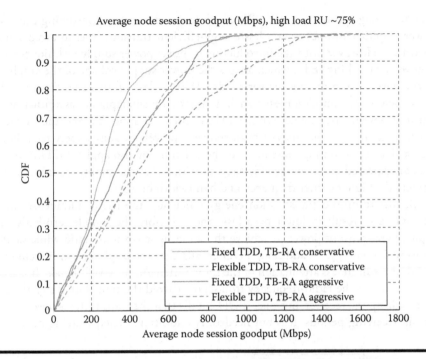

Figure 14.9 Fixed TDD versus flexible TDD performance over TB-RA scheme—high load.

reach higher peak throughputs. The important thing to note here is that when the interference conditions become severe, the TB-RA scheme has a tendency to use lower ranks for both fixed and flexible TDD schemes, reclaiming the benefit of suppressing interference independently from the source, and hence flexible TDD still retains a gain over fixed TDD, even at 75% resource use, which generally represents a case in which limited gain is expected from flexible TDD.

14.6 Energy-Saving Enablers

Energy saving is a KPI for 5G, including both smartphones and MTC devices. The latter are even expected to achieve 10 years of battery life. In the final section of this chapter, selected energy-saving enablers for 5G are reviewed to provide insight into how energy may be saved due to the 5G RAT design. The presented work is based on [25].

The proposed 5G concept has been designed to achieve better battery life than existing cellular technologies such as LTE and third generation (3G). Similarly to LTE, 5G is expected to apply OFDM in DL, but in contradiction to LTE, also in UL. This was avoided in LTE, among other reasons, due to the issue of high peak-to-average power ratio (PAPR), which forces the energy-consuming transmit power amplifier to operate at a less energy-efficient transmit level to avoid signal distortion. However, with the introduction of novel power supply techniques such as envelope tracking (ET), which adapts the power supply voltage of the amplifier to the current transmission power needs, PAPR is less of an issue. Figure 14.10 shows power consumption measurements on six LTE smartphones as a function of transmit power. The mobile terminals UE4 and UE5 apply ET and thus achieve much better energy efficiency at peak transmit power. Because supply techniques such as ET are becoming more mature, the 5G concept also applies OFDM in UL with an expected good energy efficiency, provided that the supply techniques also support the higher carrier frequency and bandwidth of 5G.

Besides improved transmission energy efficiency, the use of ultradense small cells will also result in lower path loss, and therefore the mobile terminals can apply lower transmit power to achieve the same signal-to-noise ratio while saving transmission energy. In addition, TDD entails that the duplexer can be removed and thus result in less attenuation of the transmitted and received signals. It is usually inserted in the radio-frequency front-end to avoid the transmitted signal from desensitizing the receiver, but since TDD entails that the transmitter is never active during receiving periods, it is not needed. An additional energy-saving benefit of

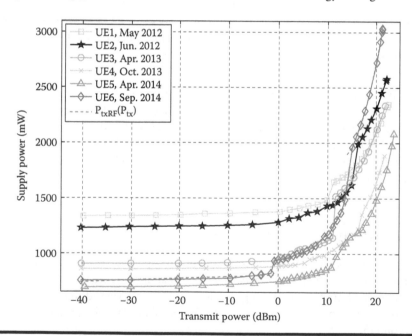

Figure 14.10 **Power consumption as a function of transmit power when using QPSK modulation. (From Lauridsen, M., Studies on mobile terminal energy consumption for LTE and future 5G, PhD thesis, Aalborg University, Aalborg, 2015, figure 2.5.)**

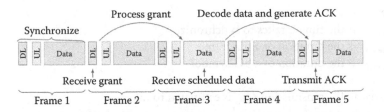

Figure 14.11   Reception of data in DL. (From Lauridsen, M., Studies on mobile terminal energy consumption for LTE and future 5G, PhD thesis, Aalborg University, Aalborg, 2015, figure 6.5.)

TDD is that in some cases, the channel may be considered reciprocal, provided the front-ends are calibrated properly. The reciprocity leads to less channel feedback, that is, less energy-consuming transmissions.

Another key feature for energy saving is the short and optimized frame structure detailed in Section 14.3.2. The new frame design is made such that a mobile terminal will receive a scheduling grant for either UL or DL in one frame, while the data will first be transferred in the following frame. The procedure for DL is illustrated in Figure 14.11, where the grant is received in Frame 2 and data are received in Frame 3. Due to this design, the mobile terminal knows whether it can apply low-power microsleep [26] in the data part of each frame. This is not possible in LTE because the data in the physical DL shared channel (PDSCH) follows directly after the scheduling information, located in the PDCCH [27]. The LTE implementation, therefore, results in reception and buffering of data that are not targeted for the mobile terminal, and thus energy is wasted. In addition to the applicability of low-power microsleep in the data part, the control channels can also be decoded in a pipelined manner, because the result is not needed until one frame later.

Besides the energy-saving features of the frame structure design, the reduced time duration also provides benefits for mobile terminal battery life. As illustrated in Figure 14.11, the mobile terminal can quickly synchronize to the channel after exiting a low-power sleep mode, because the DL control channel contains the required signals and performs a data transfer, after which it can return to a low-power sleep mode. This is an improvement compared with LTE, in which the primary and secondary synchronization signals only occur every 5 ms, which therefore will prolong the total ON time and thus the energy consumption of the device. Furthermore, the short frame structure reduces the total ON time, because a data transfer can be completed much faster than in LTE. A DL reception t_{rx5G} is estimated to be completed within

$$t_{rx5G} = t_{sync5G} + 3 \cdot t_{frame5G} + 2 \cdot t_{symb5G}$$

$$= 0.25 \text{ ms} + 3 \cdot 0.25 \text{ ms} + 2 \cdot 17.67 \text{ μs} = 1.03534 \text{ ms [s]} \qquad (14.6)$$

where:

t_{sync5G} is the time it takes to synchronize
t_{frame5G} is the duration of a 5G frame
t_{symb5G} is the duration of a 5G symbol

Similarly, a UL transmission t_{tx5G} is estimated to take

$$t_{\text{tx5G}} = t_{\text{sync5G}} + 4 \cdot t_{\text{frame5G}} + t_{\text{symb}}$$

$$= 0.25 \text{ ms} + 4 \cdot 0.25 \text{ ms} + 17.67 \text{ } \mu s = 1.26767 \text{ ms [s]} \quad (14.7)$$

This is a significant improvement as compared with TDD LTE, according to which [27] there are seven different frame configurations, and as calculated in [25], it takes 13–19 full subframes to complete a transmission, including 5 ms for synchronization results in 18–24 ms ON time. An example of such a calculation is given in Figure 14.12, where the minimum processing time for both UE and eNodeB is set to 3 ms. Note that the special subframe contains both UL and DL control channels but only DL data channels. A similar calculation for DL results in an estimated transfer time of 10–19 ms.

The battery life is calculated using the power model presented in [28] for an MTC device in 2020 for both LTE and 5G and a battery with 3 Ah at 3 V. Note that the battery self-discharge is set to 5% for the first 24 h and 2% per month thereafter, according to [29]. The result is presented in Figure 14.13 for different numbers of receptions and transmissions per second. For very low activity, the battery life is dominated by the device's sleep mode power consumption, but otherwise, the 5G concept results in 5–15 times better battery life when using discontinuous reception (DRX) and discontinuous transmission (DTX) due to the proposed short and optimized frame structure. In addition, the shorter 5G frame entails that it

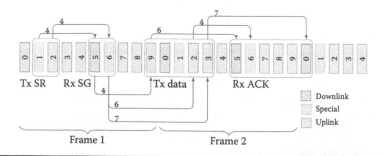

Figure 14.12 Performing a UL transmission in TDD LTE using frame configuration 0. SR and SG denote Scheduling Request and Grant, respectively. (From Lauridsen, M., Studies on mobile terminal energy consumption for LTE and future 5G, PhD thesis, Aalborg University, Aalborg, 2015, figure 6.7.)

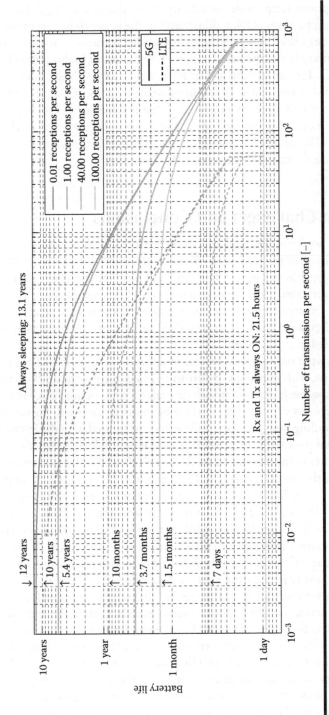

Figure 14.13 Battery life for LTE and 5G as a function of number of receptions and transmissions per second. DRX and DTX are applied.

can apply low-power sleep mode for activity levels that are so high that an LTE transceiver would be always ON, as illustrated in the lower right part of the figure.

This concludes a brief review of the energy-saving features in 5G, which targets an improved battery life as compared with LTE and support for both mobile broadband and MTC. The use of novel power amplifier supply techniques in combination with OFDM is envisioned as one improvement, while TDD will also result in lower RF front-end attenuation and less channel feedback. Finally, the short and optimized frame structure provides benefits not only in terms of the use of microsleep and DRX and DTX, but also for fast synchronization and data transfer, which will reduce the energy-consuming ON time.

14.7 Open Challenges and Conclusions

In this chapter, we have presented our vision for a 5G cmWave concept targeting ultradense deployment of small cells. We have presented our key technologies for achieving multigigabits per second data rates and ultralow latency in such strongly interference-limited scenarios. The design of a novel 5G optimized frame structure, a flexible TDD channel access mode, an interference-aware rank adaptation algorithm, and a discussion on the energy-saving benefits of our envisioned concept have been detailed in this chapter.

The present work only presents an introductory note for a unified 5G cmWave small cell concept, with numerous opportunities for future work. In the domain of flexible TDD, possible future works include evaluating its performance in the presence of multiple users per cell. The presence of multiple users will introduce some interesting scheduling problems, especially when confronted with the interaction of flexible TDD and the specific needs of different classes of devices. Another possible line of study relates to the comparison of flexible TDD and full duplex communication [30]. While the benefits of flexible TDD can only be exploited whenever there is data in either UL or DL, the benefits of full duplex can only be exploited whenever there is data in both UL and DL. It is, therefore, expected that the full duplex gain at low load will be quite limited, while at high load, there are several possibilities to obtain performance benefits that are unobtainable with flexible TDD. While attractive, the usage of full duplex at high load will induce more severe interference conditions, potentially introducing another dimension of challenge.

Future energy-consumption challenges for 5G include support for aperiodic, low-latency DL traffic and the high bandwidth and data rates. The aperiodic, low-latency traffic may occur in certain MTC scenarios when a device is asked by the AP to respond with a measurement within a certain time frame. It is difficult to accommodate this type of traffic with DRX due to the necessity of establishing a trade-off between delay and power consumption. The second challenge is to support the high bandwidth and data rate without imposing a significant power consumption penalty on the device. From a hardware perspective, the RF front-end and the converters are estimated to be able to handle the increased requirements with reasonable power consumption, but the turbo decoding and bandwidth-related tasks such as fast Fourier transform, channel estimation, and equalization may lead to a total power consumption of more than 3 W in 2020 [31]. However, it is estimated that the 5G receiver will be on a par with the LTE receiver of 2014 in 2027, but this may improve if, for example, the forward error correction coding can be made more efficient and the high bandwidth can be supported with lower processing complexity.

References

1. P. Mogensen, K. Pajukoski, B. Raaf, E. Tiirola, E. Lahetkangas, I. Z. Kovacs, G. Berardinelli, L. G. U. Garcia, L. Hu, and A. F. Cattoni, Beyond 4G local area: High level requirements and system design, in *IEEE Globecom International Workshop on Emerging Technologies for LTE-Advanced and Beyond-4G*, Anaheim, CA, IEEE, pp. 613–617, 2012.
2. P. Mogensen, K. Pajukoski, B. Raaf, E. Tiirola, E. Lahetkangas, I. Z. Kovacs, G. Berardinelli, L. G. U. Garcia, L. Hu, and A. F. Cattoni, Centimeter-wave concept for 5G ultra-dense small cells, in *IEEE 79th VTC Spring Workshop on 5G Mobile and Wireless Communication System for 2020 and Beyond (MWC2020)*, Seoul, IEEE, pp. 1–6, 2014.
3. A. Osseiran, F. Boccardi, V. Braun, K. Kusume, P. Marsch, M. Maternia, O. Queseth, et al., Scenarios for 5G mobile and wireless communications: The vision of the METIS project, *Communications Magazine, IEEE* 52(5): 26–35, 2014.
4. J. G. Andrews, S. Buzzi, W. Choi, S. V. Hanly, A. Lozano, A. C. K. Soong, and J. C. Zhang, What will 5G be? *IEEE Journal on Selected Areas in Communications* 32(6): 1065–1082, 2014.
5. E. Dahlman, G. Mildh, S. Parkvall, J. Peisa, J. Sachs, and Y. Selén, 5G radio access. *Ericsson Review*, 2014.
6. E. Hossain, M. Rasti, H. Tabassum, and A. Abdelnasser, Evolution toward 5G multi-tier cellular wireless networks: An interference management perspective, *Wireless Communications, IEEE* 21(3): 118–127, 2014.
7. M. Paolini, Mobile Data Move Indoors, 2011. Available from: http://www.senzafiliconsulting.com/Blog/tabid/64/articleType/ArticleView/articleId/59/Mobile-data-move-indoors.aspx.
8. N. H. Mahmood, G. Berardinelli, K. Pedersen, and P. Mogensen, A distributed interference aware precoding scheme for 5G dense small cell networks, in *2015 IEEE International Conference on Communication (ICC) Workshop: Smallnets*, London, IEEE, pp. 119–124, 2015.
9. F. M. L. Tavares, G. Berardinelli, N. H. Mahmood, T. B. Sørensen, and P. Mogensen, On the potential of interference rejection combining in B4G networks, in *VTC Fall, 2013 IEEE 78th*, Las Vegas, NV, IEEE, pp. 1–5, 2013.
10. A. Ghosh, T. A. Thomas, M. C. Cudak, R. Ratasuk, P. Moorut, F. W. Vook, T. S. Rappaport, G. R. Maccartney, S. Sun, and S. Nie, Millimeter-wave enhanced local area systems: A high-data-rate approach for future wireless networks, *IEEE Journal on Selected Areas in Communications* 32(6): 1152–1163, 2014.

11. E. Ben-Dor, T. S. Rappaport, Y. Qiao, and S. J. Lauffenburger, Millimeter-wave 60 GHz outdoor and vehicle AOA propagation measurements using a broadband channel sounder, in *2011 IEEE Global Communications Conference (Globecom)*, Austin, TX, IEEE, pp. 1–6, 2011.
12. S. G. Larew, T. A. Thomas, M. Cudak, and A. Ghosh, Air interface design and ray tracing study for 5G millimeter wave communications, in *Globecom Workshops: International Workshop on Emerging Techniques for LTE-Advanced and Beyond 4G*, Atlanta, GA, IEEE, pp. 117–122, 2013.
13. G. P. Fettweis, 5G—what will it be: The tactile Internet, in *IEEE International Conference on Communication (ICC)*, Budapest, IEEE, 2013.
14. P. Mogensen, K. Pajukoski, B. Raaf, E. Tiirola, E. Lahetkangas, I. Z. Kovacs, G. Berardinelli, L. G. U. Garcia, L. Hu, and A. F. Cattoni. 5G small cell optimized radio design, in *IEEE Globecom Workshops on Emerging Technologies for LTE-Advanced and Beyond-4G*, Atlanta, GA, IEEE, pp. 111–116, 2013.
15. H. Holma and A. Toskala, *LTE for UMTS: Evolution to LTE-Advanced*, New York: Wiley, 2011.
16. 3GPP, Long Term Evolution, 2009. Available from: http://www.3gpp.org/article/lte.
17. E. Lähetkangas, K. Pajukoski, E. Tiirola, G. Berardinelli, I. Harjula, and J. Vihriälä, On the TDD subframe structure for beyond 4G radio access network, in *Future Network and Mobile Summit (FutureNetworkSummit), 2013*, Lisbon, IEEE, pp. 1–10, 2013.
18. I. Poole, TDD FDD Duplex Schemes. Available from: http://www.radio-electronics.com/info/cellulartelecomms/cellular_concepts/tdd-fdd-time-frequency-division-duplex.php, December 2015.
19. G. Berardinelli, K. Pajukoski, E. Lahetkangas, R. Wichman, O. Tirkkonen, and P. Mogensen. On the potential of OFDM enhancements as 5G waveforms, in *Vehicular Technology Conference (VTC Spring), 2014 IEEE 79th*, Seoul, IEEE, pp. 1–5, 2014.
20. B. Farhang-Boroujeny, OFDM versus filter bank multicarrier, *Signal Processing Magazine, IEEE* 28(3): 92–112, 2011.
21. N. H. Mahmood, G. Berardinelli, F. M. L. Tavares, M. Lauridsen, P. Mogensen, and K. Pajukoski, An efficient rank adaptation algorithm for cellular MIMO systems with IRC receivers, in *IEEE 79th Vehicular Technology Conference (VTC-Spring)*, Seoul, IEEE, pp. 1–5, 2014.
22. IEEE. 802.16-2004 IEEE Standard for Local and Metropolitan Area Networks Part 16: Air Interface for Fixed Broadband Wireless Access Systems, 2004. Available from: http://www.wimaxforum.org/index.htm.
23. Z. Shen, A. Khoryaev, E. Eriksson, and X. Pan, Dynamic uplink-downlink configuration and interference management in TD-LTE, *IEEE Communications Magazine* 50(11): 51–59, 2012.
24. N. H. Mahmood, G. Berardinelli, F. M. L. Tavares, and P. Mogensen, A distributed interference-aware rank adaptation algorithm for local area MIMO systems with MMSE receivers, in *ISWCS*, Barcelona, IEEE, pp. 697–701, 2014.
25. M. Lauridsen, Studies on mobile terminal energy consumption for LTE and future 5G, PhD thesis, Aalborg University, Aalborg, 2015.
26. M. Lauridsen, A. Jensen, and P. Mogensen, Fast control channel decoding for LTE UE power saving, *VTC Spring, 75th*, Yokohama, IEEE, pp. 1–5, 2012.
27. 3GPP, Physical channels and modulation. *TS 36.211 V8.9.0*, 2010.

28. T. Tirronen, A. Larmo, J. Sachs, B. Lindoff, and N. Wiberg, Machine-to-machine communication in long-term evolution with reduced device energy consumption, *Transactions on Emerging Telecommunications Technologies* 24(4): 413–426, 2013.

29. Cadex Electronics Inc., *Bu-802b: Elevating Self-Discharge*, 2014.

30. N. H. Mahmood, G. Berardinelli, F. M. L. Tavares, and P. Mogensen, On the potential of full duplex communication in 5G small cell networks, in *2015 IEEE 81st Vehicular Technology Conference: VTC2015-Spring*, Glasgow, IEEE, pp. 1–5, 2015.

31. M. Lauridsen, P. Mogensen, and T. B. Sørensen. Estimation of a 10 Gb/s 5G receiver's performance and power evolution towards 2030, *IEEE Vehicular Technology Conference Proceedings*, 2015.

CM AND MM WAVE FOR 5G

Chapter 15

Millimeter-Wave Communications for 5G Wireless Networks

Turker Yilmaz and Ozgur B. Akan

Contents

15.1 Introduction

The demand on mobile communications is continuously increasing. The first stage of this trend was cellular telephone subscriptions. The number of subscribers rose from about 1 billion in 2001 [1] to nearly 7 billion at the end of 2014 [2]. However, the expansion rate has reduced as the penetration rate has increased to around 90%. According to the most recent Ericsson Mobility Report, June 2015 edition, worldwide mobile subscription numbers are expected to reach 9.2 billion by 2020, resulting in a compound annual growth rate (CAGR) of only 5% [3].

Currently, the second stage is ongoing, which is mobile broadband subscriptions. As tabulated by the International Telecommunication Union (ITU), in 2007 there were 268 million mobile broadband users, corresponding to a 4% penetration rate. These figures increased to 2.9 billion and 40% as of 2014, respectively, and are expected to escalate with a CAGR of 20% by 2020, attaining 7.7 billion [3]. Changes in the data usage per mobile device make these values even more important. Fourth-generation (4G) links support, and correspondingly expend, more data than preceding generations, and smart devices, which not only include smartphones but also mobile-connected tablets and laptops, generate much more traffic than rudimentary cellular phones. Since all these parameters, that is, mobile broadband subscriptions, 4G connections, and the number of smart devices, are on the rise, the anticipated network capacity increase from beyond 4G (B4G) systems becomes clearer.

Throughput development is just as crucial as capacity. Radio interface requirements for third-generation mobile telecommunication systems, which are called International Mobile Telecommunications (IMT)-2000 by the ITU Radiocommunication Sector (ITU-R) [4], were identified in 1997, with 20 megabits per second (Mb/s) as the highest envisioned data rate for the very high level of data service information type [5]. Furthermore, minimum downlink (DL) peak spectral efficiency for the next generation, IMT-Advanced, was set at 15 b/s/Hz in 2008, leading to a maximum data rate of 1500 Mb/s as 100 MHz-wide bandwidth is promoted to be used by mobile network operators [6]. However, DL spectral efficiency was defined assuming 4×4 multiple-input multiple-output (MIMO) antenna configuration, of which industry is trailing behind even in 2015. Therefore, obtaining the necessary data rate leap expected from a new mobile telecommunication generation solely from developments in the spectral efficiency domain seems unattainable. As a side note, in [6], uplink peak spectral efficiency is specified to be 6.75 b/s/Hz under 2×4 antenna configuration assumption.

Combining all of these, the most realistic approach to arrive at the peak data rates and total network capacity that are expected from the fifth generation (5G) of mobile communication systems emerge as increasing the operation bandwidth. Accordingly, research efforts in the *millimeter-wave (mm-wave) band*, which term classifies the electromagnetic (EM) waves that have a wavelength of between 1 and 10 mm, with a frequency range from 30 to 300 GHz, have been mounting too. The 60 GHz industrial, scientific, and medical (ISM) band has been selected as the initial target, and since December 2008, ECMA-387 [7], the Institute of Electrical and Electronics Engineers (IEEE) 802.15.3c [8] and IEEE 802.11ad [9] standards have been ratified.

Until recently, the terahertz (THz) band, which is defined to cover the frequency spectrum between 0.3 and 10 THz, was labeled as *terahertz gap* due to both the nonexistence of inexpensive signal sources that radiate in the band, and the extensive utilization of neighboring spectrums by means of electronic and optical technologies. However, with the recent emergence of silicon (Si) complementary metal-oxide semiconductor (CMOS) signal sources and front-end circuitries that can radiate in the low end of the THz band with adequate power, interest in the THz band communication has intensified. Standardization activities in the THz band also began in 2008 with the initiation of the IEEE 802.15 wireless personal area network (WPAN) THz Interest Group (IG THz).

In line with the stated arguments, this chapter provides a general overview on the mm-wave from the perspective of 5G implementation. The chapter begins with standardization actions and is followed by discussions on the EM wave propagation properties together with channel characteristics. Device technologies, specific circuitries useful for mm-wave communication systems, and an indoor access network architecture concept are presented next. Finally, conclusions are given before considerations on open issues and future research directions.

15.2 Millimeter-Wave Standardization Activities

While a comprehensive explanation of all the standards designed for the 60 GHz ISM band is available in [10], the decade-long, industry-led process can be reviewed as follows. Video is both the largest generator of mobile data traffic and application with the highest growth rate expectation [11]. Additionally, an ever popular use case for wireless communications is replacing data cables. Therefore, it should come as no surprise that in 2006, the first special interest group that was formed to target 60 GHz band, WirelessHD, focused its specification on video [12]. The second such group, Wireless Gigabit Alliance, was founded in 2009 and unlike WirelessHD, its support grew over time and it pursued a more collaborative approach by contributing its specification to the IEEE 802.11ad.

The first institutional 60 GHz standard was published by Ecma International in 2008, with the second and current edition issued two years later [7]. The highest data rate ECMA-387 supports for WPANs over a single channel with 2.1 GHz bandwidth is 6.35 Gb/s, employing a modulation and coding scheme (MCS) of 16-ary quadrature amplitude modulation (16-QAM) and Reed–Solomon encoding with a convolutional code of rate 1. The only other 60 GHz WPAN standard is IEEE 802.15.3c [8]. Three different physical layer (PHY) modes, that is, single carrier (SC), high-speed interface, and audio/visual, are identified for various needs, and 5.775 Gb/s is possible through one of the orthogonal frequency-division modulation (OFDM) MCSs of 64-QAM and low-density parity check (LDPC) (672, 504). IEEE 802.11ad, on the other hand, is the single 60 GHz wireless local area network (WLAN) standard at this time. Approved 3 years later than IEEE 802.15.3c in 2012, the PHY modes of IEEE 802.11ad are SC and OFDM, in addition to the control modulation. The maximum data rate is again achievable using an OFDM MCS with 64-QAM and LDPC (672, 546), which results in 6756.75 Mb/s [9].

The preliminary standards are set. Still, 60 GHz standardization efforts are far from final. The spectrum allocated to ISM applications around 60 GHz ranges from 57 to 66 GHz in Europe [13] and from 57 to 64 GHz in the United States [14]. However, China has allocated only 5 GHz between 59 and 64 GHz, which allows communication in only two channels of the 2.16 GHz bandwidth, as defined by IEEE 802.11ad. To address this possibility of limited use, the China Millimeter Wave Study Group (SG) was formed in January 2012 under IEEE 802.11 to both adapt IEEE 802.11ad to the Chinese spectrum and further make operational the unlicensed channels between 42.3–47 and 47.2–48.4 GHz, which are available in China [15]. Following the approval of the project authorization request document in August 2012, the SG was transformed into Task Group AJ (TGaj) with the title "Enhancements for Very High Throughput to support Chinese millimeter wave frequency bands." Another IEEE 802.11 task group, TGay, titled "Enhanced throughput for operation in license-exempt bands above 45 GHz" has also been in effect since March 2015. The aim is to improve the rate target of 20 Gb/s.

As a consequence of the listed standardization efforts, 60 GHz ISM band can be seen as the decided mm-wave frequency for 5G adoption. However, the general direction of mm-wave research activities can be outlined as follows:

- Initial WPAN and WLAN standardization activities on the 60 GHz ISM band were successfully concluded in 2012, with commercial devices expected in 2016.
- Urban propagation measurements up to 73 GHz are already being performed [16].
- The main motivation for the 60 GHz utilization is the available wide and continuous bandwidth, which is even more abundant within the unallocated spectrum above 275 GHz [17].
- Transmission characteristics of the 60 GHz band are significantly different from the sub 6 GHz bands, while those are considerably similar to the low-THz band [18].
- WPAN standardization activities began in 2008 for bands above 275 GHz, with the current active groups being IEEE 802.15 WPAN IG THz and TG3d.

Combining all these points, the low-THz band emerges as the other candidate for 5G systems. Frequencies up to 275 GHz are already assigned to numerous services by the ITU [19]. However, the rest of the spectrum is unallocated apart from some passive applications. Extrapolating the conditions that resulted in the 60 GHz utilization efforts arises the low end of the THz band as another solution for the required throughput and network capacity improvements. Accordingly, in 2014, the standardization of wireless switched point-to-point applications branched out under IEEE 802.15 TG3d. IG THz is also active for further THz band investigations.

15.3 Millimeter-Wave Channel Properties

Mm-wave is a very broad spectrum, much larger than the current frequency bands that electronic technology is successfully providing for. Predictably, EM wave propagation characteristics in mm-wave differ from the sub 6 GHz bands and also vary within. The first such change is increased attenuation by atmospheric gases. Gaseous attenuation is essentially nonexistent in conventional bands. However, it increases with frequency. Specific attenuation due to atmospheric gases up to 1 THz is examined in detail in [18], and in Figure 15.1 attenuation for the frequencies between 30 GHz and the first local maximum in the terahertz band, 325.178 GHz, is presented. The recommendation ITU-R P.676-9 [20] is used with the standard ground-level atmospheric conditions [21].

The local maxima of the atmospheric attenuation plot in Figure 15.1 are due to resonance lines of either oxygen or water vapor molecules. As can also be observed from the dry air and water vapor lines, the first two attenuation peaks at 60.83 and 118.77 GHz are caused by oxygen and those at 183.37 and 325.18 GHz originate from water vapor. The figure also includes attenuation due to rain for three different rain rates. The calculations are made for vertical polarization according to the recommendation ITU-R P.838-3 [22]. Rain attenuation also rises with the rain rate. The

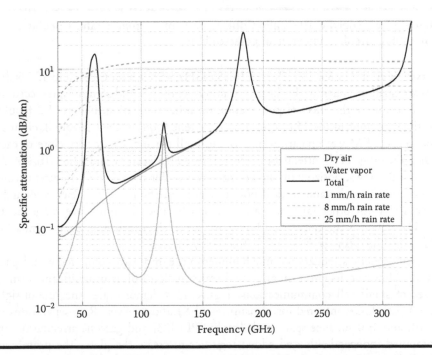

Figure 15.1 Specific attenuation due to atmospheric gases and rain, calculated between 30 and 325.178 GHz at 1 MHz intervals under standard ground-level atmospheric conditions and for rain rates of 1, 8, and 25 mm/h.

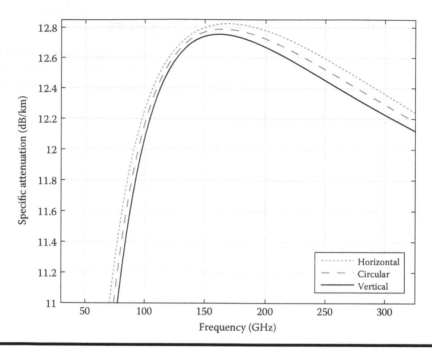

Figure 15.2 Specific attenuation due to rain of 25 mm/h rate, calculated for horizontal, circular, and vertical polarizations.

curves peak at 224.85, 186.32, and 161.85 GHz with 1.648, 6.154, and 12.75 dB/km attenuation values for the illustrated rain rates of 1, 8, and 25 mm/h, respectively.

To examine the effect of polarization on rain attenuation, Figure 15.2 focuses on the variations in the 25 mm/h rain rate attenuation for the horizontal, circular, and vertical EM wave polarizations. Because all three plots are very close to each other, the range of the attenuation axis is limited between 11 and 12.85 dB/km. As the figure shows, horizontally polarized waves suffer most from rain attenuation, followed by circular and vertical polarizations. Furthermore, this fact is virtually always true for other rain rates and extends beyond the mm-wave range up to 1 THz, which is the calculation boundary set in the overseeing recommendation [22]. The lines also get closer as the frequency increases.

Atmospheric attenuation is much higher in the mm-wave compared with legacy bands. However, its effect on an actual transmission link is insignificant in most cases of small cell communications. Figure 15.3 demonstrates the line-of-sight (LoS) path losses inflicted for transmitter (TX) and receiver (RX) separations of 1, 10, and 100 m. Free-space path loss (FSPL) [23] and gaseous attenuation are computed independently and added together to create the plots. The prominent parabolic shapes visible in the curves are due to the FSPL, and the irregularities, which are only apparent in the 100 m plot, are caused by atmospheric attenuation. Gaseous attenuation never exceeds 30 dB/km in the mm-wave range, and is much

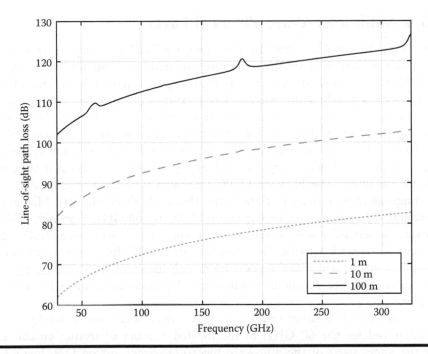

Figure 15.3   LoS path loss, composed of FSPL and gaseous attenuation calculated between 30 and 325.178 GHz and for transmission distances of 1, 10, and 100 m.

lower for most of the band. Therefore, compared with the inherently present FSPL, it can be concluded that atmospheric attenuation does not create an additional problem for mm-wave utilization in access networks.

Non-line-of-sight (NLoS) transmission is governed by three fundamental propagation mechanisms, namely, reflection, diffraction, and scattering. Since respective theories are widely covered in the literature, only their impacts on mm-wave communication are discussed henceforth. In [24], changes in the absorption coefficients and refractive indices of a number of materials from 100 GHz to 1 THz are described. In summary, refractive indices stay essentially constant, whereas absorption coefficients increase at varying rates with frequency. The latter causes higher material absorption and less transmitted power for mm-wave systems, which would impair coverage of access points (APs) beyond the rooms that they are located in. Furthermore, as the wavelengths are between 1 and 10 mm, surfaces can no longer be assumed smooth and reflection coefficients should be multiplied by the Rayleigh roughness factor for correct quantities [25]. Because this factor is smaller than 1 for mm-wave frequencies, reflected wave power densities are further weakened in the mm-wave, which is also the case for scattering [26]. Diffraction, however, is shown to be absent in 60 and 300 GHz except for a very specific situation [27]. Therefore, all NLoS propagation mechanisms decline in the mm-wave band, an outcome that has to be averted through PHY techniques, and a related coverage area study is available in [28].

15.4 Millimeter-Wave Physical Layer

The ample bandwidth offered by the mm-wave band can support very high data rates, which increases the importance of source coding and compression techniques, especially for applications requiring high data volume too. Many compression algorithms that are developed for the conventional bands are also suitable for the mm-wave channels for high spectral efficiency [29]. Moreover, the mm-wave band suffers from higher propagation losses and is very susceptible to changes in the communication channel. In line with these, a suitable mm-wave source code should be capable of efficiently adjusting to changes in the data rate, and adaptive codes based on LDPC are shown to be effective [30], resulting in compression performances close to the Slepian–Wolf bound [31]. Moreover, convolutional code is also compared with LDPC for 60 GHz. However, LDPC provided a better frame error rate than the employed convolutional code [32]. In general, research activities for a source code suitable for mm-wave communication systems, which can support both high data rates and traffic, and has high error correction capability in addition to low operation complexity are still ongoing.

Modulation is another important PHY topic. While up to 64-QAM is standardized for the 60 GHz, in the limited number of studies conducted for the upper parts of the mm-wave spectrum, either analog [33], low efficiency digital modulation methods such as amplitude shift keying or 16-QAM are mostly used [34–36]. Transmission is much faster in the mm-wave band. However, due to higher losses, the bit error ratio becomes even more important for successful reception. Therefore, low constellation digital modulation techniques, whose effect of low spectral efficiency can also be offset using a larger bandwidth, are more suitable for initial low-THz band applications, as demonstrated in [37].

Mm-wave channels can alter very rapidly. Communication links rely more heavily on the LoS component compared with sub 6 GHz bands and therefore, instantaneous obstructions such as humans walking by or doors opening also affect the transmission to the degree of complete blockage. For this reason, it is important to employ adaptive MCSs for stable connections. There are several such studies available in the literature. However, MCS should be considered together with beam forming and steering in the mm-wave band, an area that is yet to be investigated in depth. The effect of channel prediction on a beam forming and adaptive modulation system is analyzed in [38]. A method to accomplish adaptive modulation with beam forming is presented in [39]; however, assumptions such as single antenna and perfect channel state information at the RX are made, which are unrealistic for mm-wave systems. Overall, to the best of the authors' knowledge, there are no publications on adaptive MCSs, which are formed specifically to address the properties of mm-wave channels, and this also is the case for rate control among other PHY subjects.

15.5 Millimeter-Wave Devices

Si CMOS circuitries are the most suitable type for widespread adoption due to their lower cost, though the foremost needs of an mm-wave communication system are signal sources, which are capable of high output power and frequency tuning. The sole performance of state-of-the-art voltage-controlled oscillators (VCOs) fall below sufficient levels at around 100 GHz and onward. Adding frequency multipliers to the VCO blocks is a possible solution; however, they are not ideal since they intensify the phase noise. Consequently, the two primarily researched spectrum within the mm-wave band, namely, the 60 GHz ISM and low-THz bands, encounter different availability of devices and technologies. Hardware operating in the 60 GHz band is expected to be on sale to the general public as soon as 2016. Accordingly, integrated Si transceiver (TRX) chipsets, such as those from Hittite Microwave [40,41], have been on the market for a considerable amount of time. Also, many surveys and books have been published covering the 60 GHz devices in great depth [42–45]. For all these reasons, in the remainder of the section, the small number of Si CMOS circuitries developed for functioning in the low-THz band are covered.

A recent Si CMOS source implementation is reported to generate a peak output power of −1.5 dBm at 288 GHz [46]. More impressively, when the source is packaged with an Si lens on an FR-4 board, the radiated power still remains at −4.1 dBm. The circuit is designed in a 65-nm CMOS process with two 3-push NFET oscillators that are differentially locked and thus combine third-harmonic signals at the common output. With a direct current (dc) power consumption of 275 mW, dc to radio-frequency conversion efficiencies are 0.26% and 0.14% and the total die areas are 120×150 and 500×570 μm^2, for the oscillators and package as a whole, respectively.

An example of a complete Si front-end for low-THz band is tunable from 276 to 285 GHz [47]. All signal generation, frequency multiplication, filtering and radiation are performed by distributed active radiators (DARs), which are unique to the study and designed by means of the inverse design approach, where active and passive elements are designed after the metal surface currents are formulated. The DARs are combined in radiating arrays of 4×4 and realized in 45-nm CMOS Si on an insulator process. The device can steer beams up to slightly less than $\pi/2$ radians in both azimuth and elevation, and outputs a maximum effective isotropic radiated power (EIRP) of 9.4 dBm by using 16 dBi directivity. The chip area is 2.7×2.7 mm^2. The total dissipated power without directivity is 190 μW and the dc power consumption is 820 mW, thus the dc-to-RF conversion efficiencies for the TX are found to be 0.023% and 1.06%, excluding and including directivity, respectively.

The two main requirements from the mm-wave antenna systems are beamforming capability and high radiation efficiency, which are possible using either microstrip antennas or rectangular waveguides. While both have lower losses in general, they are expensive and hard to manufacture and occupy a larger die area. One popular alternative structure is substrate integrated waveguide (SIW). SIW is a transmission line technology [48], which primarily benefits from low leakage loss and cost, and compatibility to planar circuits [49]. High-performance SIW implementations of most types of mm-wave active and passive electronic circuits, including filters and oscillators [50,51], on top of beam-forming antennas and complete receivers [52,53] are already available in the literature. To summarize, like other circuit elements, mm-wave antennas are also a growing research field with multiple options competing for implementation.

15.6 Millimeter-Wave Indoor Access Network Architecture

Utilizing the mm-wave band for 5G communication will require a new access network concept together with pioneering architectural design to overcome the intrinsic additional losses and fulfill all the technology, user, and capacity requirements expected from the new generation of mobile communication systems. Developing the main ideas of the network architecture that is to be adopted at least a decade from now initially requires general forethought of the state of the world and the routines of different groups of people in 10 years time. One major transformation that is likely to happen and have a great impact on basic infrastructure systems is the completion of the change already started in the public and government views on the right to Internet access. While many governments have already committed to providing universal high-speed Internet service in line with their plans of creating information societies, in 2010, the Finnish government made broadband Internet access a legal right for their citizens, making Finland the first country to do so.

The importance of the legislation comes from its direct consequence that, in Finland, the communications infrastructure will now be considered as fundamental as water and energy infrastructures and new construction projects will be built with the communication grid installed, capable of supporting throughput and data rate needs of 5G systems. Mounting wireless APs to each room of a building that are combined over a wavelength-division multiplexing passive optical network system to form a fiber to the premises might seem to be a type of last-mile network architecture that is currently both unnecessary and financially unfeasible. However, we believe that people will demand this kind of indoor access within a decade, and, therefore, we planned our access network architecture accordingly.

For example, the dimensions of the authors' laboratory are $11.65 \times 12.12 \times 3.26$ m, seating 20 researchers. While typically maximum permitted base station EIRPs are in excess of 60 dBm per carrier, since our indoor access network proposal resembles WLANs more than the cellular networks due to envisioning an AP for every room, directives for WLANs are considered. In fact, the European Telecommunications Standards Institute regulations limit the maximum output power to 20 dBm for the 2.4 GHz ISM band and recalling the currently highest reported low-THz band signal source peak output power of −1.5 dBm, the antennas of the future mm-wave band communication systems will need to offer gains around 20 dBi.

While devising antennas with such gains at the necessary frequency range is itself a problem, let alone the additional favored properties such as beam-steering, high-gain antennas also inherently suffer from very narrow beamwidths. Increasing the gain, and so the directivity, of an antenna requires energy to be focused into smaller angles, which consequently results in smaller half power beamwidths (HPBWs). Measurements on a standard gain horn antenna, which is considered appropriate for THz communications and available for purchase, are systematically conducted between 275 and 325 GHz and the maximum gain result is 18.6 dBi with HPBWs being 16.5° and 17.1° in azimuth and elevation planes, respectively [54]. To put these figures into context, if this antenna were to be used at a TX mounted right at the center of the ceiling of the authors' labora-tory, and the HPBWs were assumed to be 16.8° for both planes, the coverage area on a desk that is perpendicularly below the antenna and lies 74 cm above the floor, as in the actual case, would be a 0.44 m² circle with a radius of 37.21 cm. The larg-est coverage area by one such antenna would also occur on the floor at the corners of the room, having an area of 6.48 m² and a shape of two semi-ellipses attached on their shared minor axis.

The coverage of a single antenna with adequate gain is therefore able to serve the personal area of only one user. Installing one antenna per each inhabitant of a room is unreasonable. However, if antenna numbers are determined optimally as some small percentage of the average number of users and mm-wave access is complemented with device-to-device (D2D) links that are formed through a cellu-lar controlled short-range communication system, the connectivity needs of all user equipment can be met. Moreover, if these D2D links are set in the frequency bands that are allocated for 4G telecommunication systems, then many advantages of the already established 4G mobile communication can strategically be exploited by setting these temporary links even at will, such as reducing the number of hando-vers. An exemplary indoor access network architecture developed over the concepts described in this section is provided in Figure 15.4.

15.7 Conclusions

In this chapter, an introductory overview of mm-wave communications from the 5G utilization perspective is provided. The legacy bands are inadequate to support the expected data traffic and rates of the B4G systems. Recognizing this, initial WPAN and WLAN standards are authorized for the 60 GHz ISM band, and efforts are continuing for the low-THz band. Increased frequency affects

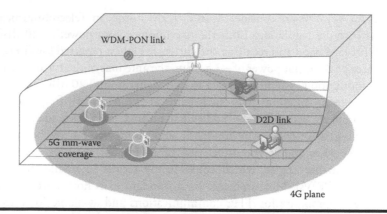

Figure 15.4 An illustrative indoor access network architecture suitable for 5G wireless communication systems in the mm-wave band.

mm-wave links mostly in FSPL; however, the reduced effects of NLoS propagation mechanisms also indicate waning multipath propagation. Research on PHY techniques for mm-wave channels are yet to intensify. Commonplace low-cost devices for the 60 GHz ISM band are about to become available, although this is not the case for mm-wave transmission windows above 100 GHz. To summarize, mm-wave is the next frontier to be employed for widespread wireless communications, and with the drive from 5G, open problems are tackled at an unmatched pace.

15.8 Future Research Directions

While exponentially increasing data traffic can be counterbalanced with the ample spectrum available in the THz band, channel measurement campaigns for various types of sites and overall modeling efforts are primarily needed, since a comprehensive low-THz band indoor channel model is not available in the literature. As nearly all types of path losses increase with frequency, THz band communication links are highly sensitive to changes in the LoS paths. One of the main sources of such changes in indoor environments is human movement, and therefore human blockage models for the THz band are also required. Given that propagation modeling is yet to be completed to satisfaction, THz band communication techniques within physical, data link, and network layers are all open for research. The high number of TRXs, which is a consequence of small cell networks, needs self-organized networking together with distributed backhaul links. Shorter wavelengths enable smaller antenna dimensions and spacings, thus MIMO and beam-forming methods can be efficiently utilized. Closely located TRXs also support network-controlled traffic offloading. All types of construction require novel femtocell in building access and heterogeneous backhaul network architectures. There also is a need for inexpensive devices, which would allow real-world network deployments. Therefore, the low-THz band is a rapidly advancing and promising field that contains many open issues for high-impact research outcomes.

Acknowledgment

This work was supported in part by the Scientific and Technological Research Council of Turkey (TUBITAK) under grant #113E962.

References

1. World Telecommunication Development Report 2002: Reinventing Telecoms. Report, International Telecommunication Union, Geneva, 2002.
2. The World in 2014: ICT Facts and Figures. Report, International Telecommunication Union, 2014.
3. Ericsson Mobility Report. Report, Ericsson, 2015.
4. Resolution ITU-R 56-1: Naming for International Mobile Telecommunications. ITU-R Resolutions, ITU, Geneva, 2012.
5. Report ITU-R M.1034-1: Requirements for the radio interface(s) for international mobile telecommunications-2000 (IMT-2000). ITU-R Recommendations and Reports, ITU, Geneva, 1997.
6. Report ITU-R M.2134: Requirements related to technical performance for IMT-Advanced radio interface(s). ITU-R Recommendations and Reports, ITU, Geneva, 2008.
7. Standard ECMA-387: High rate 60 GHz PHY, MAC and HDMI PALs, pp. 1–302, 2010.
8. IEEE Standard for Information technology—Telecommunications and information exchange between systems—Local and metropolitan area networks—Specific requirements. Part 15.3: Wireless medium access control (MAC) and physical layer (PHY) specifications for high rate wireless personal area networks (WPANs) Amendment 2: Millimeter-wave-based Alternative Physical Layer Extension. IEEE Std 802.15.3c-2009 (Amendment to IEEE Std 802.15.3–2003), pp. c1–187, 2009.
9. IEEE Standard for Information technology–Telecommunications and information exchange between systems–Local and metropolitan area networks–Specific requirements–Part 11: Wireless LAN Medium Access Control (MAC) and Physical Layer (PHY) Specifications Amendment 3: Enhancements for very high throughput in the 60 GHz Band. IEEE Std 802.11ad-2012 (Amendment to IEEE Std 802.11–2012, as amended by IEEE Std 802.11ae-2012 and IEEE Std 802.11aa-2012), pp. 1–628, 2012.
10. T. Yilmaz, G. Gokkoca, and O. B. Akan. *Millimetre Wave Communication for 5G IoT Applications*. New York: Springer, 2016.
11. Cisco visual networking index: Global mobile data traffic forecast update, 2014–2019. Report, Cisco Systems, 2015.

12. WirelessHD Consortium. WirelessHD Specification Version 1.1, 2010.

13. Commission implementing decision of 8 December 2011 amending Decision 2006/771/EC on harmonisation of the radio spectrum for use by short-range devices, 2011.

14. In the Matter of Revision of Part 15 of the Commission's Rules Regarding Operation in the 57–64 GHz Band, 2013.

15. H. Wang, W. Hong, J. Chen, B. Sun, and X. Peng. IEEE 802.11aj (45 GHz): A new very high throughput millimeter-wave WLAN system. *Communications, China* 11(6): 51–62, 2014.

16. G. R. MacCartney and T. S. Rappaport. 73 GHz millimeter wave propagation measurements for outdoor urban mobile and backhaul communications in New York City. In *Communications (ICC), 2014 IEEE International Conference on*, Sydney, NSW, IEEE, pp. 4862–4867, 2014.

17. T. Yilmaz, E. Fadel, and O. B. Akan. Employing 60 GHz ISM band for 5G wireless communications. In *Communications and Networking (BlackSeaCom), 2014 IEEE International Black Sea Conference on*, Odessa, IEEE, pp. 77–82, 2014.

18. T. Yilmaz and O. B. Akan. On the use of low terahertz band for 5G indoor mobile networks. *Computers & Electrical Engineering*, 2015.

19. The Radio Regulations, Edition of 2012, 2012.

20. Recommendation ITU-R P.676-9: Attenuation by atmospheric gases. ITU-R Recommendations, P Series Fasicle, ITU, Geneva, 2012.

21. Recommendation ITU-R P.835-5: Reference standard atmospheres. ITU-R Recommendations, P Series Fasicle, ITU, Geneva, 2012.

22. Recommendation ITU-R P.838-3: Specific attenuation model for rain for use in prediction methods. ITU-R Recommendations, P Series Fasicle, ITU, Geneva, 2005.

23. H. T. Friis. A note on a simple transmission formula. *Proceedings of the IRE*, 34(5): 254–256, 1946.

24. R. Piesiewicz, C. Jansen, S. Wietzke, D. Mittleman, M. Koch, and T. Kurner. Properties of building and plastic materials in the THz range. *International Journal of Infrared and Millimeter Waves* 28(5): 363–371, 2007.

25. P. Beckmann and A. Spizzichino. *The Scattering of Electromagnetic Waves from Rough Surfaces*. London: Artech House, 1986.

26. C. Jansen, S. Priebe, C. Moller, M. Jacob, H. Dierke, M. Koch, and T. Kurner. Diffuse scattering from rough surfaces in THz communication channels. *Terahertz Science and Technology, IEEE Transactions on* 1(2): 462–472, 2011.

27. M. Jacob, S. Priebe, R. Dickhoff, T. Kleine-Ostmann, T. Schrader, and T. Kurner. Diffraction in mm and sub-mm wave indoor propagation channels. *Microwave Theory and Techniques, IEEE Transactions on* 60(3): 833–844, 2012.

28. T. Yilmaz and O. B. Akan. Utilizing terahertz band for local and personal area wireless communication systems. In *Computer Aided Modeling and Design of Communication Links and Networks (CAMAD), 2014 IEEE 19th International Workshop on*, Athens, IEEE, pp. 330–334, 2014.

29. W. Mohr and W. Konhauser. Access network evolution beyond third generation mobile communications. *Communications Magazine, IEEE* 38(12): 122–133, 2000.

30. D. Varodayan, A. Aaron, and B. Girod. Rate-adaptive codes for distributed source coding. *Signal Processing* 86(11): 3123–3130, 2006.

31. D. Slepian and J. K. Wolf. Noiseless coding of correlated information sources. *Information Theory, IEEE Transactions on* 19(4): 471–480, 1973.

32. M. Marinkovic, M. Piz, Choi Chang-Soon, G. Panic, M. Ehrig, and E. Grass. Performance evaluation of channel coding for Gbps 60-GHz OFDM-based wireless communications. In *Personal Indoor and Mobile Radio Communications (PIMRC), 2010 IEEE 21st International Symposium on*, Instanbul, IEEE, pp. 994–998, 2010.

33. C. Jastrow, K. Munter, R. Piesiewicz, T. Kurner, M. Koch, and T. Kleine-Ostmann. 300 GHz transmission system. *Electronics Letters* 44(3): 213, 2008.

34. W. Cheng, L. Changxing, C. Qi, L. Bin, D. Xianjin, and Z. Jian. A 10-Gbit/s wireless communication link using 16-QAM modulation in 140-GHz band. *Microwave Theory and Techniques, IEEE Transactions on* 61(7): 2737–2746, 2013.

35. N. Kukutsu, A. Hirata, T. Kosugi, H. Takahashi, T. Nagatsuma, Y. Kado, H. Nishikawa, A. Irino, T. Nakayama, and N. Sudo. 10-Gbit/s wireless transmission systems using 120-GHz-band photodiode and MMIC technologies. In *Compound Semiconductor Integrated Circuit Symposium, 2009. CISC 2009. Annual IEEE*, Greensboro, NC, IEEE, pp. 1–4, 2009.

36. E. Laskin, P. Chevalier, B. Sautreuil, and S. P. Voinigescu. A 140-GHz double-sideband transceiver with amplitude and frequency modulation operating over a few meters. In *Bipolar/BiCMOS Circuits and Technology Meeting, 2009. BCTM 2009. IEEE*, Capri, IEEE, pp. 178–181, 2009.

37. T. Yilmaz and O. B. Akan. On the use of the millimeter wave and low terahertz bands for internet of things. In *Internet of Things (WF-IoT), 2015 IEEE 2nd World Forum on*, Milan, IEEE, 2015.

38. Z. Shengli and G. B. Giannakis. How accurate channel prediction needs to be for transmit-beamforming with adaptive modulation over Rayleigh MIMO channels? *Wireless Communications, IEEE Transactions on* 3(4): 1285–1294, 2004.

39. X. Pengfei, Z. Shengli, and G. B. Giannakis. Multiantenna adaptive modulation with beamforming based on bandwidth-constrained feedback. *Communications, IEEE Transactions on* 53(3): 526–536, 2005.

40. Hittite Microwave Corporation. HMC6000LP711E–60 GHz Tx with Integrated Antenna, 2013.

41. Hittite Microwave Corporation. HMC6001LP711E–60 GHz Rx with Integrated Antenna, 2013.

42. A. M. Niknejad. Siliconization of 60 GHz. *Microwave Magazine, IEEE* 11(1): 78–85, 2010.

43. A. M. Niknejad and H. Hashemi. *mm-Wave Silicon Technology: 60 GHz and Beyond.* New York: Springer, 2008.

44. T. S. Rappaport, J. N. Murdock, and F. Gutierrez. State of the art in 60-GHz integrated circuits and systems for wireless communications. *Proceedings of the IEEE* 99(8): 1390–1436, 2011.

45. S.-K. Yong, P. Xia, and A. Valdes-Garcia. *60 GHz Technology for Gbps WLAN and WPAN: From Theory to Practice.* New York: Wiley, 2011.

46. J. Grzyb, Z. Yan, and U. R. Pfeiffer. A 288-GHz lens-integrated balanced triple-push source in a 65-nm CMOS technology. *Solid-State Circuits, IEEE Journal of* 48(7): 1751–1761, 2013.

47. K. Sengupta and A. Hajimiri. A 0.28 THz power-generation and beam-steering array in CMOS based on distributed active radiators. *Solid-State Circuits, IEEE Journal of* 47(12): 3013–3031, 2012.

48. J. Hirokawa and M. Ando. Single-layer feed waveguide consisting of posts for plane TEM wave excitation in parallel plates. *Antennas and Propagation, IEEE Transactions on* 46(5): 625–630, 1998.

49. M. Bozzi, A. Georgiadis, and K. Wu. Review of substrate-integrated waveguide circuits and antennas. *Microwaves, Antennas & Propagation, IET* 5(8): 909–920, 2011.

50. Z. Cao, X. Tang, and K. Qian. Ka-band substrate integrated waveguide voltage-controlled Gunn oscillator. *Microwave and Optical Technology Letters* 52(6):1232–1235, 2010.

51. L. Gwang-Hoon, Y. Chan-Sei, Y. Jong-Gwan, and K. Jun-Chul. SIW (substrate integrated waveguide) quasi-elliptic filter based on LTCC for 60-GHz application. In *Microwave Integrated Circuits Conference, 2009. EuMIC 2009. European*, Rome, IEEE, pp. 204–207, 2009.

52. H. F. Fan, W. Ke, H. Wei, H. Liang, and C. Xiao-Ping. Low-cost 60-GHz smart antenna receiver subsystem based on substrate integrated waveguide technology. *Microwave Theory and Techniques, IEEE Transactions on* 60(4): 1156–1165, 2012.

53. A. B. Guntupalli and W. Ke. Multi-dimensional scanning multi-beam array antenna fed by integrated waveguide Butler matrix. In *Microwave Symposium Digest (MTT), 2012 IEEE MTT-S International*, Montreal, QC, IEEE, pp. 1–3, 2012.

54. S. Priebe, M. Jacob, and T. Kurner. The impact of antenna directivities on THz indoor channel characteristics. In *Antennas and Propagation (EUCAP), 2012 6th European Conference on*, Prague, IEEE, pp. 478–482, 2012.

Chapter 16

Network Architecture, Model, and Performance Based on Millimeter-Wave Communications

Yi Wang and Zhenyu Shi

Contents

16.1 Introduction

Broadening the fifth generation (5G) to millimeter-wave (mmWave) bands is an emerging hot topic that is widely discussed in industry and academia [1–4]. The widely accepted range for mmWave bands is 6–100 GHz, in comparison with the spectrum bands below 6 GHz for international mobile telephony (IMT) systems; wavelengths of 6–30 GHz are in the centimeter range. The mmWave band has the advantage of an ultrawide band available for transmission. A survey [4] shows that a total spectrum of 45 GHz is available between 6 and 100 GHz, which is tens of times the available bands below 6 GHz. Such a huge spectrum band makes it easy to achieve a data rate of tens of gigabits per second for transmission and a 1000 times throughput improvement over Long-Term Evolution (LTE) systems [5,6].

However, mmWave suffers from greater propagation fading than lower frequency bands, particularly in non-line-of-sight (NLOS) and moving scenarios. Another challenge is that the transmit power decreases as the carrier frequency increases due to the limits of front-end components such as power amplifiers [2]. Fortunately, pioneering research shows that there are lists of key technologies that make mmWave communications feasible for some scenarios [2,3,5]. Antenna array-based beam forming and tracking could partly compensate path loss; channel measurements show that 6–100 GHz can cover a range of small cells [2,7,8]; and ultradense network and self-backhauling could improve the network capacity while keeping the cost reasonable [4–6]. Recently, several companies have announced their prototype verification for mmWave communications [9–13]. Samsung has realized a peak data rate of 7.5 Gbps at 28 GHz frequency [9]; DoCoMo has realized a peak data rate of 10 Gbps at 11 GHz frequency [10]; and Huawei and Nokia have demonstrated a peak data rate of 115 Gbps [11] and 10 Gbps [12] at 72 GHz bands, respectively. This research has attracted further broad focus on 5G mmWave communications.

In parallel, industrial standards are being widely discussed to pave the way for mmWave communications. The World Radiocommunication Conference 2015 (WRC-15) has decided the spectrum for 5G in the sub-6 GHz range, and a common view is that WRC-19 will decide the spectrums above 6 GHz for 5G. The International Telecommunication Union Radiocommunication Section (ITU-R) Working Party 5D (WP5D) was initiated in 2012, aiming to standardize the IMT system that will be commercialized in the year 2020 (5G). Currently, it has finished the standard timeline plan in ITU and will output a vision paper on 5G in 2015 [14]. In March 2015, the 3rd Generation Partnership Project (3GPP) agreed to initiate 5G standards at the end of 2015 and to create a study item working on the channel model of mmWave bands [15]. A widely accepted view is that 3GPP Releases 14, 15, and 16 will be the period for standardizing 5G systems. It is expected that 5G mmWave communications will be standardized in R15 and R16, which is later than 5G sub-6 GHz standards. Regional discussion on mmWave communications include IMT2020 of China, where Huawei is chairing an mmWave communications topic; European projects such as Mobile and Wireless Communications Enablers for Twenty-Twenty Information Society (METIS) and the 5G Infrastructure Public Private Partnership (5GPPP); Association of Radio Industries and Businesses (ARIB) activities in Japan, and so on. Most of these are expected to output their research results to 3GPP and ITU.

In this chapter, the authors attempt to give an overview of research on 5G mmWave communications, including channel modeling, beam tracking, and network architecture, and investigate the key technology solutions up to the current stage. Section 16.2 discusses 5G candidate frequency bands. Section 16.3 points out the necessity of using beam forming in mmWave and provides some beam tracking techniques. Section 16.4 proposes a new channel model for beam tracking. Section 16.5 elaborates on a new network structure, and Section 16.6 illustrates the system-level performance of 72 GHz mmWave cellular networks.

16.2 Spectrum

In WRC-12, one resolution was agreed to study additional spectrum requirements and potential candidate frequency bands for IMT systems [16]. In accordance with this resolution, one agenda item in WRC-15 will consider additional spectrum allocations to the mobile service on a primary basis and the identification of additional frequency bands for future IMT or so-called 5G [17]. According to the work of Joint Task Group (JTG) 4-5-6-7, WRC-15 will mainly focus on frequency bands below 6 GHz. The following frequency ranges have been indicated as suitable for the possible future deployment of IMT: 410–430, 470–790, 1000–1700, 2025–2110, 2200–2290, 2700–5000, 5350–5470, and 5850–6425 MHz [18], as illustrated in Figure 16.1. The strategies and use for these frequency ranges may be different. For example, 470–790 MHz is suitable for providing coverage both indoors and outdoors due to its good propagation characteristics. Part of the L-band, with specific reference to the 1427–1525 and 1525–1660 MHz ranges, may provide good

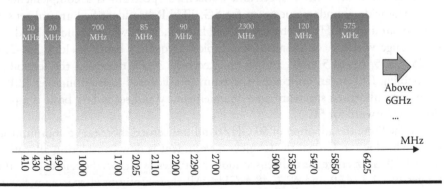

Figure 16.1   Candidate frequency bands for 5G below 6 GHz.

coverage and complement the range below 1 GHz for providing capacity. In the C-band, 3400–3800 MHz may be suitable for providing capacity to fulfill increasing traffic requirements, especially for coverage of small areas with denser network deployment.

Recently, higher frequency bands (above 6 GHz) have been identified as a good candidate spectrum for 5G. A dominant feature is that there is abundant spectrum available to support ultrahigh data rate transmission. This range of frequency bands, 3–300 GHz, is usually referred to as the mmWave band. However, not all mmWave bands can be considered for mobile communications. There are three factors that affect spectrum selection. Firstly, candidate spectrum selection crucially depends on allocations of spectrum administrations and regulators, and the primary/coprimary services for the allocations. Taking the 28 GHz band as an example, with specific reference to 24.25–29.5 GHz, it is a global allocation for mobile service on a coprimary basis except for 24.25–25.25 GHz, which is allocated in Region 3 only [16]. The situation is similar for the E-band: 71–76 and 81–86 GHz. There is a good chance of establishing a global harmonized spectrum allocation for these bands. Secondly, it is preferable to have a contiguous spectrum of several hundred megahertz, or even up to a few gigahertz. Such a continuous spectrum can provide more flexibility for administrations and regulations to manage spectrum allocation strategy. Moreover, it can provide more flexibility for the mobile network operator (MNO) to use its spectrum resource. Thirdly, the propagation characteristics of the candidate spectrum should be friendly to carrying mobile communication services. Since the propagation characteristics of mmWave bands can be quite different compared with frequency bands below 6 GHz, the candidate spectrum selection should consider multiple channel propagation issues, such as severe path loss, the influence of weather conditions and atmosphere, Doppler shift with even slow movement due to higher carrier frequency, NLOS channels. Preliminary studies have shown the possibility of providing NLOS coverage for cellular communications with mmWave bands [2]. Despite this significant progress, a complete characterization of the mmWave link for next-generation 5G mobile broadband remains elusive.

In addition to licensed spectrum, unlicensed spectrum is a complementary way to provide abundant spectrum for 5G. ITU-R has identified an unlicensed band of around 1 GHz (in the sub-100 GHz frequency), which could be used for short-range wireless communications. Typical frequency bands include 2.4–2.5, 5.725–5.875, 61–61.5 GHz, and so on. Recently, 3GPP discussed the feasibility of enabling LTE in an unlicensed spectrum in a licensed-assisted manner [2,19]. Using an unlicensed spectrum on a complementary and secondary basis is a possible way forward for 5G.

To facilitate the adoption of 60 GHz band unlicensed short-range communication technology in China, the Chinese government is now considering allocating the 40–50 GHz range for unlicensed mobile services, with specific reference to the ranges of 42.3–47.0 and 47.2–48.4 GHz. All these unlicensed spectrum bands could be potential complements to alleviate the spectrum shortage for 5G.

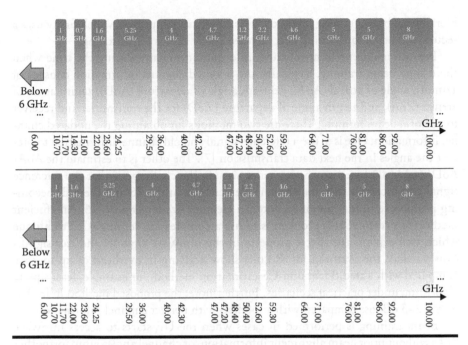

Figure 16.2   Candidate frequency bands for 5G research above 6 GHz.

WRC-15 in [20] has taken into account the frequency bands above 6 GHz for IMT-2020 systems. The following bands are allocated to the mobile service on a primary basis: 24.25–27.5, 37–40.5, 42.5–43.5, 45.5–47, 47.2–50.2, 50.4–52.6, 66–76, and 81–86 GHz. The following bands may require additional allocations to the mobile service on a primary basis: 31.8–33.4, 40.5–42.5, and 47–47.2 GHz. Further resolution and decision of the candidate bands will be reviewed in the WRC-19 conference.

16.3 Beam Tracking

The air interface of mmWave communications features antenna array–based beam forming and tracking. Both mmWave base stations (mBSs) and user equipment (UE) use an antenna array to compensate the large path loss of mmWave propagation. Such a scheme with high-gain narrow beams brings challenges for algorithm design to align the narrow beams between mBSs and multiple UE. Overhead cost, complexity, and tracking ability are the key criteria to evaluate performance. A widely accepted solution for beam alignment consists of two beam phases: beam training and beam tracking. Beam training performs a rough beam alignment whereby both quasi-omnidirectional beams and wide beams can be used for training. Since exhaustive beam search might involve a high cost in designing pilots, there are potential methods to reduce the beam training period and overhead. The hierarchical beam training method [21] is an efficient approach, which firstly uses a sector-level beam for training and then uses wide beams for searching.

Beam tracking performs channel information updates during the time when there is no beam training. Typically, the updated channel information consists of azimuth angle of arrival (AoA), azimuth angle of departure (AoD), zenith angle of arrival (ZoA), and zenith angle of departure (ZoD) for the transmitter and receiver to perform beam forming. There are two methods for obtaining the updated channel information. One is to use a reference signal and old channel information to predict the angles in the next data transmission [7]. The other is to estimate the AoA/AoD/ZoA/ZoD based on the reference signal. The challenge is that the reference signal cost increases with the antenna element size in the BS and the UE, becoming prohibitive in the case of an antenna element size larger than 16. An efficient method to overcome this problem is to use a compression-sensing (CS) technique which well exploits the sparse property of the mmWave channel and can significantly reduce the overhead [8]. Results show a saving in overhead of up to 75% compared with a traditional non-CS estimation method, for example, least square algorithm, under practical scenarios. Furthermore, the proposed method in [8] has only a 2–3 dB loss compared with a method with perfect channel information.

Beam training is performed in cases when the UE starts to access mmWave links without prior beam alignment information, or channel angles jump to another direction, which may cause failure of data transmission, such as from line-of-sight (LOS) to NLOS. Such cases can be modeled by using different drops in the 3GPP spatial channel model (SCM) model, which is beyond the scope of this chapter. This chapter investigates the case of a channel in which UEs are moving over a short range. BSs and UEs can maintain data communication at lower-order modulation due to moving UEs. Beam tracking can be used to correct beams at BSs and UEs. Hence, to study and evaluate beam-tracking techniques, the requirements for a new channel model include variant AoA/AoD/ZoA/ZoD and consistent spatial channel. Spatial consistence is needed to keep large-scale fading invariant so that we can focus on small-scale fading to evaluate beam tracking.

16.4 Channel Model with Variant Angles

In this section, we extend the 3GPP 3-D channel model to meet beam tracking requirements in mmWave bands. The 3GPP 3-D channel model is an extension of SCM from two to three dimensions. SCM is a geometry-based stochastic model in which the locations of scatters are not explicitly specified. The channel parameters for individual snapshots are determined stochastically based on statistical distributions extracted from channel measurements. Channel realizations are generated through the application of the geometrical principle by summing contributions of rays with specific small-scale parameters, such as delay, power, and angles.

Superposition results in correlation between antenna elements and temporal fading with a geometry-dependent Doppler spectrum.

An important concept of SCM is "drop." A drop is defined as one simulation run over a certain short time period in which the random properties of the channel remain constant except for the fast fading caused by the changing phases of rays. The constant properties during a single drop include power, delays, and directions of rays. Thus, large-scale propagation (e.g., path loss) is constant, which keeps spatial consistency in a drop. Multiple drops can be simulated to average the large-scale properties in a given area. But consecutive drops are independent; that is, both large-scale and small-scale parameters are independently generated at each drop duration.

There are two approaches to extending the SCM model to introduce variant angles (AoA/AoD/ZoA/ZoD). The first approach is to obtain variant angles by consecutive drops. The effort is to study methods that keep spatial consistence between drops, where both large-scale fading and angles are changing continuously in space. An efficient solution is to estimate the locations of scatters and reconstruct the channel impulse response based on the fixed scatters [22]. However, such an approach may change the SCM framework from a stochastically based to a scatter-based model, which is unacceptable to 3GPP standards. The second approach is to introduce variant angles in a drop in which spatial consistency is well maintained. Few researchers are working on this approach to the authors' knowledge. One constraint is that a drop duration has a short time period, normally about 1000 transmission time intervals (TTIs) in LTE simulations, which is equal to 1 s in time. In such a short time, an LTE channel has negligible channel changes, particularly when a small number of antennas are used. However, the situation is very different when a large-scale antenna array is used in mmWave communications. Firstly, a large-scale antenna array may form a beam with a very narrow width. For example, a plate with a size of 10 × 10 cm may accommodate 1024 antenna elements at the E-band, which can form a 3 dB beam width as narrow as 3°. UE moving over a short range may cause a large bias in beam pairs. Secondly, the mmWave channel suffers large channel variance due to its much smaller wavelength. Moving environments, such as moving cars, trees, and scatters around the UE may cause larger channel changes compared with lower-frequency bands (e.g., 2 GHz in LTE).

We assume that UE is moving with a speed v in direction (θ_v, ϕ_v), where θ_v and ϕ_v are the vertical and horizontal directions, respectively, in the global coordination system (GCS), and the BS is located at the center. The 3GPP SCM model is described by a number of clusters with different delays. A number of rays constitute a cluster, in which all rays are diffused in space, in delay or angle domains or in both. Consider the nth cluster from the antenna element u at the transmitter to the antenna element s at the receiver. The corresponding channel impulse response in SCM is extended to be

$$H_{u,s,n}(t) = \sqrt{P_n/M} \sum_{m=1}^{M} \begin{bmatrix} F_{rx,u,\theta}\left(\theta_{n,m,\mathrm{ZoA}}(t), \phi_{n,m,\mathrm{AoA}}(t)\right) \\ F_{rx,u,\phi}\left(\theta_{n,m,\mathrm{ZoA}}(t), \phi_{n,m,\mathrm{AoA}}(t)\right) \end{bmatrix}^T$$

$$\begin{bmatrix} \exp\left(j\Phi_{n,m}^{\theta\theta}\right) & \sqrt{\kappa_{n,m}}^{-1} \exp\left(j\Phi_{n,m}^{\theta\phi}\right) \\ \sqrt{\kappa_{n,m}}^{-1} \exp\left(j\Phi_{n,m}^{\phi\theta}\right) & \exp\left(j\Phi_{n,m}^{\phi\phi}\right) \end{bmatrix}$$

$$\begin{bmatrix} F_{tx,s,\theta}\left(\theta_{n,m,\mathrm{ZoD}}(t), \phi_{n,m,\mathrm{AoD}}(t)\right) \\ F_{tx,s,\phi}\left(\theta_{n,m,\mathrm{ZoD}}(t), \phi_{n,m,\mathrm{AoD}}(t)\right) \end{bmatrix}$$

$$\exp\left(j2\pi\lambda_0^{-1}\left(\hat{r}_{rx,n,m}^T(t).\bar{d}_{rx,u}\right)\right)\exp\left(j2\pi\lambda_0^{-1}\left(\hat{r}_{tx,n,m}^T(t).\bar{d}_{tx,s}\right)\right)$$

$$\exp\left(j2\pi v_{n,m}(t)t\right)$$

$$(16.1)$$

where:

$F_{rx,u,\theta}(t)$ and $F_{rx,u,\phi}(t)$	are the receive antenna element u field patterns in the direction of the spherical basis vectors,…, respectively
$F_{tx,u,\theta}(t)$ and $F_{tx,u,\phi}(t)$	are the transmit antenna element s field patterns in the direction of the spherical basis vectors,…, respectively
n	denotes a cluster
m	denotes a ray within cluster n
$d_{rx,u}$	is the location vector of the uth receive antenna element
$d_{tx,s}$	is the location vector of the sth transmit antenna element
$k_{n,m}$	is the cross-polarization power ratio in linear scale
λ_0	is the wavelength of the carrier frequency
$r_{rx,n,m}(t)$	is the spherical unit vector with azimuth arrival angle $\phi_{n,m,AoA}(t)$ and elevation arrival angle $\theta_{n,m,ZoA}(t)$, given by

$$\hat{r}_{rx,n,m}(t) = \begin{bmatrix} \sin\theta_{n,m,\mathrm{ZoA}}(t)\cos\phi_{n,m,\mathrm{AoA}}(t) \\ \sin\theta_{n,m,\mathrm{ZoA}}(t)\sin\phi_{n,m,\mathrm{AoA}}(t) \\ \cos\theta_{n,m,\mathrm{ZoA}}(t) \end{bmatrix} \qquad (16.2)$$

$r_{tx,n,m}(t)$ is the spherical unit vector with azimuth departure angle $\phi_{n,m,AoD}(t)$ and elevation departure angle $\theta_{n,m,ZoA}(t)$, given by

$$\hat{r}_{tx,n,m}(t) = \begin{bmatrix} \sin\theta_{n,m,\mathrm{ZoD}}(t)\cos\phi_{n,m,\mathrm{AoD}}(t) \\ \sin\theta_{n,m,\mathrm{ZoD}}(t)\sin\phi_{n,m,\mathrm{AoD}}(t) \\ \cos\theta_{n,m,\mathrm{ZoD}}(t) \end{bmatrix} \qquad (16.3)$$

where:

n	denotes a cluster
m	denotes a ray within cluster n
$\bar{d}_{rx,u}$	is the location vector of the receive antenna element
u and $\bar{d}_{tx,s}$	is the location vector of the transmit antenna element
s, $\kappa_{n,m}$	is the cross-polarization power ratio in linear scale
λ_0	is the wavelength of the carrier frequency

If polarization is not considered, the 2×2 polarization matrix can be replaced by the scalar $\exp(j\Phi_{n,m})$, and only vertically polarized field patterns are applied. The Doppler frequency component $v_{n,m}$ is calculated from the arrival angles (AoA, ZoA), UE velocity vector $\bar{v}$ with speed v, travel azimuth angle ϕ_v, and elevation angle θ_v, and is given by

$$v_{n,m} = \frac{r_{rx,n,m}^T \cdot \bar{v}}{\lambda_0} \tag{16.4}$$

where $\bar{v} = \left[v\sin\theta_v \cos\phi_v, \ v\sin\theta_v \sin\phi_v, \ v\cos\theta_v \right]^T$.

The difference from the SCM model is that the angles $\theta_{n,m,\text{ZoA}}(t)$, $\theta_{n,m,\text{ZoD}}(t)$, $\phi_{n,m,\text{AoA}}(t)$, and $\phi_{n,m,\text{AoD}}(t)$ in Equation 16.1 vary with time, whereas SCM keeps fixed angles.

In the following, the cluster and ray index (n, m) is omitted for simplicity. Figure 16.3 demonstrates AoA/AoD/ZoA/ZoD in the GCS, where the BS is located at the center; h_{BS} and h_{UE} are the heights of BS and UE, respectively; d' denotes the

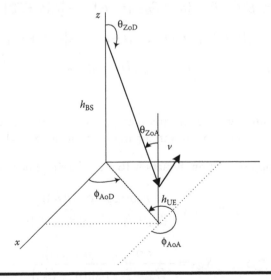

Figure 16.3   Angles of geometry coordination system.

projection of the distance between BS and UE on the x,y field, assuming that UE is only moving toward the horizontal direction (ϕ_v) over a small range. In the case of LOS, AoD and ZoD at time t by derivation are expressed by

$$\theta_{ZoD}(t) = \pi + \arctan\left(\tan\left(\theta_{ZoD}(t_0)\right) - \frac{vt\cos\left(\phi_v - \phi_{AoD}(t_0)\right)}{h_{BS} - h_{UE}} \right)$$

$$\phi_{AoD}(t) = \arctan\left(\frac{d'\sin\left(\phi_{AoD}(t_0)\right) + vt\sin\left(\phi_v\right)}{d'\cos\left(\phi_{AoD}(t_0)\right) + vt\cos\left(\phi_v\right)} \right) \quad (16.5)$$

From the relationship shown in Figure 16.3, AoA/AoD and ZoD/ZoA are related by

$$\theta_{ZoA}(t) = \pi - \theta_{ZoD}(t) \quad \text{and} \quad \phi_{AoA}(t) = \pi + \phi_{AoD}(t) \quad (16.6)$$

An accurate method to model the variant angles is to calculate the angles at each location using information on the geometry of BS and UE. The cost is that the computational complexity is high, which is unacceptable for a fast simulation. A simple way is to assume that the angles are changing linearly with time. Since the moving range is much smaller than the distance between BS and UE, the angles are expected to show little change, and hence, linear approximation is an efficient method. Assume that a linear model for variant angles is given by

$$\theta_{ZoA}(t) = \theta_{ZoA}(t_0) + S_{ZoA}.(t - t_0), \quad t \in [t_0, t_0 + T_m]$$

$$\theta_{ZoD}(t) = \theta_{ZoD}(t_0) + S_{ZoD}.(t - t_0), \quad t \in [t_0, t_0 + T_m]$$

$$\phi_{AoA}(t) = \phi_{AoA}(t_0) + S_{AoA}.(t - t_0), \quad t \in [t_0, t_0 + T_m]$$

$$\phi_{AoD}(t) = \phi_{AoD}(t_0) + S_{AoD}.(t - t_0), \quad t \in [t_0, t_0 + T_m] \quad (16.7)$$

where:

S_{ZoA} and S_{ZoD} are the slopes of variant angles in the vertical direction

S_{AoA} and S_{AoD} are the slopes of variant angles in the horizontal direction

Using linear approximation, Equation 16.5 can be simplified to

$$S_{ZoD} = -S_{ZoA} = \frac{v\cos\left(\phi_v - \phi_{AoD}(t_0)\right)}{\left(h_{BS} - h_{UE}\right)/\cos\left(\theta_{ZoD}(t_0)\right)}$$

$$S_{AoD} = S_{AoA} = -\frac{v\sin\left(\phi_v - \phi_{AoD}(t_0)\right)}{\left(h_{BS} - h_{UE}\right)\tan\left(\theta_{ZoD}(t_0)\right)} \quad (16.8)$$

Notice that the four slopes are fixed in a drop period T_m, although they can be extended to a time-varying version, but at the cost of higher computational complexity.

Consider an NLOS case where there is one reflection ray. Given the angle of the reflection surface ϕ_{RS}, a virtual UE is introduced, which is an image of UE to the reflection surface. The virtual is moving toward a direction with angle $\phi_{v'}$. The relationship between the virtual UE moving angle $\phi_{v'}$ and the original UE moving angle ϕ_v is

$$\phi_{v'} = \frac{\pi}{2} + \phi_{RS} - \phi_v \qquad (16.9)$$

The virtual UE can be seen as LOS to BS. Thus, by substituting Equation 16.9 into Equation 16.5, the function of AoD and ZoD are

$$\theta_{ZoD}(t) = \pi + \arctan\left(\tan\left(\theta_{ZoD}(t_0)\right) + \frac{vt \sin\left(\phi_v + \phi_{AoD}(t_0)\right) - \phi_{RS}}{h_{BS} - h_{UE}} \right)$$

$$\phi_{AoD}(t) = \arctan\left(\frac{d' \sin\left(\phi_{AoD}(t_0)\right) + vt \cos\left(\phi_v - \phi_{RS}\right)}{d' \cos\left(\phi_{AoD}(t_0)\right) + vt \sin\left(\phi_v - \phi_{RS}\right)} \right) \qquad (16.10)$$

with

$$\theta_{ZoA}(t) = \pi - \theta_{ZoD}(t) \text{ and } \phi_{AoA}(t) = 2\phi_{RS} + \pi - \phi_{AoD}(t) \qquad (16.11)$$

Similarly, the simplified versions in the case of NLOS are

$$S_{ZoD} = -S_{ZoA} = -\frac{v \sin\left(\phi_v + \phi_{AoD}(t_0)\right) - \phi_{RS}}{(h_{BS} - h_{UE})/\cos\left(\theta_{ZoD}\right)(t_0)}$$

$$S_{AoD} = -S_{AoA} = -\frac{v \cos\left(\phi_v + \phi_{AoD}(t_0)\right) - \phi_{RS}}{(h_{BS} - h_{UE})\tan\left(\theta_{ZoD}(t_0)\right)} \qquad (16.12)$$

Therefore, the channel impulse response of each cluster can be obtained via Equations 16.1 through 16.12. Following the procedure of SCM in [23], complete multiple-in multiple-out (MIMO) channel impulse responses are available.

16.5 UAB Network Architecture

A unified access and backhaul (UAB) network with macro base stations (MBSs) and mmWave mBSs is proposed in this section, in which MBSs and mBSs are connected with each other via backhauls, as illustrated in Figure 16.4. The first layer is made up of mMBs, which are working in both sub-6 GHz frequency carriers and higher-frequency (higher than 6 GHz) carriers. An MBS can communicate with all the mBSs in the covered area through lower frequency. Since the sub-6 GHz frequency bandwidth is relatively narrow and has lower propagation loss, mmWave in an MBS is used to deliver user-plane data to UEs as well as to backhaul data to neighboring mBSs. Although lower frequency and mmWave frequency share the same site in the MBS, their antenna and remote radio-frequency (RF) units are fully separated and may be installed at different heights, depending on coverage requirements.

The second layer consists of mBSs, which are much denser than MBSs and current LTE small cell networks. Considering their high propagation loss, the density of mBSs may range from 6 to 500 mBSs per square kilometer, which corresponds to a cell radius of 25–200 m. Part of the mBSs may function as anchor evolved Node B (eNodeB) in a way similar to MBSs. The data from or to the core network are communicated through anchor eNodeBs (ABs) to mBSs and UEs. Each mBS installs mmWave frequency for both backhaul and radio access. In the UAB architecture, both radio access and backhaul share the same platform, including antenna array, intermedium and radio frequency (IRF), and baseband unit. The advantage of UAB is that backhaul and radio access can be jointly managed to schedule radio resource and antenna resource, and hence, resources can be used in a more efficient way. For example, when backhauling has a larger load than radio access, more frequency band or antenna beams can be allocated to backhauls rather than radio access, and vice versa.

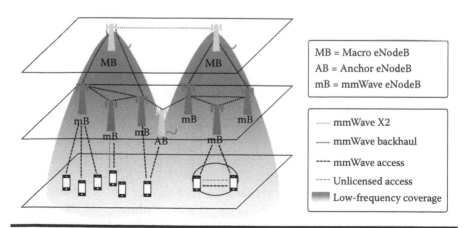

Figure 16.4 Hybrid network architecture.

The third layer is radio access, whereby UEs could access mBSs through lower frequency and mmWave, or to mBSs through mmWave. Unlicensed spectrum, in the network architecture, is used only for radio access to deliver information that is less important, because links over unlicensed spectrum might suffer from unexpected interference. mmWave is also suitable for device-to-device (D2D) communication. Multiple UEs with D2D connections can perform joint transmitting and receiving to improve transmission.

16.5.1 *Load-Centric Backhauling (LCB)*

The UAB network enables adaptive backhauling. Beams can be adaptively generated to adjust backhauls. Adaptive self-backhauling is particularly important when traffic loads are varying in different areas. Since the geographic distribution of a traffic load may be nonuniform [24], it is expected that the backhaul network can be adaptively adjusted to track traffic load changes in the network, so-called load-centric backhauling.

Therefore, we propose to use hierarchical radio resource management (RRM) architecture to realize load-centric networking. RRM for backhauling (BH-RRM) performs the function of allocating radio resource for all the backhauls between nodes. The function is located in the MBS, which can communicate with all its covered mBSs via the sub-6 GHz frequency. Each mBS performs the RRM of radio access (RA-RRM) as well as the function of executing backhauling at the resource and configuration given by the BH-RRM. Note that the RA-RRM function is the same as that of the RRM in LTE eNodeB, which allocates radio resource to local users.

The function splitting between BH-RRM and RA-RRM is crucially dependent on system architecture. If there is enough bandwidth between MBS and mBS, RA-RRM can actually move to the MBS, and a powerful RRM in the MBS may perform scheduling for both backhaul and radio access. Such a centralized structure is also suitable for cooperative communications such as coordinated multipoint (CoMP). Joint transmitting and joint receiving among multiple distributed mBSs can be successfully implemented in the centralized unit. Another extreme scenario is that BH-RRM is combined with RA-RRM and located in every mBS. The network then becomes similar to a mesh network, and every node performs scheduling in a distributed way. The advantage of the distributed architecture is that the centralized node is not required, which simplifies network deployment. However, the cost of this approach is that it is hard to perform adaptive backhauling and cooperative communication.

Last, let us exemplify the performance of LCB. Consider a network with three clusters of small cells. Each cluster has an anchor mBS (AB) in the center, and two layers of small cells surrounding each AB, as illustrated in Figure 16.5. Let us first consider the case with fixed backhauling. Suppose each small cell has a traffic load of 1; then there is a total traffic load of 18 in a cluster. The centered cell covered

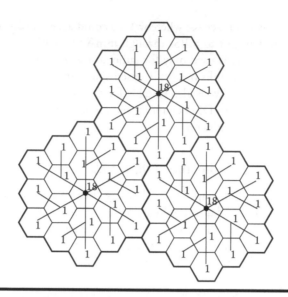

Figure 16.5 **Network topology considered in simulations.**

by AB is used only for backhaul aggregation and does not have its own traffic. We further assume that the traffic capability of each backhaul is 3, which is the load delivered between AB and its neighboring mBSs. Figure 16.5 illustrates the optimal routing, in the sense that all three ABs can deliver total traffic loads of 54, which is the sum load of all small cells.

However, traffic loads are always distributed with a very nonuniform geometry. Most traffic may concentrate on a small number of base stations. For example, if a cluster concentrates 80% of its traffic in 20% nodes, a network with fixed routing, as in Figure 16.5, can only deliver a total traffic load of 24. With LCB, each mBS can select any of the neighboring mBSs for backhauling. Here, each node selects the backhaul with the largest reserved capability. If there are backhauls with equal reserved capability, it selects the backhaul that is approaching the closest AB. Figure 16.6 shows the network capacity gain using the LCB technique. The network capacity is the sum of the traffic load delivered through the three ABs.

It is assumed that 60%–90% of traffic is concentrated in 10% of mBSs. We can see that network capacity gains by using LCB are 9.60%, 21.50%, 56.00%, and 159.70% in the case of 60%, 70%, 80%, and 90% traffic load, respectively, concentrated in 10% of nodes. The trend is that, as traffic loads have a less uniform geometric distribution, adaptive backhauling may achieve a greater capacity gain than fixed backhaul.

16.5.2 *Multifrequency Transmission Architecture*

The UAB network supports multiple-frequency carriers to deliver ultrahigh data rates (tens of gigabits per second). This is similar to the carrier aggregation (CA) technique being discussed in 3GPP LTE. A dominant difference from LTE is that

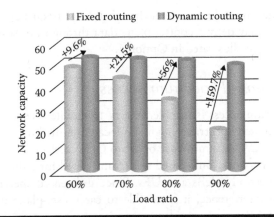

Figure 16.6 Network capacity.

it may cover much larger interval bands, such as E-band, Ka-band, and V-band carriers. Different frequency carriers have totally different propagation characteristics.

16.5.2.1 C/U Splitting Technique

The control and user planes have different requirements for data transmission. The control-plane data from the mobile management entity (MME) is always important information and requires a lower error rate than data from the user plane, whereas the user plane has a much higher data rate to be delivered than the control plane. This motivates the control and user plane (C/U) splitting technique [25].

Figure 16.7 shows the C/U splitting schemes. Current solutions in which both control and user planes route the same links to UEs are illustrated in Option 1, while in Options 2 and 3, control-plane data are delivered through sub-6 GHz frequency by MBSs, and user-plane data are delivered through mmWave from mBSs.

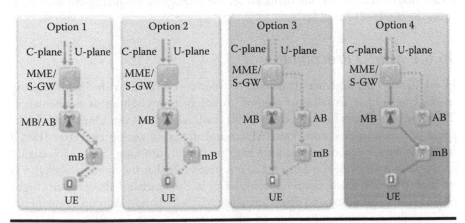

Figure 16.7 C/U splitting schemes.

When an MBS has limited resource in the sub-6 GHz frequency to transmit control-plane data, it may deliver control-plane data through mmWave to desired or neighboring mBSs, as illustrated in Option 4.

16.5.2.2 Unlicensed Spectrum Access

Unlicensed spectrum access is complementary to mobile access on licensed spectrum [26]. Unlicensed spectrum might be located at all kinds of eNodeBs and provide additional CA with mmWave or sub-6 GHz carriers.

In the hybrid network, unlicensed spectrum is used for radio access to deliver information between eNodeBs and UEs. Since unlicensed spectrum may suffer from unexpected interference, it is suitable to carry user-plane data rather than control-plane data. Based on regulatory requirements for coexistence and radiation safety, the transmit power of unlicensed spectrum is limited (e.g., 250 mW at 5.17–5.33 GHz in the United States), and its coverage is smaller than that of mBSs.

16.6 System-Level Capacity

In this section, we evaluate the system performance of 72 GHz and 28 GHz systems. The bandwidths of 72 GHz and 28 Hz are 2.5 Hz and 500 MHz, respectively. This is because the total available bandwidths are 10 GHz and 2 GHz for 72 GHz and 28 GHz, respectively. Consider a 3GPP heterogeneous network (HetNet) topology with an MBS with radius of 500 m and three mBSs distributed in a macro cell, each mBS having six cells with a radius of 50 m. The antenna apertures are 66 × 66 mm at each cell of the mBS and 16 × 16 mm at the UE. All antenna elements have half wavelength separation. Exhaustive beam training is applied to align beams in the mBSs and the UEs. Phase noise with Wiener model and error vector magnitude (EVM) are included for the orthogonal frequency-division multiplexing (OFDM)-based systems.

16.6.1 MIMO Precoding

In addition to the antenna array used in front-end, MIMO with digital steering in the baseband is an efficient method to enhance throughput or performance. There are two factors limiting the MIMO realization, however. One is that high-speed analog-to-digital converter (ADC) and digital-to-analog converter (DAC) for mmWave are expensive and have high power consumption. Since each digital chain needs a set of ADCs and DACs, it is preferable that the number of chains is no higher than 4. Computational complexity in the baseband also prevents high-order MIMO precoding and detection, particularly for bandwidths up to gigahertz.

In multiuser communications, an efficient solution is to separate users in space by beams and multiuser MIMO to improve transmission performance. Each user may own one or multiple data streams depending on channel status. The performance of multiuser MIMO plus beam forming has been analyzed. An interesting result is that with the same antenna aperture in mBS and UE, higher frequency may suffer less interference from neighboring beams, cells, and sites. System-level performance is compared at 72 and 28 GHz, deploying 3GPP HetNet topology with a small cell radius of 50–150 m. The 72 GHz system uses beams with half power beamwidth (HPBW) of 4 and 13 in mB and UE, whereas the 28 GHz system uses 10 and 21, respectively. Each cell is assumed to support four beams for simultaneous transmission. It is found in Figure 16.8 that the interference in a 28 GHz system may degrade the average signal-to-noise ratio (SNR) by 30 dB, whereas the degradation in a 72 GHz system is 5 dB. An important reason is that the wider the HPBW, the greater the interference it may produce. Wide sidelobes

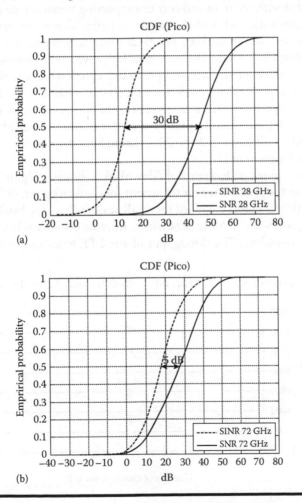

Figure 16.8   Interference analysis in (a) 28 GHz and (b) 72 GHz.

also play important roles in causing interference. Hence, we conclude that MIMO precoding and complex MIMO detection are not mandatory if the beam is narrow enough to avoid interference. An interference-mitigation technique such as MIMO precoding may be not necessary for 72 GHz, but it is mandatory for 28 GHz.

16.6.2 Performance Evaluation

In this section, we consider downlink only. Perfect channel estimation and perfect channel quality indicator (CQI) feedback are assumed. Radio resources for multiple users are scheduled in space and time dimensions. There is no further division in the frequency domain because lesser granularity, such as a physical resource block (PRB) in the LTE system, results in only 6.8% performance gain [5], which is small compared with the front-end cost of supporting frequency division. So, each user will occupy a slot with a whole frequency band. This is suitable for mmWave communications, since an analog antenna can form a beam pattern at a time that applies to all frequency bands. The authors of [27] have studied different scheduling algorithms to reduce interference and presented signal-to-leakage-plus-noise ratio (SLNR)-based and signal-to-interference-plus-noise ratio (SINR)-based proportional fair (PF) algorithms. The idea is to select the beams with smaller interference with each other while maintaining fairness. In this chapter, we use PF for reasons of simplicity, and it will result in 20%–30% throughput degradation compared with an SINR-based algorithm [27].

System throughput performance of a downlink is shown in Figure 16.9. The path loss is based on the preliminary measurement for outdoors given in [5]. The 3GPP urban micro (UMi) model is applied for small-scale fading. The baseline is an LTE system configured with 20 MHz bandwidth, intersite distance (ISD) = 500 m, and 4×2 MIMO downlink. The throughput of the LTE baseline is 0.69 Gbps/km².

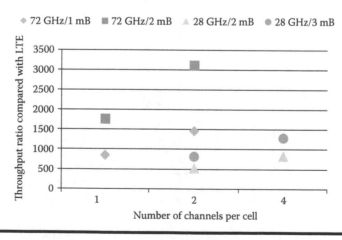

Figure 16.9   System throughput compared with LTE, 72 GHz, and 28 GHz.

We investigate the cases with one, two, and three mBSs per macro cell, and one, two, and four channels per mBS cell, respectively. For fairness, the 72 GHz system has a bandwidth of 2.5 GHz, and the 28 GHz system has 500 MHz bandwidth. To reach a 1000 times increase in throughput over LTE, it is shown that the 72 GHz system needs one channel per mBS cell and two mBSs per macro cell, or two channels per mBS cell and one mBS per macro cell. However, 28 GHz needs four channels per mBS cell and three mBSs per macro cell. Compared with 72 GHz, the 28 GHz system needs to increase node density and channels six times. Therefore, we conclude that 72 GHz can achieve a 1000 times throughput enhancement with fewer channels and sparser node density, hence reducing the capital and operating expenditure (CAPEX and OPEX) in networking.

References

1. W. Tong, 5G goes beyond smartphone, keynote speech at IEEE Globecom '14, Austin, TX, 2014.
2. T. Rappaport, S. Sun, R. Mayzus, H. Zhao, Y. Azar, K. Wang, G. N. Wong, J. K. Schulz, M. Samimi, and F. Gutierrez, Millimeter wave mobile communications for 5G cellular: It will work! *IEEE Access* 1: 335–349, 2013.
3. W. Roh, J. Y. Seol, J. H. Park, B. Lee, J. Lee, Y. Kim, J. Cho, and K. Cheun, Millimeter-wave beamforming as an enabling technology for 5G cellular communications: Theoretical feasibility and prototype results, *IEEE Transactions on Communications* 52(2): 106–113, 2014.
4. Y. Wang, J. Li, L. Huang, J. Yao, A. Georgakopoulos, and P. Demestichas, 5G Mobile: Spectrum broadening to higher-frequency bands to support high data rates, *IEEE Vehicular Technology Magazine* 9(3): 39–46, 2014.
5. Z. Shi, Y. Wang, and L. Huang, System capacity of 72GHz mmWave transmission in hybrid networks, submitted to IEEE Globecom'15, San Diego, CA, 2015.
6. J. Zhang, A. Beletchi, Y. Yi, and H. Zhuang, Capacity performance of millimeter wave heterogeneous networks at 28GHz/73GHz, in *Proceedings of the IEEE Globecom'14 Workshop on Mobile Communications with Higher Frequency Bands*, Austin, TX, pp. 405–409, 2014.
7. J. He, T. Kim, H. Ghauch, K. Liu, and G. Wang, Millimeter wave MIMO channel tracking systems, in *Proceedings of the IEEE Globecom '14 Workshop on Mobile Communications in Higher Frequency Bands*, Austin, TX, pp. 414–419, 2014.
8. H. Huang, K. Liu, R. Wen, Y. Wang, and G. Wang, Joint channel estimation and beamforming for millimeter wave cellular system, accepted for IEEE Globecom '15, San Diego, CA, 2015.
9. Samsung, Samsung achieves 7.5Gbps transmission speeds on a 5G data network. Available from: http://www.sammobile.com/2014/10/16/samsung-achieves-7-5gbps-transmission-speeds-on-a-5g-data-network/.
10. S. Suyama, J. Shen, Y. Oda, H. Suzuki, and K. Fukawa, DOCOMO and Tokyo Institute of Technology achieve world's first 10 Gbps packet transmission in outdoor experiment, 2013. Available from: https://www.nttdocomo.co.jp/english/info/media_center/pr/2013/0227_00.html.

11. Huawei, Huawei named key member of new 5G association, announces faster than 100 Gbps speed achievement at Mobile World Congress 2014. Available from: http://pr.huawei.com/en/news/hw-328622-ictmwc.htm.

12. Nokia, Nokia networks showcases 5G speed of 10Gbps with NI at the Brooklyn 5G summit, 2015. Available from: http://networks.nokia.com/news-events/press-room/press-releases/nokia-networks-showcases-5g-speed-of-10gbps-with-ni-at-the-brooklyn-5g-summit.

13. Y. Wang, L. Huang, Z. Shi, H. Huang, D. Steer, J. Li, G. Wang, and W. Tong, An introduction to 5G mmWave communications, accepted for *Proceedings IEEE Globecom 2015 Workshop*, San Diego, USA, December 2015.

14. ITU-R WP 5D, Document 5D/TEMP/548(Rev.3), Preliminary draft new recommendation ITU-R M, 2015.

15. RP-150483, RAN Chairman, Getting ready for 5G, 3GPP TSG RAN #67, Shanghai, China, 2015.

16. ITU-R Resolution 233, Studies on frequency-related matters on International Mobile Telecommunications and other terrestrial mobile broadband applications, 2012.

17. ITU-R Administrative Circular CA/201, To administrations of member states of ITU and radiocommunication sector members: Preparation of the draft CPM Report to WRC-15, 2013.

18. ITU-R Joint Task Group Document 4-5-6-7/393-E, Annex 3 to joint task group 4-5-6-7 chairman's report working document towards preliminary draft CPM text for WRC-15 agenda item 1.1, 2013.

19. 3GPP RP-131635, Introducing LTE in unlicensed spectrum, Qualcomm, 3GPP TSG RAN Meeting #62, Busan, Korea, 2013.

20. WRC-15 Document 462-E, Report on the results of the discussion on the pending issues under agenda item 10, 25 November 2015, Geneva, Switzerland.

21. Wireless LAN medium access control (MAC) and physical layer (PHY) specifications, IEEE std 802.11ad, 2012.

22. Z. Zhu, Y. Zhu, T. Zhang, and Z. Zeng, A time-variant MIMO channel model based on the IMT-Advanced channel model, in *Proceedings of the International Conference on Wireless Communications & Signal Processing (WCSP), 2012*, vol. 1, no. 5, pp. 25–27, 2012.

23. 3GPP TR 25.996, 3GPP TSGRAN, spatial channel model for multiple input multiple output (MIMO) simulations (Release 11), 2012–09.

24. Z. Niu, TANGO: Traffic-aware network planning and green operation, *IEEE Transactions on Wireless Communications* 18(5): 25–29, 2011.

25. H. Ishii, Y. Kishiyama, and H. Takahashi, A novel architecture for LTE-B: C-plane/U-plane split and phantom cell concept, in *Proceedings of the IEEE Globecom '12 Workshops*, Anaheim, CA, pp. 624–630, 2012.

26. T. Nihtila, V. Tykhomyrov, O. Alanen, M. A. Uusitalo, A. Sorri, M. Moisio, S. Iraji, R. Ratasuk, and N. Mangalvedhe, System performance of LTE and IEEE 802.11 coexisting on a shared frequency band, in *Proceedings of the IEEE Wireless Communications and Networking Conference (WCNC) 2013*, Shanghai, pp. 1038–1043, 2013.

27. H. Li, L. Huang, and Y. Wang, Scheduling schemes for interference suppression in millimeter-wave cellular network, in *Proceedings of IEEE PIMRC 2015*, Hong Kong, pp. 46–50, 2015.

Chapter 17

Millimeter-Wave (mmWave) Radio Propagation Characteristics

Joongheon Kim

Contents

17.1 Introduction

Today, millimeter-wave (mmWave) wireless communication technologies are considered as one of the major elements of fifth-generation (5G) wireless cellular network evolution. This is because mmWave wireless systems can provide an extremely ultrawide channel bandwidth and therefore, a linear increase in achievable data rates with the ultrawide bandwidth.

Even though mmWave 5G wireless networks have many benefits based on the ultrawide bandwidth, the propagation of mmWave wireless links is high directional and is also highly attenuated due to its high carrier frequency (from around 30 to 300 GHz). To quantify the directionality and attenuation factors, this chapter provides an extensive summary of the International Telecommunication Union (ITU) standard documents for carrying out research on mmWave radio wave propagation characteristics. The summary includes ITU-standardized antenna radiation patterns, path loss models, mmWave-specific attenuation factors in mmWave wireless systems, and so forth. Based on the given models and parameters, a link budget calculation is performed to identify how much distance is achievable with given threshold data rates in mmWave wireless propagation links. Note that this chapter mainly pays attention to 28, 38, and 60 GHz mmWave wireless channels, which are the most investigated for 5G cellular and peer-to-peer wireless access networks.

The remainder of this chapter is organized as follows. Section 17.2 gives an overview of mmWave characteristics, including high directionality and background noise calculation. Section 17.3 presents propagation models and parameters, including path loss models and mmWave-specific attenuation factors. Section 17.4 presents the link budget calculation results, both theoretical and practical using IEEE 802.11ad. Finally, Section 17.5 concludes the chapter.

17.2 Propagation Characteristics

Even though the use of mmWave radio technologies is attractive due to their large bandwidth availability, they have high directionality, which is positive in terms of mmWave network device densification (due to spatial reuse) but negative in terms of beam-tracking overheads (harmful in terms of mobility support [1]). Therefore, it is necessary to quantify the beamwidth of directional beams. In Section 17.2.1, the directionality of mmWave beams is determined based on an ITU recommendation. In addition, the mmWave system is noise limited, whereas conventional cellular systems are interference limited. Thus, background noise in 28, 38, and 60 GHz mmWave systems is identified in Section 17.2.2.

17.2.1 High Directionality

The directionality of wireless radio propagation depends on antenna types and corresponding parameters. Without loss of generality, this chapter considers ITU-standard antenna radiation patterns. The ITU-recommended reference antenna radiation patterns for sharing studies from 400 MHz to about 70 GHz are presented in an ITU recommendation as follows [2]:

$$G(\varphi,\theta)=\begin{cases} G_{max}-12|x|^2 & 0\leq x<1 \\ G_{max}-12-15\ln|x| & 1\leq x \end{cases}$$

where:

$G(\varphi,\theta)$ is the antenna gain

φ and θ are azimuth and elevation angles, where $-180°\leq\varphi\leq180°$ and $-90°\leq\theta\leq90°$

x is defined as

$$x=\frac{\aleph}{\aleph_\alpha}$$

where $\aleph$ and $\aleph_\alpha$ can be formulated as

$$\aleph=\cos^{-1}(\cos\varphi\cdot\sin\theta)$$

$$\aleph_\alpha=\begin{cases} \dfrac{1}{\sqrt{\left(\dfrac{\cos\alpha}{\varphi_{bw}}\right)^2+\left(\dfrac{\sin\alpha}{\theta_{bw}}\right)^2}} & 0°\leq\aleph\leq90° \\[20pt] \dfrac{1}{\sqrt{\left(\dfrac{\cos\theta}{\varphi_{3m}}\right)^2+\left(\dfrac{\sin\theta}{\theta_{bw}}\right)^2}} & 90°\leq\aleph\leq180° \end{cases}$$

where:

φ_{bw} and θ_{bw} are the half power beamwidth (HPBW) in azimuth and elevation planes

α $=\tan^{-1}(\tan\theta/\sin\varphi)$

φ_{3m} is the equivalent HPBW in the azimuth plane for adjustment of horizontal gains (degrees)

Thus, it can be calculated as follows:

$$\varphi_{3m} = \begin{cases} \varphi_{bw} & 0° \le |\varphi| \le \varphi_{th} \\ \dfrac{1}{\sqrt{\left[\dfrac{\cos\left(\dfrac{|\varphi|-\varphi_{th}}{180-\varphi_{th}}\cdot 90\right)}{\varphi_{bw}}\right]^2 + \left[\dfrac{\sin\left(\dfrac{|\varphi|-\varphi_{th}}{180-\varphi_{th}}\cdot 90\right)}{\theta_{bw}}\right]^2}} & \varphi_{th} \le |\varphi| \le 180° \end{cases}$$

where φ_{th} is defined as the boundary azimuth angle (degrees), that is, $\varphi_{th} = \varphi_{bw}$. φ_{bw} and θ_{bw} can be calculated as follows [2]:

$$\theta_{bw} = \frac{31{,}000 \cdot 10^{-G_{max}/10}}{\varphi_{bw}}$$

and we assume $\varphi_{bw} \approx \theta_{bw}$, that is,

$$\varphi_{bw} \approx \theta_{bw} \approx \sqrt{31{,}000 \cdot 10^{(-G_{max}/10)}}$$

Based on these given models, the HPBW values for various G_{max} values are summarized in Table 17.1.

Based on the models presented in this section, the ITU-standard antenna radiation pattern can be plotted. Figures 17.1 and 17.2 present azimuth plane plotting and elevation plane plotting, respectively.

Table 17.1 Beam Directionality

G_{max} (dBi)	HPBW
10	55.67764363
15	31.3098399
20	17.60681686
25	9.901040726
30	5.567764363
35	3.13098399
40	1.760681686
45	0.990104073
50	0.556776436

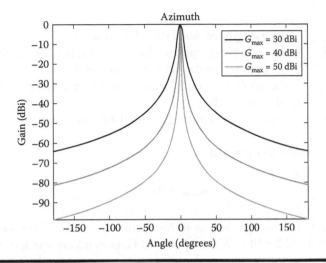

Figure 17.1   Azimuth plane plotting.

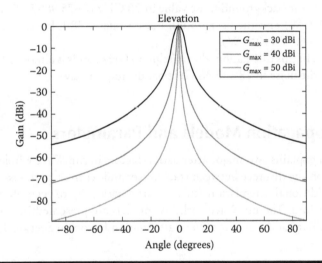

Figure 17.2   Elevation plane plotting.

17.2.2 Noise-Limited Wireless Systems

The performance of wireless systems with a large channel bandwidth can be affected by background noise levels in the system. In 60 GHz wireless standards (such as IEEE 802.11ad and IEEE 802.15.3c), the channel bandwidth is defined as 2.16 GHz. With a bandwidth of 2.16 GHz, the background noise can be calculated as follows [3]:

$$n_{\text{dBm}} = k_B T_e + 10 \log_{10} BW + L_{\text{implementation}} + n_F$$

where:

n_{dBm} is the background noise on a decibel scale

$k_B T_e$ is the noise power spectral density, which is -174 dBm/Hz

BW is the channel bandwidth (i.e., 2.16 GHz)

$L_{\text{implementation}}$ is the implementation loss, assumed by the IEEE 802.11ad standard to be 10 dB

n_F is a noise figure, assumed by the IEEE 802.11ad standard to be 5 dB

Then, $n_{\text{dBm}} = -65.6555$ dBm and

$$n_{m\text{watt}} = 10^{(n_{\text{dBm}}/10)}$$

where $n_{m\text{watt}}$ is the background noise on a milliwatt scale. Therefore, the background noise is 2.72×10^{-10} W. In 28 and 38 GHz mmWave wireless systems, the background noise values can be calculated in the same way under the assumption that the channel bandwidths in 28 and 38 GHz are 200 and 500 MHz, respectively. Finally, the background noise value in 28 GHz is -75.9897 dBm (equivalent to 2.52×10^{-11} W) and the background noise value in 38 GHz is -72.0103 dBm (equivalent to 6.29×10^{-11} W).

As shown in this calculation, the noise in 60 GHz bands is almost 10 and 4 times higher than the noise in 28 and 38 GHz bands, respectively.

17.3 Propagation Models and Parameters

This section explains two major attenuation factors in mmWave wireless channels depending on the distance between transmitter and receiver: path loss models and auxiliary additional attenuation (such as attenuation by oxygen absorption and rain attenuation). The mmWave path loss models are presented in Section 17.3.1 and the auxiliary additional mmWave attenuation factors in Section 17.3.2.

17.3.1 Path Loss Models

Free-space basic transmission (i.e., line-of-sight [LOS]) loss is determined as a function of the distance between transmitter and receiver [4] on a decibel scale:

$$PL(d_{\text{km}}) = 92.44 + 20\log_{10} f + n \cdot 10\log_{10} d_{\text{km}}$$

where:

d_{km} is the distance between transmitter and receiver (kilometers)

f stands for the carrier frequency (gigahertz)

n is the path loss coefficient, equal to 2.2 when $f \geq 10$ [5]

This equation is defined by the ITU.

The measurement-based 28 and 38 GHz path loss models are derived as summarized in [6]. The fundamental equation is

$$PL(d) = 20\log_{10}\left(\frac{4\pi d_0}{\lambda}\right) + n \cdot 10\log_{10}\left(\frac{d}{d_0}\right) + X_\sigma$$

where:
  d is the distance between transmitter and receiver (meters)
  d_0 is the close-in free-space reference distance (set to $d_0 = 5$ m)
  λ is the wavelength (10.71 mm in 28 GHz and 7.78 mm in 38 GHz)
  n is the average path loss coefficient over distance and all pointing angles
  X_σ is a shadowing random variable, which is represented as a Gaussian random variable with zero mean and σ standard deviation
  n and σ are summarized in Table 17.2 [7,8]

The measurement-based 60 GHz path loss models are presented in IEEE 802.11ad standard documents. As defined in [9], a 60 GHz mmWave IEEE 802.11ad LOS path loss model is

$$PL(d) = A + 20\log_{10} f + n \cdot 10\log_{10} d$$

on a decibel scale, where $A = 32.5$ dB. This value is specific for the selected type of antenna and beam-forming algorithms, which depend on the antenna beamwidth,

Table 17.2 Path Loss Exponent (*n*) and Standard Deviations of Shadowing Random Variables (σ)

Configuration		n	σ
25 dBi antenna at 38 GHz	LOS	2.20	10.3
	NLOS	3.88	14.6
13.3 dBi antenna at 38 GHz	LOS	2.21	9.40
	NLOS	3.18	11.0
24.5 dBi antenna at 28 GHz	LOS	2.55	8.66
	NLOS	5.76	9.02

Source: Y. Azar et al. 28 GHz propagation measurements for outdoor cellular communications using steerable beam antennas in New York City, in *Proceedings of IEEE International Conference on Communications (ICC)*, Budapest, IEEE, 2013.

but for the considered beam range from 60° to 10°, the variance is very small, less than 0.1 dB. In this equation, n refers to the path loss coefficient, which is set to $n=2$, and f stands for a carrier frequency on a gigahertz scale, set to $f=60$. Note that there is no shadowing effect in the LOS path loss model as presented in [9].

The non-line-of-sight (NLOS) model of the 60 GHz mmWave IEEE 802.11ad standards is defined as [9]

$$PL(d) = A + 20\log_{10} f + n \cdot 10 \log_{10} d + X_\sigma$$

on a decibel scale, where $A = 51.5$ dB is the value for the selected type of antenna and beam-forming schemes. This value depends on the antenna beamwidth, and the variance is very small, less than 0.1 dB in the considered beam range from 80° to 10°. In this model, $n=0.6$ and $f=60$, as previously defined. Finally, X_σ stands for the shadowing effects due to NLOS, which can be calculated by Gaussian distribution with zero mean and standard deviation σ, where $\sigma = 3.3$ dB. The 60 GHz mmWave IEEE 802.11ad path loss model in NLOS has a randomness of X_σ.

LOS and NLOS path loss plotting in 60 GHz mmWave IEEE 802.11ad wireless systems is shown in Figure 17.3.

17.3.2 Millimeter Wave-Specific Attenuation Factors

As stated in [5], there are two mmWave-specific auxiliary attenuation factors: oxygen attenuation and rain attenuation.

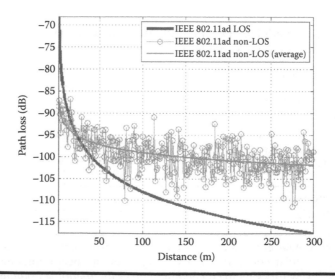

Figure 17.3 Path loss comparison in 60 GHz mmWave IEEE 802.11ad standards.

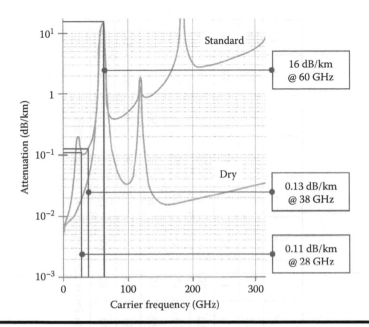

Figure 17.4 Oxygen attenuation factors in mmWave channels.

17.3.2.1 Oxygen Attenuation

The signal attenuation in wireless mmWave radio propagation due to oxygen absorption is significant, and it cannot be ignored. Figure 17.4 shows experimental results for wireless radio wave propagation attenuation in mmWave channels. The oxygen attenuation in 28, 38, and 60 GHz is 0.11, 0.13, and 16 dB/km, respectively. As shown in Figure 17.4, the performance degradation in terms of oxygen attenuation in 60 GHz bands is extremely poor. This is the main reason why 60 GHz mmWave bands are left unlicensed [10].

17.3.2.2 Rain Attenuation

The signal attenuation in wireless mmWave radio propagation due to rainfall is also significant and cannot be ignored. From the table of "Rain Climatic Zones" in [11], rain rate information in millimeters per hour can be obtained for each segmented area. For example, Northern California, Oregon, and Washington in the United States are in ITU Region D. In addition, the heaviest rain area is ITU Region Q (including the middle of Africa). The table is reproduced in this chapter as Table 17.3. It shows that the rain rates in ITU Region D with 1% outage (i.e., 99% availability) and 0.1% outage (i.e., 99.9% availability) are 2.1 and 8 mm/h, respectively, while the rain rates in ITU Region Q with 1% outage and 0.1% outage are 24 and 72 mm/h, respectively.

Table 17.3 Rain Rates Depending on Rain Climatic Zones

Percentage of Time	A	B	C	D	E	F	G	H	J	K	L	M	N	P	Q
1.0	<0.1	0.5	0.7	2.1	0.6	1.7	3	2	8	1.5	2	4	5	12	24
0.3	0.8	2	2.8	4.5	2.4	4.5	7	4	13	4.2	7	11	15	34	49
0.1	2	3	5	8	6	8	12	10	20	12	15	22	35	65	72
0.03	5	6	9	13	12	15	20	18	28	23	33	40	65	105	96
0.01	8	12	15	19	22	28	30	32	35	42	60	63	95	145	115
0.003	14	21	26	29	41	54	45	55	45	70	105	95	140	200	142
0.001	22	32	42	42	70	78	65	83	55	100	150	120	180	250	170

Source: ITU Recommendation, Characteristics of precipitation for propagation modelling, *ITU-R PN.837-1*, 1994.

Note: Rainfall intensity exceeded (mm/h) (Figures 17.1 through 17.3).

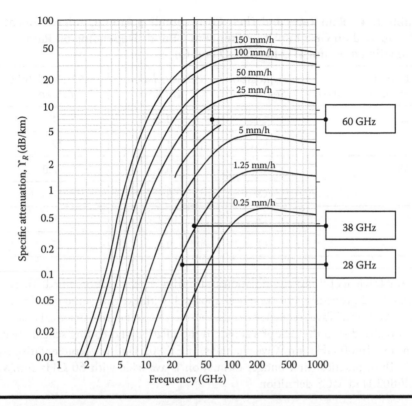

Figure 17.5   Attenuation by rain rates. (Federal Communications Commission (FCC), Office of Engineering and Technology, *Bulletin Number 70*, 1997.)

Based on this rain rate information, rain attenuation factors can be obtained [12], and measurement-based curves for the specific attenuation in each frequency depending on the rain rate can be obtained, as shown in Figure 17.5 [12]. Rain attenuation factors in decibels per kilometer can also be obtained (Figure 17.5, Table 17.4).

The actual impacts due to oxygen and rain attenuation factors are quantified in Section 17.4 in terms of link budget analysis.

17.4 Link Budget Analysis

Based on propagation characteristics, path loss models, and mmWave-specific auxiliary attenuation factors in terms of oxygen absorbance and rain rates, wireless system designers should be able to define the achievable performance. This is why link budget calculation is essential in mmWave systems engineering. In this section, two different types of link budget estimation procedures are presented. In Section 7.4.1, the Shannon capacity equation is used for estimating achievable data rates with the computation of signal-to-noise ratio (SNR). However, the Shannon capacity is only

Table 17.4 Rain Rates and Their Corresponding Attenuation Factors at 28, 38, and 60 GHz mmWave Frequency Bands Depending on Rain Climatic Zones (for ITU Regions D and Q)

Carrier Frequency (GHz)	ITU Region Segment	99% Availability (dB/km)	99.9% Availability (dB/km)
28	D	0.25	1.4
	Q	4	12
38	D	0.6	2.0
	Q	6	17
60	D	1.2	3.5
	Q	9	25

achievable when optimum modulation and coding schemes are assumed. Therefore, Section 7.4.2 presents a more practical approach with existing standards. In the 28 and 38 GHz mmWave frequency bands, there are no standards; practical analysis is not available due to the lack of a standard modulation and coding scheme (MCS) definition. In 60 GHz mmWave channels, IEEE 802.11ad is a representative standard. Thus, practical link budget estimation is available with 60 GHz mmWave IEEE 802.11ad MCS definition.

17.4.1 Shannon Capacity–Based Calculation with Signal-to-Noise Ratio Computation

Based on the well-known Shannon capacity equation, achievable data rates between transmitter and receiver can be calculated as

$$C(d) = BW \cdot \log_2\left(\frac{P_{mwatt}^{RX}(d)}{n_{mwatt}} + 1\right)$$

where:

$C(d)$ is the achievable rate when d is the distance between the transmitter and the receiver

BW is the channel bandwidth (200 MHz at 28 GHz, 500 MHz at 38 GHz, and 2.16 GHz at 60 GHz)

n_{mwatt} is the background noise, calculated as in Section 7.2.2

$P_{mwatt}^{RX}(d)$ is the receive signal strength at the receiver when d is the distance between the transmitter and the receiver

$P_{mwatt}^{RX}(d)$ can be calculated as

$$P_{mwatt}^{RX}(d) = 10^{\left(P_{dBm}^{RX}(d)/10\right)}$$

and $P_{dBm}^{RX}(d)$ is the receive signal strength at the receiver (when the distance between the transmitter and the receiver is d) in decibels, and this can be calculated as

$$P_{dBm}^{RX}(d) = \text{EIRP} - PL(d) - O(d) - R(d)$$

where $PL(d)$, $O(d)$, and $R(d)$ stand for path loss (refer to Section 17.3.1), oxygen attenuation (refer to Section 17.3.2.1), and rain attenuation (refer to Section 17.3.2.2), respectively, depending on the distance d. In addition, equivalent isotropically radiated power (EIRP) can be calculated as

$$\text{EIRP} = P_{dBm}^{TX} + G_{dBi}^{TX}$$

where P_{dBm}^{TX} and G_{dBi}^{TX} are the transmit power and transmit antenna gain, respectively. In this study, fundamental upper bounds will be observed, that is, EIRP limits are considered. In outdoor point-to-point links, the EIRP limit is defined as 82 dBm at 60 GHz mmWave bands, whereas the EIRP limit is 43 dBm at 60 GHz mmWave bands in other applications [13,14].

The achievable rate computation results are plotted as shown in Figures 17.6 and 17.7 at 60 GHz mmWave bands for 0–1500 and 0–200 m, respectively. Even though this section presents the achievable rates only in 60 GHz bands, the link budget calculation with the Shannon capacity equation can be performed for 28 GHz and 38 GHz mmWave bands in the same way.

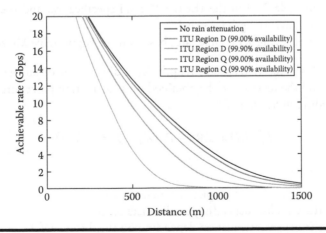

Figure 17.6 Achievable rates in LOS outdoor point-to-point 60 GHz links.

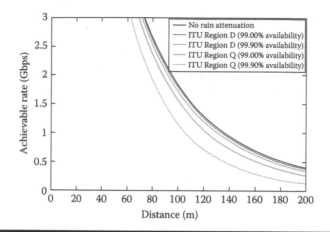

Figure 17.7 Achievable rates in LOS general 60 GHz links.

17.4.2 IEEE 802.11ad Baseband-Based Calculation in 60 GHz mmWave Channels

The achievable rates in the previous section, that is, by Shannon capacity equation–based link budget calculation, can only be obtained when optimum modulation formats and coding schemes are available. Therefore, the Shannon capacity–based approach is a theoretical upper bound. In this section, practical achievable data rates are calculated based on 60 GHz mmWave IEEE 802.11ad baseband parameters (i.e., MCS set). For the IEEE 802.11ad MCS-based link budget calculation, the following three steps are required:

- Step 1: Computing received signal strength.
- Step 2: Finding supportable MCS levels by comparing the receiver sensitivity values in table 21-3 in the IEEE 802.11ad specification and the computed received signal strength in Step 1.
- Step 3: Retrieving achievable rates based on the supportable MCS levels.

For Step 1, the received signal strength depending on the distance between the transmitter and the receiver can be obtained by a calculation procedure equivalent to the procedure in Section 17.4.1:

$$P_{dBm}^{RX}(d) = \mathrm{EIRP} - PL(d) - O(d) - R(d)$$

where:
EIRP is the equivalent isotropically radiated power
$PL(d)$ is the path loss depending on the distance d
$O(d)$ is the oxygen attenuation depending on the distance d
$R(d)$ is the rain attenuation depending on the distance d

Table 17.5   Receiver Sensitivity Values and MCS Matching

Receiver Sensitivity (dBm)	MCS Index (Mbps)	Supportable MCS
−78	MCS0 (27.5)	MCS0
−68	MCS1 (385)	MCS1
−66	MCS2 (770)	MCS2
−65	MCS3 (962.5)	MCS3
−64	MCS4 (1155)	MCS4
−63	MCS6 (1540)	MCS6
−62	MCS5 (1251.25), MCS7 (1925)	MCS7
−61	MCS8 (2310)	MCS8
−59	MCS9 (2502.5)	MCS9
−55	MCS10 (3080)	MCS10
−54	MCS11 (3850)	MCS11
−53	MCS12 (4620)	MCS12

For Step 2, the calculated received signal strength values in Step 1 should be compared with the receiver sensitivity values in table 21-3 in the IEEE 802.11ad specification. Table 17.5 shows this matching.

As presented in Table 17.5, if the received signal strength is about −70 dBm, for example, MCS1 is not supportable, because the value is less than the receiver sensitivity value in MCS1 (i.e., −70 < −68 dBm). Therefore, only MCS0 is supportable. When the received signal strength is −61.5 dBm, there are two choices: MCS5 and MCS7. In this case, the MCS that can support the higher data rate, MCS7, will be selected. Note that Table 17.5 is organized with single-carrier MCS values, which are mandatory features in IEEE 802.11ad. This standard also defines orthogonal frequency multiple duplexing (OFDM)-based MCS and low-power MCS (from MCS13 to MCS24); however, these are optional features and are not included in Table 17.5. Then, final link budget calculation results are plotted as shown in Figures 17.8 and 17.9. Similarly to the plotting in Section 17.4.1, the MCS-based link budget calculation can be performed only at 60 GHz bands, because there are no standards yet at 28 and 38 GHz mmWave bands.

In Step 3, supportable data rates based on the selected MCS values can be directly obtained from Table 17.5.

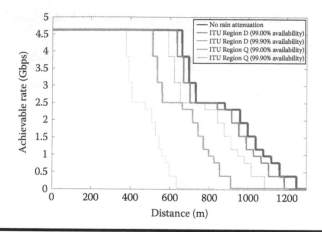

Figure 17.8 MCS-based rates in LOS outdoor point-to-point 60 GHz links.

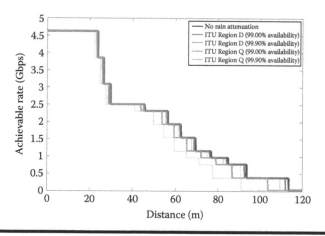

Figure 17.9 MCS-based rates in LOS general 60 GHz links.

Under the additional consideration of the optional OFDM-based MCS and low-power MCS features in addition to the mandatory single-carrier MCS values (from MCS0 to MCS12), Table 17.5 can be revised as shown in Table 17.6, with a similar approach in Step 2; corresponding data rates can be derived by Step 3 and plotted as shown in Figures 17.10 and 17.11.

17.5 Concluding Remarks

This chapter summarizes the major characteristics of 28, 38, and 60 GHz mmWave wireless radio wave propagation. The directionality of the propagation is numerically formulated and simulated based on standard ITU models. In addition, path loss

Table 17.6   Receiver Sensitivity Values and MCS Matching (Including Optional OFDM-Based MCS)

Receiver Sensitivity (dBm)	MCS Index (Mbps)	Supportable MCS
−78	MCS0 (27.5)	MCS0
−68	MCS1 (385)	MCS1
−66	MCS2 (770 , MCS13 (693)	MCS2
−65	MCS3 (962.5)	MCS3
−64	MCS4 (1155), MCS14 (866.25), MCS25 (626)	MCS4
−63	MCS6 (1540), MCS15 (1386)	MCS6
−62	MCS5 (1251.25), MCS7 (1925), MCS16 (1732.5)	MCS7
−61	MCS8 (2310)	MCS8
−60	MCS17 (2079), MCS26 (834)	—
−59	MCS9 (2502.5)	MCS9
−58	MCS18 (2772)	MCS18
−57	MCS27 (1112), MCS28 (1251), MCS29 (1668), MCS30 (2224), MCS31 (2503)	—
−56	MCS19 (3465)	MCS19
−55	MCS10 (3080)	—
−54	MCS11 (3850), MCS20 (4158)	MCS20
−53	MCS12 (4620), MCS21 (4504.5)	MCS12
−51	MCS22 (5197.5)	MCS22
−49	MCS23 (6237)	MCS23
−47	MCS24 (6756.75)	MCS24

models are presented in LOS, and NLOS situations are provided in the mmWave channels. As well as path loss, mmWave wireless channels also include additional mmWave-specific oxygen and rain attenuation. Based on the mmWave propagation models and parameters provided, a link budget calculation is performed to identify what data rates can be obtained depending on the distance between the transmitter

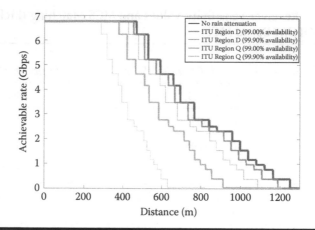

Figure 17.10 MCS-based (for both single-carrier and OFDM MCS) rates in LOS outdoor point-to-point 60 GHz links.

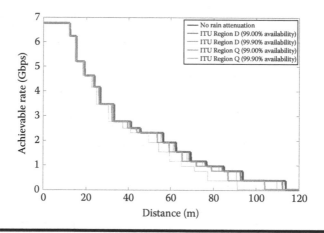

Figure 17.11 MCS-based (for both single-carrier and OFDM MCS) rates in LOS general 60 GHz links.

and the receiver of mmWave wireless propagation links. The link budget calculation is performed in two ways: the Shannon capacity equation and IEEE 802.11ad MCS-based practical data rate estimation.

References

1. J. Kim and A. F. Molisch, Fast millimeter-wave beam training with receive beamforming, *IEEE/KICS Journal of Communications and Networks* 16(5): 512–522, 2014.

2. ITU Recommendation, Reference radiation patterns of omnidirectional, sectoral and other antennas for the fixed and mobile services for use in sharing studies in the frequency range from 400 MHz to about 70 GHz, *ITU-R F.1336-4*, 2014.
3. A. F. Molisch, *Wireless Communications*, 2nd Edn, New York: Wiley, 2011.
4. ITU Recommendation, A general purpose wide-range terrestrial propagation model in the frequency range 30 MHz to 50 GHz, *ITU-R P.2001-1*, 2013.
5. ITU Recommendation, Propagation data and prediction methods for the planning of short-range outdoor radiocommunication systems and radio local area networks in the frequency range 300 MHz to 100 GHz, *ITU-R P.1411-7*, 2013.
6. J. Kim and A. F. Molisch, Quality-aware millimeter-wave device-to-device multi-hop routing for 5G cellular networks, in *Proceedings of IEEE International Conference on Communications (ICC)*, Sydney, IEEE, 2014.
7. Y. Azar, G. N. Wong, K. Wang, R. Mayzus, J. K. Schulz, H. Zhao, F. Gutierrez, Jr., D. D. Hwang, and T. S. Rappaport, 28 GHz propagation measurements for outdoor cellular communications using steerable beam antennas in New York City, in *Proceedings of IEEE International Conference on Communications (ICC)*, Budapest, IEEE, 2013.
8. T. S. Rappaport, F. Gutierrez, Jr., E. Ben-Dor, J. N. Murdock, Y. Qiao, and J. I. Tamir, Broadband millimeter wave propagation measurements and models using adaptive beam antennas for outdoor urban cellular communications, *IEEE Transactions on Antenna and Propagation* 61(4): 1850–1859, 2013.
9. A. Maltsev, E. Perahia, R. Maslennikov, A. Lomayev, A. Khoryaev, and A. Sevastyanov, Path loss model development for TGad channel models, *IEEE 802.11-09/0553r1*, 2009.
10. ITU Recommendation, Attenuation by atmospheric gases, *ITU-R P.676-10*, 2013.
11. ITU Recommendation, Characteristics of precipitation for propagation modelling, *ITU-R PN.837-1*, 1994.
12. Federal Communications Commission (FCC), Office of Engineering and Technology, *Bulletin Number 70*, 1997.
13. FCC, Operation of unlicensed devices in the 57–64 GHz band, *FCC 13-112*, 2013.
14. J. Kim, Y. Tian, S. Mangold, and A. F. Molisch, Joint scalable coding and routing for 60 GHz real-time live HD video streaming applications, *IEEE Transactions on Broadcasting* 59(3): 500–512, 2013.

Chapter 18

mmWave Communication Characteristics in an Outdoor Environment

Péter Kántor and János Bitó

Contents

18.1 Introduction

The amount of data traffic in mobile networks has increased exponentially in recent years due to new technical developments. However, this evolution indicates a global bandwidth shortage for mobile operators. The 5000-fold mobile data traffic increase that is projected by the year 2030 can be met through increased performance, spectrum availability, and massive densification of small cells [1–3]. Recent advances in air interface design provide spectral efficiency performance that is very close to the Shannon capacity limit [2,3]. To overcome this challenge, wireless service providers will need to use the higher-frequency millimete-wave (mmWave) spectrum and apply highly directional beam-forming or beam-steering antennas in fifth-generation (5G) wireless networks [1–4]. Since the available spectrum is a limited resource, it is also clear that increasingly high frequencies must be used in the future [4]. If the trend of rapidly growing demand for mobile data continues in the following years, then future mobile networks (such as 5G mobile networks) in urban areas will have to employ very small cells and much higher frequencies in addition to an overlaying macro layer operating at currently used frequencies, as shown by Rangan et al. and other researchers [4–7].

Radio propagation at millimeter-wavelength frequencies increases the available bandwidth by several orders of magnitude [4–8]. For instance, the unlicensed spectrum at 60 GHz offers a 10–100-fold increase in the available spectrum range compared with what is currently available for industrial, scientific, and medical (ISM) bands [9]. Moreover, the entire bandwidth from 300 MHz to 10 GHz—which covers all present-day cellular systems—is more than seven times less than the bandwidth from 30 to 100 GHz. Therefore, for 5G networks, the target is above 6 GHz, and the research in this field covers electromagnetic field aspects, link budgets, propagation issues, and channel model descriptions, as shown by Karjalainen et al. [10].

Figure 18.1 depicts the mmWave bands of interest in the 20–50 GHz range along with their currently allocated use. As can be observed in Figure 18.1, several contiguous bandwidth segments up to 4–5 GHz can be found in this frequency range [10].

Furthermore, the lower frequencies are currently shared by very many applications. Currently, the International Telecommunication Union (ITU) envisages mobile bands in the 20–50 GHz range as coprimary bands together with other coprimary services, for example, fixed satellite services and navigation, while these

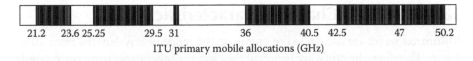

21.2	23.6 25.25	29.5 31	36	40.5 42.5	47	50.2

ITU primary mobile allocations (GHz)

Figure 18.1   Existing allocations for 20–50 GHz mm wave spectrum, with ITU coprimary mobile bands in black.

large chunks of mmWave spectrum could also be allocated for cellular usage. Therefore, the coexistence of mmWave communication with other existing systems needs to be carefully considered [10].

Having access to such large spectrum blocks makes it possible to trade off spectral efficiency for bandwidth. Therefore, high data rates can be achieved even with low-order modulation schemes requiring lower powers and lower complexity and cost [1], since the wider spectrum range makes it possible to achieve higher data rates [9]. Therefore, 5G networks applying the mmWave spectrum present a new opportunity to use channel bandwidths of 1 GHz or more. The spectrum at 28, 38, and 70–80 GHz looks especially promising for next-generation cellular systems due to the propagation characteristics [9], as will be discussed in Section 18.2.

For the abovementioned reasons, mmWave frequency ranges for the 5G standards have started to attract attention within the wireless industry [11–13] and are also on the research agenda of the EU's Horizon 2020 5G Infrastructure Public Private Partnership (5GPPP) initiative [14].

Understanding radio channels and finding accurate channel models are fundamental to developing future mmWave access systems as well as adequate backhaul techniques. However, at the time of writing, outdoor, outdoor–indoor, and vehicular mmWave channel models are limited to trials in both the lower and higher ends of the mmWave spectrum [15]. Channel parameters relating to path loss coefficients, path loss exponent (PLE), root mean square (RMS) delay spread, and angular spread need to be identified based on the new channel models, which are introduced in Sections 18.2.2 and 18.2.3.

These models are highly important for developing and testing the required physical and higher-layer solutions. Moreover, to perform link and system-level feasibility studies and to investigate regulatory issues (such as interference risks and coexistence, and coprimary exploitation of the mmWave spectrum), the newly developed mmWave channel models are also vital [1]. With a firm technical understanding of the mmWave channel, researchers may explore new methods for the air interface, multiple access, and architectural approaches that include cooperation and interference mitigation and other signal-enhancement techniques [9].

18.2 mmWave Channel Characteristics

Millimeter waves are considered to comprise the range of wavelengths from 10 to 1 mm. Therefore, the mmWave region of the electromagnetic spectrum corresponds to the radio-frequency range of 30–300 GHz, and is also called the *extremely high-frequency* (EHF) range [16].

At lower frequencies, radio waves can bend around objects and do not require line-of-sight (LOS) propagation conditions. In contrast, at millimeter wavelengths, the effects of diffraction are relatively minor [9]. Therefore, if a mobile device is behind an obstacle, it is possible that no connection can be established even if the mobile device is close to the base station—thus, no distinct coverage areas exist [9].

There are other unfavorable qualities of mmWave communication links compared with current commercially available, lower-frequency access links, such as increased free-space path loss, increased atmospheric loss, increased signal penetration through obstacles, increased surface scattering, and attenuation due to precipitation. As frequencies increase, the wavelengths become shorter, and the reflective surface appears rougher. This results in more diffused reflection as opposed to specular reflection, as described by Huang and Wang [16]. Recent works presented by Rappaport et al. and other researchers have found that urban environments provide a rich multipath, especially reflected and scattered energy at or above 28 GHz [9–11,15]. In an outdoor environment of 5G access and backbone links operating in the mmWave range, high precipitation can cause outage of a considerable part of the network, which has to be taken into account during network planning [17]. The effects of precipitation on mmWave propagation are explained in detail in Section 18.2.2.2.

In summary, the main factors that affect mmWave propagation are

- Atmospheric gases attenuation
  - Water vapor absorption
  - Oxygen absorption
- Precipitation attenuation
  - Rain, sleet, fog
- Penetration loss
- Foliage blockage
- Scattering effects
  - Diffused reflections
  - Specular reflections
- Diffraction (bending)

The characteristics of large- and small-scale fading caused by these effects need to be described by appropriate channel models. Rappaport et al. [9,15] showed that a fundamental concern of mmWave propagation is the feasibility of non-line-of-sight (NLOS) links or of LOS links exceeding the 100–200 m range in outdoor applications. As will be explained in Section 18.3, the rich multipath environment can be exploited to increase received signal power in NLOS propagation conditions [15].

Due to the research being conducted in the field of mmWave applications, it is foreseeable that mmWave will find its way into 5G vehicular applications in the coming years. Therefore, it is important to model mmWave radio propagation characteristics in intervehicle communications as well. The characteristics of mmWave in vehicular environments and the results of measurements will be introduced in the following in Section 18.2.4.

As will be explained in Section 18.3, mmWave propagation characteristics, despite the challenges, make the mmWave frequency range useful for a variety of applications, including the transmission of large amounts of data.

18.2.1 Free-Space Propagation

The antennas at each end of a path direct the electromagnetic energy toward each other. As with all propagating electromagnetic waves, for mmWaves in free space, the power falls off as the square of the range. This effect is due to the spherical spreading of the radio waves as they propagate [16]. Therefore, with a wavelength of about 5 mm, the free-space propagation loss at 60 GHz is 28 dB higher than at 2.4 GHz [18].

The frequency and distance dependence of the path loss (*PL*) between two isotropic antennas can be expressed by Equation 18.1:

$$PL_{\text{freespace}}^{[dB]} = 20 \cdot \log_{10}\left(4 \cdot \pi \cdot \frac{d}{\lambda}\right) \tag{18.1}$$

where:

$PL_{\text{freespace}}$ is the free-space path loss in decibels
d is the distance between the transmitting and receiving antennas
λ is the operating wavelength [19]

Equation 18.1 describes LOS wave propagation in free space and indicates that the free-space loss increases with the increase of the frequency [19]. Therefore, the mmWave spectrum is best used for short-distance communication links.

For directive antennas, and also taking the system losses into account, the path loss can be expressed by the Friis transmission equation [20]. This accounts more completely for all the factors contributing to the received power P_{RX} [16], according to Equation 18.2 [21]:

$$P_{RX} = P_{TX} \cdot G_{RX} \cdot G_{TX} \frac{\lambda^2}{(4 \cdot \pi \cdot d)^2 \cdot L} \tag{18.2}$$

where:
  G_{TX} is the transmitting antenna gain
  G_{RX} is the receiving antenna gain
  L is the system loss factor (≥ 1)

The transmission loss equation is often expressed in decibels [19] according to Equation 18.3:

$$PL^{[dB]} = 20 \cdot \log_{10}(d \cdot f) + 92.45 - G_{TX}^{[dB]} - G_{RX}^{[dB]} + L^{[dB]} \qquad (18.3)$$

where:
  PL is the path transmission loss
  G are the antenna gains in decibels relative to an isotropic antenna (dBi)
  d is distance in kilometers
  f is the frequency in gigahertz
  L is the system loss in decibels

18.2.2 Large-Scale Fading

An understanding of the propagation media and the circumstances that cause the fading of average received signal levels is required for a characterization of the mmWave wireless channels. To describe this behavior, one needs to consider that the path between transmitter and receiver is time, space, and frequency variant. The variations of the channel can be classified into large-scale fading and small-scale fading, depending on how fast the received power fluctuates [19,21].

Small-scale fading is the rapid fluctuations in received signal level, which appear in two different forms: time-spreading of the signal due to multipath propagation and time-variant behavior of the channel caused by the Doppler effect. Large-scale fading occurs in the received signal over a long period of time or a longer distance (on the order of a wavelength) [21].

As presented by Misra [21], the effects of large-scale fading are usually described by a path loss model that applies some sort of power law. Large-scale fading is typically characterized by path loss $PL(d)$, which is defined as a local average of the received signal power as a function of transceiver–receiver (TX–RX) distance d [22]. Moreover, the variation of the averaged received power is called *shadowing*, and it is often modeled as a zero mean Gaussian random variable X_σ with a standard deviation σ. The formula for modeling $PL(d)$ is typically expressed on a decibel scale, according to Equation 18.4:

$$PL(d) = PL(d_0) + 10 \cdot n \cdot \log_{10}\left(\frac{d}{d_0}\right) + X_\sigma \qquad (18.4)$$

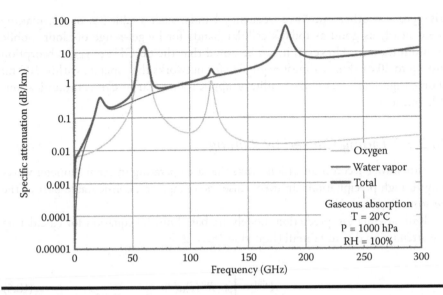

Figure 18.2 Clear air specific attenuation vs. frequency in the UHF, superhigh-frequency (SHF), and extremely high-frequency (EHF) bands. (From Crane, R. K., *Propagation Handbook for Wireless Communication System Design*, CRC LLC, Boca Raton, FL, 2003.)

where:

n denotes the PLE

$PL(d_0)$ is the initial path loss (path loss intercept) at a reference distance d_0 [16]

18.2.2.1 Fading Caused by Atmospheric Gases

As presented by Crane [19], oxygen and water vapor in the lower atmosphere significantly affect path attenuation at higher frequencies, in the introduction of Section 18.2. As an example, specific attenuation for a location at the Earth's surface is depicted in Figure 18.2 for a temperature of 20°C and 100% relative humidity. The oxygen curve gives the specific attenuation for 0% relative humidity.

18.2.2.1.1 Impact on 5G Applications

As reported by Crane in [19], the frequency bands are named *atmospheric windows* between the attenuation peaks around about 22, 60, 118, and 183 GHz due to the relatively low specific attenuation. In the frequency window below the water vapor absorption line at 22.3 GHz, the specific attenuation increases with frequency and may be more than 10 times higher at 15 GHz than at 2 GHz. Therefore, long-distance terrestrial microwave links are possible at the lower frequencies in this window [19]. Also, the 20 dB/km oxygen attenuation at 60 GHz disappears at other mmWave frequency bands, such as 28, 38, or 72, making them nearly as good as today's cellular bands for longer-range outdoor mobile communications. However, it can be noted that the 60 GHz oxygen absorption loss up to 20 dB/km is almost negligible for networks that operate within 100 m: short-range mesh networks, military applications, car-to-car communications [9], and so on.

18.2.2.2 Fading Caused by Raining

For mmWave telecommunications links that are operating in the millimeter wavelength, high precipitation can even cause the outage of a considerable part of the network [17,19].

Rain attenuation prediction models are based on the equation for calculating the attenuation due to rainfall (Equation 18.5) [23]:

$$A^{[dB]} = \int_0^d k \cdot R^\alpha(l) dl \tag{18.5}$$

where:
k and α are frequency- and polarization-dependent empirical coefficients
$R(l)$ is the value of the point rain intensity in millimeters per hour along the path at distance l
d is the path length of the link

Rain events are highly inhomogeneous in time and space; therefore, the rain rate and hence the attenuation may vary significantly along longer paths, and for practical use a path-average value can be considered. Therefore, the most important task for modeling rain attenuation is statistically describing the distribution of the rain rate across space and time.

The most commonly used rain attenuation prediction model is the ITU-R P.530-16 model [23], which does not use the full rain rate distribution but only one parameter, $R_{0.01}$, representing the rain rate exceeded for 0.01% of an average year (with an integration time of 1 min).

The model assumes an equivalent rain cell with exponential spatial rainfall distribution (the EXCELL model), which can represent the effect of the nonuniform rainfall rate along the propagation path. The ITU gives the following empirical equation (Equation 18.6) for $A_{0.01}$ (dB) rain attenuation level, which is exceeded with a probability of $p = .01\%$, where d (in kilometers) is the path length and r is the distance factor [23].

$$A_{0.01}^{[dB]} = k \cdot R_{0.01}^\alpha \cdot d_{eff} = k \cdot R_{0.01}^\alpha \cdot d \cdot r \tag{18.6}$$

Assuming that the EXCELL equivalent rain cell may intercept the link at any position with equal probability, the variation of the rain intensity along the path of the link is taken into account by the distance factor r (7), where f (in gigahertz) is the frequency. The ITU also states that the value of r should be capped at a maximum of 2.5 [23].

$$r = \frac{1}{0.477d^{0.633} \cdot R_{0.01}^{0.073*\alpha} \cdot f^{0.123} - 10.579(1 - e^{-0.024d})} \qquad (18.7)$$

To calculate the attenuation exceeded at other percentages of time A_p between $p = 1\%$ and .001%, an extrapolation formula is used (the parameters of C1, C2, and C3 are given in [23]):

$$A_p = C_1 \cdot p^{-(C_2 + C_3 \log_{10} p)} \cdot A_{0.01} \qquad (18.8)$$

The theoretical complementary cumulative distribution function (CCDF) of rain attenuation based on the ITU-R P.530-16 model for different link lengths is shown in Figure 18.3. The parameters of the calculation were as follows: $f = 38$ GHz, $R_{0.01} = 42$ mm/h, and horizontal polarization was assumed. It can be observed that considerable additional attenuation can be detected due to rainfall even at relatively short link lengths.

18.2.2.2.1 Impact on 5G Applications

The maximum tolerable path loss in such a 5G network can be obtained using Equation 18.9 [24]:

$$PL_{\max} = PL(d_{\max}) = PL(d_0) + 10 \cdot n \cdot \log_{10}\left(\frac{d_{\max}}{d_0}\right) \qquad (18.9)$$

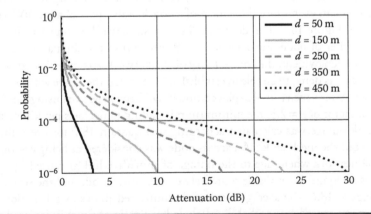

Figure 18.3 Theoretical CCDF of rain attenuation for different link lengths according to the ITU-R P.530-16 model.

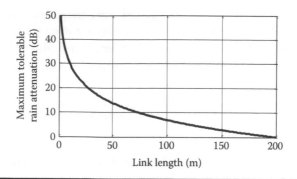

Link length (m)

Figure 18.4   Maximum tolerable rain attenuation as a function of the link length (From Kántor, P., et al., Precipitation modelling for performance evaluation of ad-hoc microwave 5G mesh networks, *2015 9th European Conference on Antennas and Propagation [EuCAP2015]*, Lisbon, IEEE, 2015.).

where d_{max} is the maximum applicable TX–RX separation [15].

The maximum tolerable additional attenuation caused by rain for a link with a length of $d \leq d_{max}$ can be calculated according to Equation 18.10 [17]:

$$A_{rain}(d) \leq PL_{max} - PL(d) = 10 \cdot n \cdot \log_{10}\left(\frac{d_{max}}{d}\right) \qquad (18.10)$$

The maximum tolerable rain attenuation is depicted in Figure 18.4 as a function of the link length for $n = 2.30$, $d_{max} = 200$ m, and $d_0 = 5$ m (at $f = 38$ GHz)—which are the parameters measured by Rappaport et al. in [15]. The corresponding channel model is discussed in detail in Section 18.3.

In [17], simulations were conducted by Kántor et al. to investigate whether high precipitation can even cause the outage of a considerable part of an mmWave mesh network. In Figure 18.5, a random mmWave mesh network is depicted, in which a certain number of nodes are deployed randomly within the simulation area. After the random deployment, the connectivity of the network was investigated [17]. Two nodes were considered to be able to establish a LOS link if they were within 200 m of each other. After a scenario of deployed nodes had been generated using the 38 GHz LOS PLE factor of $n = 2.3$, the minimum rain fading in decibels per kilometer that causes link outage was calculated according to Equation 18.7 for every mmWave link between the mesh nodes. An access node was considered as being disconnected from the network when, due to the outage of mmWave link(s) caused by raining, there was no longer an available route between the given node and the sink node.

In Figure 18.6, the average ratio of disconnected nodes as a function of the rain intensity exceedance probability (labeled as a ratio of time) is depicted assuming rerouting as the only technique to improve network resilience [17]. It can be observed that by applying a random node deployment, the average ratio of nodes

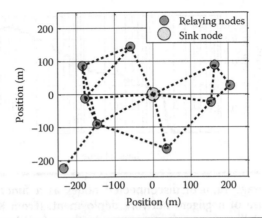

Figure 18.5 An example of a random node deployment with 10 nodes comprising a mesh network graph. (From Kántor, P., et al., Precipitation modelling for performance evaluation of ad-hoc microwave 5G mesh networks, *2015 9th European Conference on Antennas and Propagation [EuCAP2015]*, Lisbon, IEEE, 2015.)

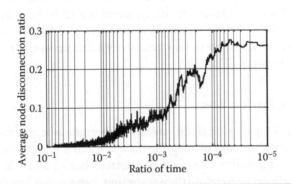

Figure 18.6 Average ratio of disconnected nodes as a function of the ratio of time over a randomly deployed mmWave mesh network. (From Kántor, P., et al., Precipitation modelling for performance evaluation of ad-hoc microwave 5G mesh networks, *2015 9th European Conference on Antennas and Propagation [EuCAP2015]*, Lisbon, IEEE, 2015.)

that cannot reach the sink node is higher than 10% with a probability of 6.48×10^{-4}. This means that for 5.5 h cumulative time duration in an average year, at least 10% of the nodes are disconnected.

Moreover, in Figure 18.7, it can be observed that an unfavorable network deployment (depicted in Figure 18.5) may seriously affect the network performance. In such an unfavorable node deployment, the ratio of disconnected nodes can be significantly higher than in the averaged deployment cases. The average ratio of nodes that cannot reach the sink node through any other node is higher than 20% with

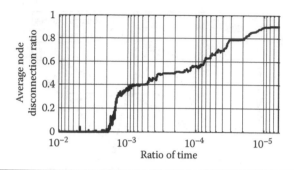

Figure 18.7 **Average ratio of disconnected nodes as a function of the ratio of time in the case of negligent network deployment. (From Kántor, P., et al., Precipitation modelling for performance evaluation of ad-hoc microwave 5G mesh networks,** *2015 9th European Conference on Antennas and Propagation* **[*EuCAP2015*], Lisbon, IEEE, 2015.)**

a probability of 1.5×10^{-3}. This probability is more than 10 times higher than the one obtained by assuming random node deployment. This means that in an average year, at least 20% of the nodes are disconnected for 13 h. Moreover, in an average year, at least 50% of the nodes are disconnected for 2.5 h. It is evident that an unfavorable network deployment may excessively affect network performance in an mmWave mesh network [17].

18.2.2.3 Penetration Loss

As described by Rappaport et al. [9], another important aspect of mmWave propagation is the greater amount of rough-surface scattering due to the smaller wavelength, especially for walls made of rough concrete, bricks, and other construction materials. In addition to rough-surface scattering, penetration loss due to building material also has to be considered, since mmWaves have a weak ability to diffract around obstacles with a size significantly larger than their wavelength [25].

As described by Niu et al. [25], typical relative permittivity and conductivity for different building materials are given in ITU-R 1238-8 and other studies [26,27], while a comparative study between 5.8 and 62.4 GHz is given in [28] by Cuinas et al. The ITU-R recommendation gives an expression for the conductivity σ of building materials. This gives a value of 0.0326 at 1 GHz in contrast to 0.908 at 60 GHz, leading to higher penetration loss.

18.2.2.3.1 Impact on 5G Applications

Cuinas et al. [28] reported that the penetration loss into buildings at 60 GHz is on the order of 3.44 dB for a plastic partition with 0.8 cm thickness, 6.09 dB for

0.8 cm plywood, 9.24 dB for 1.8 cm wood board, and 4 dB for 0.7 cm tempered glass when both the transmit and receive antennas are vertically polarized. Zhao et al. [29] conducted penetration and reflection measurements at 28 GHz in New York City, and found that tinted glass and brick pillars have high penetration losses of 40.1 and 28.3 dB, respectively.

The frequency dependence of penetration loss has also been reported to vary from 18.9 dB at 900 MHz, to 26 dB at 11.4 GHz, to 36.2 dB at 28.8 GHz for a three-wall partition between antennas [1]. These additional losses at the higher frequencies will require compensation through higher effective radiated powers. As described by Niu [25], for the reflection measurement, outdoor materials have higher reflection coefficients, and indoor materials have lower reflection coefficients.

18.2.3 Small-Scale Fading

Small-scale fading describes the rapid fluctuation in signal amplitude and phase that occurs in the received signal over a short period of time or a short distance (on the order of a wavelength) [21]. These types of fluctuations appear in two different forms: time-spreading of the signal due to multipath propagation and time-variant behavior of the channel caused by the Doppler effect.

18.2.3.1 Multipath Propagation

Fading due to multipath propagation is caused by the presence of multiple copies of the transmitted signal that occur at the receiver due to reflections against obstacles, as shown by Misra [21]. This phenomenon occurs because waves traveling along different paths may be completely out of phase when they reach the antenna, as described by Crane [19]. Therefore, these multiple replicas of the transmitted signal are superposed in either a constructive or a destructive manner depending on the phase of each partial wave [21]. These additions may create a fading notch in the received signal power and distort the frequency response characteristics of the transmitted signal. However, these distortions are linear and need to be combined at the receiver by applying equalization and diversity [21].

Xu et al. [30] showed that to measure channel time dispersion, the bandwidth of the channel sounder must exceed the channel coherence bandwidth. This ensures that all significant multipath components can be resolved and recorded in a power delay profile (PDP). Important and generally known parameters can be extracted from the PDP and used to evaluate the delay dispersion of the mmWave channel [22].

Huang et al. [16] describe that for antennas with a narrow beamwidth, a notch appears in the frequency response of channel measurement. Moreover, for antennas with a broad beamwidth, the notch in the frequency response becomes severe. The notch step is affected by the delay time, while the notch depth is affected by the difference in the path gains (or losses). In addition, the notch position is affected by the difference in the lengths of the propagation paths [16].

The maximum excess delay τ_{max} is the maximum delay value, after which all the power levels of the multipath components are below some threshold value [22]. The square root of the second central moment of the PDP is called the RMS delay spread τ_{RMS}, according to Equation 18.11:

$$\tau_{RMS} = \sqrt{\frac{\sum_{i=1}^{N} (\tau_i - \overline{\tau})^2 \cdot P_i}{\sum_{i=1}^{N} P_i}} \tag{18.11}$$

where τ_i and P_i are the excess delay and the power level of the ith multipath component of the PDP, respectively, as described by Kyrö [22].

18.2.3.2 Angular Distribution of the Propagation Channel

The angle spread is defined as the standard deviation of the direction of rays [31]. The angular characteristics of the propagation channel can be measured by rotating directive antennas, as presented in [22,32–37]. However, rotating the TX and RX antennas increases the measurement time significantly and limits the channel measurements to static channels only, as performed, for example, by Rappaport et al. [15].

Another way to estimate the angle of arrival (AOA) and angle of departure (AOD) parameters is to use multiple-input multiple-output (MIMO) measurements and beam forming or other estimation methods [22]. At mmWave frequencies, MIMO measurements are usually performed with virtual antenna arrays, as presented by Ranvier et al. in [24,38].

18.2.3.3 Weather Effects on Multipath Characteristics

Wideband measurements showed significant changes in multipath characteristics during certain weather events. Xu et al. presented measurements in [30] in which although no multipath component was detected during clear conditions, a multipath component was detected about 16 dB below the direct path during light rainfall and 12 dB below the direct path during moderate rain before a hailstorm. Two hypotheses may explain the presence of these multipath components. According to the first one, multipath components that occurred right before and after a hailstorm may have been caused by the sharp edge of the hailstorm cell [30]. This hypothesis is supported by measurements presented by de Wolf and Ligthart in [39], where it was shown that multipath components can occur at the edges of very intense and compact rain cells, because pressure, temperature, and rain can alter the refractivity of the atmosphere, thus creating varying propagation paths and propagation delays. The second hypothesis is based on the change of the electromagnetic properties of the surface or the formation of standing water surfaces during rain. Therefore, if the surface becomes wet or a standing water surface forms during a rain event, the reflected power in the specular direction would increase. In the measurement presented in [30], multipath components remained after rain, which seems to support the second hypothesis.

18.2.3.3.1 Impact on 5G Applications

The fundamental concern, due to the characteristics of the small-scale fading of mmWave, is the feasibility of NLOS links or of LOS links exceeding the 100–200 m range in outdoor applications [9,15]. However, as will be explained in Section 18.3.2, the rich multipath environment can be exploited to increase received signal power in NLOS propagation conditions [15].

18.2.4 *mmWave Characteristics in Vehicular Environment*

Takahashi et al. showed in [40] that an abrupt and substantial increase in path loss due to interruption, curves, and traveling in different lanes has been a major concern in intervehicle communications. Although the vehicles move quickly on the ground, the relative speed between the communicating vehicles is low. Thus, direct communications between such mobile stations is similar to that of conventional fixed stations in low frequency ranges. In contrast, mmWave-based communication may be seriously affected by even small amounts of relative movement between vehicles, due to the short wavelength, as presented by Kato et al. [41].

Moreover, the influence of the multipath effect from the surrounding terrestrial features near the road may be relatively small, as the antenna will have a relatively narrow beamwidth. However, the signal reflection from the road surface is not negligible, and very strong fading may occur due to interference between the direct waves and the reflected waves from the road surface [41]. In this way, the radio wave–propagation phenomenon observed during mmWave intervehicle communication will differ significantly from that of conventional mobile communications. Therefore, it is important to clarify the propagation mechanism and establish a propagation model for the practical use of such a 5G system. Such models are presented in [40–43].

Several realistic radio propagation models have been proposed, and the characteristics of path loss variation have been derived [40]. Kato et al. [41] suggest Equation 18.12 to obtain received power P_{RX}:

$$P_{RX} = \frac{P_{TX} \cdot G_{RX} \cdot G_{TX}}{L(d)} \cdot \left(\frac{\lambda}{2 \cdot \pi \cdot d}\right)^2 \cdot \sin^2\left(\frac{2 \cdot \pi \cdot h_{RX} \cdot h_{TX}}{\lambda \cdot d}\right) \qquad (18.12)$$

considering only the direct wave and the reflected wave when a transmitter is held at a height h_{TX} and a receiver at a height h_{RX}. The path lengths of the direct wave and the wave reflected from the road surface are considered to be identical (d); $L(d)$ is the absorption factor [44]; λ is the wavelength of the carrier wave; and G_{TX} and G_{RX} are the gains of the boresight of the transmitter and the receiver, respectively.

18.2.4.1 Impact on 5G Applications

Measurements of radio propagation in vehicular environments are important, because the path loss in intervehicle communications may be site dependent [44]. Takahashi et al. [40] also give figures to demonstrate the relationship between the short-term median value of received power and the distance between vehicles described by Equation 18.12 for a center frequency of 59.1 GHz with different antenna heights. The curve in Figure 18.8 represents the received power predicted by a model using Equation 18.12 for the individual intervehicle distances, modeled in [41]. It can be observed that the calculated curve implies that the level of received power changes along with the distance between vehicles due to interference between direct waves and reflected waves.

Kato et al. [41] indicated through measurements that the level of received power may also fluctuate significantly, even when the distance between vehicles remains constant. This is probably because the actual height of the vehicle changes as it travels on the road. Under the given conditions, the received power fluctuates even if the distance between vehicles remains unchanged.

Moreover, Takahashi et al. [40] observed additional losses of 15 dB for highways and 5 dB for regular roads when the intervehicle distance was more than approximately 30 m. The results for highways showed that the path loss in a quasistatic environment can be accounted for by interference between two dominant

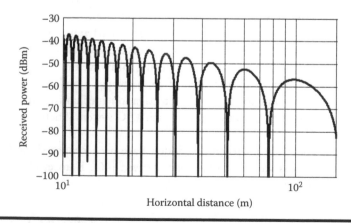

Figure 18.8 Calculated results for the short-term median value of received power in relation to the horizontal distance.

radio waves. Their results also suggested that the path loss in the Rayleigh amplitude distribution was caused by the vertical fluctuation of the cars.

Nevertheless, an abrupt and substantial increase in path loss due to interruption, curves, and traveling in different lanes is a major concern in intervehicle communications.

18.3 mmWave Propagation Models

As mentioned in Section 18.2, due to the high propagation loss explained in the previous subsections and the low power budget at mmWave bands, the feasibility of NLOS links or of LOS links exceeding the 100–200 m range is a fundamental concern [9]. Therefore, such models are especially important, since the application of mmWave transmission to multigigabit wireless systems adopting advanced massive MIMO and beam-forming techniques is envisaged [1,15].

Nowadays, there are different approaches to modeling the 5G mmWave propagation channel. Geometry-based stochastic channel models (GSCM) provide path-loss models by fitting a least-squares linear regression best line fit to the measured path losses [45]. Another approach is the so-called close-in free-space reference path-loss model [9]. This model applies a so-called reference distance—for a definition, see Equation 18.4—over which free-space propagation is assumed. In contradiction to stochastic approaches, ray-tracing methods also are commonly applied. There are also path loss and shadowing models exploiting the combination of these methods.

As stated in Rappaport et al. [9], path loss and shadowing models for narrow antenna beamwidth are of vital importance for the design of mmWave radio links. Moreover, propagation models need to account for time and the angle dispersion of the signal, as explained in Section 18.2.3.

18.3.1 Geometry-Based Stochastic Channel Models

The WINNER II/ITU IMT-Advanced/3GPP 3-D propagation model provides omnidirectional path loss models by fitting measurements with the smallest standard deviation between the regression line and the measured data in a least-squares linear regression [46]. This is called a *floating-intercept model*.

These path loss models are typically described in the form of Equation 18.13:

$$PL = A \cdot \log_{10}(d^{[m]}) + B + C \cdot \log_{10}\left(\frac{f^{[GHz]}}{5}\right) + X \qquad (18.13)$$

where:

d is the distance between the transmitter and the receiver in meters
f is the system frequency in gigahertz

A   is the fitting parameter, which includes the *PLE*
B   is the intercept
C   is a path loss frequency-dependent parameter
X   is an optional, environment-specific term (e.g., wall attenuation) [45,47]

The models can be applied in the frequency range from 2–6 GHz and for different antenna heights [45]. However, in principle, a similar floating-intercept model can be applied for higher frequencies as well (such as for 28 and 73 GHz in [46]).

18.3.2 Close-In Free-Space Reference Path Loss Models

The parameters used in a close-in free-space reference model provide physical insight into channel propagation as compared with free-space propagation, unlike the models applying GSCM methods, which provide a best minimum error fit to collected path losses, as stated by MacCartney et al. [46].

Rappaport et al. [15] concluded from measurements at 28 and 73 GHz that mmWave channels are more directional at both the TX and the RX than conventional microwave (ultra high-frequency [UHF]) channels. Moreover, Jämsä et al. discussed in [47] that in contrast to the omnidirectional path loss models that are commonly applied (i.e., GSCM), beam-forming and beam-combining technologies require mmWave directional path loss models that allow one to estimate the power level received by narrow-beam antennas in a given direction.

Therefore, Rappaport et al. [15,48] conducted the 28 GHz urban propagation campaign in New York City, where steerable directional antennas were applied. The distance between the transmitter and the receiver ranged from 75 to 125 m. By employing highly directional steerable horn antennas to simulate an antenna array, they were able to obtain AOA and AOD data, which are necessary to determine the multipath angular spread. Samimi et al. [49] also conducted AOA and AOD measurement in outdoor urban environments in New York City. They found that New York City has a rich multipath when using highly directional steerable horn antennas, and at any receiver location.

Due to the highly reflective outdoor environment, PDPs displayed numerous multipaths with large excess delay for both LOS and NLOS environments. The average number of resolvable multipath components in an LOS environment was 7.2 with a standard deviation of 2.2 for a TX–RX separation of less than 200 m [15]. The results show that the LOS PLE is 2.55 (see Equation 18.4), resulting from all the measurements acquired in New York City [15].

NLOS measurements with TX–RX separation less than 100 m showed that the number of average received multipath components is 6.8, with a standard deviation equal to that of a LOS case. However, the average PLE in the NLOS situations was increased to 5.76 [15].

Azar et al. [50] reported an outage study conducted in Manhattan, New York. They found that the signal acquired by the RX for all cases was within 200 m;

however, beyond 200 m, 57% of locations were in outage due to obstructions. The maximum coverage distance was shown to increase with increasing antenna gains and a decrease of the PLE.

Similar results were reported by Akdeniz et al. [7], derived from measurements of channels at 28 and 73 GHz in New York. The parameters of the applied channel model obtained from their measurements include the PLE, the number of spatial clusters, the angular dispersion, and outage. It was found that even in highly NLOS environments, strong signals can be detected 100–200 m from potential cell sites, and spatial multiplexing and diversity can be supported at many locations, with multiple path clusters received.

Rappaport et al. [51–53] conducted 38 GHz cellular propagation measurements in Austin, Texas, at the University of Texas main campus. The LOS PLE was measured to be 2.30, while the NLOS PLE was 3.86. Based on an outage study, it was found that base stations of lower heights have better close-in coverage, and most of the outages occurred at locations beyond 200 m from the base stations.

As a conclusion from several field test measurements reported by Rappaport and his team, the maximum coverage distance achieved is 200 m in a highly obstructed outdoor environment. Furthermore, the outage probability is greatly affected by the transmitted power and antenna gains as well as the propagation environment [15].

In addition to PLE, PDPs and RMS delay spread characteristics are also important to accurately describe the mmWave channel [54]. As Rappaport et al. presented in [15], most LOS measurements had minimal RMS delay spread, on the order of 1.1 ns. The NLOS measurements exhibited higher and more varied delay spreads, with a mean of around 14 ns. More than 80% of the NLOS links had RMS delay spreads under 20 ns.

As a conclusion, it can be stated that the delay spreads with each antenna were nearly identical in distribution, despite the discrepancy in NLOS path loss.

Moreover, directional beam-steering antennas such as antenna arrays may also be used to reduce the RMS delay spread. Murdock et al. [51] showed that RMS delay spread increases with combined angle, as steeper angles are correlated with a higher number of signal bounces from the transmitter to the receiver. At small angles, both antennas exhibit low variance in delay spread. RMS delay spread is inversely proportional to the transmitter separation distance and directly proportional to the combined off-boresight angle. The results also show that angle spread (AOAs) occurs mostly when the RX azimuth angle is between −20° and +20° about the boresight of the TX azimuth angle.

Ben-Dor et al. [35] conducted wideband propagation measurements in cellular peer-to-peer outdoor environments and in-vehicle scenarios. Path loss data for 10 measured courtyard locations were calculated with respect to a $d_0 = 3$ m free-space reference distance (defined in Equation 18.4) for 60 GHz measurements.

A minimum mean square error (MMSE) best-fit path loss model for LOS courtyard measurements with antennas pointing at each other yielded a PLE of

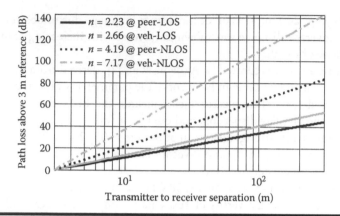

Figure 18.9   Path loss for peer-to-peer and vehicle environments.

$n = 2.23$ ($\sigma = 1.87$ dB) and 2.66 ($\sigma = 5.4$ dB) for peer-to-peer and vehicle environments, as shown in Figure 18.9. NLOS antenna pointing conditions, in which transmitter and receiver antennas were each pointed to find reflective objects, provide $n = 4.19$ ($\sigma = 9.98$ dB) and 7.17 ($\sigma = 23.8$ dB), respectively, for the courtyard and into a vehicle.

Figure 18.9 shows that high delay spreads are less likely at large TX–RX separations (>50 m). This is explained by the high absorption of energy at 60 GHz, which reduces the number of observable multipath components when directional antennas are used [35].

18.3.3 Ray-Tracing Simulations

Another alternative approach for obtaining channel characteristics is a ray-tracing simulation, which has shown good agreement with real measurement results from various papers [31]. Ray-tracing models represent an appropriate choice for deterministic channel modeling of mmWave propagation. By ray-tracing prediction methods, path loss models can be derived [55,56], or multidimensional channel characterization can be performed directly—often in combination with measurements [57,58]. From the ray-tracing simulation, both large-scale and small-scale channel statistics, such as delay spread and angle spread, can be derived [31].

In deterministic modeling, it is important to know the material parameters of different building materials. Measured reflection coefficients for different wall materials in the 60 GHz frequency range have been presented in [59,60].

Nguyen et al. [61] conducted a wideband propagation measurement campaign using rotating directional antennas at 73 GHz at the New York University (NYU) campus, and based on these results, they presented an empirical ray-tracing model to predict the propagation characteristics at the 73 GHz E-Band.

Hur et al. presented in [31] a full 3-D ray-tracing simulation based on geometrical optics and the uniform theory of diffraction. At each receiver point, several rays were collected in descending order, and then a channel impulse response including signal power, phase, propagation time, and AOD and AOA for both azimuth and elevation were calculated. They compared different path loss models by ray-tracing simulations and demonstrated similarities among the corresponding results in dense urban scenarios, even though the transmitter antenna height introduced different slope and intercept into the models.

18.3.4 Combined Methods

The Mobile and Wireless Communications Enablers for the Twenty-Twenty Information Society (METIS) channel modeling approach consists of a map-based model and a number of alternatives to stochastic models. The objective of METIS is to prepare a frequency-dependent path loss model for a huge frequency range, from 0.45 to 86 GHz [62].

The METIS channel modeling approach is depicted in Figure 18.10 [47]. The METIS path loss model is still emerging, because it is challenging to reach this goal due to the limited measurement data. It is also planned to implement ray tracing–based path loss models in the METIS channel model.

18.4 Summary

With the potential for greater capacity, exceeding the capacity of current communication systems by orders of magnitude, mmWave communications have become a promising candidate for 5G mobile networks. This is supported by the availability of unallocated spectrum and the possibility of exploiting the cognitive spectrum management in the mmWave band. This feature offers an opportunity to provide high data rates to enable immersive user experiences currently unachievable with cellular radio networks.

The application of these mmWave frequencies to cellular users presents a challenge due to shadowing effects and the need for adaptive beam forming in a high-mobility environment. In addition, path loss and shadowing effects will impact the data rate, and there is a need for extensive measurements to obtain full radio propagation characterization. Therefore, gigabit data transmission in these bands requires accurate channel models verified by sophisticated channel sounders to provide precise angular spread and time delay characterization of the multipath components. Furthermore, in outdoor environments, the effects of precipitation will also have to be taken into account.

mmWave channel models are still emerging. The parameters of channel models applicable for 5G network planning should be based on large measurement databases. Furthermore, other aspects of mmWave propagation over short-range

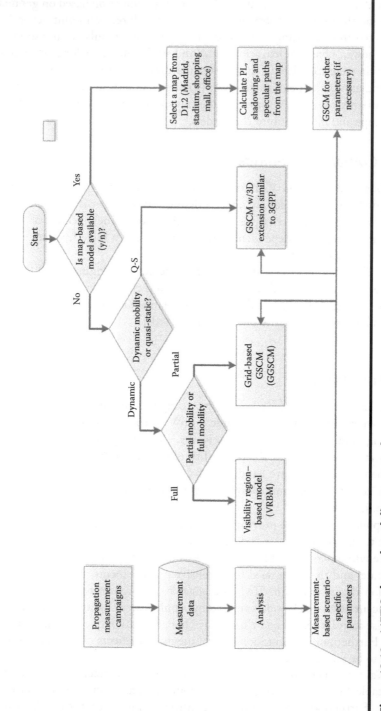

Figure 18.10 METIS channel modeling approach.

5G links, such as foliage blockage and attenuation caused by sleet, need to be further investigated.

However, recent research has shown that even in highly NLOS environments, strong signals can be detected 100–200 m from potential cell sites, and spatial multiplexing and massive MIMO can be supported at many locations, with multiple path clusters received.

References

1. S. Salous, V. Degliesposti, M. Nekovee, and S. Hur, Millimeter-wave propagation characterization and modelling towards 5G systems, *10th IC1004 MC and Scientific Meeting*, contribution TD(14)10091, Aalborg, 2014.
2. Cisco Systems, The zettabyte era: Trends and analysis, White Paper, Cisco Visual Networking Index (VNI), 2014.
3. Cisco Systems, Global mobile data traffic forecast update 2014–2019, White Paper, Cisco Visual Networking Index (VNI), 2015.
4. P. Zhouyue and F. Khan, An introduction to millimeter-wave mobile broadband systems, *IEEE Communications Magazine* 49(6): 101–107, 2011.
5. S. Rangan, T. S. Rappaport, and E. Erkip, Millimeter wave cellular wireless networks: Potentials and challenges, *Proceedings of the IEEE* 102(3): 366–385, 2014.
6. F. Boccardi, R. W. Heath, A. Lozano, T. L. Marzetta, and P. Popovski, Five disruptive technology directions for 5G, *IEEE Communications Magazine* 52(2): 74–80, 2014.
7. M. R. Akdeniz, L. Yuanpeng, M. K. Samimi, S. Sun, S. Rangan, T. S. Rappaport, and E. Erkip, Millimeter wave channel modeling and cellular capacity evaluation, *IEEE Journal on Selected Areas in Communications* 32(6): 1164–1179, 2014.
8. A. Ghosh, T. A. Thomas, M. C. Cudak, R. Ratasuk, P. Moorut, F. W. Vook, T. S. Rappaport, G. R. MacCartney, S. Sun, and S. Nie, Millimeter-wave enhanced local area systems: A high-data-rate approach for future wireless networks, *IEEE Journal on Selected Areas in Communications* 32(6): 1152–1163, 2014.
9. T. Rappaport, R. W. Heath, R. Daniels, and J. Murdock, *Millimeter Wave Wireless Communications*. Westford, MA: Pearson Education, 2014.
10. J. Karjalainen, M. Nekovee, H. Benn, W. Kim, J. Park, and H. Sungsoo, Challenges and opportunities of mm-wave communication in 5G networks, *2014 9th International Conference on Cognitive Radio Oriented Wireless Networks and Communications (CROWNCOM)*, Oulu, pp. 372–376, 2014.
11. W. Roh, J.-Y. Seol, J. Park, B. Lee, J. Lee, Y. Kim, J. Cho, K. Cheun, and F. Aryanfar, Millimeter-wave beamforming as an enabling technology for 5G cellular communications: Theoretical feasibility and prototype results, *IEEE Communications Magazine* 52(2): 106–113, 2014.
12. Nokia Solutions and Networks, White Paper—Looking Ahead to 5G—Building a Virtual Zero Latency Gigabit Experience, White Paper, 2014.
13. IWPC, White Paper on 5G: Evolutionary and disruptive visions towards ultra-high capacity networks, *International Wireless Industry Consortium*, White Paper, 2014.
14. Advanced 5G Network Infrastructure for the Future Internet—Public Private Partnership in Horizon 2020, *Horizon 2020 5GPPP Initiative*, 2013.

15. T. S. Rappaport, R. S. Shu, H. Z. Mayzus, Y. Azar, K. Wang, G. N. Wong, J. K. Schulz, M. Samimi, and F. Gutierrez, Millimeter wave mobile communications for 5G cellular: It will work! *IEEE Access* 1: 335–349, 2013.

16. K-C. Huang and Z. Wang, *Millimeter Wave Communication Systems*, Piscataway, NJ: Wiley-IEEE, 2011.

17. P. Kántor, L. Csurgai-Horváth, Á. Drozdy, and J. Bitó, Precipitation modelling for performance evaluation of ad-hoc microwave 5G mesh networks, *2015 9th European Conference on Antennas and Propagation (EuCAP2015)*, Lisbon, IEEE, paper: 1570048329, 2015.

18. G. R. MacCartney and T. S. Rappaport, 73 GHz millimeter wave propagation measurements for outdoor urban mobile and backhaul communications in New York City, *2014 IEEE International Conference on Communications*, Sydney, NSW, IEEE, pp. 4862–4867, 2014.

19. R. K. Crane, *Propagation Handbook for Wireless Communication System Design*, Boca Raton, FL: CRC LLC, 2003.

20. H. T. Friis, A note on a simple transmission formula, *Proceedings of the IRE* 34(5): 245–256, 1946.

21. S. Misra, *Selected Topics in Communication Networks and Distributed Systems*, New York: World Scientific, 2009.

22. M. Kyrö, Radio wave propagation and antennas for millimeter-wave communications, Doctoral dissertation, Aalto University, 2012.

23. ITU Recommendations P.530-16, *Propagation Data and Prediction Methods Required for the Design of Terrestrial Line-of-Sight Systems*, 2015.

24. S. Ranvier, J. Kivinen, and P. Vainikainen, Millimeter-wave MIMO radio channel sounder, *IEEE Transactions on Instrumentation and Measurement* 56(3): 1018–1024, 2007.

25. Y. Niu, Y. Li, D. Jin, L. Su, and A. V. Vasilakos, A survey of millimeter wave (mmWave) communications for 5G: Opportunities and challenges, *Wireless Networks* 21(8): 2657–2676, 2015.

26. L. M. Correia and P. O. Françês, Estimation of materials characteristics from power measurements at 60 GHz, *IEEE International Symposium on Personal, Indoor, Mobile Radio Communications*, The Hague, IEEE, pp. 510–513, 1994.

27. J. Lu, D. Steinbach, P. Cabrol, P Pietraski, and R. V. Pragada, Propagation characterization of an office building in the 60 GHz band, *8th European Conference on Antennas and Propagation (EuCAP 2014)*, The Hague, IEEE, pp. 809–813, 2014.

28. I. Cuinas, J-P. Pugliese, A. Hammoudeh, and M.G. Sanchez, Comparison of the electromagnetic properties of building materials at 5.8 GHz and 62.4 GHz, *Vehicular Technology Conference, 2000. IEEE 52nd VTS Fall*, Boston, MA, IEEE, vol. 2, pp. 780–785, 2000.

29. H. Zhao, R. Mayzus, S. Shu, M. Samimi, J. K. Schulz, Y. Azar, K. Wang, G. N. Wong, F. Gutierrez, and T. S. Rappaport, 28 GHz millimeter wave cellular communication measurements for reflection and penetration loss in and around buildings in New York City, in *Proceedings of the IEEE International Conference*, Budapest, IEEE, pp. 5163–5167, 2013.

30. H. Xu, T. S. Rappaport, R. J. Boyle, and J. H. Schaffner, Measurements and models for 38-GHz point-to-multipoint radiowave propagation, *IEEE Journal on Selected Areas in Communications*, 18(3): 310–321, 2000.

31. S. Hur, Y. Chang, S. Baek, Y. Lee, and J. Park, mmWave propagation models based on 3D ray-tracing in urban environments, *10th IC1004 MC and Scientific Meeting*, contribution TD(14)10054, Aalborg, IEEE, 2014.

32. I. Sarris and A. Nix, Power azimuth spectrum measurements in home and office environments at 62.4 GHz, in *Proceedings of the IEEE 18th Int. Symp. Personal, Indoor and Mobile Radio Communications (PIMRC'07)*, Athens, IEEE, pp. 1–4, 2007.
33. N. Moraitis, P. Constantinou, and D. Vouyioukas, Power angle profile measurements and capacity evaluation of a SIMO system at 60 GHz, *Proceedings of the IEEE 21th Int. Symp. Personal, Indoor and Mobile Radio Communications (PIMRC'10)*, Istanbul, IEEE, pp. 1027–1031, 2010.
34. Z. Muhi-Eldeen, L. Ivrissimtzis, and M. Al-Nuaimi, Modelling and measurements of millimetre wavelength propagation in urban environments, *IET Microwaves, Antennas and Propagation* 4(9): 1300–1309, 2010.
35. E. Ben-Dor, T. S. Rappaport, Y. Qiao, and S. J. Lauffenburger, Millimeterwave 60 GHz outdoor and vehicle AOA propagation measurements using a broadband channel sounder, *Proceedings of the IEEE Global Telecommunications Conference (GLOBECOM 2011)*, Houston, TX, IEEE, pp. 1–6, 2011.
36. Y. Shoji, H. Sawada, C.-S. Choi, and H. Ogawa, A modified SV-model suitable for line-of-sight desktop usage of millimeter-wave WPAN systems, *IEEE Transactions on Antennas and Propagation* 57(10): 2940–2948, 2009.
37. H. Xu, V. Kukshya, and T. S. Rappaport, Spatial and temporal characteristics of 60-GHz indoor channels, *IEEE Journal on Selected Areas in Communications* 20(3): 620–630, 2002.
38. S. Ranvier, M. Kyrö, K. Haneda, T. Mustonen, C. Icheln, and P. Vainikainen, VNA-based wideband 60 GHz MIMO channel sounder with 3-D arrays, in *Proceedings of the IEEE Radio and Wireless Symposium (RWS'09)*, San Diego, CA, pp. 308–311, 2009.
39. D. de Wolf and L. Ligthart, Multipath effects due to rain at 30–50 GHz frequency communication links, *IEEE Transactions on Antennas and Propagation* 41(8): 1132–1138, 1993.
40. S. Takahashi, A. Kato, K. Sato, and M. Fujise, Distance dependence of path loss for millimeter wave inter-vehicle communications, *IEEE 58th Vehicular Technology Conference, VTC 2003-Fall*, IEEE, pp. 26–30, 2003.
41. A. Kato, K. Sato, and M. Fujise, Technologies of millimeter-wave inter-vehicle communications—Propagation characteristics, *Journal of the Communications Research Laboratory* 48(4): 100–110, 2001.
42. Y. Karasawa, Multipath fading due to road surface reflection and fading reduction by means of space diversity in ITS vehicle-to vehicle communications at 60 GHz, *Transactions on IEICE* 83(4): 518–524, 2000.
43. K. Tokuda, Y. Shiraki, K. Sekine, and S. Hoshina, Analysis of millimeter-wave band road surface reflection fading (RSRF) in vehicle-to vehicle communications, *Technical Report of IEICE AP98-134*, 1999.
44. A. Kato, K. Sato, M. Fujise, and S. Kawakami, Propagation characteristics of 60-GHz millimeter waves for ITS inter-vehicle communications, *IEICE Transactions on Communications* E84-B(9): 2530–2539, 2001.
45. P. Kyösti, J. Meinilä, L. Hentilä, X. Zhao, T. Jämsä, C. Schneider, M. Narandzić, et al., WINNER II channel models, *IST-4-027756 WINNER II, D1.1.2 V1.2*, 2008.
46. G. R. MacCartney, M. K. Samimi, and T. S. Rappaport, Omnidirectional path loss models in New York City at 28 GHz and 73 GHz, *IEEE 2014 Personal Indoor and Mobile Radio Communications (PIMRC)*, Washington, DC, IEEE, 2014.

47. T. Jämsä, T. Rappaport, G. R. MacCartney Jr, M. K. Samimi, J. Meinilä, and T. Imai, Harmonization of 5G path loss models, *10th IC1004 MC and Scientific Meeting*, contribution TD(14)10073, Aalborg, IEEE, 2014.

48. G. R. MacCartney Jr., M. K. Samimi, T. S. Rappaport, and T. Jämsä, Path loss models in New York City at 28 GHz and 73 GHz, *10th IC1004 MC and Scientific Meeting*, contribution TD(14)10072, Aalborg, IEEE, 2014.

49. M. K. Samimi, K. Wang, Y. Azar, G. N. Wong, R. Mayzus, H. Zhao, J. K. Schulz, S. Sun, F. Gutierrez, Jr., and T. S. Rappaport, 28 GHz angle of arrival and angle of departure analysis for outdoor cellular communications using steerable beam antennas in New York City, *Proceedings of the IEEE VTC*, New York, IEEE, pp. 1–6, 2013.

50. Y. Azar, G. N. Wong, K. Wang, R. Mayzus, J. K. Schulz, H. Zhao, F. Gutierrez, D. Hwang, and T. S. Rappaport, 28 GHz propagation measurements for outdoor cellular communications using steerable beam antennas in New York City, in *Proceedings of the IEEE International Conference on Communications*, New York, IEEE, pp. 1–6, 2013.

51. J. N. Murdock, E. Ben-Dor, Y. Qiao, J. I. Tamir, and T. S. Rappaport, A 38 GHz cellular outage study for an urban campus environment, *Proceedings of the IEEE Wireless Communications and Networking Conference*, Paris, France, pp. 3085–3090, 2012.

52. T. S. Rappaport, Y. Qiao, J. I. Tamir, J. N. Murdock, and E. Ben-Dor, Cellular broadband millimeter wave propagation and angle of arrival for adaptive beam steering systems (invited paper), *Proceedings of the IEEE Radio Wireless Symposium*, Santa Clara, CA, IEEE, pp. 151–154, 2012.

53. T. S. Rappaport, E. Ben-Dor, J. N. Murdock, and Y. Qiao, 38 GHz and 60 GHz angle-dependent propagation for cellular and peer-to peer wireless communications, *Proceedings of the IEEE International Conference on Communications*, Ottawa, ON, IEEE, pp. 4568–4573, 2012.

54. S. Geng, J. Kivinen, X. Zhao, and P. Vainikainen, Millimeter-wave propagation channel characterization for short-range wireless communications, *IEEE Transactions on Vehicular Technology* 58(1): 3–13, 2009.

55. Y. Chang, S. Baek, S. Hur, Y. Mok, and Y. Lee, A novel dual slope mmWave channel model based on 3D ray-tracing in urban environments, *IEEE 2014 Personal Indoor and Mobile Radio Communications (PIMRC)*, Washington, DC, IEEE, pp. 222–226, 2014.

56. M. Jacob, S. Priebe, T. Kurner, M. Peter, M. Wisotzki, R. Felbecker, and W. Keusgen, Extension and validation of the IEEE 802.11ad 60 GHz human blockage model, *7th European Conference on Antennas and Propagation (EuCAP2013)*, Gothenburg, IEEE, pp. 2806–2810, 2013.

57. W. Peter, W. Keusgen, and R. Felbecker, Measurement and ray-tracing simulation of the 60 GHz indoor broadband channel: Model accuracy and parameterization, *2nd European Conference on Antennas and Propagation (EuCAP'07)*, Edinburgh, IEEE, pp. 1–8, 2007.

58. D. Dupleich, F. Fuschini, R. Mueller, E. Vitucci, C. Schneider, V. Degli Esposti, and R. Thomä, Directional characterization of the 60 GHz indoor-office channel, *31th URSI General Assembly and Scientific Symposium*, Beijing, pp. 1–4, 2014.

59. K. Sato, T. Manabe, T. Ihara, H. Saito, S. Ito, T. Tanaka, K. Sugai, et al., Measurements of reflection and transmission characteristics of interior structures of office building in the 60-GHz band, *IEEE Transactions on Antennas and Propagation* 45(12): 1783–1792, 1997.

60. B. Langen, G. Lober, and W. Herzig, Reflection and transmission behaviour of building materials at 60 GHz, in *Proceedings of the IEEE 5th International Symposium on Personal, Indoor and Mobile Radio Communications (PIMRC'94)*, The Hague, IEEE, vol. 2, pp. 505–509, 1994.
61. H. C. Nguyen, G. R. Maccartney, T. Thomas, T. S. Rappaport, B. Vejlgaard, and P. Mogensen, Evaluation of empirical ray-tracing model for an urban outdoor scenario at 73 GHz E-band, in *2014 IEEE 80th Vehicular Technology Conference (VTC Fall)*, Vancouver, BC, pp. 1–6, 2014.
62. T. Jämsä, P. Kyösti, and K. Kusume (Eds), Initial channel models based on measurements, *ICT-317669-METIS Deliverable D1.2 V1.0*, 2014.

Chapter 19

Millimeter-Wave (mmWave) Medium Access Control: A Survey

Joongheon Kim

Contents

As millimeter-wave (mmWave) radio wave propagation is highly directional, new medium access control (MAC) mechanisms are required for directional mmWave wireless systems. Therefore, directional beam management is required in mmWave MAC. This chapter summarizes the beam management schemes in academic literatures and industry standards in mmWave systems. In addition, mmWave-specific scheduling and relaying features are discussed. Video streaming and cellular network–related MAC features are also introduced in this chapter.

19.1 Introduction

One of the fundamental roles of medium access control (MAC) in wireless and computing networking is "collision and interference management." One of the most famous and successful random access schemes in wireless networking is carrier sensing multiple access with collision avoidance (CSMA/CA), and this also coordinates wireless medium access with the concept of collision avoidance.

However, collision and interference management is no longer a key role in millimeter-wave (mmWave) wireless networks because of the high directionality of mmWave wireless beams/links [1]. If the azimuth and elevation beamwidths are assumed to be φ and θ, respectively, the probability of interference existence is theoretically analyzed as $(\theta/2\pi)$ $(\varphi/2\pi)$ [2]. Therefore, the probability of interference existence is around $7.7 \times 10^{-2}\%$ when the azimuth and elevation beamwidths are 10°; that is, interference management is no longer an important element in mmWave wireless networks [1].

On the other hand, "directional beam management" has become one of the key research topics in high-directional mmWave wireless networks such as "beam training and tracking." Based on this research direction, the Institute of Electrical and Electronics Engineers (IEEE) 802.11ad standard, which is one of the best-known 60 GHz mmWave wireless standards, contains a detailed description of "Beamforming and Training." If cellular base stations (BSs) want to use mmWave technologies for next-generation fifth-generation (5G) cellular networks, the directional mmWave antennas in the BS should be able to rapidly track mobile cellular users. Otherwise, mobility support is no longer possible. Therefore, fast beam training and tracking are essential for using mmWave in mobile cellular systems.

The remainder of this chapter is organized as follows. Section 19.2 gives an overview of mmWave beam management schemes in IEEE standards and academic literatures. Section 19.3 presents an overview of scheduling and relay selection methodologies for mmWave wireless systems. Section 19.4 presents the video applications of mmWave wireless systems and corresponding MAC features. Section 19.5 presents the MAC design considerations of next-generation 5G mmWave wireless cellular networks. Finally, Section 19.6 concludes the chapter.

19.2 Directional Beam Management in mmWave MAC Design

As explained in Section 19.1, directional beam management schemes are important in mmWave wireless systems. Therefore, this section presents various beam-training schemes in IEEE standards and academic publications.

19.2.1 Exhaustive/Brute-Force Search

The general beam-forming and training procedure using transmit beam forming (TXBF) is illustrated in Figure 19.1, as explained in [3]. In Figure 19.1, each beam-training initiator (BI) and beam-training responder (BR) has N beam directions.

First of all, the brute-force search with transmit beam forming works as follows. To initiate the beam-training procedure, the BI sweeps through all beam directions, transmitting one training packet in each direction. During this time, the BR receives the packets with an omnidirectional antenna pattern. At the end of this period, the BR can figure out which beam direction of the BI resulted in the highest signal-to-noise ratio (SNR) at the BR. Subsequently, the BI and the BR exchange their roles and repeat the procedure, allowing the BI to determine the direction of the BR leading to the highest SNR. A last step of exchanging feedback packets allows both sides to learn their own optimal directions.

A variant of this approach uses receive beam forming (RXBF) instead of TXBF, as illustrated in Figure 19.1a. Each node, BI and BR, has N beam directions. The BI transmits packets in the omnidirectional mode, while the BR scans through all directions; then the BI and the BR exchange roles. The two nodes then know their best beam directions without further exchange of feedback packets. In addition,

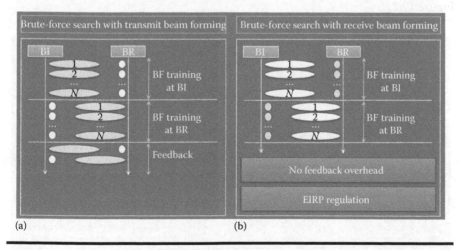

(a)   (b)

Figure 19.1 Brute-force beam training (a) with transmit beam forming and (b) with receive beam forming.

the brute-force search with receive beam forming is not impacted by constraints on equivalent isotropically radiated power (EIRP), but only on transmitted absolute power.

As presented in [4,5], the beamwidth of commercial mmWave high-gain horn and Cassegrain antennas is near 1°, and similar values can be achieved with adaptive antennas of realistic size. Thus, in the worst case, N should be $360°/1° = 360$ for two-dimensional beam geometry and $360°/1° \times 1800°/1° \approx 6.5 \times 10^4$ for three-dimensional beam geometry. Consequently, the beam-training procedures can require a significant overhead.

19.2.2 Two-Stage Beam Training in IEEE Standards

This section presents an overview of currently existing standardized mmWave beam-training schemes. In IEEE, there are two standards for 60 GHz mmWave wireless networks, IEEE 802.15.3c WPAN and IEEE 802.11ad WLAN, as explained in [4,5].

In the IEEE 802.11ad WLAN and IEEE 802.15.3c WPAN beam forming and training, the standards use a two-stage beam-forming and training operation: coarse-grained beam training (called *sector sweeping* in IEEE 802.11ad and *low-resolution (L-Re) beam training* in IEEE 802.15.3c) and fine-grained beam training (called *beam refinement* in IEEE 802.11ad and *high-resolution (H-Re) beam training* in IEEE 802.15.3c [6–8]).

If the standards consider TXBF, BF and BI determine the optimum coarse-grained beam according to the exhaustive-search protocol described in Section 19.2.1. In the next stage, fine-grained beam training, the same type of operation is performed to identify the best beam in each coarse-grained beam. Similar principles hold when RXBF is considered. This procedure is illustrated in Figure 19.2.

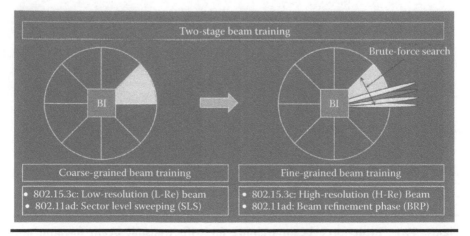

Figure 19.2   Two-stage beam training in IEEE standards.

Even if both standards have their own specific beam-forming and training protocols, the protocols are fundamentally based on two-stage beam training. While this can accelerate the beam forming, it is still slow, as shown by simulation results in [5].

In addition, the numbers of coarse-grained and fine-grained beam-training search spaces also have an impact on the performance of beam-training speed. The proposed algorithm in [9] finds the numbers of coarse-grained and fine-grained beam-training search spaces that minimize the overall number of control signal transmissions. This reduces the beam-training time as well as the number of transmitted control signals. This is good for fast link configuration, and is additionally beneficial in terms of energy awareness, as discussed in [9].

19.2.3 Interactive Beam Training

The fundamental reason why brute-force search is inefficient lies in the fact that even when a BI or a BR finds a fairly good beam direction, it cannot stop in the middle of the brute-force search operation, because it has to search all possible beam directions. Of course, to find the globally optimum direction, a complete search is necessary. However, it is often sufficient to find a "good enough" direction that can maintain the mmWave wireless communications. Therefore, beam-training overhead can be reduced by letting the beam search stop when both the BI and the BR find acceptable beam directions. This is the main design philosophy of interactive beam training, and details are presented in [4,5].

As illustrated in Figure 19.3, the BI and the BR change their communication mode between transmitter (TX) and receiver (RX) after every training packet transmission. Thus, after sending a training packet in an omnidirectional TX (Omni-TX), the device, either the BI or the BR, updates its communication mode as a beam-formed Rx (BF-RX) to receive the training packet from

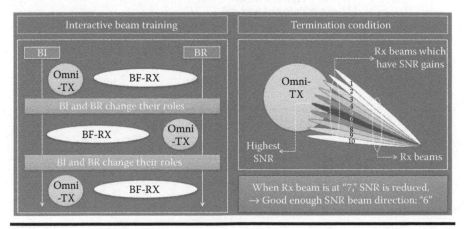

Figure 19.3 Interactive beam training.

the given direction of the opposite side via RXBF. Having identified a beam direction with "sufficient quality" (i.e., sufficient SNR), the RX will continue the search till it can be sure of having found a local optimum, that is, until it has determined that the SNR is worse on both sides of the "sufficiently good" direction. This is done to increase the robustness of the received scheme, and in light of the fact that finding the local optimum does not impose a significant increase in training overhead. This concept is illustrated in Figure 19.3 (termination condition).

If either the BI or the BR finds an acceptable beam direction in a BF-RX mode, it can piggyback this information on the next training packet. When both the BI and the BR have found their acceptable beam directions, this beam-training procedure immediately stops.

The performance of interactive beam training is well studied, and the plotting is shown in Figure 19.4 [5]. As presented in this figure, if the link configuration time is less than the session reinitiation thresholds of voice over IP (VoIP) and video services, the link can be reconnected and serve the corresponding VoIP or video services without any disconnection. The exhaustive search with RXBF shown in Figure 19.4 (i.e., brute-force search with RXBF) cannot serve VoIP and video services even if the beamwidth is near 10°. In the case of average performance, if the beamwidth is larger than 5.3°, VoIP service can be served (i.e., VoIP service sessions can be reconnected before the session threshold expires) even though the service user is moving. Similarly, if the beamwidth is larger than 9°, video streaming service can be served (i.e., reconnection of video service sessions before the

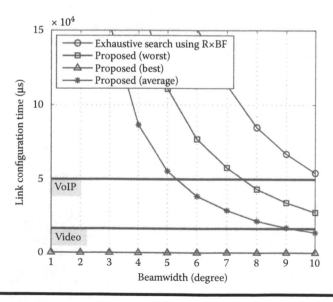

Figure 19.4 Performance of interactive beam training.

session threshold expires) even though the service user is moving. In the case of worst performance, if the beamwidth is larger than 7.5°, VoIP service can be served (i.e., VoIP service sessions can be reconnected before the session threshold expires) even though the service user is moving.

19.2.4 Prioritized Sector Search Ordering

To accelerate the average search speed, the order of RX beam directions to be searched can be prioritized. For this purpose, this proposed prioritized sector search ordering (PSSO) orders the segmented spaces in terms of network association request/response (NAR) statistics. Note that the term *segmented spaces* is equivalent to "sectors" in IEEE 802.11ad and "low-resolution (L-Re) beams" in IEEE 802.15.3c.

This PSSO is quite useful in mmWave wireless systems, because physical obstacles can constitute very strong attenuators, thus greatly restricting the angular range from which useful signals can come in a given room (this is especially true for walls, which can be easily penetrated by microwaves, but are impervious to mmWaves, and which might not be effective reflectors for certain geometric configurations either). The regions with the highest number of NAR statistics might thus constitute the angular regions from which radiation can physically occur, or they might be regions that are preferred by users. This operation is illustrated in Figure 19.5.

In Figure 19.5, the system has eight sectors, and each sector has its own different NAR value. The NAR of Sector 8 is the highest value, which means that the sector has the best population. Thus, the system starts beam searching from Sector 8. In the same way, the system searches the given sectors in terms of ordering by NAR statistics.

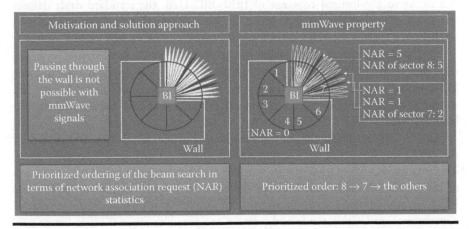

Figure 19.5   Prioritized sector search ordering.

19.3 Scheduling and Relay Selection for mmWave Systems

In conventional wireless networking systems, designing scheduling algorithms is one of the key issues in wireless MAC research. Due to the high directionality of mmWave wireless communications, network device densification is achievable by spatial reuse. However, there is discussion in the literature about scheduling schemes even in mmWave high-directional wireless communications with various optimization criteria. In addition, several relaying schemes are proposed and introduced in the 60 GHz mmWave IEEE 802.11ad standard to combat the short-distance data transmission limitation due to high attenuation in the air.

19.3.1 Scheduling

The fundamental directionality is considered on top of various currently existing channel access mechanisms. The scheme proposed in [10] considers directionality in CSMA/CA random access. Similarly, the algorithm proposed in [11] is for time-division multiple access (TDMA) under consideration of spatial reuse due to the high directionality of mmWave beams.

In addition, directional CSMA/CA can cause a deafness problem, which is clearly defined in [12], and the issue was resolved with a multihop RTS/CTS mechanism in high-directional wireless mesh networks.

Lastly, due to the high attenuation characteristics in mmWave radio wave propagation, blockage-aware robust scheduling algorithms have also been designed in [1,13].

19.3.2 Relay Selection in IEEE 802.11ad

According to the limited coverage of IEEE 802.11ad, the standard draft defines two kinds of relaying, link cooperating (LC) and link switching (LS), as explained in [14].

In LS, if the source–destination direct physical mmWave wireless propagation link is disrupted, the source redirects the mmWave wireless transmission of frames addressed to the destination via the relay. The direct link between the source and the destination can resume after the direct link between them has been recovered.

In LC, a frame transmission from the source to the destination is repeated by the relay even when the source–destination link is being used at the same time. This may increase the signal quality received at the destination by taking advantage of cooperative diversity and improve the network capacity significantly [15]. For LC, both cooperative communications with amplify-and-forward and cooperative communications with decode-and-forward are possible. Since it offers better performance than LS, we henceforth consider only LC.

Furthermore, the possibility of source and destination communicating with each other without relaying noncooperative communications needs to be taken into account.

Interestingly, constructing relay networks are required for both indoor and outdoor applications, but the fundamental reasons are different. In indoor applications, the relay deployment is needed to combat non-line-of-sight (NLOS) situations, whereas in outdoor applications, the relay deployment is required for extending wireless communication coverage.

19.4 Video Streaming

19.4.1 Uncompressed Video Streaming Indoors

Since the year 2000, mmWave wireless systems have attracted a lot of attention, because an mmWave system was used for uncompressed high-definition (HD) wireless video transmission, and thus the WirelessHD consortium was established to define 60 GHz mmWave wireless technologies for this point-to-point stationary video streaming. In addition, the major use case scenarios of 60 GHz IEEE 802.11ad are for indoor video streaming over 60 GHz mmWave wireless channels.

In the WirelessHD and IEEE 802.11ad standards, CSMA/CA is also defined; however, the standards consider reserved/scheduled time allocation (with TDMA) for this wireless HD video streaming.

In a 1080p HD video stream, one frame consists of 1080×1920 pixels, each of which is represented by $3 \times 8 = 24$ bits (8 bits red, green, and blue [RGB]). Thirty frames of image data are transmitted per second in a standard mode. Thus, the required data rate to transmit uncompressed 1080p HD video is approximately 1.5 Gbps ($1080 \times 1920 \times 24 \times 30$). In enhanced mode, the number of frames per second is doubled, and thus a data rate of 3 Gbps is required. For the format of YCbCr 4:2:0 (instead of RGB), the number of bits in a frame is half as many as for a frame of RGB; that is, 0.75 and 1.5 Gbps are required for uncompressed 1080p HD video streaming in standard and enhanced modes, respectively.

The 60 GHz mmWave IEEE 802.11ad standard includes four subchannels with a bandwidth of 2.16 GHz for each; thus, uncompressed 1080p HD video wireless transmission can be achieved in ideal channel conditions.

19.4.2 Real-Time Video Streaming Outdoors

In outdoor video streaming, most applications are for longer-distance scenarios compared with indoor applications. As calculated in [16,17], the achievable distance when the target threshold is set to 1 Gbps is about 200–300 m even if high-gain Cassegrain and horn antennas are used. This means that mmWave wireless links are not suitable for long-distance outdoor video delivery. To overcome this disadvantage, it is necessary to construct relay networks. A well-studied example is given in [16,17]. The authors construct two-hop relay networks and then design an algorithm for joint relay selection and video stream allocation.

As illustrated in Figure 19.6, each source (a wireless video camera) is located at the top of the target network. Each source records video signals, which are delivered to relays and eventually arrive at the destination D (i.e., the broadcasting center). In this architecture, the authors [16,17] designed an optimization framework for joint source coding and video stream distribution.

As illustrated in Figure 19.7, HD video cameras record the scene using the embedded camera. Then, the recorded signals travel to a scalable video coding (SVC) encoder, and the bit streams are reorganized as layered information (one basement layer and multiple enhancement layers for video quality enhancement). If the mmWave channel condition is not good, the source needs to compress more (select a lower number of enhancement layers) for transmitting video signals in a real-time manner. On the other hand, if the channel condition is quite good, the source node can transmit more enhancement layers for better video quality.

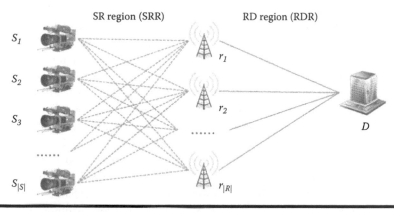

Figure 19.6 Two-hop outdoor mmWave streaming networks.

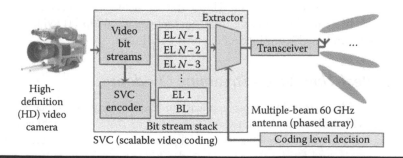

Figure 19.7 Source devices in outdoor mmWave streaming platforms.

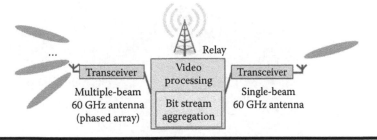

Figure 19.8   Relay devices in outdoor mmWave streaming platforms.

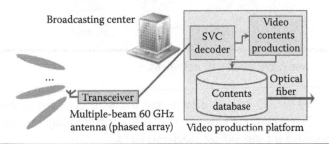

Figure 19.9   Broadcasting center in outdoor mmWave streaming platforms.

As illustrated in Figure 19.8, each relay receives streams from its connected sources. Then, each relay aggregates the streams and sends them to the final destination (the broadcasting center).

As illustrated in Figure 19.9, the broadcasting center is wirelessly connected with all deployed relays. Then, the broadcasting center aggregates all signals from end-hop wireless HD video cameras; it generates multimedia contents; and forwards the contents to customers.

19.5 Next-Generation Wireless Cellular Network MAC

As presented in Figures 19.10 and 19.11, two types of cellular networking architectures are considered for mmWave cellular networks.

Figure 19.10   mmWave cellular networks.

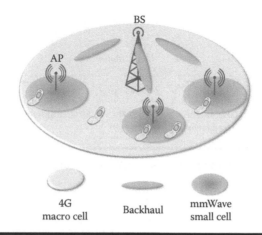

Figure 19.11   mmWave small cell networks.

In Figure 19.10, 5G BSs are directly talking with deployed mobile users via mmWave wireless access. For this purpose, fast beam-training and tracking algorithms are required for supporting mobile services.

However, deploying mmWave BSs in entire areas is not possible because of the cost. Because mmWave beams are directional, the mmWave BSs should be densely deployed, which is worse in terms of deployment cost. Therefore, deploying mmWave (APs) in required areas can be considered in terms of cost-effective design. Figure 19.11 illustrates the deployment of mmWave small cells. Service providers can deploy mmWave small cells only in the required hot-spot areas. In addition, the backhaul links between the APs and the BS can be designed with mmWave channels to achieve high capacity.

Lastly, direct communication between two mobile users (so-called device-to-device communications) can be performed with mmWave wireless technologies, because most device-to-device applications are for social network–based video delivery; that is, a large bandwidth is required for higher data rates. In [18], a device-to-device routing algorithm under the consideration of video quality maximization is proposed.

For these three major mmWave cellular access technologies (broadband mmWave cellular access, small cell mmWave cellular access, and device-to-device cellular access), the following considerations are required for cellular MAC protocol design.

For broadband and small cell mmWave cellular access technologies (as in Figures 19.10 and 19.11), fast beam-training and tracking algorithms are required for mobile service support. In addition, reliable technologies to support mobile users who are suffering from blockage and NLOS situations are required.

For device-to-device cellular access technologies, fully distributed MAC should be additionally designed, because there is no centralized network component that can make adequate scheduling decisions. One good example of fully distributed MAC mechanisms for cellular networks is FlashLinQ, which was designed by Qualcomm. Because the key component of FlashLinQ is signal-to-interference ratio (SIR)-based scheduling, the implementation of FlashLinQ for mmWave wireless channels will be simpler, because interference in mmWave wireless systems is rare [19].

19.6 Summary

This chapter discusses MAC issues in mmWave wireless systems. Because mmWave radio wave propagation is highly directional, interference is no longer a major consideration in MAC. Rather, managing high-directional beams has become a major consideration in mmWave MAC design. Therefore, fast mmWave beam-training and tracking algorithms have been discussed in mmWave research. This chapter summarizes beam-training and tracking algorithms in academic literatures and IEEE standards (including IEEE 802.11ad and IEEE 802.15.3c). Then, fundamental scheduling and relaying technologies are introduced. In addition, video streaming in indoor and outdoor scenarios in mmWave wireless systems is discussed. Lastly, various mmWave cellular architectures (broadband, small cell, and device-to-device networks) are presented, and corresponding design issues are addressed.

References

1. S. Singh, F. Ziliotto, U. Madhow, E. Belding, and M. Rodwell, Blockage and directivity in 60 GHz wireless personal area networks: From cross-layer model to multihop MAC design, *IEEE Journal on Selected Areas in Communications* 27(8): 1400–1413, 2009.
2. S. Singh, R. Mudumbai, and U. Madhow, Interference analysis for highly directional 60-GHz mesh networks: The case for rethinking medium access control, *IEEE/ACM Transactions on Networking* 19(5): 1513–1527, 2011.
3. F. Dai and J. Wu, Efficient broadcasting in ad hoc wireless networks using directional antennas, *IEEE Transactions on Parallel and Distributed Systems* 17(4): 335–347, 2006.
4. J. Kim, Elements of next-generation wireless video systems: Millimeter-wave and device-to-device algorithms, PhD Dissertation, University of Southern California, Los Angeles, CA, 2014.
5. J. Kim and A. F. Molisch, Fast millimeter-wave beam training with receive beamforming, *IEEE/KICS Journal of Communications and Networks* 16(5): 512–522, 2014.
6. E. Perahia, C. Cordeiro, M. Park, and L. L. Yang, IEEE 802.11ad: Defining the next generation multi-Gbps Wi-Fi, in *Proceedings of IEEE Consumer Communications and Networking Conference (CCNC)*, Las Vegas, NV, IEEE, pp. 1–5, 2010.
7. T. Baykas, C.-S. Sum, Z. Lan, J. Wang, M. A. Rahman, H. Harada, and S. Kato, IEEE 802.15.3c: The first IEEE wireless standard for data rates over 1 Gb/s, *IEEE Communications Magazine* 49(7): 114–121, 2011.

8. J. Kim, A. Mohaisen, and J-K. Kim, Fast and low-power link setup for IEEE 802.15.3c multi-gigabit/s wireless sensor networks, *IEEE Communications Letters* 18(3): 455–458, 2014.

9. J. Kim and S-N. Hong, Dynamic two-stage beam training for energy-efficient millimeter-wave 5G cellular systems, *Telecommunication Systems* 59(1): 111–122, 2015.

10. M. X. Gong, D. Akhmetov, R. Want, and S. Mao, Directional CSMA/CA protocol with spatial reuse for mmWave wireless networks, in *Proceedings of the IEEE Global Telecommunications Conference (GLOBECOM)*, Miami, FL, IEEE, pp. 1–5, 2010.

11. C.-S. Sum, L. Zhou, M. A. Rahman, J. Wang, T. Baykas, R. Funada, H. Harada, and S. Kato, A multi-Gbps millimeter-wave WPAN system based on STDMA with heuristic scheduling, in *Proceedings of the Global Telecommunications Conference (GLOBECOM)*, Honolulu, HI, IEEE, pp. 1–6, 2009.

12. R. R. Choudhury and N. F. Vaidya, Deafness: A MAC problem in ad hoc networks when using directional antennas, in *Proceedings of the IEEE International Conference on Network Protocols (ICNP)*, pp. 283–292, 2004.

13. Y. Niu, Y. Li, D. Jin, L. Su, and D. Wu, Blockage robust and efficient scheduling for directional mmWave WPANs, *IEEE Transactions on Vehicular Technology* 64(2): 728–742, 2015.

14. J. Kim, Y. Tian, A. F. Molisch, and S. Mangold, Joint optimization of HD video coding rates and unicast flow control for IEEE 802.11ad relaying, in *Proceedings of IEEE International Symposium on Personal Indoor and Mobile Radio Communications (PIMRC)*, Toronto, IEEE, pp. 1109–1113, 2011.

15. J. N. Laneman, D. N. C. Tse, and G. W. Wornell, Cooperative diversity in wireless networks: Efficient protocols and outage behavior, *IEEE Transactions on Information Theory* 50(12): 3062–3080, 2004.

16. J. Kim, Y. Tian, S. Mangold, and A. F. Molisch, Quality-aware coding and relaying for 60 GHz real-time wireless video broadcasting, in *Proceedings of IEEE International Conference on Communications (ICC)*, Budapest, IEEE, pp. 5148–5152, 2013.

17. J. Kim, Y. Tian, S. Mangold, and A. F. Molisch, Joint scalable coding and routing for 60 GHz real-time live HD video streaming applications, *IEEE Transactions on Broadcasting* 59(3): 500–512, 2013.

18. J. Kim and A. F. Molisch, Quality-aware millimeter-wave device-to-device multi-hop routing for 5G cellular networks, in *Proceedings of IEEE International Conference on Communications (ICC)*, Sydney, IEEE, pp. 5251–5256, 2014.

19. X. Wu, S. Tavildar, S. Shakkottai, T. Richardson, J. Li, R. Laroia, and A. Jovicic, FlashLinQ: A synchronous distributed scheduler for peer-to-peer ad hoc networks, *IEEE/ACM Transactions on Networking* 21(4): 1215–1228, 2013.

Chapter 20

Millimeter-Wave
MAC Layer Design

Busra Gozde Bali and Ozgur Baris Akan

Contents

20.1 Introduction

The demand for wireless spectrum is increasing rapidly. In recent years, there has been extensive research on increasing spectrum efficiency and spectrum reuse to accommodate the number of wireless applications that use the wireless spectrum. However, despite all optimization, the growing demand will soon surpass the bandwidth that is available in bands conventionally used for wireless communication, that is, frequencies lower than a few gigahertz. Millimeter-wave (mmWave) bands are a promising solution to the bandwidth scarcity problem. Millimeter-wave communication systems use the 30–300 GHz range of the electromagnetic spectrum, which corresponds to wavelengths of between 10 and 1 mm. Thus, mmWave communication offers an extensive amount of additional bandwidth. However, it also has unique challenges and considerable differences compared with lower-frequency bands [1]. Therefore, there is an imminent need to develop new communication schemes in all communication layers.

In this chapter, the medium access control (MAC) protocol issues and developments for mmWave communications are presented. We focus on three major issues. First, the design challenges in the MAC layer to providing high channel throughput and packet transmission are discussed. Then, design guidelines for developing algorithms in the MAC layer and layout classifications of MAC protocols along with their performance analysis are presented. Lastly, this chapter reviews standardization in mmWave communications, with a specific design goal covered in each section.

Some of the major challenges in mmWave communications can be listed as blockage, deafness, concurrent transmission, and synchronization. To meet these challenges, enhanced MAC layer protocols and algorithms have been proposed. First, we take a brief look at each of these challenges. By blockage, we mean the obstacles on the propagation path. In mmWave communications, path loss is higher than in conventional wireless channels. Therefore, directional antennas are generally used. In such low-beamwidth transmissions, blockage can be a major problem in communication [2]. Another problem with directional antennas is the so-called deafness problem. This occurs when a beam-formed wave does not reach the intended receiver [3]. The MAC layer solutions in mmWave should address this problem [2,4]. We discuss the limited work in the literature on MAC protocols that aim to overcome deafness in detail in this chapter. As a result of the aforementioned challenges, synchronization is also encountered as a major problem in mmWave communications. Due to highly directional transmission, temporal link failures due to blockage, deafness, and so on cause uncorrected clock drifts [3,4].

Standardization activities have been initiated for mmWave communications. There are ongoing standardization efforts on MAC for personal and local area networks. Currently, there is no identified standardization activity for mmWave in cellular networks, though several research projects, such as FP7 EU Project Mobile and Wireless Communications Enablers for the Twenty-Twenty Information Society (METIS) (2012–2015) [5], are working on standardization activities for mmWave cellular networks. In this chapter, we lay out the current state of these standardization activities for the mmWave MAC layer. For wireless personal area networks (WPANs), three different standards are presented: IEEE 802.15.3c [6], Wireless HD [7], and ECMA-387 [8]. The MAC layer designs for wireless communications of each standard are described in Section 20.7. These standards have different network technologies, and these differences are highlighted in detail. IEEE 802.11ad [9] and Wireless Gigabit Alliance (WiGig) [10] are offered in two different standards for wireless local area network (WLAN) [11]. IEEE 802.11ad and WiGig add modifications to the IEEE 802.11 and IEEE 802.11ad, respectively. Additionally, prestandardization activity has been initialized for mmWave in cellular networks. For instance, research projects, such as FP7 EU Project METIS [5], are addressing the standardization for fifth-generation (5G)/beyond-fourth-generation (4G) networks. An important feature is that these standards enable devices to get information about the channel.

In addition, MAC design guidelines focus on algorithms and protocols that explain the classifications of MAC protocols. These protocols focus on aspects of the well-known problems of neighbor discovery, blockage, and deafness, as well as delay characteristics, and how directional antennas can be applied in the MAC to provide reliable packet transmission throughput efficiently [4,12]. This chapter provides insight into the main challenges of mmWave communications for the MAC layer, classifies the existing standards for mmWave communications, and gives in-depth overviews of MAC protocols and algorithms, as well as related current projects. We conclude with open issues and future research directions.

20.2 Key Design Challenges and Directions for MAC Layer

With the increasing need for large data quantities, mmWave communication faces many challenges, such as deafness, blockage, and high attenuation. There are several challenges in the implementation of MAC layer solutions to increase network throughput. However, there are also properties of mmWave communications that can be exploited to obtain higher performance. Some of these ideas have been implemented in various projects. In this section, we identify the main key design problems of mmWave communications in terms of the MAC layer, identify MAC design guidelines, and summarize MAC design opportunities and the challenges of mmWave communications in the following subsections.

20.2.1 Directivity

Efficient MAC protocols should provide high link quality and minimize collisions. The use of directional transmission for the MAC layer is the most appropriate solution mentioned in the literature. To explain the importance of directional transmission for the MAC layer, we consider examples given in [2,13]. In [13], the authors explain that the propagation loss of 60 GHz signals is 22 dB higher than that of 2.4 GHz signals in free space. A directional antenna beam achieves high gain in a specific direction and has low gain in other directions. As such, a directional or beam-forming antenna achieves higher gain than omnidirectional antennas. Since a directional antenna has low power in certain directions, it reduces interference to other nodes.

20.2.2 Blockage

Blockage is one of the most critical challenges of mmWave communication. It refers to high attenuation due to obstacles. The wavelength is 5 mm at 60 GHz [14]. Highly directional beam forming could lead to network sensitivity to blockage. Channels can be blocked by obstacles, either human or material, as illustrated in Figure 20.1. The human body can attenuate mmWave signals by 35 dB [15], and materials such as brick attenuate them by 80 dB [16–19]. Thus, human movement in a room may cause significant blockage for mmWave networks. Unlike WLAN and WPAN systems, cellular networks allow non-line-of-light (NLOS) communications [6,9]. In mmWave cellular networks, network utility is optimized to overcome blockages.

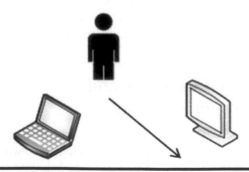

Figure 20.1   Human blockage.

20.2.3 CSMA Problems in MAC Layer

Several MAC protocols are designed for system requirements. For example, packet transmission and delay are critical. Therefore, the Carrier Sense Multiple Access with Collision Avoidance (CSMA/CA) algorithm becomes fundamental to the MAC design. CSMA/CA-based wireless networks suffer seriously from the hidden terminal problem and the exposed terminal problem. In the following subsections, we summarize deafness and the hidden/exposed terminal problem for the MAC layer.

20.2.3.1 Deafness

Deafness is a major problem in mmWave communications. Since the beams of the transmitter and the receiver do not point to each other, communication cannot be established. The signal strength of the third device is very low. Thus, new MAC protocols are needed to provide effective third devices in the network. The authors of [20] propose that the deafness problem can be easily solved by using a piconet structure. If the beams of the transmitter and the receiver only face each other, the system will be resistant to interference coming from outside. Hence, this condition reduces interference [21].

20.2.3.2 Hidden/Exposed Terminal Problem

MAC protocols aim to provide high transmission in mmWave networks. Since the exposed node problem restricts transmission capabilities, algorithms based on this problem should be redesigned to exploit spatial reuse. Due to a neighboring transmitter in wireless networks, a device is hindered from sending packets to other devices. This is the exposed node problem.

In wireless networks, the transmission range of the station is usually short. Therefore, not all the terminals located in the network can hear each other. Data transmission can only provide the location of transmitters or receivers in the transmission range of the terminal. In large-scale wireless networks, data transmission involves multihops. This causes the hidden terminal problem in the network.

20.4 Spatial Reuse

Highly directional transmission is one of the most important means of interference reduction. Transmitters and receivers are able to send data simultaneously, which is termed *spatial reuse*. If possible, new MAC protocols should lead to concurrent transmission. Interference is one of the major problems for concurrent transmissions in mmWave communications. To solve the interference problem, directional transmission is proposed. Another possible solution is using a coordinator, which is already provided by IEEE 802.15.3c. An interference-free scheme is proposed in [19]. Compared with single transmission [17], this scheme has a high network throughput. However, it causes complexity in the network. Thus, new MAC protocols should be investigated in terms of optimizing the interference problem.

20.5 Comparison of MAC Protocols for mmWave Communications

MAC layer design challenges are mentioned in Section 20.2. Several MAC protocols are proposed to overcome these challenges; directional MAC protocols, beam-forming protocols, and resource allocation play the most important roles in mmWave communications. We summarize these MAC protocols in the following subsections.

20.5.1 Resource Allocation

mmWave communication standards, including IEEE 802.11ad and IEEE 802.15.3c, play a very important role in high-throughput transmission and delay constraints. These standards allocate resources at the MAC layer for multiple users. mmWave communication is a promising technology for 5G networks. To provide a high data rate, the authors of [22] propose optimization techniques for resource allocation with IEEE 802.11ad or IEEE 802.15.3c. These optimization techniques (local descending discrete scaling [LDDS] algorithm [23], rate allocation game, Nash bargaining solution, and particle swarm optimization [PSO]) have convex functions. In addition to this, the authors use channel time allocation PSO (CTA-PSO) to solve the resource allocation problem, where the presence of the blockage problem proves that alternative resource allocation optimization techniques must be discovered.

Another MAC protocol proposes a resource allocation algorithm using mmWave in smart home networks [24]. The authors have worked on the resource allocation problem and propose a novel multichannel MAC protocol that is based on IEEE 802.15.3c, adhering to mmWave channelization. To deliver high throughput and provide aggregate network utility, multiple CTAs are preferred. Furthermore, the authors propose utility functions for battery-constrained multimedia devices. In smart home networks, simulation results prove that the proposed MAC protocol has better aggregate network utility than the existing IEEE 802.15.3c protocol.

In IEEE 802.15.3c MAC, resource allocation schemes are not specified. An enhanced MAC (EMAC) protocol that depends on a resource allocation scheme is proposed in [25]. Simulation results of the EMAC protocol show that throughput and delay characteristics are improved.

20.5.2 Transmission Scheduling

A wide range of protocols and algorithms for wireless networks have been proposed for transmission scheduling.

Several protocols [26–28] are based on time-division multiple access (TDMA). In order to clarify exclusive region (ER), [26] authoritize concurrent transmissions to investigate the spatial multiplexing gain of wireless networks. The results provide important guidelines for scheduling schemes. A multihop concurrent transmission scheme (MHCT) is proposed by [27]. To improve time-slot use, MHCT focuses on exploiting spatial capacity and time-division multiplexing (TDM). A virtual time-slot allocation (VTSA) scheme implements a multi-Gbps TDMA in a mmWave environment for throughput enhancement, which is revealed in [28]. The proposed virtual time-slot allocation (VTSA) scheme shows that system throughput improved.

Interference management and deafness have become the key design aspects for MAC layer design in mmWave networks. Therefore, memory-guided directional MAC protocol (MDMAC), aimed at finding solutions to network discovery and deafness, was proposed in [29]. This frame-based protocol is intended to achieve approximate TDM schedules without resource allocation.

To reduce blockage, high-gain directional antennas have become a hot topic. The authors of [30] have developed a blockage-robust and efficient directional MAC protocol (BRDMAC), which copes with the blockage problem through relaying. BRDMAC focuses on relay selection and transmission scheduling algorithms. Compared with the existing standards, BRDMAC has been demonstrated to perform better in terms of delay and throughput.

20.5.3 Concurrent Transmission

Concurrent transmission has a very important role in MAC layer design. Hence, several concurrent transmission algorithms have been suggested. In [31], concurrent transmission scheduling algorithms are proposed, but high propagation loss and the use of directional antennas are not taken into consideration by these algorithms.

Resource use efficiency has a significant effect on high data rate mmWave networks. To ensure this, [32] proposes a concurrent transmission scheduling algorithm. The network performance with regard to flow throughput, path loss, and a directional antenna is considered. With this heuristic algorithm, mmWave networks can support a greater number of users.

20.5.4 Blockage and Directivity

Human blockage has been a significant area of research in mmWave communications. A wide range of protocols and algorithms have been proposed for blockage [27,33]. There are several works on blockage in the literature [13,31,34,35]. These approaches are proposed to reduce the effect of human blockage and use a minimum number of hops for data transmission.

Directivity is another important challenge for MAC design. Directional antennas use higher gain and also reduce interference. Several MAC protocols improve the performance of the MAC layer using directional antennas, including [36,37]. Selection and orientation of antennas are an important factor for MAC layer design. To solve this problem, which occurs by directional transmission, coordination mechanisms must be developed.

20.5.5 Beam-Forming Protocols

Beam-forming protocols aim to maximize the transmission rate by multiple antennas. Beam forming confirms a beam toward a certain direction to maximize the transmission rate. The antenna gains of the transmitter and the receiver play an important role in transmission data rates. Therefore, beam-forming protocols should be organized adhering to the choice of metrics [38,39] for the best transmission.

There are a variety of beam-forming protocols for the MAC layer [40–42]. Moreover, [43–46] have interesting ideas for overcoming the directivity problem in mmWave networks.

To reduce the total setup time and the complexity of beam forming, [46] propose a concurrent beam-forming protocol. In order to achieve high system throughput and high energy efficiency, this protocol focused on the beam-forming problem.

20.6 MAC Design Guidelines

The MAC design guidelines focus on algorithms and protocols. These protocols focus on aspects of the well-known problems of neighbor discovery, blockage, and deafness, as well as delay characteristics, and how directional antennas can be applied in the MAC to provide reliable packet transmission throughput efficiently [4,12]. This section gives in-depth overviews of MAC protocols and algorithms as well as related current projects.

One of the main functions of MAC design is to present resource allocation for MAC layer throughput. Slotted ALOHA protocol determines that time is slotted and a packet can be transmitted at the beginning of a slot. Therefore, it can reduce the collision duration. Slotted ALOHA and TDMA are the most widely used multiple access schemes for resource allocation. Currently, mmWave MAC layer design approaches are not focused on resource allocation in terms of collision probability and throughput, which have been given in [47].

The design of synchronization plays an important role in mmWave networks because of the highly directional transmissions. Not only are directional transmissions responsible for data transmission, but synchronization is also a critical aspect for directional transmission. To establish communication, synchronization should be considered in the MAC layer design.

Interference problems are seen as one of the most important factors that reduce network performance, and should be resolved by interference management [21]. Interference management can be provided by transmission coordination and power control.

The selection of antenna is one of the important factors for the MAC layer design. To solve the problem of deafness, which occurs due to directional transmission, coordination mechanisms must be developed. Thus, spatial reuse becomes key to MAC design to ensure high network capacity [21]. There are different ways to design a MAC protocol for directional antennas. One solution is to divide the network into piconets, each having one piconet controller. This approach is used in IEEE 802.15.3c.

20.7 Standardization in mmWave Communications

Several international organizations have developed standards. There are currently five international standards for mmWave WLAN and WPAN applications, including IEEE 802.15.3 Task Group 3c (TG3c) [6], IEEE 802.11ad [9] Standardization Task Group, ECMA-387 [8], WiGig [10], and the WirelessHD Consortium [7]. We summarize these standards in this section.

20.7.1 Local Area Networks

20.7.1.1 IEEE 802.11ad

The IEEE develops IEEE 802.11ad standards for WLANs. To ensure mmWave communications, the IEEE added some modifications to the IEEE 802.11 MAC layer. IEEE 802.11ad, for the MAC layer [9] in mmWave communications, provides wireless applications, including wireless synchronization, Internet access, and high-definition (HD) multimedia transmissions.

IEEE 802.11ad covers several notable characteristics of the MAC layer: relaying, link adaptation, security, beam forming, and multigigabit access. IEEE 802.11ad operates in a similar manner to IEEE 802.15.3c, except for the resource allocation periods.

A basic service set (BSS) is a set of stations consisting of a personal BSS (PCP) and non-port control protocol devices (DEVs). The PCP provides the basic timing for the BSS, and manages the medium access according to transmission requests from the DEVs. The PCP also schedules channel access during beacon intervals (BIs). In IEEE 802.11ad, channel access time is divided into BIs. Each BI is subdivided into four portions: the beacon transmission interval (BTI), the association beam-forming training (A-BFT), the announcement transmission interval (ATI),

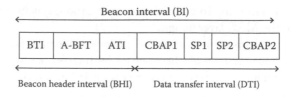

Figure 20.2   IEEE 802.11ad frame structure.

and the data transfer interval (DTI), which are reviewed in this section and illustrated in Figure 20.2.

- BTI: The access point transmits beacons to each sector.
- A-BFT: The beam-forming training period is reserved between the PCPs and the DEVs. During the BTI, it transmits beacon frames.
- ATI: During the ATI, service periods (SPs) and contention-based access periods (CBAPs) are allocated by PCP.
- DTI: After initial beam-forming training in A-BFT, peer-to-peer communications between DEVs take place in the DTI.

20.7.1.2 Wireless Gigabit Alliance

The WiGig was developed to promote multigigabit wireless communications technology. The WiGig alliance supported IEEE 802.11ad [9], announced in 2010. Devices have wireless communication at multigigabit speeds with WiGig. WiGig has many applications [10]: wireless transmission of audio data and uncompressed video from a digital camera to a HDTV, projector, or monitor.

20.7.2 Personal Area Networks

20.7.2.1 IEEE 802.15.3c

This section provides a brief summary of the IEEE 802.15 MAC. More comprehensive and detailed information on this standard is also available in [6].

The establishment of mmWave WPANs is based on IEEE 802.15.3c, that is, piconet. IEEE 802.15.3c supports several applications, including high-speed Internet access, video on demand, and HD. The piconet is the fundamental topology for the IEEE 802.15.3c WPANs. The piconet is composed of a piconet coordinator (PNC) and several DEVs, including the transmitter and the receivers. One of the DEVs, which is able to provide piconet synchronization and management, can be selected as a PNC, as shown in Figure 20.3. The PNC has a variety of functions, such as allocating channel resources for DEVs, and managing the security and authentication process. The role of the superframe is to control channel time among DEVs in the piconet. A superframe consists of three major parts: the beacon, the

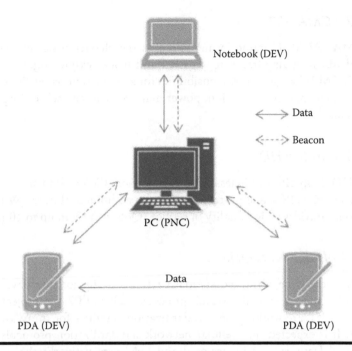

Figure 20.3   Piconet structure.

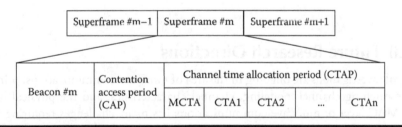

Figure 20.4   IEEE 802.15.3c MAC superframe structure.

contention access period (CAP), and the channel time allocation period (CTAP), as shown in Figure 20.4. The major responsibilities of each part are as follows:

- Beacon: Timing piconet and piconet management information with the broadcasting beacons in the piconet are provided by the PNC.
- CAP: The CAP is used for asynchronous data transmission between DEVs in the piconet. When DEVs associate with the piconet, they use CAP. CSMA/CA is the medium access method in the CAP.
- CTAP: The CTAP is composed of one or more channel time allocation blocks (CTAs) and management CTAs (MCTAs). CTAs are allocated by the PNC. TDMA and the slotted ALOHA protocol are the medium access methods in the CTAs and the MCTAs, respectively.

20.7.2.2 ECMA-387

The ECMA-387 MAC frame format contains the characteristics of many wireless standards, including frame aggregation and block acknowledgments [8]. The ECMA-387 MAC service has responsibilities for a reservation-based channel access mechanism, secure communication, power management, and scheduling of frame transmission.

20.7.2.3 WirelessHD

WirelessHD [7] specifies a wireless video area network (WVAN) to provide connectivity between CE, PCs, and portable devices. In addition to this, the WirelessHD specification provides a high quality of service (QoS) in a room up to 10 m.

20.7.3 Cellular Networks

Currently, there is no identified standardization activity for mmWave in cellular networks, though several research projects, such as FP7 EU Project METIS (2012–2015) [5], are working on standardization activities for mmWave cellular networks. It is expected that cellular network standardization protocols will use the standard features used in personal and local area networks that have been mentioned in Sections 20.7.1 and 20.7.2.

20.8 Future Research Directions

Several investigations are ongoing in the area of mmWave communications, primarily researching channel modeling, channel characteristics, and the physical layer. The MAC layer of mmWave communications has many challenges requiring new protocols and algorithms. Some future research directions such as directivity, beamforming protocols, resource allocation, and antenna design need to be addressed to support the development of MAC layer design for mmWave communications.

20.9 Conclusions

The characteristics of mmWave bands bring many challenges and opportunities for MAC protocol design. In this chapter, MAC design guidelines have focused on MAC algorithms and protocols; classifications of MAC protocols have been explained, and possible solutions have been highlighted. These protocols focus on aspects of the well-known problems of neighbor discovery, blockage, and deafness, on delay characteristics, and on how directional antennas can be applied in the MAC to provide reliable packet transmission throughput efficiently. MAC layer design for wireless communications of each standard has been described. These standards are different from each other, and these differences have been highlighted in detail. This chapter has provided insight into the main challenges of mmWave communications for the MAC layer, classified existing standards for mmWave communications, provided in-depth overviews of MAC protocols and algorithms as well as related current projects, and concluded with open issues and future research directions.

Acknowledgment

This work was supported in part by the Scientific and Technological Research Council of Turkey (TUBITAK) under grant #113E962.

References

1. T. Rappaport, S. Sun, R. Mayzus, H. Zhao, Y. Azar, K. Wang, G. Wong, J. Schulz, M. Samimi, and F. Gutierrez, Millimeter wave mobile communications for 5G cellular: It will work! *IEEE Access* 1: 335–349, 2013.
2. T. S. Rappaport, R. W. Heath Jr, R. C. Daniels, and J. N. Murdock, *Millimeter Wave Wireless Communications*. New York: Pearson Education, 2014.
3. R. R. Choudhury, X. Yang, R. Ramanathan, and N. F. Vaidya, On designing MAC protocols for wireless networks using directional antennas, *IEEE Transactions on Mobile Computing* 5(5): 477–491, 2006.
4. J. Qiao, X. Shen, J. W. Mark, and Y. He, MAC-layer concurrent beamforming protocol for indoor millimeter-wave networks, *IEEE Transactions on Vehicular Technology* 64(1): 327–338, 2014.
5. FP7 EU Project METIS (Online). Available from: http://wirelessgigabitalliance.org/.
6. IEEE Std. 802.15.3-2003 ed. Wireless medium access control (MAC) and physical layer (PHY) specifications for high rate wireless personal area networks (WPANs), Piscataway, NJ, 2006.
7. WirelessHD: WirelessHD specification overview, 2009.
8. Standard ECMA-387 2nd Edition: High Rate 60 GHz PHY, MAC and HDMI PAL, 2010.
9. IEEE P802.11ad, Part 11: Wireless LAN Medium Access Control 5 (MAC) and Physical Layer (PHY) Specifications. Amendment 3: Enhancements for Very High Throughput in the 60 GHz Band, 2013.
10. Wireless Gigabit Alliance (Online). Available from: https://www.metis2020.com.
11. R. Maslennikov and A. Lomayev, Implementation of 60 GHz WLAN Channel Model. IEEE doc. 802.11-10/0854r3, 2010.
12. Z. Xun, C. A. O. Ya-Nan, and Z. Qiang-Wei, New medium access control protocol of terahertz ultra-high data-rate wireless network, *Journal of Computer Applications* 33(11): 3019–3023, 2013.
13. J. Wang, R. V. Prasad, and I. G. M. M. Niemegeers, Enabling multi-hop on mm Wave WPANs, in *Proceedings of the IEEE ISWCS08*, Reykjavik, IEEE, pp. 371–375, 2008.

14. S. Sushil, F. Ziliotto, U Madhow, E. Belding, and M. Rodwell, Blockage and directivity in 60 GHz wireless personal area networks: From cross-layer model to multihop MAC design, *IEEE Journal on Selected Areas in Communications* 27(8): 1400–1413, 2009.

15. J. Lu, D. Steinbach, P. Cabrol, and P. Pietraski, Modeling the impact of human blockers in millimeter wave radio links, *ZTE Communication Magazine* 10(4): 23–28, 2012.

16. S. Rangan, T. Rappaport, and E. Erkip, Millimeter wave cellular wireless networks: Potentials and challenges, *Proceedings of the IEEE* 102(3): 366–385, 2014.

17. K. C. Allen, N. DeMinco, J. Hoffman, Y. Lo, and P. Papazian, Building Penetration Loss Measurements at 900 MHz, 11.4 GHz, and 28.8 MHz, U.S. Department of Commerce, National Telecommunications and Information Administration Rep, pp. 94–306, 1994.

18. A. V. Alejos, M. G. Sanchez, and I. Cuinas, Measurement and analysis of propagation mechanisms at 40 GHz: Viability of site shielding forced by obstacles, *IEEE Transactions on Vehicular Technology* 57(6): 3369–3380, 2008.

19. H. Zhao, R. Mayzus, S. Sun, M. Samimi, J. K. Schulz, Y. Azar, K. Wang, G. N. Wong, F. Gutierrez, and T. S. Rappaport, 28 GHz millimeter wave cellular communication measurements for reflection and penetration loss in and around buildings in New York City, in *Proceedings of the IEEE International Conference on Communications (ICC)*, Budapest, IEEE, pp. 5163–5167, 2013.

20. M. X. Gong, R. Stacey, D. Akhmetov, and S. Mao, A directional CSMA/CA protocol for mmWave wireless PANs, Wireless Communications and Networking Conference (WCNC), 2010 IEEE, Santa Clara, CA, IEEE, 2010.

21. S. Singh, R. Mudumbai, and U. Madhow, Interference analysis for highly directional 60-GHz mesh networks: The case for rethinking medium access control, *IEEE/ACM Transactions on Networking* 19(5): 1513–1527, 2011.

22. S. Scott-Hayward and E. Garcia-Palacios. Multimedia resource allocation in mmWave 5G networks, *Communications Magazine, IEEE* 53(1): 240–247, 2015.

23. J. W. Lee, R. R. Mazumdar, and N. B. Shroff, Nonconvexity issues for Internet rate control with multiclass services: Stability and optimality, *Proceedings of the 23rd Annual IEEE INFOCOM*, 1: 1–12, 2004.

24. B. Ma, B. Niu, Z. Wang, and V. W. S. Wong, Joint power and channel allocation for multimedia content delivery using millimeter wave in smart home networks, Global Communications Conference (GLOBECOM), 2014 IEEE, Austin, TX, IEEE, 2014.

25. R. Bernasconi, I. Defilippis, S. Giordano, and A. Puiatti, An enhanced MAC architecture for multi-hop wireless networks, in M. Conti (ed.), *Personal Wireless Communications*. Berlin: Springer, pp. 811–816, 2003.

26. L. X. Cai, L. Cai, X. Shen, and M. Jon, REX: A randomized exclusive region based scheduling scheme for mmWave WPANs with directional antenna, *IEEE Transactions on Wireless Communications* 9(1): 113–121, 2010.

27. Q. Jian, Enabling multi-hop concurrent transmissions in 60 GHz wireless personal area networks, *IEEE Transactions on Wireless Communications* 10(11): 3824–3833, 2011.

28. C. Sum, Z. Lan, R. Funada, J. Wang, T. Baykas, M. A. Rahman, and H. Harada, Virtual time-slot allocation scheme for throughput enhancement in a millimeter-wave multi-Gbps WPAN system, *IEEE Journal on Selected Areas in Communications* 27(8): 1379–1389, 2009.

29. S. Singh, R. Mudumbai, and U. Madhow, Distributed coordination with deaf neighbors: Efficient medium access for 60 GHz mesh networks, in *Proceedings of the IEEE INFOCOM*, San Diego, CA, IEEE, pp. 1–9, 2010.

30. Y. Niu, Y. Li, D. Jin, and L. Su, Blockage robust and efficient scheduling for directional mmWave WPANs, *IEEE Transactions on Vehicular Technology* 64(2): 728–742, 2015.

31. S. Singh, F. Ziliotto, U. Madhow, E. M. Belding, and M. J. W. Rodwell, Millimeter wave WPAN: Cross-layer modeling and multihop architecture, in *Proceedings of the IEEE INFOCOM07*, Anchorage, AK, IEEE, pp. 2336–2240, 2007.

32. H. Kang, G. Ko, I. Kim, J. Oh, M. Song, and J. Choi, Overlapping BSS interference mitigation among WLAN systems, in *Proceedings of the IEEE 2013 International Conference on ICT Convergence*, Jeju, South Korea, IEEE, pp. 913–917, 2013.

33. H. Park, S. Park, T. Song, and S. Pack, An incremental multicast grouping scheme for mmWave networks with directional antennas, *IEEE Communications Letters* 17(3): 616–619, 2013.

34. C. W. Pyo and H. Harada, Throughput analysis and improvements of hybrid multiple access in IEEE 802.15.3c mmWave-WPAN, *IEEE Journal on Selected Areas in Communications* 27(8): 1414–1424, 2009.

35. L. X. Cai, L. Cai, X. Shen, and J. W. Mark, Capacity analysis of UWB networks in three-dimensional space, *IEEE/KICS Journal of Communications and Networks* 11(3): 287–296, 2009.

36. E. Shihab, L. Cai, and J. Pan, A distributed asynchronous directional-to-directional MAC protocol for wireless ad hoc networks, *IEEE Transactions on Vehicular Technology* 58(9): 5124–5134, 2009.

37. M. Sanchez, T. Giles, and J. Zander, CSMA/CA with beam forming antennas in multi-hop packet radio, *Proceedings of the Swedish Workshop on Wireless Ad Hoc Networks*, pp. 63–69, 2001.

38. J. Wang, Z. Lan, C-W. Pyo, B. T. Chin-Sean Sum, M. A. Rahman, J. Gao, R. Funada, F. Kojima, H. Harada, and S. Kato, Beam codebook based beamforming protocol for multi-Gbps millimeter-wave WPAN systems, *IEEE Journal on Selected Areas in Communications* 27(8): 1390–1399, 2009.

39. H. H. Lee and Y. C. Ko, Low complexity codebook-based beamforming for MIMO-OFDM systems in millimeter-wave WPAN, *IEEE Transactions on Wireless Communications* 10(11): 3607–3612, 2011.

40. B. Li, Z. Zhou, W. Zou, X. Sun, and G. Du, On the efficient beamforming training for 60 GHz wireless personal area networks, *IEEE Transactions on Wireless Communications* 12(2): 504–515, 2013.

41. S. Hur, T. Kim, D. J. Love, J. V. Krogmeier, T. A. Thomas, and A. Ghosh, Multilevel millimeter wave beamforming for wireless backhaul, in *Proceedings of the IEEE GLOBECOM Workshops*, Houston, TX, IEEE, pp. 253–257, 2012.

42. L. Zhou and Y. Ohashi, Efficient codebook-based MIMO beamforming for millimeter-wave WLANs, in *Proceedings of the IEEE PIMRC*, pp. 1885–1889, 2012.

43. S. Bellofiore, J. Foutz, R. Govindarajula, I. Bahceci, C. A. Balanis, A. S. Spanias, J. Capone, and T. M. Duman, Smart antenna system analysis, integration, and performance for mobile ad-hoc networks (Manets), *IEEE Transactions on Antennas and Propagation* 50(5): 571–581, 2002.

44. D. Lal, R. Toshniwal, R. Radhakrishna, D. Agrawal, and J. Caffery, A novel MAC layer protocol for space division multiple access in wireless ad hoc networks, *Eleventh International Conference on Computer Communications and Networks*, IEEE, pp. 614–619, 2002.

45. H. Singh and S. Singh, DOA-ALOHA: Slotted ALOHA for ad hoc networking using smart antennas, *Proceedings Vehicular Technology Conference (VTC)*, IEEE, pp. 2804–2808, 2003.

46. J. Qiao, X. Shen, J. Mark, and Y. He, MAC-layer concurrent beamforming protocol for indoor millimeter-wave networks, *IEEE Transactions on Vehicular Technology* 64(1): 327–338, 2015.

47. J. Qiao, L. X. Cai, X. Shen, and J. Mark, Enabling multi-hop concurrent transmissions in 60 GHz wireless personal area networks, *IEEE Transactions on Wireless Communications* 10(11): 3824–3833, 2011.

Index

Latency, 5G, 23–24, 50, 191–193
Latent fault determination, 160
Layered space-time code, 118
LC, *see* Link cooperating (LC)
LCB, *see* Load-Centric Backhauling (LCB)
LD, *see* Laser diode (LD)
Legacy deployment solutions, 77–80
LF-based algorithm, *see* Load fairness
        (LF)-based algorithm
License-assisted access (LAA), 83
Licensed share access (LSA), 110
Licensed spectrum, 444
Linear detection algorithm, 120–121
Linear minimum mean square error
        (LMMSE), 376
Linear precoding, 119
Line-of-sight (LOS), 86–87, 165, 484, 490,
        498, 499
Link budget analysis, 471–476
    IEEE 802.11ad baseband-based calculation,
        474–476
    Shannon capacity-based calculation,
        472–473
Link cooperating (LC), 516
Link switching (LS), 516
Linköping University, 136
LLR, *see* Log-likelihood ratios (LLR)
LMDS, *see* Network-local multipoint
        distribution service (LMDS)
LMMSE, *see* Linear minimum mean square
        error (LMMSE)
Load-Centric Backhauling (LCB), 453–454
Load fairness (LF)-based algorithm, 405
Local area networks
    IEEE 802.11ad, 531–532
    Wireless Gigabit Alliance, 532
Local caching, 15
Log-likelihood ratios (LLR), 227, 228, 229
Lognormal shadow model, 116
Long-Term Evolution (LTE), 4, 20, 23, 74, 99,
        106, 107, 191, 400, 447, 454–455
    eNB, 39
    frame structure, 38–39
    initial procedures in 4G-EPC, 39–41
    MME pool, 39
    network architecture, 35–38
    protocol stack, 39
    proxy mobile internet protocol, 45–46
    S-GW pool, 39
    S1-based handover, 42–45
    time-frequency resource grid, 379–380
X2-based handover, 41–42

Long-Term Evolution–Advanced (LTE–A),
        46, 97
Long-Term Evolution–Time-Division
        Duplexing (LTE-TDD), 391
LOS, *see* Line-of-sight (LOS)
Low-resolution (L-Re) beam training, 512
L-Re beam training, *see* Low-resolution (L-Re)
        beam training
LS, *see* Link switching (LS)
LSA, *see* Licensed share access (LSA)
LTE, for unlicensed spectrum (LTE-U), 83
LTE, *see* Long-Term Evolution (LTE)
LTE-A, *see* Long-Term Evolution–Advanced
        (LTE-A)
LTE-TDD, *see* Long-Term Evolution–Time-
        Division Duplexing (LTE-TDD)
LTE-U, *see* LTE for unlicensed spectrum
        (LTE-U)
Lund University, 136

M

MA, *see* Multiple access (MA)
MAC, *see* Medium access control (MAC)
Machine learning approach, 220–229
Machine-to-machine (M2M) networks,
        15, 103, 104, 105, 108
Machine-type communication (MTC), 5, 30
Mach–Zehnder modulator (MZM), 197
Macro base stations (MBSs), 452, 453
Macrocell, 152, 164
    backhaul failure, 170–172
    failure, 173
Macrocellular network, 77, 78, 81
Marzetta, Thomas L., 114
Massive machine-type communications
        (M-MTC), 103
Massive MIMO, 14, 85–86, 113–148
    antennas, 127–137
    beam forming, 137–148
    channel capacity, 121–124
    multiantenna transmission model, 117–121
    multiuser, 124–127
    technology and its theoretical basis,
        115–117
Matched filtering (MF)
    algorithm, 147
    demodulation, 336, 338
MATLAB*, 200
Matrix notation, 334–338, 356–357
Maximizing signal-to-interference and noise
        ratio (Max SINR), 147

9 780367 574895